Pathology of the Developing Mouse

Pathology of the Developing Mouse

A Systematic Approach

Edited by **Brad Bolon**

CRC Press
Taylor & Francis Group
Boca Raton London New York

CRC Press is an imprint of the
Taylor & Francis Group, an **informa** business

CRC Press
Taylor & Francis Group
6000 Broken Sound Parkway NW, Suite 300
Boca Raton, FL 33487-2742

First issued in paperback 2020

ISBN 13: 978-0-367-57577-9 (pbk)
ISBN 13: 978-1-4200-7008-8 (hbk)

Library of Congress Cataloging-in-Publication Data

Pathology of the developing mouse : a systematic approach / edited by Brad Bolon.
     pages cm
   Includes bibliographical references and index.
   ISBN 978-1-4200-7008-8
   1. Mice--Diseases. I. Bolon, Brad, editor.

QL737.R6P38 2015
616.02'7333--dc23   2014040187

Visit the Taylor & Francis Web site at
http://www.taylorandfrancis.com

and the CRC Press Web site at
http://www.crcpress.com

This book is dedicated to those hardy souls

who seek understanding of large truths from small canvases,

who aspire to discover universal principles within ever-changing flesh,

who strive to build a healthier tomorrow for all by willingly

paying the price of difficult learning today.

In other words, this book is devoted to all of you

who have already suffered, or will someday encounter,

the delights and debacles of mouse developmental pathology.

Contents

Section I   Introduction

Section II   Fundamentals of Mouse Developmental Biology

Section III   Experimental Methods in Mouse Developmental Pathology

Section IV   Analytical Practices for Mouse Developmental Pathology

Preface

The goal of this book is to provide, insofar as feasible, a comprehensive reference on the design, analysis, and interpretation of abnormal findings that may be detected in developing mice before and shortly after birth. In particular, this book is designed specifically not only to serve as a "how to do" manual for developmental pathology experimentation in mice—other texts already support this niche quite well—but also, more importantly, to act as a "how to interpret" resource for pathologists and other biomedical scientists faced for the first or hundredth time with defining the significance of distorted features in some fantastic murine developmental monstrosity.

The topics covered in this book include a full range of subjects encountered when building and wielding a developmental pathology tool kit:

- Baseline anatomic and physiologic traits of developing mice
- Principles of good experimental design and statistical analysis
- Procedures for anatomic pathology examinations to evaluate structural changes at the macroscopic (gross), microscopic (cells and tissues), and ultrastructural (subcellular) levels, using conventional necropsy (autopsy) based or novel noninvasive imaging techniques
- Methods for clinical pathology testing to assess the biochemical and cellular composition of tissues and fluids
- Options and protocols for *in situ* molecular pathology analysis to undertake site-specific explorations of the various mechanisms responsible for producing adverse findings (i.e., *lesions*) during development
- Appropriate terms for normal structures and processes as well as common lesions observed in developing mice
- Well-referenced and illustrated guides to the recognition and interpretation of anatomic pathology and clinical pathology changes in the animal (embryos, fetuses, neonates, and juveniles) and its support system (placenta)

In my experience, this last point encompasses the most critical need but the least available resource in developmental biological laboratories that work with mice. Moreover, if available at all, such knowledge typically must be transferred to new personnel by the lab sage, and so is vulnerable to permanent loss if this technical expert moves or retires. This book has been designed to give sages-in-training a springboard for honing their proficiency.

Although well illustrated and replete with relevant literature citations, this book is neither an atlas nor a comprehensive survey of developmental pathology lesions. Instead, this volume communicates the collective knowledge of many veterans in the mouse developmental pathology field with the hope that others can learn from our successes and avoid repeating our mistakes.

As with most such big-picture projects, the assembly of this book is several years behind schedule. My sincere apologies, especially to my publisher. I only hope that the final product is a suitable recompense for the long wait.

Best wishes for pleasant pursuits in prenatal and perinatal pathology.

Brad Bolon
The Ohio State University
Columbus, Ohio

Acknowledgments

My appreciation goes out to the numerous individuals who were instrumental in helping to bring this book to fruition.

I thank Dr. Vincent St. Omer, my master's thesis advisor, for first stimulating my mania for developmental pathology during my initial research experiences at the University of Missouri College of Veterinary Medicine, Columbia, Missouri.

I thank Drs. Kevin Morgan and Frank Welsch, my doctoral mentors; Donald Stedman and Barbara Elswick, my technical overseers in the laboratory; and Drs. David Dorman and Ketti Terry, my fellow graduate students, for assisting me in learning the "joys" of developmental pathology and toxicology during my tenure at the Chemical Industry Institute of Toxicology (CIIT) in Research Triangle Park, North Carolina.

I thank my coworkers over the years (you know who you are!) for constantly reorganizing their experiments and schedules to help me pursue my developmental pathology passion, and I thank my colleagues for indulging my ongoing crusade to instill an equal love for developmental pathology in them.

I thank the individuals who have contributed to the production of the contents that fill this book. I am grateful to my coauthors for making the content look good; my team of biomedical illustrators (independent contractors David Glenn [Chapel Hill, North Carolina] and Colin Moore [Longmont, Colorado], but particularly Tim Vogt of The Ohio State University College of Veterinary Medicine, Columbus, Ohio) for making the exceptional drawings and my photographic team (Joe Anderson, Dr. Kelli Boyd, Dr. Elizabeth R. Magden, and Stephen Kaufman) for their superb photography. This book would not have been possible without your contributions.

I thank the publication team—John Sulzycki, David Fausel, and Richard Tressider at CRC Press, and Christine Selvan at SPi Global—for the high quality of the final product and nearly infinite patience they showed during production of the manuscript.

Finally, and most importantly, I deeply appreciate the endurance of my wife, Janine, and my scintillating school of spawn—Sean, Beth, James, and Clare—during the years of late-evening writing and constant paternal distraction. This work stands as a testament to your lasting faith and love.

Editor

Brad Bolon, DVM, MS, PhD, is a Diplomate of the American College of Veterinary Pathologists (DACVP) and the American Board of Toxicology (DABT) as well as a Fellow of the Academy of Toxicological Sciences (FATS) and the International Academy of Toxicologic Pathology (FIATP). He is currently an associate professor in the Department of Veterinary Biosciences at the Columbus campus of The Ohio State University (OSU) and also serves as the associate director of the Comparative Pathology and Mouse Phenotyping Shared Resource (CPMPSR) at the OSU Comprehensive Cancer Center. He has practiced mouse developmental pathology in many venues: contract research organizations, pharmaceutical companies, private consulting, and now academia. His main professional interests are the pathology of genetically engineered mice (especially embryos, fetuses, and placentas) and toxicologic neuropathology. He has presented and published extensively on both topics for nearly 25 years.

Contributors

Brad Bolon
Department of Veterinary Biosciences
and
Comparative Pathology and Mouse Phenotyping
    Shared Resource
The Ohio State University
Columbus, Ohio

Denise I. Bounous
Drug Safety Evaluation
Bristol-Myers Squibb
Princeton, New Jersey

Kelli L. Boyd
Division of Comparative Medicine
Vanderbilt University Medical Center
Nashville, Tennessee

Vinicius Carreira
Department of Environmental Health
College of Medicine
University of Cincinnati
Cincinnati, Ohio

Sara Cole
Campus Microscopy and Imaging Facility
The Ohio State University
Columbus, Ohio

Diane Duryea (retired)
Department of Pathology
Amgen, Inc.
Thousand Oaks, California

Julie F. Foley
Laboratory of Experimental Pathology
National Institute of Environmental Health
    Sciences
Research Triangle Park, North Carolina

Oded Foreman
Department of Pathology
Genentech
South San Francisco, California

Kathleen Gabrielson
Department of Molecular and Comparative
    Pathobiology
Johns Hopkins University
Baltimore, Maryland

Stephen Kaufman (retired)
Department of Pathology
Amgen, Inc.
Thousand Oaks, California

Olga N. Kovbasnjuk
Division of Gastroenterology and Hepatology
Johns Hopkins University
Baltimore, Maryland

Krista M.D. La Perle
Department of Veterinary Biosciences
and
Comparative Pathology and Mouse Phenotyping
    Shared Resource
The Ohio State University
Columbus, Ohio

Colin McKerlie
The Hospital for Sick Children
and
University of Toronto
Toronto, Ontario, Canada

Jennifer M. Mele
Comprehensive Cancer Center
Wexner Medical Center
The Ohio State University
Columbus, Ohio

Richard Montione
Campus Microscopy and Imaging Facility
The Ohio State University
Columbus, Ohio

Susan Newbigging
The Hospital for Sick Children
and
Mount Sinai Hospital
and
University of Toronto
Toronto, Ontario, Canada

Efrain Pacheco
Department of Pathology
Amgen, Inc.
Thousand Oaks, California

Kimerly A. Powell
Small Animal Imaging Shared Resource
The Ohio State University
Columbus, Ohio

Colin G. Rousseaux
Department of Pathology and Laboratory
    Medicine
Faculty of Medicine
University of Ottawa
Ottawa, Canada

Jerrold M. Ward
Global VetPathology
Montgomery Village, Maryland

David Weinstein
Numira Biosciences, Inc.
Salt Lake City, Utah

Geoffrey A. Wood
Department of Pathobiology
University of Guelph
Guelph, Ontario, Canada

Section I

Introduction

1

The Case for Developmental Pathology and Developmental Pathologists

Brad Bolon

CONTENTS

The twenty-first century is poised to be a time of biological wonder. The knotty problems of centuries, even millennia, now can be solved in periods so short and for costs so minor that the promise of real-time *personalized* medicine based on a patient's own genomic code and protein complement seems poised to become a fact in our lifetimes. A Brave New World is upon us, indeed.

The Mouse as a Model in Developmental Biology Research

The workhorse of the modern biomedical research engine is the mouse. We are often reminded that mice are not just small, hairy humans. Despite this truism, mice are good subjects for answering many biological questions, including basic mechanisms held in common by developing organisms of many species. Mice even make valid models for many distinctly human diseases as long as the hypothesis being tested is defined clearly in advance. For example, mouse models of human-specific congenital conditions like cerebral palsy cannot replicate the complete constellation of anatomic, functional, and molecular changes that initiate and sustain these diseases. However, mice can be used to explore the biochemical and cellular effects of anoxia on developing neurons in cerebral palsy, and similar mechanisms proposed as the basis for other human conditions.

Many factors enhance the desirability of the mouse as a test subject for fundamental biology research. Major attributes include its status as a mammal, its small size, its fecundity, and its well-characterized genome. Particular genetic factors that increase the attractiveness of the mouse as a model organism are its relatively close resemblance to the human genome,[1,16] the numerous existing strains with highly defined genetic compositions (see Chapter 2), and the ease with which the mouse genome may be reconstructed (engineered) to incorporate manipulated genes (Figure 1.1) and transplanted cells.[8,15] The peak of this trend has been realized in the last 25 years with the burgeoning ability to deliberately engineer the mouse genome to contain and express human genetic material. This skill provides a useful means for *in vivo* exploration of a human gene's function, which has become integral to improving predictions of human responses to therapeutic entities.[30] The importance of genetically engineered mice (GEM) as world-changing tools for biological discovery has been showcased by presentation of the 2007 Nobel Prize in Physiology or Medicine to Mario Capecchi, Sir Martin Evans, and Oliver Smithies *"for their discoveries of principles for introducing specific gene modifications in mice by the use of embryonic stem cells."*[4] Multiple institutions currently participate in consortia dedicated to

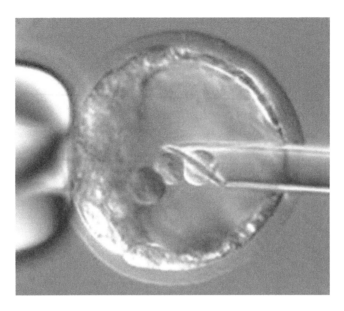

FIGURE 1.1 Photomicrograph of a mouse blastocyst being injected with embryonic stem cells. (The image is reproduced courtesy of Dr. Anne Bower and Dr. Manfred Baetscher, Oregon Health Science University, from the version posted on the website of the Genome Modification Facility, Harvard University, Cambridge, MA, http://gmf.fas.harvard.edu/.)

defining the function(s) of all mouse genes within the next decade[10,21,23,24] and providing researchers with greater access to GEM.

The ability to change the mouse genome at will has led to greater understanding of the roles played by hundreds of newly identified molecules (Figure 1.2). Many of these new GEM lines have been further investigated as novel mouse models of human disease based on the resemblance of clinical, functional, and/or pathologic findings in the mouse to those of a human disease. This concordance is

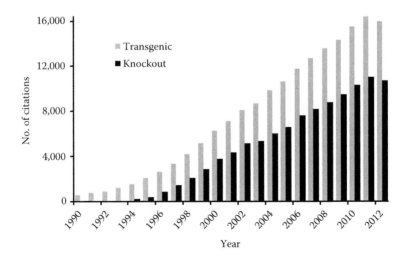

FIGURE 1.2 Graph demonstrating the rapid growth in articles related to genetic engineering technology during the last 25 years. The search was conducted via the PubMed database (http://www.ncbi.nlm.nih.gov/pubmed) at the U.S. National Library of Medicine, using the words *transgenic* and *knockout* matched with the year, on June 1, 2014.

notable given both the higher genomic heterogeneity of humans relative to the inbred mouse strains favored for many biological experiments and the tendency for human diseases to arise as polygenic conditions rather than from the catastrophic loss of a single gene such as occurs in engineered mice. These circumstances place the mouse in the winner's circle as the unquestioned mammal *par excellence* for further *in vivo* investigation of new developmental biology principles in terms of both acquiring fundamental knowledge and seeking insights for devising new therapeutic targets and treatments.

The Case for Developmental Pathology

Perhaps, though, the Brave New World is further afield than we suppose. As is ever the case with variable biological systems, research with GEM has presented researchers with many unforeseeable surprises along the way. One of the most unexpected findings is the extent to which many targeted genes being explored as potential players in adult diseases also have critical roles to play sometime during development. This inference is supported by the observation at high-throughput mouse phenotyping programs that approximately 30% of mouse lines with engineered null (knockout) mutations in which the genetic modification is present throughout all of development result in either an embryonic lethal or a perinatal lethal phenotype (Ref. [29] and Chapter 17); this outcome has filled mouse phenome databases with several thousand developmental lethal phenotypes (Table 1.1). Ensuing progress in genetic engineering technology has permitted the construction of conditional GEM, in which the pattern of gene expression or loss can be regulated to occur only at a specific location and/or at a given time.[8,15] In this manner, the gene may be removed only during adulthood, thereby avoiding the embryonic lethal consequences of abnormal gene expression during development.

Nonetheless, the crop of genes whose loss or overexpression during development leads to early lethality will keep growing (Figure 1.3) as the number of GEM projects continues to increase. This body of new genes represents a rich field for improving our understanding of the biological mechanisms that drive such fundamental processes as cell differentiation, intracellular and intercellular signaling, cell-to-cell interactions, cell migration, and many others. The knowledge of normal processes that occur from the beginning of (or even before) conception through the postnatal stages of infancy, puberty, and beyond comprises the field of developmental biology, and this information has value in its own right. However, a greater appreciation for developmental biology in mice also offers a means for improved predictions regarding the potential hazards and risks posed by exposing the developing human organism to agents—microbes, radiation, and teratogenic xenobiotics (chemicals, metals, and drugs), to name just a few—that have the capacity to upset the carefully scripted spatial and temporal waves of gene expression that must occur to ensure that development proceeds without incident. A possible bonus of increasing developmental biology expertise in the scientific community as a whole is that a better comprehension of developmental processes taking place in correctly differentiating and proliferating cells under the tight control of normal regulatory pathways will offer new hope of understanding, combating, and maybe even correcting the aberrant differentiation and proliferation events that drive such devastating, typically adult-onset conditions as degenerative diseases and neoplasia.

Developmental biology as a discipline encompasses many different investigational modalities. While many of these tools are designed to evaluate the cells, molecules, and processes that direct the appropriate assembly of a new organism, structural analysis at tissue and organ levels is an important aspect of any developmental biology study. The most common methods found in recent literature reports include whole-mount preparations of intact mouse embryos/fetuses and placentas, histopathological examination of affected organs, and *in situ* molecular methods to reveal expression of altered proteins or cell types; less frequent endpoints include acquisition of organ weights, cytological evaluation of particular cell populations, or quantification of chemical analytes in body fluids. These parameters are integral elements in standard pathology analyses, indicating that developmental pathology is an essential component of a well-designed developmental biology experiment.

TABLE 1.1

Major Classes of Developmental Pathology Phenotypes in Mice

Timing of Defect	Process Impacted by Defect	Number[a]
Embryogenesis		
Preimplantation	Zygote	17
	Morula	8
	Blastocyst	339
Implantation	Implantation errors	66
Organogenesis	Embryo anatomy	1994
	Abnormal size	847
	Patterning anomalies	924
	Embryo physiology	135
	Growth arrest	384
	Growth delay	711
Interface issues	Embryo/placental boundary	48
Placenta	Placenta anatomy	1189
	Hematopoiesis	85
	Placental physiology	51

Timing of Lethality	Penetrance of Lethality	Number[a]
Embryogenesis		
Preimplantation		215
At implantation		50
Postimplantation, pre-organogenesis		614
Organogenesis		1668
Birth		
Neonatal	Complete penetrance	1029
	Partial penetrance	455
Perinatal	Complete penetrance	449
	Partial penetrance	401
Juvenile (birth to weaning)	Complete penetrance	890
	Partial penetrance	1009
Weaning	Complete penetrance	71
	Partial penetrance	53
Premature senescence		80

Source: Data were acquired via the *Mouse Genome Informatics* database (www.informatics.jax.org) of The Jackson Laboratory, Bar Harbor, Maine, on June 1, 2014, using the *Mammalian Phenotype Browser* (http://www.informatics.jax.org/searches/MP_form.shtml) and specific search strings for abnormal embryogenesis (http://www.informatics.jax.org/searches/Phat.cgi?id = MP:0001672) and placental pathology (http://www.informatics.jax.org/searches/Phat.cgi?id = MP:0001711).

[a] Denotes the number of developmental phenotypes of a given type.

The Case for Developmental Pathologists

The case for developmental pathology is solid. What about the case for assigning this role to expert developmental pathologists?

A pathologist is an individual with formal training in the investigation, diagnosis, and interpretation of diseases. Viewed from one angle, scientists engaged in developmental biology studies in any venue are fulfilling the function of the developmental pathologist because they are employing fundamental

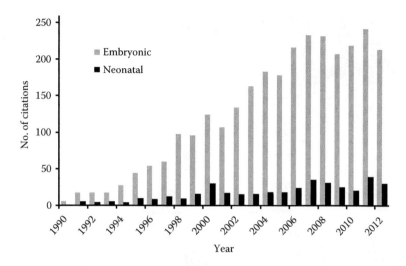

FIGURE 1.3 Graph demonstrating the steady growth in articles related to developmental lethal phenotypes during the last 25 years. The search was conducted via the PubMed database (http://www.ncbi.nlm.nih.gov/pubmed) at the U.S. National Library of Medicine, using the words *embryonic* and *neonatal* combined with the two terms, *lethality* and *mouse*, on June 1, 2014.

pathology techniques to evaluate the structure and function of organs and cell populations. This estimation has some merit given the wealth of review articles,[2,3,12,14,19,27] reference books,[5,17,18,28] and websites[9] that outline *how to do* a developmental pathology assessment coupled with the growing number of mouse developmental anatomy atlases available in the public domain.[6,7,11,13,20,22,25,26] From another angle, however, the developmental pathology tasks attending such experiments should be assigned whenever possible to an experienced developmental pathologist.

The reason for this assertion is not professional prejudice, but prior practice. Developmental pathologists have formal expertise not only in the acquisition of pathology data (or *how to do*) but also years of experience in interpreting structural data (or *how to use*) to define a diagnosis (i.e., what lesion or disease is present) and give a prognosis (i.e., what the predicted outcome of the lesion will be over time). Individuals with formal expertise in pathology have been trained as systems biologists. In other words, they have been taught to approach biological problems simultaneously at several levels—molecular, cellular, organ, and systemic—and ultimately integrate the responses of all organs in the body when interpreting the impact of an event (e.g., mutation, toxicant exposure) that might disturb development. Moreover, comparative pathologists generally have specific training in anatomy (adult and embryonic), biochemistry, and physiology of one or, ideally, many species of mammals that can permit the rational and reasonably facile translation of known biological principles and new experimental data among species. Therefore, comparative pathologists are biomedical scientists with the required education and experience to easily merge the attributes to be observed in the conventional three-dimensional body plan (characterized by length, width, and height) with the fourth dimension of time that represents the key feature of development (Figure 1.4). Pathologists are well equipped to assess the long-term impact of lesions that are superimposed on the ever-evolving anatomic landscape that comprises the developing mouse.

This book is designed for the developmental pathologist in us all. The chapters are designed to give novice researchers important information they will need to help them begin planning and performing mouse developmental pathology experiments. However, the real value of this volume will be for trained comparative pathologists and developmental biologists who already have considerable experience in manipulating specimens from developing mice. The information in these pages should help such individuals to quickly expand their expertise in this endeavor and hopefully will serve as a handy aid for interpreting the significance of pathologic changes in developing mice for years to come.

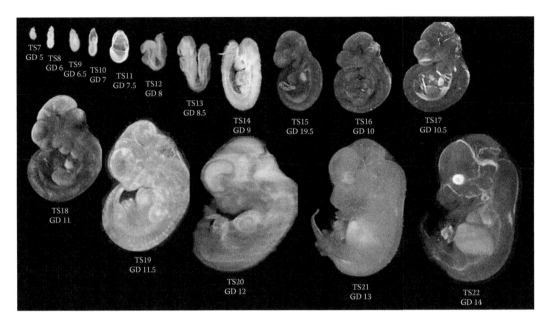

FIGURE 1.4 Composite image of embryonic mouse developmental stages from implantation (at Theiler stage [TS] 7 and gestational day [GD] 5, in upper left) throughout the rest of organogenesis (TS22 and GD 14, in lower right). (The image is reproduced courtesy of the eMOUSE Atlas Project, http://www.emouseatlas.org/emap/, as described in Richardson, L. et al., *Nucleic Acids Res.* 42(D1), D835, 2014.)

REFERENCES

1. Adams MD (2007). The mouse genome. In: *The Mouse in Biomedical Research. Vol. 1: History, Wild Mice, and Genetics* (Fox JG, Barthold SW, Davisson MT, Newcomer CE, Quimby FW, Smith AL, eds.). San Diego, CA: Academic Press (Elsevier), pp. 99–114.
2. Brayton C, Justice M, Montgomery CA (2001). Evaluating mutant mice: Anatomic pathology. *Vet. Pathol.* 38(1): 1–19. [Erratum in *Vet. Pathol.* 38(15): 568, 2001.]
3. Bronson RT (2001). How to study pathologic phenotypes of knockout mice. In: *Methods in Molecular Biology: Gene Knockout Protocols* (Tymms MJ, Kola I, eds.). Totowa, NJ: Humana, pp. 155–180.
4. Capecchi MR, Evans MJ, Smithies O (2007). The Nobel Prize in Physiology or Medicine 2007 lecture: Gene targeting to produce knockout mice. http://www.nobelprize.org/nobel_prizes/medicine/laureates/2007/popular-medicineprize2007.pdf [last accessed June 1, 2014].
5. Croy BA, Yamada AT, DeMayo FJ, Adamson SL (2014). *The Guide to Investigation of Mouse Pregnancy.* London, U.K.: Academic Press (Elsevier).
6. de Boer BA, Ruijter JM, Voorbraak FP, Moorman AF (2009). More than a decade of developmental gene expression atlases: Where are we now? *Nucleic Acids Res.* 37(22): 7349–7359.
7. EMAP (e-MOUSE Atlas Project) (2013). e-Mouse Atlas, v3.5. http://www.emouseatlas.org/emap/home.html [last accessed June 1, 2014].
8. Galbreath EJ, Pinkert CA, Bolon B, Morton D (2013). Genetically engineered animals in product discovery and development. In: *Haschek and Rousseaux's Handbook of Toxicologic Pathology*, 3rd edn., vol. 3 (Haschek WM, Rousseaux CG, Wallig MA, eds.). San Diego, CA: Academic Press (Elsevier), pp. 405–460.
9. Hill MA (2014). Mouse development. http://php.med.unsw.edu.au/embryology/index.php?title=Mouse_Development [last accessed June 1, 2014].
10. IMKC (International Mouse Knockout Consortium) (2007). A mouse for all reasons. *Cell* 128(1): 9–13.
11. Jacobwitz DM, Abbott LC (1998). *Chemoarchitectonic Atlas of the Developing Mouse Brain.* Boca Raton, FL: CRC Press.

12. Kasarskis A, Manova K, Anderson KV (1998). A phenotype-based screen for embryonic lethal mutations in the mouse. *Proc. Natl. Acad. Sci. USA* 95(13): 7485–7490.
13. Kaufman MH (1992). *The Atlas of Mouse Development*, 2nd edn. San Diego, CA: Academic Press.
14. Kulandavelu S, Qu D, Sunn N, Mu J, Rennie MY, Whiteley KJ, Walls JR et al. (2006). Embryonic and neonatal phenotyping of genetically engineered mice. *ILAR J.* 47(2): 103–117.
15. Menke DB (2013). Engineering subtle targeted mutations into the mouse genome. *Genesis* 51(9): 605–618.
16. Mouse Genome Sequencing Consortium (2002). Initial sequencing and comparative analysis of the mouse genome. *Nature* 420(6915): 520–562.
17. Nagy A, Gertsenstein M, Vintersten K, Behringer R (2003). *Manipulating the Mouse Embryo*, 3rd edn. Cold Spring Harbor, NY: Cold Spring Harbor Laboratory Press.
18. Papaioannou VE, Behringer RR (2005). *Mouse Phenotypes: A Handbook of Mutation Analysis*. Cold Spring Harbor, NY: Cold Spring Harbor Laboratory Press.
19. Papaioannou VE, Behringer RR (2012). Early embryonic lethality in genetically engineered mice: Diagnosis and phenotypic analysis. *Vet. Pathol.* 49(1): 64–70.
20. Paxinos G, Halliday G, Watson C, Koutcherov Y, Wang H (2007). *Atlas of the Developing Mouse Brain at E17.5, P0, and P6*. San Diego, CA: Academic Press (Elsevier).
21. Peters LL, Robledo RF, Bult CJ, Churchill GA, Paigen BJ, Svenson KL (2007). The mouse as a model for human biology: A resource guide for complex trait analysis. *Nat. Rev. Genet.* 8(1): 58–69.
22. Richardson L, Venkataraman S, Stevenson P, Yang Y, Moss J, Graham L, Burton N, Hill B, Rao J, Baldock RA, Armit C (2014). EMAGE mouse embryo spatial gene expression database: 2014 update. *Nucleic Acids Res.* 42(D1): D835–D844.
23. Rosenthal N, Brown S (2007). The mouse ascending: Perspectives for human-disease models. *Nat. Cell Biol.* 9(9): 993–999.
24. Rossant J, McKerlie C (2001). Mouse-based phenogenomics for modelling human disease. *Trends Mol. Med.* 7(11): 502–507.
25. Schambra U (2008). *Prenatal Mouse Brain Atlas*. New York: Springer.
26. Schambra UB (2007). Electronic prenatal mouse brain atlas. http://www.epmba.org/ [last accessed June 1, 2014].
27. Ward JM, Elmore SA, Foley JF (2012). Pathology methods for the evaluation of embryonic and perinatal developmental defects and lethality in genetically engineered mice. *Vet. Pathol.* 49(1): 71–84.
28. Ward JM, Mahler JF, Maronpot RR, Sundberg JP (2000). *Pathology of Genetically Engineered Mice*. Ames, IA: Iowa State University Press.
29. White JK, Gerdin AK, Karp NA, Ryder E, Buljan M, Bussell JN, Salisbury J et al. (2013). Genome-wide generation and systematic phenotyping of knockout mice reveals new roles for many genes. *Cell* 154(2): 452–464.
30. Zambrowicz BP, Sands AT (2003). Knockouts model the 100 best-selling drugs—Will they model the next 100? *Nat. Rev. Drug Disc.* 2: 38–51.

Section II

Fundamentals of
Mouse Developmental Biology

2

Essentials of Mouse Genetics and Nomenclature

Oded Foreman and Brad Bolon

CONTENTS

Mice have been important subjects in biomedical research for decades. The original reasons for selecting this species as an animal model undoubtedly focused on its small size, relatively docile nature, and fecundity. Thus, many animals could be bred and maintained in a confined area for little cost over a short period. More recent advances in genetic and genomic research have established mice as a major model system for modern translational medicine initiatives.[17,40] These features have made mice a favorite mammalian model for exploring basic mechanisms of health and disease.

The mouse genome is approximately equivalent to the human genome in terms of overall size, number of genes, and the order in which the genes are arranged.[1,32] The mouse genome of 2.5 Gb (gigabases, where 1 Gb equals 1 billion bases) is approximately 14% smaller than the human genome (2.9 Gb).[32] Approximately 80% of mouse genes have human orthologs (i.e., genes that developed from a common ancestral gene), and another 19% have human homologs (i.e., distinct genes with similar functions).[32] Over 90% of mouse and human genes are linked in the same order (i.e., exhibit synteny) on their respective chromosomes.[32] These facts suggest the relatively recent existence of a common mammalian ancestor for both species. Furthermore, mutations that cause human diseases often cause similar mouse diseases. Thus, genetic discoveries using mouse tissues or cells have direct applicability to understanding many gene-mediated disorders that afflict people. Extraordinary progress has been made in the ability to manipulate the mouse genome by selective breeding, and more recently by direct intervention to engineer novel mutations.[4,10,34] The mouse thus will serve as a critical component of the biomedical research community's tool chest throughout the twenty-first century.

This chapter has been assembled with several objectives in mind. The first is to review fundamental principles of mouse genetics, particularly attributes of the various house mouse (*Mus musculus*) stocks and strains that comprise the majority of research animals today. The second aim is to summarize current conventions for naming various types of mice. The final goal is to briefly review various technologies that produce or utilize developing mice as research subjects. Taken together, the present treatment of these subjects will demonstrate why mouse conceptuses and neonates are growing in importance as engines of biological discovery.

Evolution of the Mouse as a Model for Research

The subfamily Murinae harbors 4 subgenera and 38 recognized species of mice. The subgenera are *Coelomys* (shrew mice), *Mus* (house mice), *Nannomys* (African pygmy mice), and *Pyromys* (spiny mice); the subgenus *Mus* diverged from other murine lineages approximately 10 million years ago. Each subgenus contains several species, which may be distinguished by characteristic anatomic features (e.g., dentition,[5] cranial structure[38]) and, more recently, unique patterns of biochemical[6,39] and molecular[3,18,29,44] markers. Species may be separated in turn into various subspecies. For example, *Mus musculus* (*M. m.*, which comprises the bulk of all research mice) has multiple subspecies, which began segregating some 1–2 million years ago. Modern inbred strains of laboratory mice have unequal genetic contributions from all these *M. m.* subspecies.[13]

M. m. domesticus and *M. m. musculus* are both thought to have originated in central Asia before subsequently commencing independent migrations into the West.[42] *M. m. musculus* invaded southeastern Europe and the lands bordering the Mediterranean Sea approximately 10,000 years ago, while *M. m. domesticus* colonized northern and western Europe approximately 6,000 years ago. Presently, both subspecies naturally share a geographic range in central Europe, although evolutionary drift over time has rendered them reproductively isolated. Gene flow between the two subspecies is minimal in the natural setting, because all progeny of crosses between these two subspecies are sterile.

In the late nineteenth and early twentieth centuries, mouse fanciers began collecting and breeding mice. The goal of these aficionados was to perpetuate and create novel physical characteristics, such as coat color and texture. Mice were exhibited and traded throughout Europe and Asia, and eventually also in the United States. Indiscriminate breeding for unique physical traits resulted in mouse variants with unpredictable genetic contributions from many *M. musculus* subspecies. In the early twentieth century, university researchers in the northeastern United States began working with mouse fanciers (particularly Abbie Lathrop) to establish *M. musculus* as a new animal model for biological research. Shortly thereafter, Leonard Loeb (University of Pennsylvania) showed that tumors arising in Japanese waltzing mice could be transplanted to closely related mice but not to unrelated animals. In 1909, Ernest Tyzzer (Harvard University) demonstrated that tumor rejection in mice was heritable.[43] This discovery led Clarence Cook Little (Harvard University) to postulate that a genetically uniform (*inbred*) strain of mouse was required to investigate certain biological questions. Little subsequently developed the first inbred mouse strain, the Dilute Brown non-Agouti (DBA). Dozens of additional inbred strains were developed through traditional breeding programs during the remainder of the twentieth century.

Basic Concepts in Laboratory Mouse Genetics

The *M. musculus* genome is carried on 20 pairs of chromosomes (19 autosomes along with X [female] and Y [male] sex chromosomes).[32] Each chromosome is formed from two DNA strands, one from each parent. In the absence of chromosomal damage, each parental strand contains the same complement of genes, and each gene is located at the same site on each strand. This close degree of point-by-point matching is referred to as homology. The mouse genome has approximately 22,000–30,000 protein-encoding genes, which is a complement equivalent to those in humans and rats.[1,32] As in other species, multiple protein forms may be derived from a single gene via alternate splicing.

The intricate phenotypes first admired by mouse fanciers result from the unique combination of alleles borne by a given individual. An allele is a particular form of a gene, the uniqueness of which is dictated by minor variations in the DNA sequence. Some genes have only one possible allele, while others have many options. All the alleles for a given gene may be interchanged freely at the sites (called loci) on the chromosome where the gene is found. However, only one allele at a time may occupy the gene locus on each DNA strand. The most common allele in a population is termed the *wild-type* (WT) allele. Alleles that have different sequences from the WT version often still encode a fully functional protein. On the other hand, certain non-WT allele variants are *hypomorphic* alleles (encoding proteins with reduced function) or null alleles (encoding nonfunctional proteins).

Various terms are utilized to describe the complement of alleles carried by individuals for any given gene. A *homozygous* mouse bears the same allele on both homologous chromosomes, while a *heterozygous* animal has a different allele on each chromosome. This condition is applicable to all 19 autosomes and to the specific portions of the sex chromosomes that carry homologous gene loci. A *hemizygous* state is one in which a single allele specifies a trait. This situation is applicable to genes on the X chromosome for which homologous loci are missing on the truncated Y chromosome, and is also true of transgenic animals in which foreign DNA has been randomly inserted into the genome. A *dominant* allele is one that will determine the phenotype of a heterozygous animal regardless of the other allele present at the corresponding gene locus. If two dominant alleles for the same gene are present, both alleles will be expressed to produce a combined phenotype. A *recessive* allele is one that cannot specify a phenotype unless it is paired with another recessive allele.

Sex-linked alleles are a special class that directs certain male- and female-specific patterns of inheritance. The production of X-linked traits in males is specified by genes that are borne on the unpaired portion of the X chromosome. Such features are controlled by the unopposed action of the single female allele, which explains why so many X-linked diseases in males are mediated by recessive alleles. In contrast, females have two homologous X chromosomes, so these genes are inherited in females according to the same rules for expression that govern dominant and recessive autosomal genes. Those genes that are unique to the Y chromosome (Y-linked) are regulated by the single male allele.

Categories of Mice and Conventions for Nomenclature

Substantial variability exists in the genetic backgrounds of mouse research colonies that are maintained at different institutions in multiple geographic locations around the world.[4,22,24] Therefore, the utility of any given family of mice with respect to a specific research question can only be understood in terms of precise terminology used to describe its genetic attributes.[10,23,40] Shared nomenclature not only improves communication among scientists but also is an essential conceptual foundation for efforts to ensure that groups of mice have and maintain the appropriate genetic background.[41] Prior to the advent of naming conventions, considerable confusion arose as various laboratories assigned names (1) based on phenotypic characteristics that arose from the combined activities of several distinct genetic events or (2) applied multiple names to a phenotype without realizing that all the diverse components of the condition stemmed from a single genetic abnormality. Although reports in the biomedical literature are still rife with nomenclature errors, the chaos is beginning to recede as rules determined by the *International Committee on Standardized Nomenclature in Mice* (an elected oversight body) are taking hold.[9,10,28] Updated *Guidelines for Nomenclature of Mouse and Rat Strains* are maintained at http://www.informatics.jax.org/nomen/strains.shtml, while a searchable database of mouse lineages and their availability may be accessed at http://www.findmice.org//index.jsp. This section briefly introduces the fundamental genetic concepts that define different categories of mice and the current conventions for naming them (Table 2.1).

Outbred Stocks

An outbred stock is a mouse colony that is maintained by regular crossbreeding of males and female with different genetic backgrounds. This breeding scheme provides the maximum degree of heterozygosity within a defined population of mice by specifically arranging crosses so that inbreeding among closely related individuals cannot occur. At the same time, matings are kept within the colony, so no new genetic alleles are introduced. If such closed colonies are small, the genetic variation among individuals will gradually decrease over time.

TABLE 2.1

Nomenclature Conventions for Mouse Stocks and Strains

Type of Stock or Strain	Sample Symbols	Comments
Outbred stock	Crl:CD1(ICR) Hsd:ICR(CD-1)	A stock derived from a colony of Swiss mice established at the Institute for Cancer Research (ICR) in Philadelphia and now distributed from either Charles River Laboratories International, Inc. (Crl) or Harlan Laboratories, Inc. (Hsd).
Inbred strain	C57BL/6J	A C57BL/6 strain distributed from The Jackson Laboratory (J). The designation is often abbreviated as *B6*.
Inbred substrain	C57BL6/N	A C57BL/6 substrain[41] started at the U.S. National Institutes of Health (N).
Inbred substrain	C57BL/6NCrl	A C57BL6/N substrain distributed by Charles River Laboratories International, Inc. (Crl).
Inbred substrain	C57BL/6NCrlCrlj	A C57BL/6N substrain distributed by the Japanese branch of Charles River Laboratories International, Inc. (Crlj).
Mixed inbred strain	B6;129	A strain derived from two or three parental lineages, listing the host strain before the first semicolon and the donor strain following the last semicolon (in this case, a C57BL/6 blastocyst mingled with 129-derived embryonic stem [ES] cells).
Segregating inbred strain	129X1/SvJ	A 129X1/Sv strain distributed by The Jackson Laboratory (J) that is fully inbred except for heterozygous alleles at a given locus (in this case, the tyrosinase gene responsible for coat color).
Hybrid strain	B6C3F1	The progeny of a first-generation (F1) cross between fully inbred C57BL/6J females and fully inbred C3H males.
Hybrid strain	C3B6F1	The progeny of a first-generation (F1) cross between fully inbred C3H females and fully inbred C57BL/6J males, made by the laboratory of Dr. Robert W. Williams (Rww) and distributed by The Jackson Laboratory (J). Multiple recombinant inbred strains derived from the same progenitor strains are numbered sequentially; this crossing is the 49th in the series.
Hybrid strain	C3B6F2	The progeny of a second-generation (F2) cross between two C3B6F1 parents.
Recombinant inbred strain	BXD49/RwwJ	Cross (X) between C57BL/6J females and DBA/2J males to create an inbred strain (20 brother–sister matings).
Congenic strain	B6.CB17-*Prkdc^{scid}*/SzJ	An inbred strain distributed by The Jackson Laboratory (J) in which a mutant allele (the *Prkdc^{scid}* allele in this case) and a tiny segment of flanking chromosomal DNA has been transferred from the C.B-17 donor strain to the C57BL6/J recipient strain. The specific substrain was developed by Dr. Leonard Shultz (Sz).
Coisogenic strain	C57BL/6J-*Apc^{Min}*/J	A C57BL/6 strain distributed by the Jackson Laboratory (J) that is fully inbred except for one mutant allele at a given locus (in this case, the chemically induced Min [multiple intestinal neoplasia] mutation of the adenomatous polyposis coli [Apc] tumor suppressor gene).
Consomic strain	C57BL/6J-Chr 17$^{A/J}$/NaJ	A C57BL/6 strain distributed by the Jackson Laboratory (J) in which chromosome (Chr) 17 from the C57BL/6J recipient strain has been completely replaced by Chr 17 from the A/J donor strain. The specific substrain was developed by Dr. Joseph Nadeau (Na).[33]
Conplastic strain	C57BL/6J-mt$^{A/J}$/NaJ	A C57BL/6 strain distributed by the Jackson Laboratory (J) in which the entire mitochondrial (mt) genome from the C57BL/6J recipient strain has been fully replaced by corresponding mt gene complement from the A/J donor strain. The specific substrain was developed by Dr. Joseph Nadeau (Na).

In general, outbred stocks are hardier than more genetically homogeneous groupings (e.g., inbred strains). Therefore, outbred stocks live longer than inbred mice, are more resistant to infectious disease, and develop fewer spontaneous conditions. The diverse genetic background of outbred stocks mimics the high degree of heterozygosity found in typical human populations. Accordingly, some researchers prefer outbred stocks for studies for which the data set is to be extrapolated to predict possible human responses (e.g., toxicological studies).

Outbred stocks are designated by a string of coded abbreviations detailing the laboratory that bred the animal, its stock, and any associated unique genetic attributes. Punctuation marks are used to separate the different codes; a colon (:) separates the source from the stock designation, while a hyphen (-) precedes the list of mutations carried by the animals. For example, a mouse with the identification *Hsd:ND4* is a Swiss Webster mouse (code for Swiss Webster=ND4) from Harlan Laboratories, Inc. (code=Hsd). Similarly, the designation *Hsd: MF-1-Foxn1^{nu}* denotes a Harlan-derived animal from the MF-1 outbred stock that bears a mutation in the *Foxn1^{nu}* gene (forkhead box 1, a human transcription factor related to the nude [nu] gene in the mouse). By common agreement, gene names are italicized in all spaces. Mouse genes are rendered with only an initial capital letter (*Foxn1*), while their human counterparts are fully capitalized (*FOXN1*).

Inbred Strains and Substrains

An inbred strain is defined as a colony in which all individuals can be traced to a single ancestral pair. Inbred strains are genetically identical (homozygous) at essentially all alleles.[31] This consistency is attained by mating brothers and sisters for at least 20 generations (F20), at which point 98.7% of the genome will be homozygous.[4] A *fully inbred* strain (99.98% homozygous) is acquired by backcrossing for 40 generations (F40).[4,40] The genome can remain fixed for extended periods by continued brother/sister crossings, although genetic monitoring must be conducted intermittently to make sure that spontaneous mutations have not arisen. An inbred substrain is declared to exist when genetic analysis reveals a genomic variation with respect to the parental background. Substrains are produced by design when mice are removed from one colony to form a separate colony and then interbred in isolation for more than 20 generations.

Fully inbred (F20) strains and substrains are generally less hardy than outbred stocks. Common attributes of inbred strains and substrains include a heightened susceptibility to disease, smaller litter size, and shorter life spans. Constant inbreeding has also introduced numerous strain-specific diseases and conditions in many inbred strains (cancer, retinal degeneration, seizures, etc.).

Inbred strains are designated by strings of uppercase letters (e.g., AKR) or Arabic numerals (e.g., 129), or a combination of letters and numbers in which a letter comes first (e.g., C57BL). Substrain names are created from the root symbol for the original strain followed by a forward slash (/) and a unique substrain designation indicating the laboratory of origin or the creator (e.g., A/J, where A is the strain and J stands for the institute where the substrain arose [in this case, The Jackson Laboratory]). Substrains maintained at different facilities will over time give rise to new substrains that will require their own designations (e.g., C3H/HeJ, the substrain of C3H that was derived by Dr. Walter Heston at The Jackson Laboratory).

A full list of codes for inbred mice is available at the Laboratory Code Registry through the Institute for Laboratory Animal Research (http://dels.nas.edu/global/ilar/Lab-Codes). Researchers and laboratories that have created a new mouse strain or substrain should enroll with this registry before naming their colony to ensure that nomenclature conventions are followed in future scientific communications.

Segregating Inbred Strains

These specialized colonies are fully inbred strains except for a heterozygous allele for one or more gene loci. Heterozygosity must be reestablished with every generation by genetic testing of progeny and culling of animals in which the gene locus has reverted to a homozygous state. Segregating inbred strains exhibit all of the same biological attributes of regular inbred strains (e.g., increased susceptibility to disease, reduced fecundity).

The nomenclature for segregating inbred strains is comparable to that of other inbred strains, since the segregating locus for a given strain remains part of the standard genotype. An example is the SSL/LeJ mouse, an inbred strain which carries distinct piebald (*Ednrb^s^*) and piebald lethal (*Ednrb^{s-l}^*) alleles for the piebald gene. If multiple affected genes are present, the strain name must be adjusted to indicate whether or not the segregating alleles are coupled (i.e., carried on a single chromosome) or repulsed (carried on separate chromosomes). For instance, the GL/Le *Edar^{dl-J}^* +/+ *Ostm1^{gl}^* segregating inbred strain transmits the two non-WT alleles (*Edar^{dl-J}^* and *Ostm1^{gl}^*) on different chromosomes, while its GL/Le *Edar^{dl-J}^* *Ostm1^{gl}^*/++ counterpart carries both non-WT alleles on one chromosome and the two WT alleles (++) at the homologous gene loci on the other. The forward slash (/) separates the two alleles on one chromosome from the two found on the other.

Hybrid Strains

Hybrids are the first-generation (F1) progeny of two inbred strains. These offspring are heterozygous at all gene loci for which their parents had different alleles but otherwise share identical genetic and phenotypic characteristics (except for those gender-specific traits mediated by the sex chromosomes). In order to remain genetically identical, all crossings must be performed in the same direction (i.e., breeding males are always taken from one parental strain, and breeding females from the other). An additional crossing of F1 progeny to produce an F2 generation will yield hybrid offspring that all have a unique random mixture of alleles from both parental strains.

Several advantages of hybrid strains have made them popular choices for biomedical research. Their greater genetic diversity (*hybrid vigor*) permits such hybrids to live longer, reproduce more successfully, and resist disease more effectively than pure inbred strains. At the same time, the more defined genetic background of hybrids renders analysis and interpretation of gene-mediated events much simpler than is the case in outbred stocks.

The nomenclature for hybrids uses uppercase abbreviations for the two parental strains (Table 2.2) followed by the code F1. The first letter always denotes the maternal strain. For example, the CB6F1

TABLE 2.2

Approved Abbreviations for Some Common Inbred Mouse Strains

Abbreviation	Inbred Strain
129	129
A	A
AK	AKR
B	C57BL
B6	C57BL/6
B10	C57BL/10
BR	C57BR/CD
C	BALB/c
C3	C3H
CB	CBA
D1	DBA/1
D2	DBA/2
HR	HRS/J
L	C57L/J
R3	RIIIS/J
J	SJL
SW	SWR

Source: Reproduced from *Guidelines for Nomenclature of Mouse and Rat Strains* as published on the Mouse Genome Informatics (MGI) website (http://www.informatics.jax.org/mgihome/nomen/strains.shtml; last accessed June 1, 2014).

hybrid results from crossing BALB/c females with C57BL/6 males, while the B6C3F1 hybrid is the product of breeding C57BL/6 females to C3H males. The nomenclature for F2 hybrids follows a similar convention but is updated to include the new generation (e.g., CB6F2 hybrids are an intercross of two CB6F1 parents).

Recombinant Inbred Strains

Production of recombinant inbred strains requires a tiered breeding scheme. First, two inbred strains are crossed to generate an F1 hybrid. Next, F1 progeny are crossed again to produce an F2 generation. Subsequently, continuous brother–sister matings starting with a randomly chosen F2 breeding pair are carried out for 20 generations to produce a new strain. Recombinant inbred mice have an equal genetic contribution (50%) from each of the original inbred strains. However, the random recombination of the parental alleles achieved in the F2 generation means that each recombinant inbred strain will have a unique complement of alleles.

Recombinant inbred strains are designated by uppercase abbreviations of both parental strain names separated by an uppercase letter *X* (to show that two strains have been crossed) with no intervening spaces. Again, the maternal strain is always listed first. If more than one recombinant inbred strain is derived from the same original progenitor inbred strains, the different lineages are differentiated using unique serial numbers. For example, the name AXB15 describes one of at least 15 recombinant inbred strains derived by crossing A females and C57BL/6 males.

Congenic Strains

Congenic mice are identical at all gene loci except for a single mutant allele at one locus (in conjunction with a small segment of flanking chromosomal DNA). Congenic strains are made by mating mice of two genetically distinct inbred strains, only one of which (termed the donor strain) carries the mutation. The F1 progeny that bear the mutant allele are selected and then repeatedly backcrossed onto one of the parental lineages (termed the recipient strain). Offspring of subsequent matings are screened anew for the mutant allele, and carriers are used as breeders for additional backcross matings to the recipient strain. With each successive generation, the percentage of recipient alleles increases logarithmically in the mutant progeny. A strain is considered fully congenic after 10 generations (designated N10, in which the theoretical percentage of shared genes has increased to 99.9%).[4] *Speed congenics* is a breeding scheme that yields in only five generations a genetic background equivalent to that of a N10 congenic produced by ordinary backcrossing. In speed congenics, a small panel of single nucleotide polymorphism (SNP) markers is used to distinguish between the donor and recipient genetic backgrounds. Animals that bear the mutant allele and the highest percentage of recipient SNP markers are selected as breeders for the next backcrossing generation, thus accelerating the accumulation of recipient alleles.

A three-part nomenclature is used for congenic strains. The first portion of the code, which denotes the recipient strain, is separated by a period (.) from the middle part naming the donor strain. The final segment, which follows a hyphen (-), states the donated allele. An example is the B6.CB17-*Prkdc^{scid}*/SzJ congenic strain, where the genetic background (i.e., recipient strain) is C57BL/6, the donor strain was C.B-17, and *Prkdc^{scid}* is the donated allele. The forward slash (/) uses the same convention noted earlier for inbred strains to separate the congenic strain name from the substrain designation *SzJ* that communicates the investigator (Dr. Leonard Shultz) and the institution (The Jackson Laboratory). Placement of a number behind the forward slash indicates that more than one line of a particular congenic strain has been produced that has the same donated allele from the same donor strain transplanted onto the recipient background (e.g., B6.CB17-*Prkdc^{scid}*/3SzJ indicates that at least three lines of this congenic strain are in existence).

For some experiments, multiple loci are transferred to a congenic strain. In this event, the strain name will include a parenthetical phrase giving the designations (separated by a hyphen) for the most proximal and most distal known markers that were transferred. The congenic strain NOD.129-(*D19Mit10–D19Mit54*)/GseJ has placed a length of DNA from the 129 donor strain onto the NOD recipient background, where the donated DNA is bounded by the two alleles termed *D19Mit10* and *D19Mit54*.

Again, the notation behind the slash denotes the inventor (Dr. George S. Eisenbarth) and facility (The Jackson Laboratory).

If the alleles in the congenic strain derive from more than one source, the symbol Cg is added to the name. For instance, the NOD.Cg-*Prkdc^{scid}* *Il2rg^{tm1Wjl}*/SzJ congenic strain has the NOD genetic background for the recipient allelic complement, and it carries donated alleles that arose in two different donor strains: the *Prkdc^{scid}* allele from the C.B-17 strain and the IL2rg allele from the B6.129S4-*Il2rg^{tm1Wjl}* strain. The latter strain is itself a congenic strain where the allele was transferred from the donor 129S4 strain onto the recipient C57BL/6 background.

Mutations that are the result of gene targeting by homologous recombination in ES cells are given the symbol of the targeted gene with a composite superscript consisting of the symbol *tm* (to denote a targeted mutation), followed by a serial number from the laboratory of origin, and the laboratory code indicating where the mutation was produced.

Coisogenic Strains

Coisogenic strains are genetically identical except for a single gene locus where both alleles are mutated. This homozygous condition is distinct from that of the segregating inbred strain (see above), where one or more gene loci have a heterozygous set of alleles. The genetic change may have arisen spontaneously, or it may be the product of deliberate genetic engineering. The advantage of coisogenic strains for biomedical experimentation is that they are inbred except for the one mutant gene and thus should have a fairly uniform response to physiological challenges.

Coisogenic strains are named using the strain symbol followed by a hyphen (-) and the code for the mutated allele (rendered in *italic* letters). For instance, the A/J-*sunk* coisogenic strain is a lineage arising from a spontaneous mutation of the *sunk* gene in mice of the A/J inbred substrain. However, the C57BL/6-*Camk2a^{tm1Vyb}*/J coisogenic strain resulted from deliberate introduction of the Vyb targeting vector (indicated by the superscript designation *tm* [for targeted mutation, i.e., a knockout]) to specifically alter the *Camk2a* gene locus. The mutation is maintained on the C57BL/6 background by The Jackson Laboratory (J).

Consomic Strains

A consomic strain is an inbred strain in which a chromosome has been completely replaced by the homologous chromosome from another inbred strain.[16,33] Both the donor and recipient strains share the same genetic background, but one entire donor chromosome has been substituted into the recipient's chromosomal complement. If each recipient chromosome is replaced, a family of 21 consomic strains can be produced for any donor/recipient pair. A consomic strain must be backcrossed for at least 10 generations to reach completion.

Consomic strains are designated using the code for the recipient strain, a hyphen (-), the letters *Chr* (for *chromosome*), a number to indicate which chromosome has been replaced, and a superscript describing the donor strain. An example is the C57BL/6J-Chr 11^{C3H/HeJ}/J consomic strain, in which chromosome 11 of the recipient strain (C57BL/6J) has been switched with the same chromosome from the donor strain (C3H/HeJ).

Conplastic Strains

A conplastic strain harbors the nuclear genome of one inbred strain and the mitochondrial genome of another.[16,46] The nuclear genes of breeding males are sorted with the cytoplasmic (mitochondrial) gene complement of breeding females by performing an initial backcross with a female of the donor strain followed by a series of backcrosses onto males of the recipient strain. Fully conplastic strains must have been backcrossed for at least 10 generations.

Conplastic strains are named using the code for the recipient strain that is contributing the nuclear genome, a hyphen (-), the letters *mt* (for *mitochondrial genome*), and a superscript describing the donor strain supplying the cytoplasmic genome. Thus, the C57BL/6J-mt$^{A/J}$/NaJ conplastic strain couples the nuclear gene complement of the recipient C57BL/6 strain with the mitochondrial genome of the donor A/J strain.

Techniques for Manipulating Genes That Direct Mouse Development

Two general paradigms are available for discovering and investigating the functions of novel genes. Both approaches require an initial step in which the genome is manipulated to introduce one or more induced genetic events. Subsequently, the impact of these gene alterations may be explored in either the forward or reverse direction.

Forward genetics is driven by the nature of the genetic change. The genome is modified at random or at a specific gene locus of unknown function. Mice bearing the mutation(s) are bred to create a colony in which the impact of the gene (i.e., the phenotype) may be studied. This scheme is particularly valuable for evaluating the roles of known genes *in vivo*.

Methods used in forward genetics include transgenic technology (in which new genetic material is inserted at random) and gene targeting (where an endogenous mouse gene is specifically replaced [*knocked out*] using an altered homolog). Both transgenic and gene targeting constructs can be engineered to deliver one or multiple copies of foreign genes (murine or nonmurine) or an altered endogenous (murine) gene. Gene targeting constructs can also be designed to swap a nonmurine homolog in place of its murine counterpart (*knocked in*). Expression of the new genetic material may be constitutive (always turned on). The ability to use genetic engineering technology to introduce gene systems that are controlled by exogenous chemicals (i.e., not present in normal mice) and the existence of unique promoting elements for certain genes now allows investigators to limit expression to specific times (e.g., only during development, or only in adulthood) and tissues.

Reverse genetics is dependent on the induction of a new deviant phenotype, characterized by some change (anatomic, behavioral, biochemical, etc.) with respect to the normal condition of WT mice. Once a deviation is discerned and characterized, genetic approaches are used to find the mutated gene(s) responsible for the change. The capacity to detect a novel phenotype is limited only by the number of means used to seek aberrations from the norm; close scrutiny of previously characterized phenotypes utilizing new technologies commonly demonstrates one or more new phenotypes which were missed during the initial workup. Because spontaneous mutations are infrequent, exposure to mutagenic chemicals (e.g., DNA-alkylating agents) or microorganisms (e.g., viruses) is often used to accelerate the mutation rate at random gene loci, which will boost the rate at which new phenotypes are discovered.

In the last two decades, many innovative techniques have been developed to specifically manipulate the mouse genome. The most widely used methods are briefly reviewed here.

Insertional Approaches for Deliberately Engineering Mouse Genes

Direct manipulation of the mouse genome to alter the native genetic material is the realm of insertional techniques. Engineered DNA is introduced into the nuclear (or occasionally the mitochondrial) genome either randomly or at a specific site. The new gene may express a new protein or replace an endogenous gene with a new one (which can be fully functional, partially functional, or null [inactive]). On occasion, a random transgenic insertion may ablate gene function, in which case great care must be taken to define which genetic alteration (the added transgene or the lost endogenous gene) is responsible for the phenotype.[30,35,45] Trangenes encoding small hairpin RNAs and small interfering RNAs (siRNA) may also be used to specifically reduce the expression of an endogenous gene (a *knockdown*).[14,26]

Pronuclear microinjection is a common means of physically introducing transgenes into mouse cells.[7] The engineered DNA carrying the transgene is injected into the pronucleus of a zygote (one-celled embryo [i.e., E0]) using an ultrafine glass pipette. If taken up by the nuclear DNA, the transgene will be incorporated at random into the mouse genome and propagated into all future daughter cells. The impact of the transgene will depend on the site(s) at which the transgene is inserted (i.e., the complement of adjacent preexisting genes and regulatory elements) and the number of copies that have been incorporated. Thus, transgenic embryos bearing the same construct will still be genetically distinct and may develop different phenotypes, or exhibit variable degrees of the same phenotype.

Blastocyst microinjection is used to produce cells containing a specific genetic alteration—most often an engineered null mutation—into a multicelled, cavitated blastocyst (a 3.5-day-old embryo [i.e., Theiler stage 4]). This practice is a fundamental step in gene targeting (see *Targeted insertion* in the following text). The first stage of this method requires that a targeted gene be substituted for its native counterpart in embryonic stem (ES) cells *in vitro* by homologous recombination. Donor ES cells carrying the altered gene are selected and then injected into the cavity (blastocoele) of the recipient blastocyst using an ultra-fine glass pipette. The altered ES cells intermingle at random with the recipient cells to form a chimera, an organism comprised of two or more tissues that have distinct genetic origins (i.e., that arise from different ES cell lineages). If donor ES cells eventually develop into germ cells, the altered gene will be present in the gametes and can be propagated by breeding.

Viral transfection with lentiviral or other retroviral vectors can be used to deliver small DNA constructs (~10 kb) into a broad range of tissues, including mouse embryos[25,36] and placenta.[15] These viruses are able to infect noncycling cells. Infected cells will reverse transcribe the viral RNA to make DNA, which is then incorporated into the host cell's genome. Retroviral vectors have been engineered so that the RNA construct they carry is produced but their normal packaging proteins are not. Thus, each vector can deliver its package but cannot hijack the host cell to make new infectious virions.

Gene trapping uses ES cells to introduce random mutations into genes, thereby producing fusion proteins without function but which are genetically regulated in the same manner as the WT protein.[2,11,27,47] Trapping vectors insert selectable reporter genes with downstream polyadenylation (pA) sequences into functional genes. The pA sequence terminates further transcription of the disrupted gene, while the reporter simultaneously identifies cells in which a trap has succeeded.

Targeted insertion uses homologous recombination to produce site-specific integration of foreign DNA into ES cells to produce either a gain-of-function (*knockin*) or loss-of-function (*knockout*) phenotype.[8,21] The targeting vector used in the recombination step replaces a WT allele of a particular gene with the DNA of interest, without affecting any other part of the genome, by aligning identical flanking sequences in the WT gene and the vector. A fraction of the aligned sequences will exchange DNA, thereby incorporating the engineered construct into the WT locus. The inserted DNA may be from any species and may encode either a functional (fully or partially) or a nonfunctional protein. The vector also typically bears positive (e.g., antibiotic resistance) and negative (e.g., viral thymidine kinase) selection markers to allow selection of only those cells in which the correct recombination event has taken place.

Conditional targeted insertion allows expression of an engineered mutation to be confined to a specific space (target location). The impact of the induced mutation is then evaluated by mating carrier mice with an animal bearing a recombinase transgene (e.g., Cre, which excises double-stranded DNA specifically located between two loxP sites to allow recombination). The location of the enzyme activity in a Cre strain is made conditional for a specific tissue or cell type by linking the Cre gene to a site-specific promoter; many Cre strains bearing different promoters are required to evaluate function of the engineered gene in multiple locations. The orientation of the loxP sites relative to the engineered gene can determine the nature of the recombination event. This technology is particularly effective for exploring the outcome of protein inactivation during adulthood for genes which induce embryolethality if not available during development.

Inducible targeted insertion permits direct control of an engineered mutation's expression at a specific time.[12,37] This control is mediated by an inducible transcriptional activator that is linked to the promoter which controls the engineered gene. Addition of an exogenous chemical that interacts with the activator (e.g., ecdysone, tetracycline) turns expression *on* or *off*. Again, this technique is an important means of assessing the result of gene inactivation during adulthood for genes that cause embryolethality if absent during development.

Noninsertional Approaches for Randomly Engineering Mouse Genes

Spontaneous mutations are a relatively rare event in nature, but the rate of mutagenesis can be accelerated using mutagens (e.g., ethyl-*N*-nitrosourea [ENU],[20] ionizing radiation[19]). In fact, this approach is the most efficient way to generate vast numbers of novel mutations, since a mutagen can cause numerous random mutations in a single cell (e.g., ENU affects approximately 1 in 700 genes), and different patterns of random mutations in adjacent cells. The use of a mutagen produces noninsertional genetic alterations as no foreign DNA is introduced into the mouse genome. Progeny of the mutagen-treated males are evaluated for deviant phenotypes, which arise due to mutations in parental sperm cells; affected offspring are then bred to maintain the new trait so that it can be characterized. This approach has proven to be an efficient tool for discovering novel genes and revealing new functions of known genes.

Current Uses for Mice as Models for Developmental Research

In general, historical mouse strains experienced relatively little embryolethality, although nonlethal abnormalities (i.e., developmental pathology) were not infrequent. However, the pace at which new categories of mice are being made has exploded during the past two decades with the advent of the new techniques for genetic manipulation (described earlier). Developing mice from a substantial fraction of these genetically engineered strains exhibit lesions before or shortly after birth, and embryolethality and perinatal lethality are a relatively frequent finding. In fact, the need to avoid such early lethality outcomes when trying to investigate gene function during adulthood was a major impetus for developing the conditional and inducible systems for gene targeting.

Mouse developmental biology, both normal and pathologic, is a rapidly burgeoning field of investigation. Two main factors are driving this renaissance. One major cause is a desire for basic biological knowledge regarding molecular mechanisms of vertebrate, and especially mammalian, development. The relatively close genetic and physiological relationships of mice and humans make the mouse a model for questions of this nature. In addition, developmental lethality is a common outcome of mouse genetic engineering experiments performed in drug discovery and development programs. A better understanding of how genetic defects can alter developmental programming will help accelerate the pace at which such information can be applied to understand similar processes in diseased tissues (e.g., cancer, neurodegeneration).

REFERENCES

1. Adams MD (2007). The mouse genome. In: *The Mouse in Biomedical Research, Vol. 1: History, Wild Mice, and Genetics* (Fox JG, Barthold SW, Davisson MT, Newcomer CE, Quimby FW, Smith AL, eds.). San Diego, CA: Academic Press (Elsevier); pp. 99–114.
2. Aiba K, Carter MG, Matoba R, Ko MS (2006). Genomic approaches to early embryogenesis and stem cell biology. *Semin. Reprod. Med.* 24(5): 330–339.
3. Arden B, Klein J (1982). Biochemical comparison of major histocompatibility complex molecules from different subspecies of *Mus musculus*: Evidence for trans-specific evolution of alleles. *Proc. Natl. Acad. Sci. USA* 79(7): 2342–2346.
4. Berry ML, Linder CC (2007). Breeding systems: Considerations, genetic fundamentals, genetic background, and strain types. In: *The Mouse in Biomedical Research, Vol. 1: History, Wild Mice, and Genetics* (Fox JG, Barthold SW, Davisson MT, Newcomer CE, Quimby FW, Smith AL, eds.). San Diego, CA: Academic Press (Elsevier); pp. 53–78.
5. Bienvenu T, Charles C, Guy F, Lazzari V, Viriot L (2008). Diversity and evolution of the molar radicular complex in murine rodents (Murinae, Rodentia). *Arch. Oral Biol.* 53(11): 1030–1036.
6. Bonhomme F, Catalan J, Britton-Davidian J, Chapman V, Moriwaki K, Nevo E, Thaler L (1984). Biochemical diversity and evolution in the genus *Mus. Biochem. Genet.* 22(3–4): 275–303.
7. Brinster RL, Chen HY, Trumbauer ME, Yagle MK, Palmiter RD (1985). Factors affecting the efficiency of introducing foreign DNA into mice by microinjecting eggs. *Proc. Natl. Acad. Sci. USA* 82(13): 4438–4442.

8. Capecchi MR (1989). Altering the genome by homologous recombination. *Science* 244(4910): 1288–1292.

9. Committee on Standardized Genetic Nomenclature for Mice (1996). Rules for nomenclature of inbred strains. In: *Genetic Variants and Strains of the Laboratory Mouse*, 3rd edn., vol. 2 (Lyon MF, Rastan S, Brown SDM, eds.). Oxford, U.K.: Oxford University Press; pp. 1532–1536.

10. Eppig JT (2007). Mouse strain and genetic nomenclature: An abbreviated guide. In: *The Mouse in Biomedical Research, Vol. 1: History, Wild Mice, and Genetics* (Fox JG, Barthold SW, Davisson MT, Newcomer CE, Quimby FW, Smith AL, eds.). San Diego, CA: Academic Press (Elsevier); pp. 79–98.

11. Floss T, Wurst W (2002). Functional genomics by gene-trapping in embryonic stem cells. *Methods Mol. Biol.* 185: 347–379.

12. Furth PA, St Onge L, Böger H, Gruss P, Gossen M, Kistner A, Bujard H, Hennighausen L (1994). Temporal control of gene expression in transgenic mice by a tetracycline-responsive promoter. *Proc. Natl. Acad. Sci. USA* 91(20): 9302–9306.

13. Galtier N, Bonhomme F, Moulia C, Belkhir K, Caminade P, Desmarais E, Duquesne JJ, Orth A, Dod B, Boursot P (2004). Mouse biodiversity in the genomic era. *Cytogenet. Genome Res.* 105(2–4): 385–394.

14. Gao X, Zhang P (2007). Transgenic RNA interference in mice. *Physiology (Bethesda)* 22(3): 161–166.

15. Georgiades P, Cox B, Gertsenstein M, Chawengsaksophak K, Rossant J (2007). Trophoblast-specific gene manipulation using lentivirus-based vectors. *Biotechniques* 42(3): 317–318, 320, 322–315.

16. Gregorová S, Divina P, Storchova R, Trachtulec Z, Fotopulosova V, Svenson K, Donahue L, Paigen B, Forejt J (2008). Mouse consomic strains: Exploiting genetic divergence between *Mus m. musculus* and *Mus m. domesticus* subspecies. *Genome Res.* 18(3): 509–515.

17. Guénet JL, Bonhomme F (2003). Wild mice: An ever-increasing contribution to a popular mammalian model. *Trends Genet.* 19(1): 24–31.

18. Ideraabdullah FY, de la Casa-Esperón E, Bell TA, Detwiler DA, Magnuson T, Sapienza C, de Villena FP (2004). Genetic and haplotype diversity among wild-derived mouse inbred strains. *Genome Res.* 14(10A): 1880–1887.

19. Jacquet P (2004). Sensitivity of germ cells and embryos to ionizing radiation. *J. Biol. Regul. Homeost. Agents* 18(2): 106–114.

20. Justice MJ, Noveroske JK, Weber JS, Zheng B, Bradley A (1999). Mouse ENU mutagenesis. *Hum. Mol. Genet.* 8(10): 1955–1963.

21. Koller BH, Smithies O (1992). Altering genes in animals by gene targeting. *Ann. Rev. Immunol.* 10: 705–730.

22. Kruisbeek AM (2001). Commonly used mouse strains. *Curr. Protoc. Immunol.* Appendix 1: Appendix 1C.

23. Linder CC (2006). Genetic variables that influence phenotype. *ILAR J* 47(2): 132–140.

24. Linder CC, Davisson MT (2004). Strains, stocks, and mutant mice. In: *The Laboratory Mouse* (Hedrich H, Bullock G, eds.). San Diego, CA: Academic Press (Elsevier); pp. 25–46.

25. Lois C, Hong EJ, Pease S, Brown EJ, Baltimore D (2002). Germline transmission and tissue-specific expression of transgenes delivered by lentiviral vectors. *Science* 295(5556): 868–872.

26. Lykke-Andersen K (2006). Regulation of gene expression in mouse embryos and its embryonic cells through RNAi. *Mol. Biotechnol.* 34(2): 271–278.

27. Lyons GE, Swanson BJ, Haendel MA, Daniels J (2000). Gene trapping in embryonic stem cells in vitro to identify novel developmentally regulated genes in the mouse. *Methods Mol. Biol.* 136: 297–307.

28. Maltais LJ, Blake JA, Eppig JT, Davisson MT (1997). Rules and guidelines for mouse gene nomenclature: A condensed version. International Committee on Standardized Genetic Nomenclature for Mice. *Genomics* 45(2): 471–476.

29. Marshall JT (1986). Systematics of the genus *Mus. Curr. Top Microbiol. Immunol.* 127: 12–18.

30. McNeish JD, Thayer J, Walling K, Sulik KK, Potter SS, Scott WJ (1990). Phenotypic characterization of the transgenic mouse insertional mutation, *legless. J. Exp. Zool.* 253(2): 151–162.

31. Mekada K, Abe K, Murakami A, Nakamura S, Nakata H, Moriwaki K, Obata Y, Yoshiki A (2009). Genetic differences among C57BL/6 substrains. *Exp. Anim.* 58(2): 141–149.

32. Mouse Genome Sequencing Consortium (2001). Initial sequencing and comparative analysis of the mouse genome. *Nature* 420(6915): 520–562.

33. Nadeau JH, Singer JB, Matin A, Lander ES (2000). Analysing complex genetic traits with chromosome substitution strains. *Nat. Genet.* 24(3): 221–225. [Erratum in: *Nat. Genet.* 25(1): 125, 2000.]

34. Nagy A, Gertsenstein M, Vintersten K, Behringer R (2003). *Manipulating the Mouse Embryo*, 3rd edn. Cold Spring Harbor, NY: Cold Spring Harbor Laboratory Press.

35. Rijli FM, Dolle P, Fraulob V, LeMeur M, Chambon P (1994). Insertion of a targeting construct in a *Hoxd-10* allele can influence the control of *Hoxd-9* expression. *Dev. Dyn.* 201(4): 366–377.

36. Robertson E, Bradley A, Kuehn M, Evans M (1986). Germ-line transmission of genes introduced into cultured pluripotential cells by retroviral vector. *Nature* 323(6087): 445–448.

37. Schönig K, Bujard H (2003). Generating conditional mouse mutants via tetracycline-controlled gene expression. *Methods Mol. Biol.* 209: 69–104.

38. Searle JB, Jamieson PM, Gündüz I, Stevens MI, Jones EP, Gemmill CE, King CM (2009). The diverse origins of New Zealand house mice. *Proc. Biol. Sci.* 276(1655): 209–217.

39. Shimizu H, Nagakui Y, Tsuchiya K, Horii Y (2001). Demonstration of chymotryptic and tryptic activities in mast cells of rodents: Comparison of 17 species of the family Muridae. *J. Comp. Pathol.* 125(1): 76–79.

40. Silver LM (2008). *Mouse Genetics: Concepts and Applications.* Oxford, U.K.: Oxford University Press. Available online at http://www.informatics.jax.org/silver/index.shtml (last accessed October 31, 2014).

41. Sundberg JP, Schofield PN (2010). Mouse genetic nomenclature: Standardization of strain, gene, and protein symbols. *Vet. Pathol.* 47(6): 1100–1104.

42. Tucker PK (2007). Systematics of the genus *Mus*. In: *The Mouse in Biomedical Research, Vol. 1: History, Wild Mice, and Genetics* (Fox JG, Barthold SW, Davisson MT, Newcomer CE, Quimby FW, Smith AL, eds.). San Diego, CA: Academic Press (Elsevier); pp. 13–23.

43. Tyzzer EE (1909). A series of spontaneous tumors in mice with observations on the influence of heredity on the frequency of their occurrence. *J. Med. Res.* 21(3): 479–518.

44. Veyrunes F, Dobigny G, Yang F, O'Brien PC, Catalan J, Robinson TJ, Britton-Davidian J (2006). Phylogenomics of the genus *Mus* (Rodentia; Muridae): Extensive genome repatterning is not restricted to the house mouse. *Proc. Biol. Sci.* 273(1604): 2925–2934.

45. Woychik RP, Stewart TA, Davis LG, D'Eustachio P, Leder P (1985). An inherited limb deformity created by insertional mutagenesis in a transgenic mouse. *Nature* 318(6041): 36–40.

46. Yu X, Gimsa U, Wester-Rosenlöf L, Kanitz E, Otten W, Kunz M, Ibrahim SM (2009). Dissecting the effects of mtDNA variations on complex traits using mouse conplastic strains. *Genome Res.* 19(1): 159–165.

47. Zambrowicz BP, Friedrich GA (1998). Comprehensive mammalian genetics: History and future prospects of gene trapping in the mouse. *Int. J. Dev. Biol.* 42(7): 1025–1036.

3

Essential Terminology for Mouse Developmental Pathology Studies

Brad Bolon and Colin G. Rousseaux

CONTENTS

As with all specialized scientific endeavors, the fields of developmental biology and developmental pathology have built long lists of jargon to describe the essential structures, functions, and processes that are encountered in young animals. Individuals who engage in the acquisition, analysis, and interpretation of normal features and developmental lesions must have a ready command of the terminology required to ensure that information is communicated precisely to persons who seek to gain insights from developmental pathology studies. Correct terminology is also essential to permit a proper statistical analysis; calculations performed using a nonspecific term (e.g., *neural tube defect*) will require a different analysis relative to calculations made using classes defined by more discrete terms (*anencephaly*, *exencephaly*, *encephalocele*, etc.). The terms in this chapter have been drawn from multiple sources, including medical dictionaries[1,5] and published committee reports regarding preferred nomenclature.[4,6] Mouse developmental pathology researchers also should seek to glean new terms that they encounter in the scientific literature.

This chapter is divided into two parts. The first section defines appropriate nomenclature for selected structures and developmental processes in young mice. The list was devised to provide a brief introduction to fundamental concepts of importance in genetics (see Chapter 2) and developmental biology (see Chapter 4) with which mouse pathologists should be familiar. The second section presents a glossary of key terms for common lesions observed in mouse embryos, fetuses, neonates, and juveniles (see Chapter 15) as well as placenta (see Chapter 16) in which significant developmental defects have been produced by mutations, physical agents, or toxicants. Together, these lists provide a reasonable introduction to the discipline-specific lexicon needed to contribute efficiently and effectively to developmental pathology studies. Where feasible, the definitions have been cross-referenced to relevant figures in other chapters.

Terminology for Describing Normal Developmental Structures and Processes

The mouse developmental pathology literature may confuse novice researchers as two different conventions are employed for naming organs and defining their orientation with respect to other sites in the specimen. The imprecision stems from the frequent practice of applying nomenclature conventions for humans rather than the correct terms for mice. The potential for misunderstanding is implicit in the different body orientations of these two species: humans are bipedal, with a vertical (upright) body axis, while mice are quadrupeds with a horizontal body axis. These disparate body carriages dictate different naming conventions (Figure 3.1).

To avoid confusion, mouse developmental pathology reports and publications should describe organ positions using the appropriate nomenclature for this species, as set forth in the *Nomina Anatomica*

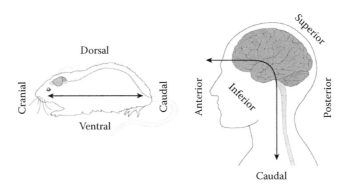

FIGURE 3.1   Schematic diagrams illustrating the most appropriate anatomic nomenclature to use for designating structure names in quadrupeds (mice) versus bipeds (humans). (The illustration was prepared by Mr. T. Vojt, The Ohio State University.)

Veterinaria.[2] Human-specific terms from the *Terminologia Anatomica*[3] (the successor to the *Nomina Anatomica*) generally should be avoided; however, a permissible compromise in naming mouse structures is to list the veterinary term for a feature and then follow it by the medical term (placed within parentheses). For example, the human anatomic term *superior cervical ganglion* should be avoided in mice and instead replaced by the recognized veterinary term *cranial cervical ganglion* or the compromise rendering *cranial cervical ganglion (superior cervical ganglion)*. In general, descriptive terms should be employed rather than eponyms (e.g., *primitive node* instead of *Henson's node* or *Spemann's organizer*) to promote clear communication of shared concepts and mechanisms across multiple species.

The other major terminology choice is the means for designating the apparent age of a developing animal (Table 6.1). After birth, the convention is to use the chronological age, setting the day of birth as postnatal day (PND [or P]) 0. Before birth, age is expressed as *days post coitum* (dpc), *embryonic day* (E), and *gestational day* (GD); these terms may be used interchangeably. Nonetheless, the timing before birth is more muddled as the day of conception is established by many investigators as GD 0 but is considered by others to be GD 1. Accordingly, users of developmental pathology data must check which convention is being used in order to correctly understand the meaning of the findings, while it is incumbent upon researchers to specifically state which option they have employed. In my experience, a simple means for communicating this point is to include a statement in the Methods section, and ideally in the abstract, indicating that the plug-positive day is set as GD 0 (or GD 1).

The following terms should be familiar to developmental pathologists. This list was chosen because the entries occur with some frequency in mouse developmental pathology reports. More importantly, these terms permit meaningful discussion among scientific collaborators with diverse scientific backgrounds.

Allantois: Plural: allantoides or allantoises. A columnar membrane extending from the primitive hind gut to the chorionic plate that serves as the main precursor of the umbilical cord (illustrated in Figures 4.34, 4.39, and 16.8).

Allele: One of several alternative forms of a gene. Mice are diploid organisms with paired chromosomes, and so have two alleles for a given gene (one on each chromosome).

Amnion: Plural: amnions or amnia. A thin, translucent, fluid-filled membrane that encloses the embryo or fetus (illustrated in Figures 4.12, 4.34, and 4.37).

Angiogenesis: Formation of new blood vessels from preexisting vessels (compare to *Vasculogenesis*).

Anlage: Plural: anlagen. The group of embryonic cells from which an organ first develops.

Apoptosis: Programmed cell death used to remodel the organization of developing structures (illustrated in Figures 5.4, 5.5, and 14.1).

Blastocyst: A multicelled, cavitated embryo (~GD 3.5 to GD 4.5) comprised of an outer wall, the trophoblast (primordial placenta), and an inner cell mass (primordial embryo) (illustrated in Figures 1.1, 4.4 through 4.6, and 7.3).

Blood island: Dispersed aggregates of hematopoietic precursor cells that form during early gestation (~GD 7.5 to GD 8.0) within the yolk sac (illustrated in Figure 4.36).

Chorion: Plural: chorions or choria. A cellular outer membrane that encloses the embryo or fetus (illustrated in Figures 4.34, 4.37, and 16.8); a portion termed the chorionic plate serves as the fusion site for the allantois (shown in Figure 4.39).

Conceptus: Plural: **concepti** or **conceptuses**. The entire mass of tissue that develops from the fertilized egg, including the embryo/fetus and the extra-embryonic membranes (i.e., placenta).

Conditional transgenic: An engineered gene regulated by elements that control when (at what development stage) and/or where (in what cells) it will be expressed.

Congenic: A rodent strain in which continual backcrossing for at least 10 generations yields a consistent and homogenous genetic constitution at all major gene loci.

Congenital: A condition that is present from birth (synonyms: *hereditary*, *inborn*).

Decidualization: Process by which uterine stromal cells differentiate into large epithelioid (*decidual*) cells (illustrated in Figure 4.34) in response to both steroid hormones and embryonic signals.

Dominant negative: A mutation whose product overrides the function of the normal wild-type gene product within the same cell (see also **Antimorph**).

Ectoderm: The outer of the three primitive germ layers in the early embryo, from which are derived the nervous system, mucous membranes of the mouth and anus, and skin.

Embryo: The unborn mouse from the time of conception until all organs are first formed (approximately GD 14.5–GD 15.0) (various stages are shown in Figure 1.4 as well as in Figures 4.1 through 4.25).

Embryogenesis: Time in development from conception until all organs have begun to form (approximately GD 14.5–GD 15.0).

Endoderm: The inner of three primary germ cell layers in the early embryo, from which are derived the epithelial lining of multiple internal organs.

Fetus: Plural: **feti** or **fetuses**. The unborn mouse from the completion of major organogenesis (approximately GD 14.5–GD 15.0) to birth (GD 19) (illustrated in Figures 4.26 through 4.29).

Gamete: A reproductive cell—spermatozoon for males, ovum for females—that contains the haploid set of chromosomes (i.e., only one allele of each gene).

Gastrulation: Portion of embryogenesis during which the mesoderm is first produced (~GD 6.5) (illustrated in Figures 4.10 and 4.11).

Gene targeting: A genetic engineering method wherein a gene is removed (*knock-out*) or replaced (*knock-in*) with specificity by homologous recombination.

Genotype: The genetic makeup of an animal.

Haploinsufficiency: The presence of a single functional gene in a cell, which by itself does not produce enough protein to sustain full function.

Haplotype: A clustered group of gene sequences at adjacent positions on a DNA strand that will be transferred together during genome replication.

Hatching: Escape of the blastocyst from within the *zona pellucida*, a necessary event in preparation for implantation.

Hemizygous: A transgenic animal into which one or more copies of a nonendogenous gene have been inserted, typically at random in a single locus; alternatively, the condition wherein an otherwise diploid cell possesses one or more unpaired genes.

Hemotrophic nutrition: Transfer of blood-borne nutrients in the definitive (late) placenta from maternal blood to embryonic blood.

Heterozygous: An animal with differing alleles at the same gene locus on both paired chromosomes (typically denoted as HET or +/−).

Histiotrophic nutrition: Transfer of nutrients from dam to embryo early in gestation via phagocytic uptake of uterine secretions by trophectoderm and trophoblast.

Homozygous: An animal with identical alleles at a gene locus (usually denoted as −/− for knock-out [KO] animals and +/+ for wild-type [WT] animals).

Humanized: An animal into which human genetic material or human cells has been inserted.

Juvenile: A young animal (in mice, often considered to extend from weaning until puberty) (illustrated in Figures 15.14 and 15.A.11).

Knock-down: An animal (sometimes abbreviated as KD) in which a gene's function has been reduced but not lost (see also **Hypomorph**).

Knock-in: An animal (often abbreviated as KI) in which a functional endogenous gene is replaced by a foreign gene (often the human homolog) to test whether or not it serves the function of the original gene.

Knock-out: An animal (generally denoted as KO or −/−) in which an endogenous gene has been replaced by a nonfunctional engineered gene (see also **Amorph**).

Locus: Plural: **loci**. The precise location of a gene on a chromosome.

Mesoderm: The middle of three primary germ cell layers in the early embryo, from which the blood, connective tissue, mesothelium, muscles, and skeletal tissues are derived.

Morula: Plural: **morulae**. A multicelled, solid embryo (illustrated in Figures 4.3 and 7.3).

Mosaic: An individual that developed from a single zygote but that possesses two or more genetically different cell populations, which are genetically distinct due to a gene mutation or chromosomal nondisjunction during early embryogenesis. The extent of the mosaic state depends on the cleavage stage at which the genetic event occurred.

Neonate: A newborn, technically PND 0–1 but often up to PND 7 (which is the time of development at which physiological evolution of mouse pups approximately matches that of human neonates) (illustrated in Figure 4.30).

Neurulation: Portion of embryogenesis during which the central nervous system is defined (~E8–E9.5).

Oocyte: A developing egg cell (ovum) at any stage before or after ovulation, but before fertilization.

Organogenesis: The portion of embryogenesis during which major organs/systems are first formed (approximately GD 8.0 to GD 14.5).

Perinatal: The time of development from soon before to soon after birth. Definitions differ among investigators, although in many instances, an exact range is not given. (A useful span is from GD 17.5 to PND 2.)

Phenotype: The sum of all functional and structural changes produced by the activity of an animal's genotype.

Placenta: Plural: **placentae** or **placentas**. A transient organ that supports the prenatal growth of the embryo from which it is derived by permitting the exchange of nutrients and wastes between the maternal and embryonic circulations.

Polar body: Each of the small cells (essentially discarded nuclear remnants) that buds from an oocyte at the two meiotic divisions so that the normal diploid gene complement is retained once the oocyte has been fertilized (illustrated in Figures 4.1, 4.2, and 7.3).

Puberty: The period during which juvenile animals reach sexual maturity and so become capable of reproduction.

Stem cell: A cell that upon division gives rise to one daughter that can differentiate into a specialized cell type and another daughter that retains stem cell properties.

Transgenic: An engineered animal into which a specific genetic modification is induced by random insertion of foreign genetic material.

Trophectoderm: The trophoblast layer after gastrulation, which is continuous with the ectoderm layer of the embryo (illustrated in Figure 4.7).

Trophoblast: The outer layer of the mammalian blastocyst.

Vasculogenesis: Formation of new blood vessels by direct differentiation from pluripotent mesenchymal cells (extra-embryonic mesoderm) (illustrated in Figures 16.7 and 16.14). (See also **Angiogenesis**.)

Vestigial: A small but recognizable remnant of something that was once larger (in the same individual, or in other members of the species).

Vitelline: Blood vessels of the yolk sac.

Yolk sac: A cellular outer membrane that encloses the embryo or fetus (illustrated in Figures 4.35, 4.36 through 4.38, 4.40, 16.2, and 16.3).

***Zona pellucida*:** A thick, protective membrane of glycoproteins that surrounds the embryo until hatching.

Zygote: A single-cell embryo (i.e., a newly fertilized ovum).

Terminology for Describing Developmental Lesions and Processes

Developmental pathologists typically employ one of two strategies in assigning a name for a lesion. The first is to use a specific term for the change, where such an entity is well recognized by the community at large. The list below gives a census of common developmental pathology terms as they have been presented previously in the developmental biology, developmental toxicology, and teratology literature. The second option, which is utilized by necessity when a more specific word is not available, is to describe features in terms of physical characteristics (color, consistency, shape, size, etc.; see Chapter 15 for more detail). Thus, a *small eye* is referred to using a specific term, microphthalmia, while an *open eye at birth* (see Figure 15.23) is described using that phrase since a more explicit word is not universally applied. The key principle to remember is that the chosen words must communicate the nature of the lesion. Even well-known terms may need to be defined in documents if the some of the readers are likely to be unfamiliar with the jargon.

Ablepharia: Absence of a fusion point for the eyelids, so that the eyes are covered with a continuous skin layer.

Abrachia: Absence of the fore limbs.

Acampsia: Rigidity or inflexibility of a joint (see also **Ankylosis**).

Acardia: Absence of the heart.

Acaudia, acaudate: Without a tail (synonym: **Anury**).

Acephalostomia: Absence of the head but with the presence of an orifice in the neck region.

Acephaly: Agenesis of the head (synonym: **Acephalia**).

Acheiria: Congenital absence of one or both hands (forepaws).

Achondroplasia: Inadequate bone formation resulting in short limbs and other defects, due to abnormal cartilage (e.g., skeletal dysplasia).

Acorea: Absence of the pupil of the eye.

Acrania: Partial or complete absence of the cranium—generally applied to the calvaria (i.e., the dorsal skull) (illustrated in Figure 15.4).

Acystia: Absence of the urinary bladder.

Adactyly: Absence of digits (illustrated in Figure 15.A.3).

Agalactosis: A lack of milk secretion.

Agenesis: Lack of development of an organ (synonym: **Aplasia**) (illustrated in Figures 15.4, 15.12, 15.13, and 15.17 as well as 15.A.3 and 15.A.7).

Aglossia: Absence of the tongue.

Agnathia: Absence of the lower jaw (mandible) (illustrated in Figure 15.4).

Amastia: Absence of the mammary glands.

Amelia: Complete absence of a limb or limbs (see also **Ectromelia** and **Phocomelia**) (illustrated in Figure 15.13).

Ametria: Absence of the uterus.

Amorph: A mutation resulting in a total loss of gene function (a *null* mutation), either by ablation of the gene (DNA amorph), failure of gene transcription (RNA amorph), or abnormal translation (protein amorph).

Anasarca: Generalized edema.

Anencephaly: Absence of the cranial vault with the brain (especially cerebrum and midbrain) either missing or greatly reduced in size (illustrated in Figure 15.4).

Anephrogenesis: Absence of kidney(s) (illustrated in Figure 15.A.7).

Aniridia: Absence of the iris.

Anisomelia: Inequality in size between paired limbs.

Ankyloglossia: Partial or complete adhesion of the tongue to the floor of the mouth.

Ankylosis: Abnormal fixation of a joint; implies bone fusion (see also **Acampsia**).

Anodontia: Absence of some or all of the teeth.

Anomaly: A feature (applied to structures rather than biochemical changes or functional alterations) that deviates from the expected (i.e., "normal") pattern.

Anonchia: Absence of some or all of the claws.

Anophthalmia: Absent eye(s) (illustrated in Figure 15.17).

Anorchism: Unilateral or bilateral absence of the testes (synonym: **Anorchia**).

Anostosis: Defective development of bone, often via failure of ossification.

Anotia: Absence of the external ear(s) (i.e., pinnae, auricles).

Anovarism: Absence of the ovaries (synonym: **Anovaria**).

Antimorph(ic): A dominant negative mutation that works counter to normal gene activity (see also **Dominant Negative**).

Anury: See **Acaudia**.

Aphakia: Absence of the eye lens.

Aphalangia: Absence of a digit or of one or more phalanges.

Aplasia: See **Agenesis**.

Apodia: Absence of one or more paws.

Apoptosis: Self-destruction of single or clustered cells in response to a nonphysiological insult rather than inherent programming (illustrated in Figures 5.4 and 14.1).

Aprosopia: Absence of the face.

Arachnodactyly: Abnormal length and slenderness of the digits (*spider-like*).

Arhinia: Absence of the nose (illustrated in Figure 15.4).

Arthrogryposis: Persistent flexure or contracture of a joint. (**Arthrogryposis multiplex congenita** is a syndrome distinguished by congenital fixation of the joints leading to secondary muscle hypoplasia.)

Ascites: Accumulation of an effusion (watery fluid) in the abdomen (outmoded synonym: dropsy).

Asplenia: Absence of the spleen.

Astomia: Absence of the opening of the mouth.

Atelectasis: Incomplete expansion of a fetal lung or a portion of the lung at birth.

Athelia: Absence of the nipple(s).

Athymism: Absence of the thymus gland (synonym: **Athymia**).

Atresia: Absence of a normally patent lumen or closure of a normal body opening (e.g., **Atresia ani**— agenesis of the anal opening).

Atrial septum defect: Postnatal communication connecting the atria (abbreviated ASD) (illustrated in Figure 15.8).

Bipartite: Division of a single structure into two parts; often applied to areas of skeletal ossification (illustrated in Figure 15.16).

Brachydactyly: Abnormal shortness of digits (alternative spelling: brachiodactyly) (illustrated in Figure 15.A.3).

Brachygnathia: Abnormal shortness of the mandible (synonym: **Hypognathia**; illustrated in Figure 15.4).

Brachyury: Abnormally short tail (synonym: **Microcauda**; illustrated in Figure 15.22).

Buphthalmia: Enlargement and distension of the fibrous coats of the eye; often a consequence of congenital glaucoma (synonym: **Buphthalmos**).

Camptodactyly: Permanent flexion of one or more digits (synonym: **Camptodactylia**).

Cardiomegaly: Large heart (often due to chronic overload of the right heart) (illustrated in Figure 15.8).

Celoschisis: See **Gastroschisis**.

Celosomia: Fissure or absence of the sternum, with visceral herniation.

Cephalocele: Protrusion of part of the brain through the cranium. (See also **Encephalocele** and **Meningocele**.)

Cheiloschisis: Cleft lip, usually affecting the upper lip (*harelip*).

Choristoma: An ectopic *rest of normal tissue* (i.e., a histologically normal tissue in an abnormal location).

Cleft lip: See **Cheiloschisis**.

Cleft palate: See **Palatoschisis** (illustrated in Figures 15.4 and 15.15).

Clinodactyly: Permanent lateral or medial deviation of one or more digits.

Club Foot: See **Talipes**.

Coarctation: A stricture or stenosis, usually of the aorta.

Coloboma: Fissure or incomplete development of the eye and/or eyelid.

Conjoined twins: Two fetuses from a single zygote that are partially fused (though to varying degrees) due to incomplete separation during cleavage.

Cranial doming: Increased rounding of the top of the head, which may or may not be related to hydrocephalus.

Craniopagus: Conjoined twins connected at the head (usually by fusion of the skull rather than the brain).

Craniorachischisis: Fissure of the skull and vertebral column (possibly with extension into the brain and spinal cord).

Cranioschisis: Cleft skull.

Craniostenosis: Malformation of the skull resulting from **Craniosynostosis**.

Craniosynostosis: Premature ossification of cranial sutures.

Cryptophthalmia: Formation of a continuous skin layer rather than eyelids over the eye(s).

Cryptorchidism: Failure of one or both testes to enter the scrotum (synonym: **Cryptorchism**).

Cyst: Any abnormal sac, usually filled with some material (often fluid).

Curled tail: Circular displacement of tail tip by at least 180° (a half circle or more) (illustrated in Figure 15.22).

Cyclopia: Fusion of the two orbits into a single orbit, often along with absence of the nose or the presence of a tubular appendage (proboscis) above the orbit.

Deformation: Structural defect caused by an extrinsic mechanical force acting to distort but not destroy a nearby normal tissue.

Degeneration: Biochemical and/or structural deterioration of one or more cells, typically leading to a reduction (transient or permanent) in function (illustrated in Figures 15.18 and 16.10).

Delayed ossification: Incomplete mineralization of an otherwise normal ossification center.

Developmental delay: Temporary deferral in attaining one or more anatomic or functional attributes (illustrated in Figure 15.6).

Dextrocardia: Abnormal displacement of the heart to the right side of the thoracic cavity.

Diaphragmatic hernia: Displacement of abdominal organs into the thoracic cavity through a fissure in the diaphragm.

Dicephalus: Conjoined twins with one body and two heads.

Dilated cerebral ventricle: Minimal to mild increase in the size of cerebral ventricles (especially the lateral ventricles), with no evidence of decrease in cortical thickness (see also **Hydrocephalus**).

Diplomyelia: Doubling of the spinal cord (partial or complete) due to a longitudinal fissure.

Diplopagus: Conjoined twins with largely complete bodies that may share one or more internal organs.

Diprosopus: Facial duplication (usually partial).

Disruption: A structural defect due to an extrinsic insult that actually destroys normal tissue.

Diverticulum: A localized pocket formed in the wall of a hollow organ that opens into its lumen.

Dysarthrosis: Malformation of a joint.

Dysgenesis: Defective generation of a structure; commonly used to mean that the abnormality began at the time the structure was first starting to form (illustrated in Figures 5.6c, 15.9, and 15.A.9).

Dysmorphogenesis: The development of ill-formed structures.

Dysplasia: Abnormal tissue development in which altered organization of cells and tissue(s) results in morphologic defects; usually employed to mean that the structure began to develop correctly but finished in an inappropriate fashion (illustrated in Figures 5.5, 5.6b, 15.20, 15.A.10, 16.10, and 16.11).

Dysraphism: Failure of fusion of a seam, especially applied to the neural folds (synonym: **Dysraphia**) (illustrated in Figures 15.3 and 15.4).

Dystocia: Prolonged or difficult delivery.

Dystrophy: A disorder in which an organ or tissue develops abnormally or, once having formed, eventually begins to regress.

Ectopia: Displacement or malposition of a cell, tissue, or organ (see also **Choristoma** and **Heterotopia**; illustrated in Figure 15.20).

Ectopia cordis: Displacement of the heart outside of the thoracic cavity due to failed closure of the ventral wall (i.e., sternum).

Ectopic pregnancy: Pregnancy occurring outside the lumen of the uterus (commonly in the oviduct or abdominal cavity).

Ectrodactyly: Absence of all or only part of one or more digits (see also **Adactyly** and **Brachydactyly**; illustrated in Figure 15.A.3).

Ectromelia: Hypoplasia or absence of one or more limbs (see also **Amelia** and **Phocomelia**; illustrated in Figure 15.13).

Ectropion: Abnormal eversion of the margin of the eyelid.

Edema: Transfer of an effusion into a tissue or cavity.

Effusion: A cell- and protein-poor fluid.

Encephalocele: Herniation of part of the brain, generally encased in meninges, through an opening in the skull (see also **Cephalocele, Meningocoele,** and **Meningoencephalocoele**) (illustrated in Figure 15.4).

Entropion: Abnormal inversion of the margin of the eyelid.

Epispadias: Absence of the upper wall of the urethra, especially in males (leading to the urethral orifice opening on the dorsal surface of the penis).

Ethmocephaly: A defect in the holoprosencephaly spectrum in which a proboscis separates two closely set, often small (microphthalmic) eyes.

Eventration: Protrusion of the intestines through a fissure in the abdominal wall (usually on the ventral surface) (illustrated in Figure 15.23).

Exencephaly: A neural tube closure defect in which a cleft in the skull exposes the brain (usually the cerebrum and midbrain) (illustrated in Figure 15.4).

Exomphalos: See **Omphalocoele** (illustrated in Figure 15.23).

Exophthalmos: Abnormal protrusion of the eye from the socket (synonym: **Proptosis**).

Exostosis: Abnormal bony growth that projects outward from the surface of a bone.

Exstrophy: Congenital eversion of a hollow organ (typically applied to the urinary bladder).

Fetal wastage: Postimplantation death of a fetus (or embryo) (synonym: **Resorption**; illustrated in Figure 7.15).

Fistula: An abnormal narrow passage between two normally unconnected hollow organs, or between an internal structure (typically hollow) and the surface of the body.

Fusion: Abnormal union of two structures (illustrated in Figure 15.24; see also Figure 16.8, demonstrating failed fusion).

Gastroschisis: Fissure of the ventral abdominal wall that does not involve the umbilicus, through which viscera may protrude (usually the intestines); extension of the fissure into the thoracic cavity is termed **Thoracogastroschisis.**

Gonadal dysgenesis: General term for anomalous gonad (ovarian or testicular) development.

Hamartoma: A benign nodular mass resulting from faulty embryonal development of cells and tissues that normally occur at the affected site.

Hemimelia: Absence or shortening of all or part of the distal portion of a limb.

Hemivertebra: Incomplete formation of one side of a vertebra.

Hemorrhage: Accumulation of extravasated (i.e., extravascular) blood (illustrated in Figure 15.10).

Hermaphrodite: An individual with both male and female gonadal tissue.

Heterotaxy: Condition where certain thoracic and abdominal organs are arranged abnormally across the left–right axis; commonly affected organs include the heart (especially the atria, which often exhibit **Isomerism**) and spleen.

Heterotopia: Development of a normal tissue in an abnormal location (see also **Ectopia**) (illustrated in Figure 15.20).

Holocardius: A monozygotic twin with no heart whose circulation depends on the heart of the more fully developed twin.

Holoprosencephaly: Failed division of the prosencephalon (the initial, unified primary cerebral vesicle) resulting in deficient midline craniofacial development with ocular **Hypotelorism** (and true **Cyclopia** in severe cases) (illustrated in Figure 15.4).

Holorachischisis: Fissure of the entire vertebral column (see also **Rachischisis**).

Hydranencephaly: Complete or almost complete absence of the cerebral hemispheres; the space where brain should have been located is filled with cerebrospinal fluid (synonym: **Hydranencephalus**).

Hydrocele: Accumulation of fluid in the tunica vaginalis of the testis or along the spermatic cord.

Hydrocephalus: Dilation of the cerebral ventricles due to fluid accumulation; may lead to **Cranial doming** (synonym: **Hydrocephaly**, technically denoting an increase of fluid inside the skull).

Hydronephrosis: Distension of the renal pelvis and calyces of the kidney with urine due to obstructed urinary outflow (illustrated in Figure 15.21).

Hydropericardium: Collection of fluid in the pericardial sac that envelops the heart (illustrated in Figure 15.A.6).

Hydroureter: Dilation of the ureter with urine because of a distal obstruction of the ureter or urethra (illustrated in Figure 15.21).

Hypermastia: Presence of one or more supernumerary mammary glands (synonym: **Polymastia**).

Hypermorph(ic): A mutation leading to an increase in gene function. Mechanisms for producing this effect include boosting either the gene dosage (a gene duplication), over-expression of mRNA and/or protein, or generating a constitutively active protein.

Hypertelorism: Abnormally great distance between two paired structures (although commonly applied to the eyes).

Hypomorphic: A mutation leading to reduction but not complete loss of gene function (see also **Knock-down**).

Hypospadias: Anomaly in which the urethra opens on the underside of the penis or the perineum in males, or into the vagina in females.

Hypotelorism: Abnormally decreased distance between paired structures (although commonly applied to the eyes).

Ichthyosis: A developmental skin disorder characterized by altered (usually excessive) skin keratinization, leading to dryness and scaling.

Iniencephaly: An abnormality of the head and neck characterized by an occipital bone defect, *spina bifida* of the cervical vertebrae, and fixed retroflexion of the head backward onto the cervical midline.

Intersex: Condition in which structural attributes of both males and females may be seen.

Ischiopagus: Conjoined twins connected at the pelves.

Isomerism: Similarity of bilateral structures that are normally dissimilar (such as cardiac atria).

Kinked tail: Localized twisting (usually a side to side undulation) of the tail (illustrated in Figure 15.22).

Kyphosis: Excessive dorsal (upward) curvature of the thoracic spine as viewed from the side (illustrated in Figure 15.14).

Levocardia: Condition in which the heart is located correctly on the left side of the body while the related vessels are on the wrong side (i.e., *situs inversus*).

Lordosis: Excessive ventral (downward) curvature of the thoracic spine as viewed from the side.

Macroglossia: Abnormally large and often protruding tongue.

Macrognathia: Enlarged or protruding jaw (synonym: **Prognathism**).

Macrophthalmia: Abnormally large eye(s).

Macrotia: Large ears (pinnae).

Malalignment: Abnormal relative position of structures (typically paired) on opposite sides of a dividing line or located around a central axis.

Malformation: A lasting structural defect arising from an intrinsically abnormal developmental process that usually is incompatible with survival or that markedly curtails postnatal function.

Malposition: Displacement to an abnormal location and/or orientation.

Malrotation: Position of a structure at an abnormal angle relative to its normal location, or less commonly the failure of a structure to rotate into its normal position.

Megencephaly: Enlarged brain, typically applied to the cerebrum.

Meningocele: Herniation of the meninges through a defect in the skull or vertebral column.

Meningoencephalocele: Herniation of meninges-covered brain tissue through a defect in the skull (see also **Cephalocele**, **Encephalocele**, and **Meningocele**).

Meningomyelocele: Herniation of meninges and spinal cord tissue through a defect in the vertebral column.

Meroanencephaly: Absence of part of the brain, usually the cerebrum and/or midbrain (synonym: **Anencephaly**) (illustrated in Figure 15.4).

Meromelia: Absence of part of a limb (see also **Micromelia** and **Phocomelia**).

Mesocardia: Malposition of the heart with the apex at the central midline (i.e., a right shift).

Microcaudia: Reduced tail length (synonym: **Brachyury**; illustrated in Figure 15.22).

Micrencephaly: Having an abnormally small brain (variant: **Microencephaly**) (illustrated in Figure 15.5).

Microcephaly: Abnormally small head (synonyms: **Microcephalus**).

Microencephaly: Variant of **Micrencephaly**.

Microglossia: Reduced size of the tongue.

Micrognathia: Abnormally small jaw (usually the mandible) (illustrated in Figure 15.4).

Micromelia: Having abnormally small or short limb(s).

Microphthalmia: Abnormal smallness of one or both eyes (synonym: **Microphthalmos**).

Microstomia: Hypoplasia of the mouth.

Microtia: Hypoplasia of the pinna (ear), with an absent or atretic (closed) external auditory meatus.

Mutagenic: The capacity for an agent to damage DNA.

Myelocele: Herniation of spinal cord through a vertebral defect.

Myeloschisis: Cleft spinal cord.

Necrosis: Death of one or more cells, often leading in developing tissues to followed by cell disintegration and tissue cavitation (illustrated in Figures 15.19 and 16.16).

Neomorph(ic): A dominant gain of gene function due to ectopic expression of mRNA or protein or to production of altered proteins that serve a new function.

Neural tube defect (or neural tube closure defect): Any fissure (i.e., **Dysraphism**) along the dorsal midline as a consequence of nonelevation and/or nonfusion of the neural folds; examples include **Anencephaly**, **Cranioschisis**, **Encephalocele**, **Exencephaly**, **Meningocele**, **Myeloschisis**, **Rachischisis**, etc.

Nevus: A circumscribed skin malformation, usually colored due to hyperpigmentation (more melanin) or with abnormal (usually increased) vascularization.

Oligodactyly: Having fewer than the normal number of digits (illustrated in Figure 15.A.3).

Oligohydramnios: Abnormally reduced quantity of amniotic fluid.

Omphalocele: Fissure of the ventral abdominal wall that involves the umbilicus, usually associated with a macroscopically visible umbilical hernia (synonym: **Exomphalos**) (illustrated in Figure 15.23).

Otocephaly: Extreme underdevelopment of the lower jaw, leading to close approximation or even union of the ears on the ventral neck.

Overriding aorta: Displacement of the aorta to the right, so that it receives blood from both ventricles.

-pagus: Suffix indicating that the developmental lesion involves conjoined twins.

Palatoschisis: Fissure or cleft in the bony palate (synonym: **Cleft palate**) (illustrated in Figure 15.15).

Patent *ductus arteriosus*: An open communication between the pulmonary trunk and the aorta that has persisted after birth (illustrated in Figure 15.A.5).

Patent *foramen ovale*: Failure or inadequate postnatal closure of the *foramen ovale*, a small portal in the atrial septum allowing interchange of blood between the atria; primary atrial septal defects are larger than a patent *foramen ovale*.

Phocomelia: Reduction or absence of the proximal part of a limb(s), leaving the distal part (i.e., paw) attached to the trunk by a small, irregularly shaped bone (see also **Meromelia** and **Micromelia**).

Placenta accreta: Abnormal implantation characterized by absence of the decidua and adherence of the embryonic placenta to the surface of the myometrium.

Placenta increta: Abnormal implantation characterized by absence of the decidua and invasion of the embryonic placenta into but not through the myometrium.

Placenta percreta: Abnormal implantation characterized by absence of the decidua and invasion of the embryonic placenta completely through the myometrium.

Placenta previa: Condition in which the placenta blocks the opening of the uterine horn, thereby preventing expulsion of the conceptuses located farther up the horn.

Placentomegaly: Enlargement of the placenta.

Plagiocephaly: Asymmetry of the cranium due to premature closure of the cranial sutures on one side.

Polydactyly: Supernumerary digits (illustrated in Figure 15.A.3).

Polyhydramnios: Abnormally increased quantity of amniotic fluid.

Polymastia: See **Hypermastia**.

Proboscis: A cylindrical, fleshy protuberance from the face which may be located anywhere from the normal location of the nose to the forehead; often linked to **Holoprosencephaly**.

Prosoposchisis: Fissure of the face and jaw (chiefly the maxilla [upper] jaw) (illustrated in Figure 15.4).

Pseudohermaphroditism: Condition in which an individual of one sex has external genitalia resembling those of the other sex.

Ptosis: Drooping of the upper eyelid because of abnormal ocular muscle or nerve development.

Rachipagus: Conjoined twins connected at the vertebral column.

Rachischisis: Fissure of the vertebral column.

Resorption: A conceptus that implanted successfully but later died, leading to removal of the embryonic tissue (but retention of extra-embryonic tissue comprising the placenta) (synonym: **Fetal wastage**; illustrated in Figure 7.15).

Retinal fold: Usually focal undulation of the retinal layers.

Rhinocephaly: Craniofacial malformation featuring a proboscis-like nose above partially or completely fused eyes (see also **Cyclopia**).

Runt: A normally developed fetus, neonate, or juvenile that is significantly smaller than its littermates (synonym: **Stunting**; illustrated in Figures 15.9, 15.14, and 15.A.6).

Scoliosis: Lateral deviation of the vertebral column as viewed from above.

Sirenomelia: Having fused lower limbs (synonym: **Symmelia**).

Situs inversus: Lateral transposition of the viscera to the mirror position across the midline.

Spina bifida **(S.B.):** Localized defective closure of the vertebral arches, through which the spinal cord and/or meninges may protrude.

Spina bifida aperta: S.B. in which the neural tissue is exposed.

Spina bifida cystica: S.B. with intact skin and no herniated neural tissue.

Spina bifida occulta: S.B. with herniation of a cystic swelling containing the meninges (**Meningocele**), spinal cord (**Myelocele**), or both (**Meningomyelocele**).

Supernumerary: Having more than the usual or expected number (common misspelling: supernumery) (illustrated Figure 15.A.3).

Sympodia: Fusion of the (hind) paws.

Syndactyly: Webbing or fusion between adjacent digits (illustrated in Figures 5.5 and 15.A.3).

Syndrome: A group of clinical signs and/or structural abnormalities that occur together.

Synpolydactyly: A malformation featuring both too many digits and fusion of some digits.

Talipes: Deformity of the foot, which is twisted out of shape or position.

Teratothanasia (alternatively **terathanasia**): Death (*miscarriage*) of a malformed embryo or fetus.

Teratogen: An agent that can cause abnormal development of an embryo or fetus.

Tetralogy of Fallot: A combination of cardiac defects consisting usually of pulmonary stenosis, ventricular septal defect, overriding aorta, and right ventricular hypertrophy.

Thoracopagus: Conjoined twins connected at the thorax, often with sharing of a single heart (see also **Diplopagus** and **Holocardius**).

Thoracoschisis: Fissure of the thoracic wall.

Thoracostenosis: Narrowing of the thoracic region.

Tracheoesophageal fistula: Abnormal channel between the lumens of the trachea and esophagus.

Transposition of great vessels: Cardiac malformation in which the aorta originates from the right ventricle while the pulmonary trunk arises from the left ventricle.

Umbilical hernia: See **Omphalocele** (illustrated in Figure 15.23).

Ventricular septal defect: Persistent communication between the cardiac ventricles (illustrated in Figure 15.8).

Wavy ribs: Extra bends in one or more ribs.

Xenobiotic: An exogenous chemical substance within an organism that is not normally expected to be found within that organism.

REFERENCES

1. Dorland (2006). *Dorland's Illustrated Medical Dictionary*, 32nd edn. Philadelphia, PA: Elsevier.
2. International Committee on Veterinary Gross Anatomical Nomenclature (ICVGAN) (2012). *Nomina Anatomica Veterinaria,* 5th edn. http://www.wava-amav.org/Downloads/nav_2012.pdf (last accessed June 1, 2014).
3. International Federation of Associations of Anatomists (IFAA) (1998*). Terminologia Anatomica*. http://www.unifr.ch/ifaa/ (last accessed June 1, 2014).
4. Makris SL, Solomon HM, Clark R, Shiota K, Barbellion S, Buschmann J, Ema M, Fujiwara M, Grote K, Hazelden KP, Hew KW, Horimoto M, Ooshima Y, Parkinson M, Wise LD (2009). Terminology of developmental abnormalities in common laboratory mammals (version 2). *Birth Defects Res B Dev Reprod Toxicol* **86** (4): 227–327.
5. Stedman TL (2006). *Stedman's Medical Dictionary*, 28th edn. Baltimore, MD: Lippincott, Williams & Wilkins.
6. Wise LD, Beck SL, Beltrame D, Beyer BK, Chahoud I, Clark RL, Clark R, Druga AM, Feuston MH, Guittin P, Henwood SM, Kimmel CA, Lindstrom P, Palmer AK, Petrere JA, Solomon HM, Yasuda M, York RG (1997). Terminology of developmental abnormalities in common laboratory mammals (version 1). *Teratology* **55** (4): 249–292.

4

Anatomy and Physiology of the Developing Mouse and Placenta

Brad Bolon and Jerrold M. Ward

CONTENTS

Comparative pathologists who engage in mouse developmental pathology studies must possess a thorough understanding of normal anatomic features in order to effectively find and characterize lesions in developing mice. Most investigators will possess a fair degree of assurance in their ability to assess structural traits in near-term fetuses (gestational day [GD] 17.5 or older), neonates, and juveniles (up to postnatal day [PND] 35). Confidence in this regard is founded in the close structural resemblance of most cells, organs, and tissues during these late developmental stages to the appearance of the corresponding components in adult mice. However, developmental anatomy is not confined merely to consideration of the normal three dimensions of length, width, and height that constitute the usual focus of clinicians, pathologists, and other biomedical scientists. Instead, proficiency in the art and science of developmental anatomy necessarily includes comprehension of structural changes in the fourth dimension of time as well (see Figure 1.4). The transient, sometimes weird, constantly shifting morphology of the embryo and the extraembryonic membranes early during gestation affords a particular challenge to comparative pathologists tasked with unraveling the cause of embryonic lethal phenotypes.

A comprehensive review of mouse development is far beyond the scope of this volume. Interested readers will find more information on the topic in many books,[36,47,66,73,87,91,114,131,165,166,197,217] book chapters,[13,45,53,62,89,90,129,130,177] developmental anatomy atlases,[52,83,88,142,174,175,199] review articles,[12,29,125,163,172,215,218] and websites.[9,24,52,70,190] Accordingly, this chapter has been designed with two aims in mind. The first is to briefly review features of normal embryonic and fetal development in the mouse. This objective will be pursued using a brief narrative supplemented with selected photographic and schematic illustrations as well as tables of key developmental milestones, arranged by both day and organ system (see the Appendix, Tables 4.A.1 through 4.A.16) to showcase the key structures and events that should be easily identifiable by developmental pathologists. The second aim is to give a detailed summary of macroscopic (gross) and microscopic (histologic) anatomy of the extraembryonic membranes as they evolve over time. In our experience, the analysis of mouse placenta is especially difficult for many comparative pathologists because of its complex anatomical organization and the dearth of systematic morphological descriptions for the cells and regions that comprise this unique, transient organ.

Stages in the Anatomic Evolution of the Developing Mouse

In developing vertebrates, the stages of development generally are defined in terms of one or more major events or features that occur during that period. Anatomic similarities offer a ready means of comparing the development of structure and function across species. In mammals, the principal developmental periods that occur prior to birth are conception, preimplantation, implantation, organogenesis, and fetal growth, while those taking place after birth are the neonatal, juvenile, and adult phases. Each period then may be subdivided using additional events or features. Specific developmental *ages* in mice generally are assigned using the Theiler staging (TS) system[18,52,70,88,91,124] (Table 4.1). Similar schemes developed for use in other species include Carnegie staging for human beings[18,47,69] and the Witschi system for rats.[18,71]

TABLE 4.1

In Utero Growth of the *Average* Mouse Embryo/Fetus

Prominent Feature of Embryo/Fetus	Developmental Stage		Approximate Peak Dimension[a] (mm)
	Theiler Stage (TS)	**Gestational Day (GD)**	
Zygote	1	0.5–1.0	0.1
Morula	2, 3	1.5–2.5	0.1
Blastocyst	4, 5	3.0–4.0	0.1
Blastocyst (implanted)	6	4.5–5.0	0.1
Egg cylinder	7, 8	5.0–6.0	0.1–0.15
Primitive streak	9, 10	6.5–7.0	0.15–0.2
Neural plate	11	7.0–7.5	0.2–0.25
Somite pairs nos. 1–7	12	8.0–8.5	1.0–1.2
Somite pairs nos. 8–12	13	8.5–9.0	1.3–1.5
Somite pairs nos. 13–20	14	8.5–9.0	1.5–1.8
Somite pairs nos. 21–29	15	9.0–9.7	2.0–2.2
Somite pairs nos. 30–34	16	9.7–10.2	2.5–3.0
Somite pairs nos. 35–39	17	10.0–10.5	3–4
Somite pairs nos. 40–44	18	10.5–11.0	4–5
Somite pairs nos. 45–47	19	11.0–11.5	6–7
Somite pairs nos. 48–51	20	11.5–12.0	7–8
Somite pairs nos. 52–55	21	12.5–13.0	8–9
Somite pairs nos. 56–60+	22	13.5–14.0	9–10
Eyelids closed	23	14.5–15.0	11–12
Umbilical hernia closing	24	15.5–16.0	12–14
Eyelids closed	25	16.5–17.0	14–17
Pinna covers auditory meatus	26	17.5–18.0	17–20
		18.5–19.0	19–22
Birth	27	19.0	23–27

Note: Data adapted from values reviewed in Refs. [52,91,124,130,166,198].

[a] Peak (i.e., maximum) dimension in one direction of the embryo or fetus—where feasible, this represents the crown-to-rump length.

These schemes provide criteria for recognizing major developmental milestones based on morphological modifications that occur over a relatively short span of time. For some purposes, however, even shorter interstage intervals may be necessary. In such cases, Theiler stages are subdivided or replaced using other anatomic benchmarks, including altered contours or dimensions (e.g., crown-rump length, body weight)[46,144,206]; changing cell or somite counts[14,60,206]; or the appearance of new features (e.g., cranio-facial elements,[119] foot pads,[121] limb buds,[213] ossification centers in specific appendicular bones,[141] or tooth eruption[144]). Somites make a particularly good scale for partitioning Theiler stages as the genesis of each new somite pair in mouse embryos has been calculated to require 2 h.[60] In general, the relevance of staging criteria must be confirmed by each laboratory for its mouse colonies as time-dependent differences in development amounting to delays of 12 h or more exist among animals having different genetic backgrounds.[110,200]

Many conventions for timing, such as days *post coitum* (dpc) and embryonic day (E), are used by various researchers in communicating the timing of developmental events. The common options are reviewed in Chapter 6. In the current chapter, the timing is given using the gestational day (GD) and postnatal day (PND) conventions. This choice acknowledges the fact that gestation in the mouse includes distinct embryonic and fetal periods, where the extended embryonic phase involves the initial specification of the body plan and organs while the short fetal phase (commencing with the mineralization of bone at approximately GD 15) encompasses their subsequent growth; admittedly, the length of fetal development *in utero* is short in mice (4 days, representing approximately 20% of gestation) relative to the fetal period in humans (6 months, or about 67% of gestation). The continuing structural and functional changes that occur in developing mice during the 2 weeks from the end of organogenesis (GD 15.0) until approximately PND 10 are comparable in most respects to developmental events that take place in human fetuses during the 6 months spanning the second and third trimesters, especially in terms of neural development.[24,25]

Conception

Mouse development proper commences only after fertilization. Prior to this event, the haploid genomes of the female and male gametes (ova and spermatozoa, respectively) are incapable of launching an independent embryo. The act of conception merges the haploid genomes of a male and female gamete to provide each new embryo with a complete diploid genome.

Conception typically can occur once during any estrus cycle in which the uterus does not already contain one or more embryos. A mature female mouse spontaneously releases multiple ova from each ovary (usually 4–10 per side, depending on the strain or stock) during each estrus cycle throughout the course of her early adult life (which extends from approximately 2 to 8 months of age [Table 4.2]). If the female is mated with a fertile male, the flotilla of ova will encounter a wave of migrating sperm in the ampulla (proximal oviduct) shortly after ovulation; each ovum will fuse with one sperm. This process of fertilization is the event that embodies conception. The outcome of conception is a zygote (a one-celled embryo; Figure 4.1).

Conception happens on average at approximately 0.5 dpc. The age difference between the oldest and youngest embryos in a mouse litter is reported to range from 10[88] to 16 h[166] to as much as 24 h[200] (see Figure 6.3). Thus, early in organogenesis (i.e., the period in development during which blueprints for all major organs are being established), the morphological features of the oldest embryos will place them at a different Theiler stage than their youngest littermates. The variance in age among embryos in a litter is influenced by the genotype, as male (XY) embryos develop more rapidly and grow larger than female (XX) embryos during the preimplantation period.[16,118,202] Individual rates of embryonic development can be altered by epigenetic factors such as subtle variations in maternal homeostasis (e.g., total body weight,[132,170] extrauterine inflammation,[54] malnutrition,[94] and stress[17,208]) and/or intrauterine conditions (e.g., fluid composition,[65] overcrowding,[15] sex of neighboring embryos[169,207])—without inducing any lasting adverse consequences on the viability of the animal. Each embryo has the innate ability to modulate the extent to which such epigenetic factors can override its genetic programming.[21,42,102,184] Certain paternal traits, including semen biochemistry,[161] gene activity,[168,201] and nutrition,[117] also may modify maternal and/or embryonic physiological responses during this phase of gestation.

TABLE 4.2

Physiological Parameters of the Average Mouse

Parameter	Typical Values (Potential Range)
Life span (natural)	18–24 months (fed *ad libitum*)—varies by strain/line/stock
Breeding age (most efficient)	60 days (male)—generally to 10–12 months 55 days (female)—usually 2–8 months (after 4 or 5 litters)
Estrus cycle length	4–5 days
Ovulation stimulus	Spontaneous
Uterus conformation	Bicornuate
Litter size: average (range)	8–10 (4–12) in natural litters—varies by strain/line/stock ~30 (15–45) following superovulation
Gestation length	19 days (18.5–20)
Birth weight	~1.2 g (0.8–1.5 g)
Intake of solid food	Postnatal day (PND) 10 (PND 9–12)
Weaning weight	
Male	~21 g (18–25 g)
Female	~18 g (16–24 g)

Note: Adapted in part from Refs. [111,124,151,156,166].

Preimplantation

Mouse embryos spend several days (usually 4–5) developing within the lumen of the reproductive tract before they implant in the uterine wall. The first recognizable event is the equal division of the zygote to produce an embryo comprised of two identical daughter cells (Figure 4.2). This cleavage occurs in the proximal oviduct at approximately GD 1.0. Further rounds of division in the distal oviduct take place from GD 1.0 to GD 2.5, thereby resulting in the formation of a morula: an expanding, solid, compacted mass of 4, 8, 16, or occasionally 32 stem cells (Figure 4.3). The equipotent nature of these original stem cells (termed blastomeres) is shown by their essentially identical cytoarchitectural, functional, and

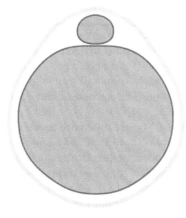 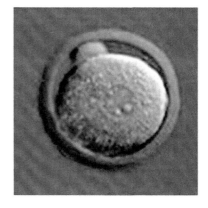

FIGURE 4.1 A Theiler stage (TS) 1 mouse conceptus (gestational day [GD] 0.5) is a one-celled embryo (or *zygote*) characterized by a large, round, cytoplasm-rich central cell with nucleus; a small, peripheral nucleus (the *first polar body*, a genomic remnant from a meiotic division in the ovary); and a thick, smooth outer shell of glycoprotein (termed the *zona pellucida*) that prevents entry of additional sperm. Bar = 50 μm. (The diagram has been adapted by Mr. Tim Vojt, The Ohio State University, from the EMAP (e-MOUSE Atlas), v3.5, http://www.emouseatlas.org/emap/home.html, last accessed December 1, 2014, by courtesy of the Medical Research Council of the United Kingdom. The photomicrograph was kindly provided by Mr. Joe Anderson, Amgen, Inc., Thousand Oaks, CA.)

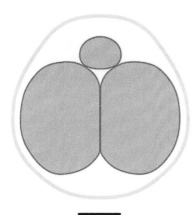
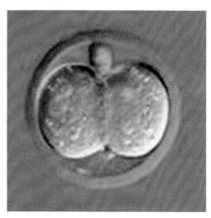

FIGURE 4.2   A Theiler stage 2 mouse conceptus (GD 1.0) is a two-celled embryo characterized by two oval, cytoplasm-rich central cells; a small, peripheral second polar body (a genomic remnant from the second meiotic division, which occurs after fertilization); and a *zona pellucida*. Bar = 50 µm. (The diagram has been adapted by Mr. Tim Vojt, The Ohio State University, from the EMAP (e-MOUSE Atlas), v3.5, http://www.emouseatlas.org/emap/home.html, last accessed December 1, 2014, by courtesy of the Medical Research Council of the United Kingdom. The photomicrograph was kindly provided by Mr. Joe Anderson, Amgen, Inc., Thousand Oaks, CA.)

molecular characteristics. Thus, damage leading to loss of one blastomere usually can be repaired completely by a surviving neighbor.

Embryos begin their regional anatomic specialization as they exit the distal oviduct through the uterotubal junction into the lumen of the uterus. This transfer occurs at about GD 2.5. Once in the uterus, the solid mass that comprises a morula begins to form a cavity (termed a blastocoele) at one pole. This event typically transpires upon accretion of 32 or 64 cells, which occurs at approximately GD 3.0. Dilation of the cavity is accompanied by differentiation of two initial embryonic cell types: the trophectoderm, which forms the outer layer of the entire embryo and is the progenitor of the extraembryonic membranes (i.e., placenta), and the inner cell mass (ICM), a group of pluripotent stem cells that form a crescent-shaped cap along the interior surface of the trophectoderm at one pole that over time develops into the embryo proper (Figure 4.4). The existence of zonal differences in the blastocyst is indicative of a biochemical polarity that is involved in establishing the body axis and specifying the location of major cell populations and organelles; the unequal division of molecular constituents among cells in the animal (i.e., ICM) and vegetal (i.e., placental) poles is regulated by maternal factors that act beginning at the four-cell stage.[51] Interestingly, the four-cell stage also represents the time at which the embryonic genome is activated and at which those cells that do not follow a correct developmental program will begin to prune themselves spontaneously by apoptosis.[55] The ICM continues to amass cells by equal stem cell cleavage divisions, and the blastocoele enlarges as the blastocyst migrates down the uterus. Ultimately, the *zona pellucida* is shed by the blastocyst in a process termed *hatching* (Figure 4.5), and the ICM exhibits a small wave of apoptosis (postulated to remove excess elements in advance of the next developmental surge[137]) at GD 4.0 to prepare for embryo implantation. The mouse blastocyst consists of approximately 200 cells at the end of preimplantation development.[62]

Structural evolution from morula to blastocyst is accompanied by significant changes in embryonic chemistry. Development after conception is characterized by a low metabolic rate (termed *quietness*) in embryonic cells.[103] Energy metabolism shifts from pyruvate-oriented aerobic glycolysis in earlier preimplantation embryos to glucose-based oxidative phosphorylation in nascent blastocysts.[112] This transition is controlled by stage-specific upregulation of genes associated with initial lineage specification of cells. The shift to glucose as the chief fuel source for embryonic cells is accompanied by a spike in oxygen utilization.[75] Alterations in the intrauterine environment that stress the embryo may disrupt these normal patterns, thereby leading to a loss of quietness and a decrease in embryonic viability.[103] Similarly, xenobiotic agents that impact the efficiency of energy metabolism may reduce the proliferation of embryonic

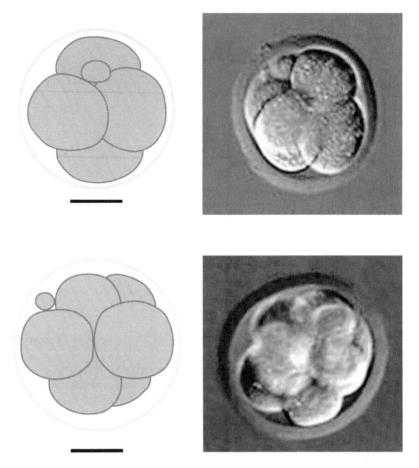

FIGURE 4.3 A Theiler stage 3 mouse conceptus (GD 2.0) is a *morula*: a multicelled, compact mass of equipotent stem cells (or *blastomeres*) surrounded by a *zona pellucida*. Embryos at this stage have 4, 8, 16, or 32 cells. Bar = 50 µm. (The diagram has been adapted by Mr. Tim Vojt, The Ohio State University, from the EMAP (e-MOUSE Atlas), v3.5, http:// www.emouseatlas.org/emap/home.html, last accessed December 1, 2014, by courtesy of the Medical Research Council of the United Kingdom. The photomicrograph was kindly provided by Mr. Joe Anderson, Amgen, Inc., Thousand Oaks, CA.)

cells, which may lead to long-term reductions in embryonic and/or fetal growth.[117] Blastocysts of many but not all mouse strains possess functional enzymes (e.g., cytochromes p450 [CYP 450]) capable of xenobiotic metabolism. Strain-specific differences in enzyme expression (embryonic and maternal) are thought to be responsible for strain-specific patterns of embryonic vulnerability.[143]

Implantation

The uterine wall has been primed to host embryos through the shifting concentrations of circulating estrogen and progesterone,[77,138,139] which lead to *decidualization* (or *decidual reaction*) of the endometrium.[155] The conceptus also participates in controlling this decidual reaction, acting through the trophoblast giant cells that form the outer layer of embryonic tissue near the uterine wall.[6] Major changes in the decidua-rich uterine wall that promote both embryo attachment and survival include stromal cell expansion,[155] enhanced blood vessel numbers and permeability,[20,21] and increased secretory capability by glandular epithelium. The decidual reaction provides the chief means for supporting the conceptuses until they are able to form their own placentae. These changes appear to be controlled by maternal leukocytes, including uterine dendritic cells and uterine natural killer (uNK) cells.[10] At GD 4.5, the enlarging *zona*-free blastocyst (Figure 4.6) forms a loose connection with the endometrial surface, usually after

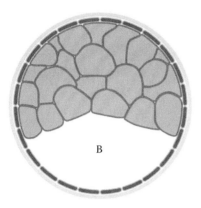
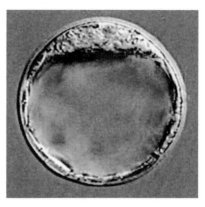

FIGURE 4.4 A Theiler stage 4 mouse conceptus (GD 3.0) is an early (32- or 64-cell) *blastocyst*: a multicelled embryo with a crescent-shaped inner cell mass (ICM [orange cells]) at one pole and a cavity (blastocoele [B]) at the opposite pole, both encompassed by the *zona pellucida* (outer, continuous, pale blue ring). The ICM represents the pluripotent stem cells that will give rise to the embryo, while the mural trophectoderm (inner, dashed, dark blue ring) that comprises the wall of the blastocoele is the earliest portion of the placenta. Bar = 50 μm. (The diagram has been adapted by Mr. Tim Vojt, The Ohio State University, from the EMAP (e-MOUSE Atlas), v3.5, http://www.emouseatlas.org/emap/home.html, last accessed December 1, 2014, by courtesy of the Medical Research Council of the United Kingdom. The photomicrograph was kindly provided by Mr. Joe Anderson, Amgen, Inc., Thousand Oaks, CA.)

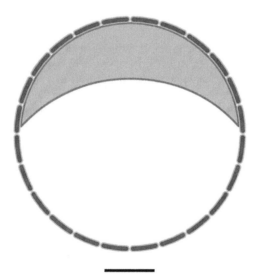

FIGURE 4.5 A Theiler stage 5 mouse conceptus (GD 4.0) is a blastocyst that has shed its *zona pellucida*. Bar = 100 μm. (The diagram has been adapted by Mr. Tim Vojt, The Ohio State University, from the EMAP (e-MOUSE Atlas), v3.5, http://www.emouseatlas.org/emap/home.html, last accessed December 1, 2014, by courtesy of the Medical Research Council of the United Kingdom.)

coming to rest in a uterine gland. The orientation of the ICM is random at first, but in short order signaling between uterine epithelial cells and the embryo causes the ICM to reorient to rest as close as possible to the mesometrial wall of the uterus. Trophoblast begins to proliferate, invades the endometrium, and differentiates into syncytiotrophoblasts (which abut the maternal tissue) and cytotrophoblasts (which envelop the ICM).

Successful implantation is a team effort requiring cooperation between the maternal tissue and the embryo. The hormonal fluctuations that follow conception induce uterine epithelial cells to express

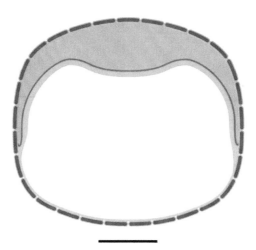

FIGURE 4.6 A Theiler stage 6 mouse conceptus (GD 4.5) is an implanting blastocyst in which the ICM exhibits the first evidence of distinct cell populations: an *epiblast* (large outer region of columnar cells, representing the future embryo) and a *hypoblast* (thin inner layer of cuboidal cells, which will form the future yolk sac). The blastocoele is round and cavernous and is partially lined (near the ICM) by a thin layer of cuboidal cells (the parietal endoderm [thick, pale blue arc]). At this early stage of postimplantation development, the trophectoderm (dashed, dark blue ring) represents the major portion (70%–80% of cells) of the conceptus. Bar = 100 μm. (The diagram has been adapted by Mr. Tim Vojt, The Ohio State University, from the EMAP (e-MOUSE Atlas), v3.5, http://www.emouseatlas.org/emap/home.html, last accessed December 1, 2014, by courtesy of the Medical Research Council of the United Kingdom.)

molecules that promote blastocyst attachment.[153] Expression of prosurvival factors by the blastocyst (i.e., autocrine control) and the uterus (i.e., paracrine cross-talk) is necessary to sustain pregnancy.[133,186] Decidual cells accumulate glycogen and lipids, presumably as stored energy to help nourish the embryo. A few decidual cells near the embryo, especially on the mesometrial side, will undergo apoptosis in normal implantation sites as a means of increasing the capacity of maternal blood channels available to circulate blood near the expanding placenta.[137] However, the decidual response will regress locally in those regions at a distance from implantation sites if the litter size is small (i.e., if the horn holds only one or two widely spaced embryos).

Uterine secretions also play important roles in aiding and sustaining implantation. Many molecules are conveyed in the fluid that facilitate implantation (e.g., integrins and other adhesion molecules, matrix proteins like fibronectin and laminin) or minimize uterine contractions that might expel the blastocyst (e.g., hormones like prolactin and relaxin). Secretions also harbor many signaling molecules (e.g., growth factors, steroid hormones) to promote survival and growth of established embryos. A related function of uterine fluid is to carry vital micronutrients and proteins (e.g., albumin) needed for embryonic expansion. Together these materials ensure that most viable embryos that come into contact with a properly primed uterine wall will have an opportunity to achieve implantation.

The implanted blastocyst undergoes a conformational change so that the cavernous blastocoele begins to adopt an elongate, cylindrical profile (Figure 4.7) by GD 5.0. The resulting *egg cylinder* (Figure 4.8) exhibits such identifiable anatomic features as the ectoplacental cone (EPC, a wedge of extraembryonic tissue that invades the decidua lining the mesometrial side of the uterine wall); a complete layer of parietal endoderm lining the inner surface of the mural trophectoderm; and differentiation of the ICM into distinct zones: the *epiblast* (i.e., the future embryo), a thick layer of columnar cells located near the trophectoderm, and the *hypoblast* (i.e., the future yolk sac [YS]), a thin layer of cuboidal cells that covers the surface of the ICM adjacent to the blastocoele.[225] The epiblast will give rise to all the embryonic germ layers (ectoderm, mesoderm, and endoderm) as well as some extraembryonic ectoderm and mesoderm, while the hypoblast will produce the YS endoderm and some extraembryonic mesoderm. Generation of ectoderm and endoderm from the epiblast begins at approximately GD 6.0. The formation of the *primitive streak* (Figure 4.9) on the caudal half of the embryo's dorsal surface

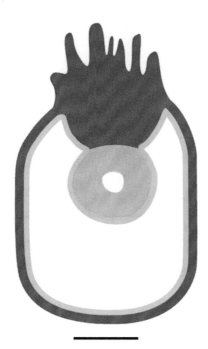

FIGURE 4.7 A Theiler stage 7 mouse conceptus (GD 5.0) is an elongated *egg cylinder* with a distinct layer of cuboidal cells (the parietal endoderm) lining the whole inner surface of the mural trophectoderm (i.e., the cells forming the outer layer of the blastocoele wall). Ongoing proliferation of trophectoderm on the mesometrial side of the ICM produces a roughly triangular ectoplacental cone (EPC), while other trophectodermal cells stop dividing and enlarge by endoreduplication to become trophoblast giant cells. Bar = 100 μm. (The diagram has been adapted by Mr. Tim Vojt, The Ohio State University, from the EMAP (e-MOUSE Atlas), v3.5, http://www.emouseatlas.org/emap/home.html, last accessed December 1, 2014, by courtesy of the Medical Research Council of the United Kingdom.)

at approximately GD 6.5 marks the midline plane of the longitudinal body axis as well as the site at which mesoderm (the third germ layer) will begin to form through subduction of surface ectoderm (a process termed *gastrulation*; Figure 4.10). The primitive node and pit, which develop at the cranial end of the primitive streak, serve as the origin of cells that migrate cranially to form the notochord. The continuation of gastrulation leads to production of a trilaminar embryo associated with several distinct extraembryonic cavities and membranes by approximately GD 7.5.[46]

The outcome of these early postimplantation events are that the embryo has produced the necessary building blocks for constructing all cell populations, tissues, and organs that will be necessary to sustain survival and growth throughout the remainder of its life. The next phase of embryogenesis is to begin assembling these building blocks into the *anlagen* (plural of *anlage*, the formal term for an embryonic primordium) for the various organs and systems while sequentially remodeling the body plan in four dimensions: length, breadth, width, and time.

Organogenesis

The span during embryogenesis in which organs and systems first are configured is termed organogenesis. During this time, the anlagen of various organs are formed by differing mixtures of tissues derived from the three germ layers produced by the epiblast: ectoderm, mesoderm, and endoderm. Organogenesis extends from approximately GD 7.0 to GD 7.25 through GD 15.0. Each organ, as well as the specific cell populations within distinct regions of the organ, possesses its own *critical period* during which the principal developmental events required to initiate and expand the structure take place (see Chapter 5). Thus, the critical period for the development of most organs and systems extends over some days (see Table 4.1 and also Appendix Tables 4.A.1 through 4.A.14). The evolving conformations

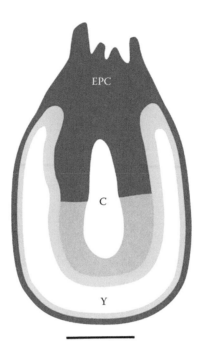

FIGURE 4.8 A Theiler stage 8 mouse conceptus (GD 6.0) is a differentiating egg cylinder in which the embryonic and extraembryonic tissues are accompanied by formation of the proamniotic cavity (C), yolk sac cavity (Y), and Reichert's membrane (a thick, noncellular membrane located between the outer trophectoderm (thin, outer, dark blue line) and the inner parietal endoderm [thin, inner, pale blue line]). The ectoplacental cone (EPC) holds abundant maternal blood (i.e., nonnucleated erythrocytes), and many trophoblast giant cells invade the decidua. Bar = 100 μm. (The diagram has been adapted by Mr. Tim Vojt, The Ohio State University, from the EMAP (e-MOUSE Atlas), v3.5, http://www.emouseatlas.org/emap/home.html, last accessed December 1, 2014, by courtesy of the Medical Research Council of the United Kingdom.)

of major organs result in the progressive adjustment of the embryonic/fetal profile throughout the post-implantation period.[88,201] The appearance and organ proportions of the conceptus begin to attain the adult pattern only as the organogenesis period is concluded. The importance of critical periods is that they represent times of heightened vulnerability to developmental disruption arising from exposure to endogenous and exogenous insults.[43,61,158,216] An oft-forgotten fact is that critical periods occur not only in gestation but also after birth (Figure 5.2).[162,204]

Many critical processes take place during organogenesis, all of which are essential to the proper differentiation of cells, tissues, organs, and systems.[165] For instance, *histogenesis*—the differentiation of diverse tissues—results from selective rejection by stem cells of some differentiation pathways in favor of other options as development continues. The initial totipotent stem cells, which can express all genes at need and so may become any cell type, evolve progressively into pluripotent stem cells with a wide array of fate options; oligopotent stem cells with only a few options; and at last into terminally differentiated cells, which in ordinary situations can no longer become another cell type but instead are dedicated in perpetuity to their own particular function. Induction occurs when a cell population (the inducer) affects the development of a second population. This influence may occur during initial specifi-cation of the body plan (*primary induction*) or during the many subsequent cascades of events needed to control cell interactions during formation of a particular organ (*secondary induction*). Migration (*mor-phogenetic movement*) in a reproducible direction and pattern is a stereotypical response of induced embryonic cells. Apoptosis (*programmed cell death*) to remodel solid structures, produce hollow ones, or prune suboptimal features is a recurring theme during vertebrate development. Patterns of apoptosis differ by location and over time; these regional variations must be recognized as *within normal limits* and not as manifestations of developmental damage, although altered apoptosis (either enhanced[49,235] or reduced[97,107]) may be a manifestation of some malformation syndromes and types of toxicant-induced

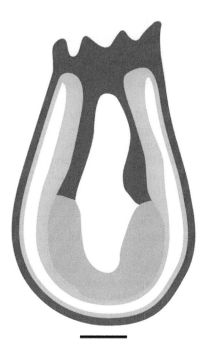

FIGURE 4.9 A Theiler stage 9 mouse conceptus (GD 6.5) is characterized by the initial formation of the primitive streak, a thin groove on the dorsal surface at the caudal pole of the embryo. The streak is the earliest structural evidence indicating the existence and position of the embryonic axis. Bar = 100 μm. (The diagram has been adapted by Mr. Tim Vojt, The Ohio State University, from the EMAP (e-MOUSE Atlas), v3.5, http://www.emouseatlas.org/emap/home.html, last accessed December 1, 2014, by courtesy of the Medical Research Council of the United Kingdom.)

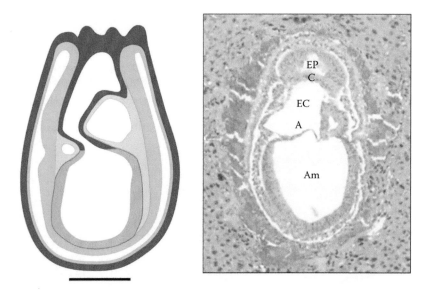

FIGURE 4.10 A Theiler stage 10 mouse conceptus (GD 7.0) is engaged in *gastrulation*, a process during which the middle germ layer (i.e., the mesoderm) is formed by migration of cells from the edges of the primitive streak down and under the surface ectoderm. The chorion (C) separates the ectoplacental cavity (EP) from the exocoelomic cavity (EC), while the amniotic fold (A) caps the amniotic cavity (Am). Bar = 100 μm. (The diagram has been adapted by Mr. Tim Vojt, The Ohio State University, from the EMAP (e-MOUSE Atlas), v3.5, http://www.emouseatlas.org/emap/home.html, last accessed December 1, 2014, by courtesy of the Medical Research Council of the United Kingdom.)

embryonic injury. All these and also other developmental processes are controlled by direct cell-to-cell contact (e.g., between adhesion molecules and extracellular matrix proteins) and gradients of diffusible substances (termed *morphogens*, as they dictate structural organization). Expression of these control mechanisms may cycle up and down one or more times over the course of development. Some molecules that have signaling functions in adult animals serve as morphogens during the developmental period (e.g., classic neurotransmitters[99]).

The neural plate (i.e., the primordium of the central nervous system) is the first embryonic organ to be specified (at approximately GD 7.5) (Figure 4.11). Soon thereafter, the rostral portion of the embryo exhibits a cardiac tube (the heart primordium) and the rise of the head folds (Figure 4.12), which are dominated by the tissue that will become the brain. The lateral edges of the neural plate begin to elevate (at GD 7.75 to GD 8.0) and begin fusing to form the neural tube (Figure 4.13); the first fusion occurs in the thoracic region, after which the folds fuse in zipper-like waves extending both cranially and caudally. At the same time, the YS mesoderm begins to produce hematopoietic precursors,[50] while the cardiac tube begins to loop (Figure 4.14),[172,190,191] connects with newly formed blood vessels throughout the body and within the nascent placenta and YS, and starts to contract. Dramatic increases in embryonic heart rate, arterial flow rate, and cardiac output occur as organogenesis progresses.[84,92,109,194,228] These changes may reflect the gradual transition from large primitive (*fetal*) erythrocytes (produced by the YS) to smaller definitive (*adult*) erythrocytes[100] as well as the increasing size and septation of the heart over time. Cardiovascular and erythropoietic abnormalities that limit the effective circulation of blood in the embryo and/or placenta are key causes of embryonic lethality.

The body plan undergoes five major changes at this point in gestation. First, the axis becomes segmented, resulting in visible condensation of paraxial mesoderm that brackets the neural tube into somites (Figure 4.13). Other axial structures develop a similar scheme for partitioning, which is visible grossly as distinct bulges or grooves (e.g., the five secondary brain vesicles and their margins) or bilateral formation of anatomic landmarks (e.g., dorsal root ganglia and nerve roots associated with the spinal cord). Segmentation is indistinct in some structures (e.g., the intermediate mesoderm and lateral plate mesoderm, which reside at a distance from the axis) even though regional differences in molecular

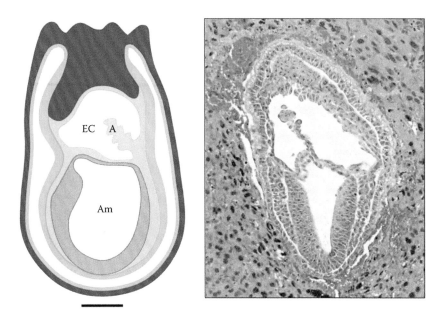

FIGURE 4.11   A Theiler stage 11 mouse conceptus (GD 7.5) features a thickened ectodermal cell bed, termed the neural plate; an elongated allantois (A); and distinct amniotic (Am) and exocoelomic (EC) cavities. Bar = 100 μm. (The diagram has been adapted by Mr. Tim Vojt, The Ohio State University, from the EMAP (e-MOUSE Atlas), v3.5, http://www.emouseatlas.org/emap/home.html, last accessed December 1, 2014, by courtesy of the Medical Research Council of the United Kingdom.)

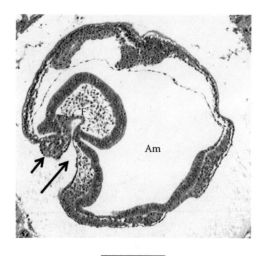

FIGURE 4.12 A Theiler stage 11 mouse conceptus (GD 7.75) features elevated head folds (F), a cardiac tube (short arrow), and the stomodeum (primordial oral cavity [long arrow]). A, amnion; Am, amniotic cavity.

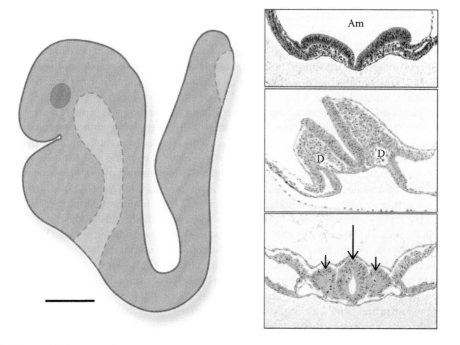

FIGURE 4.13 A Theiler stage 12 mouse conceptus (GD 8.0) is characterized by the progressive elevation of the neural plate (top) to form folds (middle), the tips of which eventually meet at the midline and fuse (bottom) to generate the initial portion of the neural tube (long arrow). Somites (short arrows) become visible as condensations of mesoderm adjacent to the neural tube. Am, amniotic cavity; D, paired dorsal aortae. Bar = 200 μm. (The diagram has been adapted by Mr. Tim Vojt, The Ohio State University, from the EMAP (e-MOUSE Atlas), v3.5, http://www.emouseatlas.org/emap/home.html, last accessed December 1, 2014, by courtesy of the Medical Research Council of the United Kingdom.)

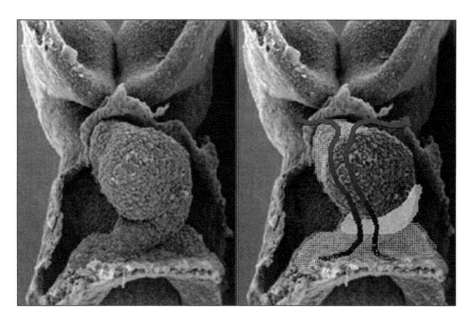

FIGURE 4.14   A Theiler stage 13 mouse conceptus (GD 8.5) is distinguished by initial expansion (in length and diameter), looping, and regional specification of the cardiac tube. Blood from the body (carried by the common cardinal vein), immature placenta (via the umbilical vein), and yolk sac (by the vitelline vein) circulates sequentially (red lines) through the sinus venosus (green), common atrium (blue), primitive left ventricle (purple), and bulbus cordis (the right ventricle precursor; yellow). (These scanning electron micrographs, including annotations, are reproduced here courtesy of Dr. Kathleen Sulik, University of North Carolina, Chapel Hill, NC, from an online tutorial in mouse embryology, Embryo images: Normal and abnormal mammalian development, http://www.med.unc.edu/embryo_images/, last accessed December 1, 2014.)

expression reveal the continuation of regional specialization. The second main development at this time of gestation is *turning*, a process unique to the mouse embryo in which the caudal portion of the animal rotates so that the tail moves from a position behind the head to a location in front and just to the right of the individual (Figures 4.15 and 4.16). The third major shift is continued looping of the heart tube to the right so that its original orientation approximately along the mid-sagittal plane (Figure 4.14) rotates into a transverse (i.e., more adult-like) position (Figure 4.17). The evolving cardiac morphology may be observed readily at this period (GD 8.0–GD 10.5) due to the translucent natures of the pericardium and thoracic wall, and may be followed progressively in whole embryo cultures; alternatively, the structural evolution of the cardiovascular system may be highlighted using contrast media[44] or three-dimensional reconstruction of thin tissue slices produced by noninvasive imaging (see Chapter 14) or histological sectioning. The fourth prominent change is gradual development of an umbilical hernia, beginning at about GD 10.5, as the rapidly expanding abdominal viscera take up residence outside the narrow confines of the abdominal cavity. Finally, various embryonic structures (other than blood vessels) begin to display color, the most obvious of which are the liver primordium (located within the bulging abdomen just caudal to the heart) and the retina (which appears as a black circular background within the optic vesicle).

Subsequent stages in organogenesis lead to continued evolution of external features and internal organs. The changes that define any two stages typically are quite subtle, but the cumulative effect is to convert the flat, featureless neural plate (GD 7.5) into a late embryo (GD 14.5) that in most respects now exhibits the adult body plan. Important external landmarks that occur during the middle of gestation include the complete closure of the neural tube (Figures 4.16, 4.18, and 4.19) and the appearance of the forelimb bud (Figure 4.18), hind limb bud (Figure 4.20), and nasal primordia (Figures 4.19 and 4.21). The divisions of the brain expand greatly (Figure 4.22) and the limb buds develop regional sculpting that culminates in the formation of indentations on the distal margin of the limb buds (Figures 4.23 and 4.24), first on the forepaw (about GD 12.3) and shortly thereafter on the hind paw (GD 12.8). These indentations presage the development of actual digits on the forepaw at GD 14 (Figure 4.25) and on the hind paw at GD 15 (Figure 4.26).

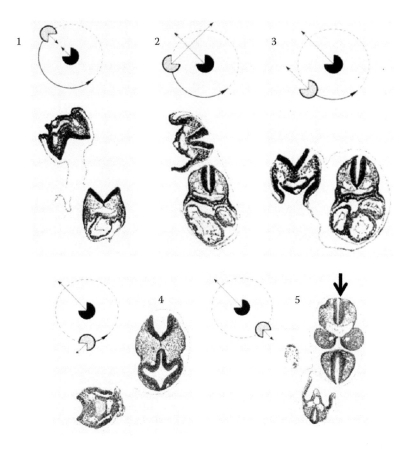

FIGURE 4.15 A Theiler stage 13 mouse conceptus engages in *turning* (shown as viewed from above, with steps moving consecutively in numerical sequence) from approximately GD 8.5 to GD 9.0 to reposition the tail (pale hemicircle) from its original site behind the head (black hemicircle) to a new location to the front of the animal. Note that orientation of the head (demonstrated by the arrow extending from the groove in the black hemicircle) remains constant. The curved arrow shows the direction of the turn, which usually is directed to the right. Note that turning takes place concurrently with progressive closure of the neural tube, especially in the cranial region (thick arrow in panel 5). (Reproduced from Kaufman, M.H., *The Atlas of Mouse Development*, 2nd edn., Academic Press, San Diego, CA, Copyright 1992, with permission from Elsevier.)

Fetal Development

The fetal period (beginning at GD 15.0 and extending to term) is so truncated in mice that events during this period often are considered together with those that occurred during embryogenesis. In fact, many mouse developmental biologists do not use the term *fetus* for mice but merely continue to refer to conceptuses as embryos until birth. We have chosen to distinguish the fetal period specifically to highlight the biological differences between the events that occur in embryos (e.g., organ specification) and fetuses (organ growth). The beginning of the fetal period coincides with the end of organogenesis and the time when ossification centers (initial foci of bony mineralization) are evident in the bones of the pelvis and all the large bones of the hind limbs.

With some minor exceptions, mouse fetuses have external and (to a lesser degree) internal features that closely resemble those of adult mice. This similarity means that many investigators choose to begin their developmental pathology studies at this time during gestation because they have at least some familiarity with the adult body plan and adult-like cell and organ architectures, and thus can extrapolate their knowledge to successfully understand fetal traits. The 4-day fetal phase is characterized chiefly by the expansion of committed cells and tissues to remodel organs and systems.

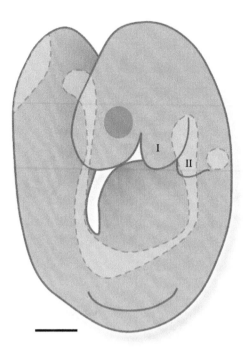

FIGURE 4.16   A Theiler stage 14 mouse conceptus (GD 9.0) is characterized by complete closure of the rostral (anterior) neuropore, initial formation of the caudal (posterior) neuropore, successful conclusion of the turning process, initial formation of the linear alimentary tract (pale central zone with dashed border), and a prominent heart. Branchial arches (I) and (II) are visible as small, bilateral protuberances in the cervical region, and bilateral condensations of tissue on the body wall that span somites 8–12 delineate the future sites of the forelimb buds. Bar = 250 μm. (The diagram has been adapted by Mr. Tim Vojt, The Ohio State University, from the EMAP (e-MOUSE Atlas), v3.5, http://www.emouseatlas.org/emap/home.html, 2013, last accessed December 1, 2014, by courtesy of the Medical Research Council of the United Kingdom.)

Nonetheless, some distinctive structural changes do occur in this time frame. The pinnae (external ears) grow to cover the external auditory meatuses (ear canals) by approximately GD 16 (Figure 4.27). The umbilical hernia begins to recede at GD 15 or so as the organs take up residence in the now more capacious abdomen; this process is completed by GD 17. The eyelids fuse in most animals by GD 16–GD 17 (Figure 4.28). Anatomic changes on the last gestational day (Figure 4.29) and the day of birth (Figure 4.30) are striking mainly because the increase in body size is substantial. Pups may be sexed with reasonable ease at this time because the lack of fur permits ready evaluation of the ano-genital distance (AGD, which is twice as long in males) and observation of the bulbourethral gland[28,183] (which forms a large dark spot in the superficial subcutis between the anus and genital tubercle[178,183]; see Figure 7.19). The AGD is increased in female neonates exposed *in utero* to androgens produced by adjacent male littermates.[169,207]

Postnatal Development

At birth, *perinatal* (a term used by scientists who study carcinogenesis in mice) or neonatal mice (or *pinkies* or *pups*) are observed to be essentially hairless, nearly immobile, pale pink cylinders. A healthy pup normally will display a white spot on the left flank soon after birth, which is an indication that it has a milk-filled stomach. The anatomic and physiological development of neonates during the early post-natal period (PND 0 through PND 7–10) encompasses a set of events and processes that are equivalent to those that occur during the last two trimesters of human pregnancy. Initial instinctive behaviors like geotaxis (i.e., the tendency to reorient the head so that the nose is at the most elevated position when placed on an inclined plane) and suckling are rudimentary at birth but become progressively rapid and smooth in the first days after birth. Such physiological changes are accompanied by profound changes

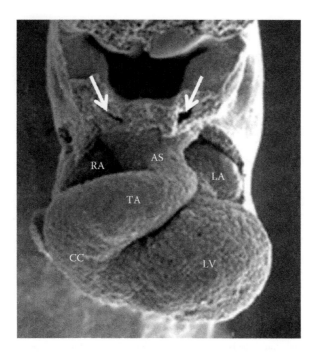

FIGURE 4.17 A Theiler stage 14 mouse conceptus (approximately GD 9.25) exhibits characteristic cardiac morphology resulting from progressive looping and ballooning needed to shift the heart tube into a more transverse (i.e., adult-like) orientation. The common atrium has moved from its original location—caudal to the left ventricle (LV)—both dorsally and cranially to form paired left (LA) and right (RA) atria. The bulbus cordis (i.e., right ventricle primordium, seen here as the conus cordis [CC]) has rotated to the right and migrated caudally. The truncus arteriosis (TA) is the precursor of the ascending aorta and pulmonary arterial trunk) and thus delivers blood to the aortic sac (AS), from whence it enters the dorsal aortae (arrows). (These scanning electron micrographs, including annotations, are reproduced here courtesy of Dr. Kathleen Sulik, University of North Carolina, Chapel Hill, NC, from an online tutorial in mouse embryology entitled Embryo images: Normal and abnormal mammalian development, http://www.med.unc.edu/embryo_images/, last accessed December 1, 2014.)

in the central nervous system. Major findings include an increase in the sizes of the cerebellar lobules (Figure 4.31) as well as the white matter tracts connecting brain centers either to each other or to the spinal cord. Early movements are stilted because many neural cell populations have yet to reach their final locations (Figure 4.32) and myelination does not commence extensively until at least a week after birth. Similarly, hematopoiesis shifts progressively during the early postnatal period from a hepatocentric process toward the more adult pattern of blood cell production in the bone marrow; this reshuffling involves the temporary regression of certain blood cell populations (see Ref. [166] and Chapter 8). Other pronounced anatomic shifts in this period include the initial eruption of the incisor teeth at PND 10 and renewed patency of various surface structures, including the external auditory meatus (at PND 11), eyelids (at PND 12[56]), and finally the vagina (at approximately PND 30–35). The timing of important pubertal events in female mice (e.g., vaginal opening, first vaginal cornification, and the onset of estrus cyclicity) is dependent on the mouse strain.[127]

Anatomic changes during postnatal development obviously will be attended by far-reaching evolution of physiological processes as well. A detailed review of developmental events during the neonatal and juvenile periods is beyond the scope of this book. However, superb review articles on physiological advances during postnatal development have been crafted recently for many systems in rodents (mice and/or rats): central nervous system,[224] gastrointestinal system,[212] heart,[68] immune system,[74] kidney,[230] lung,[231] reproductive tract (male[113] and female[8]), and skeleton.[232] Additional morphological descriptions for this phase of development may be obtained from reference texts[166,182] and by evaluating relevant websites from research institutions[52,81,82] and commercial mouse vendors.

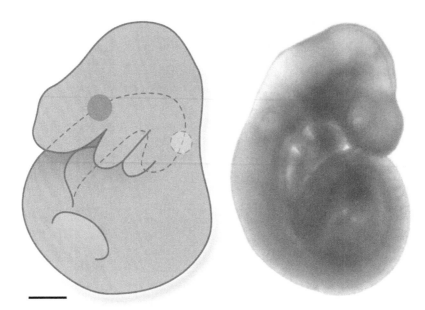

FIGURE 4.18 A Theiler stage 15 mouse conceptus (GD 9.5) is recognized by narrowing or complete closure of the caudal neuropore and distinct forelimb buds. The head becomes more bulbous as the forebrain vesicle (termed the prosencephalon) partitions to become the telencephalon (cerebral hemispheres) and diencephalon (thalamus and hypothalamus). Bar = 500 μm. (The diagram has been adapted by Mr. Tim Vojt, The Ohio State University, from the EMAP (e-MOUSE Atlas), v3.5, http://www.emouseatlas.org/emap/home.html, last accessed December 1, 2014, by courtesy of the Medical Research Council of the United Kingdom.)

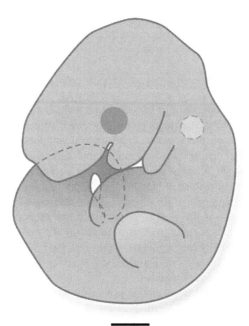

FIGURE 4.19 A Theiler stage 16 mouse conceptus (GD 10.0) showing full closure of the caudal neuropore, prominent forelimb buds, and a short tail. The nasal processes form bilaterally on the rostroventral portion of the head in front of the optic pits. Bar = 1 mm. (The diagram has been adapted by Mr. Tim Vojt, The Ohio State University, from the EMAP (e-MOUSE Atlas), v3.5, http://www.emouseatlas.org/emap/home.html, last accessed December 1, 2014, by courtesy of the Medical Research Council of the United Kingdom.)

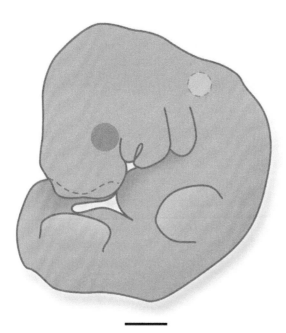

FIGURE 4.20 A Theiler stage 17 mouse conceptus (GD 10.5) demonstrates deepening of the lens pits, subdivision of branchial arch I into dorsal and ventral regions (the primordial maxilla and mandible, respectively), hind limb buds located bilaterally at somites 23–28, and an elongating tail. Bar = 1 mm. (The diagram has been adapted by Mr. Tim Vojt, The Ohio State University, from the EMAP (e-MOUSE Atlas), v3.5, http://www.emouseatlas.org/emap/home.html, last accessed December 1, 2014, by courtesy of the Medical Research Council of the United Kingdom.)

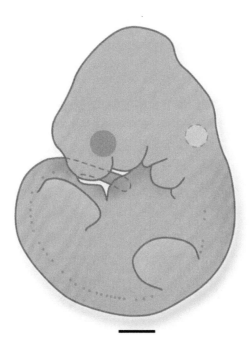

FIGURE 4.21 A Theiler stage 18 mouse conceptus (GD 11.0) is recognized externally by closure of the lens vesicle, initial formation of the nasal pits, and disappearance of the somites in the cervical regions. Bar = 1 mm. (The diagram has been adapted by Mr. Tim Vojt, The Ohio State University, from the EMAP (e-MOUSE Atlas), v3.5, http://www.emouseatlas.org/emap/home.html, last accessed December 1, 2014, by courtesy of the Medical Research Council of the United Kingdom.)

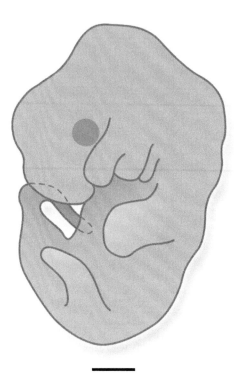

FIGURE 4.22   A Theiler stage 19 mouse conceptus (GD 11.5) is discerned by total separation of the lens vesicle from the surface ectoderm, sharp definition of the eye margins, and sculpting of the forelimb bud to yield a narrow proximal column (the future shoulder and arm) and a distal paddle (the nascent paw). Bar = 1 mm. (The diagram has been adapted by Mr. Tim Vojt, The Ohio State University, from the EMAP (e-MOUSE Atlas), v3.5, http://www.emouseatlas.org/emap/home. html, last accessed December 1, 2014, by courtesy of the Medical Research Council of the United Kingdom.)

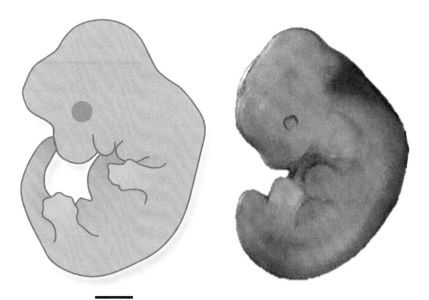

FIGURE 4.23   A Theiler stage 20 mouse conceptus (GD 12.0) is characterized by the appearance of digital rays to define the digits on the forepaw (at approximately GD 12.3) and hind paw (GD 12.8). Bar = 1 mm. (The diagram has been adapted by Mr. Tim Vojt, The Ohio State University, from the EMAP (e-MOUSE Atlas), v3.5, http://www.emouseatlas.org/emap/ home.html, last accessed December 1, 2014, by courtesy of the Medical Research Council of the United Kingdom.)

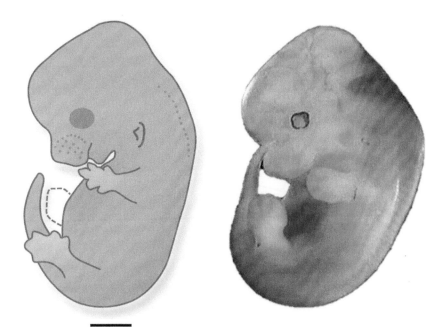

FIGURE 4.24 A Theiler stage 21 mouse conceptus (GD 13.0) features multiple indentations of the apical ridge of the paw plate on the forelimb (denoting the positions where the webs between digits will regress) and the conspicuous appearance of vibrissae follicles in rows on the snout. The pinna is beginning to become elevated. An umbilical hernia (dashed brown line) comprised of intestinal loops is evident. Bar = 1 mm. (The diagram has been adapted by Mr. Tim Vojt, The Ohio State University, from the EMAP (e-MOUSE Atlas), v3.5, http://www.emouseatlas.org/emap/home.html, last accessed December 1, 2014, by courtesy of the Medical Research Council of the United Kingdom.)

Stages in the Anatomic Evolution of the Mouse Placenta

The placenta (*plakuos* [Greek] = *flat cake*) is a temporary organ formed *in utero* to sustain the developing offspring. Placenta is the first organ created by a newly implanted embryo (GD 5.0), although its construction necessitates that extraembryonic membranes from the embryo work cooperatively with the maternal decidua (a hormone-primed cell population derived from endometrial stroma) to complete the task. The core function of the placenta and its membranes is to serve as an interface for exchanging nutrients and wastes between the embryonic and maternal circulations and to provide an environment suitable for supporting growth of the embryo. Additional roles include production of hormones that regulate maternal hemodynamic and metabolic activities to ensure that the pregnancy is sustained. The relatively nonvascular decidua serves the additional purpose of erecting a physical barrier between maternal blood and embryonic trophoblast,[223] thereby preventing premature mingling of potentially immunogenic elements. The morphology and gene expression of the mouse placenta is modified extensively during gestation in response to the embryo's evolving metabolic demands.[59,85]

General Features of Placental Anatomy and Physiology

Basic Placental Structure

The genesis of the mouse placenta has been well characterized.[30–33,36,59,76,126,145,154,163,164,184,214,218] The mouse placenta (as well as that of rats and humans) forms a discoid, hemochorial, chorioallantoic-derived interface with the uterine lining (Figure 4.33). The *discoid* type means that the exchange of materials between conceptus and dam occurs via a regionally localized plaque (or disc) rather than over the entire placental surface, while the *hemochorial* configuration indicates that maternal capillaries

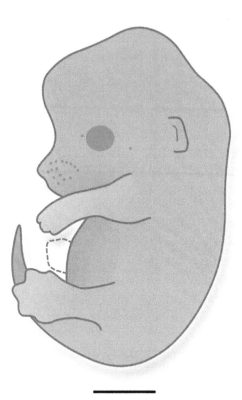

FIGURE 4.25   A Theiler stage 22 mouse conceptus (GD 14.0) exhibits separation of the distal digits on the forelimb and multiple indentations of the apical ridge of the paw plate on the hind limb. The outer edge of the elevated pinna is beginning to fold rostrally. Hair follicles are prominent on the trunk. Bar = 1 mm. (The diagram has been adapted by Mr. Tim Vojt, The Ohio State University, from the EMAP (e-MOUSE Atlas), v3.5, http://www.emouseatlas.org/emap/home.html, last accessed December 1, 2014, by courtesy of the Medical Research Council of the United Kingdom.)

rupture to form sinusoids (i.e., blood-filled channels that lack an endothelial lining) that are separated from endothelial-lined embryonic capillaries by embryonic trophoblasts. In fact, the mouse placenta uses a *hemotrichorial* arrangement in which embryonic capillaries are enveloped by three bands of trophoblasts: two inner, continuous layers of multinucleated syncytiotrophoblasts, and a single outer, discontinuous stratum of mononuclear trophoblasts, which abut the sinusoids. This structure prevents most maternal blood cells from crossing the placenta, thereby limiting any maternal immune response while enhancing gas and nutrient exchange by minimizing the width of the tissue across which materials must be transferred. The *chorioallantoic* derivation means that the embryonic contribution to the final placenta stems from two cell lineages: the chorionic plate (extraembryonic ectoderm) and the allantois (extraembryonic mesoderm).[59]

Morphological Characteristics of the Yolk Sac Placenta

Shortly after implantation begins, the placental primordium consists chiefly of the trophectoderm (the outer cell layer of the blastocyst), with a small contribution of embryonic endoderm that serves as a lining for the inner surface of the blastocoele (Figure 4.6). The trophectoderm represents the first terminally differentiated cell type within the conceptus, although specialized trophectoderm subpopulations do have different roles. The polar trophectoderm forms a cap over the ICM that eventually will differentiate into the extraembryonic ectoderm (EEE, the source for labyrinthine trophoblast cells) and the EPC ([Figure 4.7], the source for the spongiotrophoblast cells). The mural trophectoderm, which forms the wall of the blastocoele, is the source of most trophoblast giant cells (TGC) that invade

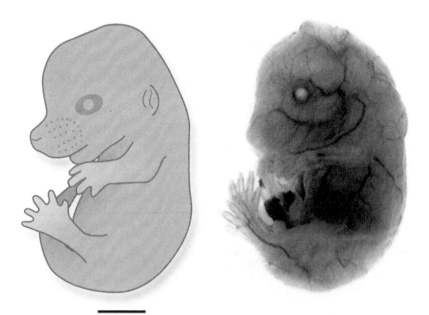

FIGURE 4.26 A Theiler stage 23 mouse conceptus (GD 15.0) possesses fully separated but divergent digits (i.e., splayed rather than parallel) on all four limbs. The forward-directed pinna covers about half of the external auditory meatus. Hair follicles now cover the head as well. Bar = 1 mm. (The diagram has been adapted by Mr. Tim Vojt, The Ohio State University, from the EMAP (e-MOUSE Atlas), v3.5, http://www.emouseatlas.org/emap/home.html, last accessed December 1, 2014, by courtesy of the Medical Research Council of the United Kingdom.)

FIGURE 4.27 A Theiler stage 24 mouse conceptus (GD 16.0) displays nearly parallel digits on the forelimb and visible claw primordia. The eyelids are mostly or completely fused, and the pinna covers essentially all of the external auditory meatus. The umbilical hernia is reduced in size as the intestinal loops relocate into the abdominal cavity. Bar = 2 mm. (The diagram has been adapted by Mr. Tim Vojt, The Ohio State University, from the EMAP (e-MOUSE Atlas), v3.5, http://www.emouseatlas.org/emap/home.html, last accessed December 1, 2014, by courtesy of the Medical Research Council of the United Kingdom.)

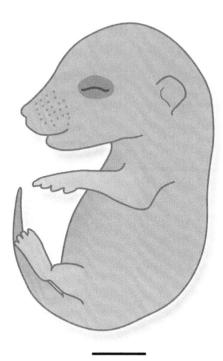

FIGURE 4.28 A Theiler stage 25 mouse conceptus (GD 17.0) has thickened skin with prominent folds, resulting in greatly reduced prominence of the subcutaneous blood vessels. The eyelids are completely fused, and short vibrissae have recently erupted. Bar = 2 mm. (The diagram has been adapted by Mr. Tim Vojt, The Ohio State University, from the EMAP (e-MOUSE Atlas), v3.5, http://www.emouseatlas.org/emap/home.html, last accessed December 1, 2014, by courtesy of the Medical Research Council of the United Kingdom.)

the adjacent decidua (Figure 4.34). The TGC are formed by persistent endoreduplication of DNA in the absence of cell division. As gestation proceeds, the expansion of the EEE differentiates further to form the chorionic epithelium, the cells of which over time aggregate to form the thick chorionic plate (i.e., the future base of the definitive placenta). The allantois arises as a narrow bridge of mesoderm from the caudal end of the embryo. As gestation progresses, the allantois (i.e., precursor of the umbilical cord) projects toward the chorion; the fusion of these two organs at approximately GD 8.5 produces the chorioallantois. Completion of this connection is essential if successful creation of an efficient system for gas and nutrient exchange (i.e., definitive placentation) is to become functional later in gestation.

The yolk sac (YS) is the first functional placental structure in the mouse and remains the only source of gas and nutrient exchange until the elements of the definitive placenta begin to assemble (approximately GD 10.0–GD 10.5).[86] The YS is a translucent, fluid-filled pouch connected to the ventral surface of the embryonic plate (Figure 4.35). Early in gestation, the YS is seen in histological sections as a thin membrane formed of cuboidal epithelium subtended by a thin layer of mesoderm; the mesodermal layer is punctuated at intervals by aggregates of hematopoietic stem cells (*blood islands*), which are the main source of erythrocytes during early pregnancy (approximately GD 8–GD 11) (Figure 4.36). Later in gestation, the YS appears microscopically as a thin, prominently folded membrane that no longer has blood islands (Figure 4.36). The YS is lined by extraembryonic endoderm (a derivative of the hypoblast); the endoderm has two components, an outer layer lining the blastocoele (the parietal endoderm) and an inner layer attached to the embryo (the visceral endoderm). The visceral endoderm is supported in part on a mesenchyme bed derived from extraembryonic mesoderm. Early after implantation, the ICM sinks away from the EPC to form the elongated *egg cylinder* (Figure 4.8). This repositioning brings the outer (parietal) and inner (visceral) walls of the YS into proximity with each other in a distinctive process termed *inversion* (Figures 4.8 and 4.37). Gases move across the YS by diffusion, but

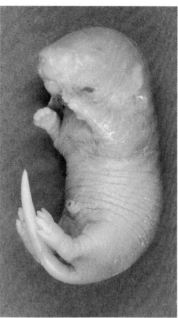

FIGURE 4.29 A Theiler stage 26 mouse conceptus (GD 18.0) exhibits long vibrissae and long pinnae that completely obscure the external auditory meatuses. The fetus on the right was fixed by immersion in Bouin's solution. Bar = 3 mm. (The diagram has been adapted by Mr. Tim Vojt, The Ohio State University, from the EMAP (e-MOUSE Atlas), v3.5, http://www.emouseatlas.org/emap/home.html, last accessed December 1, 2014, by courtesy of the Medical Research Council of the United Kingdom.)

macromolecules are transferred by the process of *histiotrophic nutrition*,[59,150,181,233] whereby YS phagocytes engulf secretions from uterine glands and pass them to the embryo. Ablation of transcription factors that control expression of genes whose products are needed to bind or transport micronutrients leads to early embryonic lethality.[22,48] The YS vascular tree is established by GD 8.0–GD 8.5 but continues to branch until between GD 12 and GD 14, suggesting that the YS continues to fill a histiotrophic role even after the definitive placenta has taken over as the principal means of embryonic support (which occurs at approximately GD 12.5).[122]

Morphological Characteristics of the Definitive Placenta

Later in gestation, the main procurer of nutrients for embryonic use switches from the YS to the definitive placenta. This change is essential because the growing metabolic needs of the embryo cannot be served by histiotrophic nutrition alone. Accordingly, placental structure evolves to include a greatly expanded interface for exchange of fluids, gases, and macromolecules between the maternal and embryonic vascular systems, a process termed *hemotrophic nutrition*. The inability to switch from histiotrophic to hemotrophic nutrition will cause early embryonic lethality between GD 10.5 and GD 12.5.

At the macroscopic level, the definitive placenta forms a dark red, discoid mass that caps the mesometrial pole of the YS (Figures 4.33 and 4.38). The thin inner membrane that surrounds the embryo (after turning has been completed) is the amnion. The central, dark red circle at the base of the disc is the chorionic plate, representing the site at which the allantois extending from the embryo will fuse to complete formation of the umbilical cord. The brown ring encircling the chorionic plate is the labyrinth, a hemispherical tissue mass derived mainly from embryonic endothelia and trophoblasts but containing parallel arrays in which embryonic capillaries abut maternal sinusoids. The pale outer ring is decidua

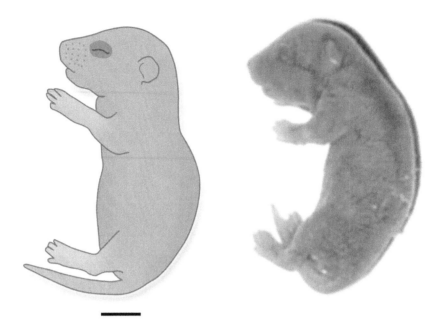

FIGURE 4.30 A Theiler stage 27 mouse (GD 19 or GD 20) is a neonate. The appearance is similar to that of a near-term fetus except that the individual is larger—and no longer housed within the uterus! Bar = 5 mm. (The diagram has been adapted by Mr. Tim Vojt, The Ohio State University, from the EMAP (e-MOUSE Atlas), v3.5, http://www.emouseatlas.org/emap/home. html, 2013, last accessed December 1, 2014, by courtesy of the Medical Research Council of the United Kingdom. The gross photograph is reprinted from Klinger, S., Turgeon, B., Lévesque, K., Wood, G.A., Aagaard-Tillery, K.M., and Meloche, S., Loss of Erk3 function in mice leads to intrauterine growth restriction, pulmonary immaturity, and neonatal lethality, *Proc. Natl. Acad. Sci. USA*, 106(39), 16710–16715. Copyright [2009] National Academy of Sciences, U.S.A.)

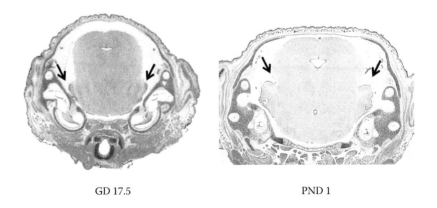

GD 17.5   PND 1

FIGURE 4.31 Postnatal development in the cerebellum is associated with pronounced structural changes and not merely a progressive increase in size. During the fetal (here at GD 17.5) and early neonatal (here at postnatal day [PND] 1) periods, the cerebellum is distinguished as a pair of small primordia (arrows) anchored to the sides of the metencephalon. These primordia do not expand into a recognizable cerebellum until several days after birth. Mouse, CD-1 stock.

derived from the uterine wall. In the absence of a well-formed definitive placenta, embryos will begin to degenerate between GD 9 and GD 9.5 and ultimately will die sometime between mid-gestation (GD 12.5) to the end of organogenesis (GD 15.0) (see Chapter 16 for more details on major placental lesions). In most cases, expiration of the embryo occurs approximately 36–60 h after a defect in the placenta first leads to insufficient transfer of micronutrients and oxygen.

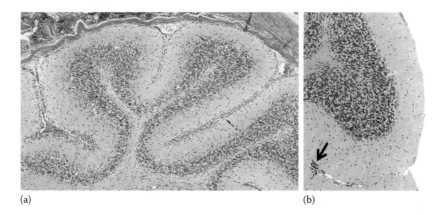

(a)   (b)

FIGURE 4.32 Postnatal development in the cerebellum is associated with migration of precursor cells to the various neuronal layers. At postnatal day (PND) 14 (a), the cerebellum features a thin but uniform external layer of granule cells on the meningeal surface of the folia, and the borders between the cerebellar layers are irregular; together, these features indicate that some cells are still migrating to reach their final positions. In an adult (approximately 120-day-old) animal, rare and small ectopic rests (arrow, b) beneath the meninges represent the residue of the transient external granule cell layer, and the boundaries between the cerebellar layers are sharp. Mouse, FVB strain.

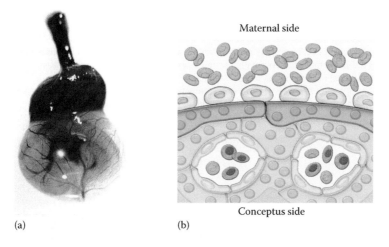

Maternal side

Conceptus side

(a)   (b)

FIGURE 4.33 Organization of the mouse placenta. Macroscopically (a), the discoid definitive placenta forms a bright red, soft mass atop the translucent, highly vascularized yolk sac. At the microscopic level (b), capillaries of embryonic origin (white lumen) contain a mixture of primitive (nucleated) and definitive (nonnucleated) erythrocytes and are lined by a continuous layer of endothelial cells. These capillaries are separated from the maternal sinusoids (pale yellow lumen) by three chorial cell layers: two inner continuous strata of multinucleated syncytiotrophoblasts and one outer discontinuous layer of mononuclear trophoblast (which forms the lining of the sinusoid). Sinusoids are large channels filled with maternal blood in which no layers of maternal-origin tissue (i.e., epithelium, connective tissue, or endothelium) remain from the original uterine wall. The direct contact of maternal blood with embryonic tissue is a hemochorial arrangement (or more specifically a hemotrichorial plan in mice since there are three trophoblast layers).

At the microscopic level, the late-gestation placenta is a complex organ arranged in strata.[59] Each part may be identified by its location and its distinctive cytoarchitectural features. The amnion is a thin protective membrane that envelops and closely conforms to the turned embryo. The amniotic fluid within it serves to cushion the embryo from trauma. The next envelope out is the YS, with its visceral wall positioned next to the amnion and its parietal wall associated with the acellular Reichert's membrane (RM). An interrupted stratum of trophoblast giant cells is situated between the RM and the decidua along the antimesometrial (bottom) and side walls of the conceptus. The definitive placenta evolves as

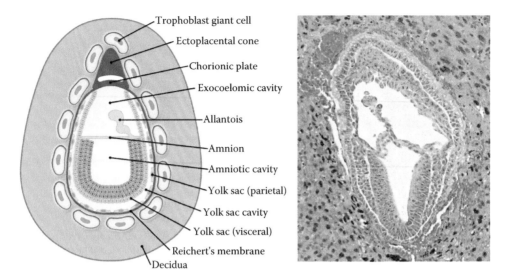

FIGURE 4.34 The primordial placenta (shown here at GD 7.5 as both a schematic diagram and matched H&E-stained histological section) has many easily discerned anatomic features. The trophectoderm (i.e., the outer wall of the blastocyst) gives rise to the trophoblast cells, which is the first terminally differentiated cell type in the mouse embryo. The polar trophectoderm establishes the ectoplacental cone on the side of the conceptus closest to the major maternal blood vessels, while on the opposite (antimesometrial) side, the mural trophectoderm cells form a ring of trophoblast giant cells that eventually encircle the entire conceptus. The allantois originates from mesoderm at the caudal end of the embryo and projects toward the chorionic plate, eventually making contact at about GD 8.5 to form the umbilical cord. The amnion is a protective bilayered membrane that eventually encompasses the mouse embryo once it has completed turning. The yolk sac is the first functional placenta in the mouse. (Adapted from Bolon, B., Pathology analysis of the placenta, in *The Guide to Investigations of Mouse Pregnancy*, Croy, B.A., Yamada, A.T., DeMayo, F., Adamson, S.L., eds., Academic Press [Elsevier], San Diego, CA, pp. 175–188, Copyright 2014, with permission from Elsevier.)

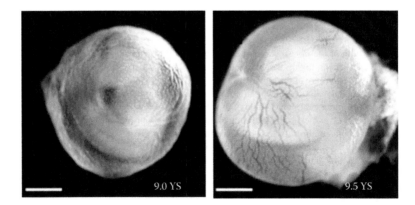

FIGURE 4.35 The yolk sac (YS) appears macroscopically as a thick, translucent membrane that encircles the embryo. The YS is most prominent prior to completion of the definitive placenta, when the YS serves to absorb nutrients, to excrete waste products, and to establish the initial cardiovascular and hematopoietic components critical for embryonic survival beyond E10.5. The YS surface has narrow, branching cords of vitelline blood vessels that over time become more highly ramified and engorged. The numbers denote the gestational day. (Reproduced from Rhee, S. et al., *PLoS ONE*, 8(3), e58828, 2013. With permission.)

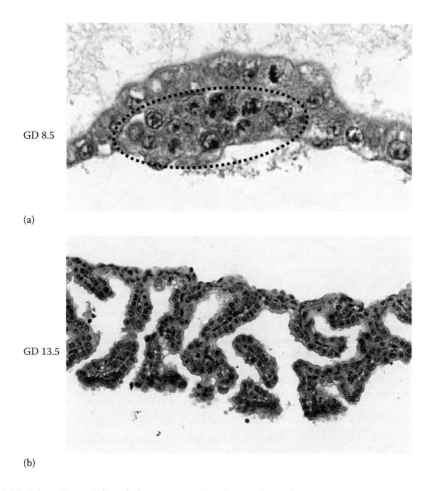

GD 8.5

(a)

GD 13.5

(b)

FIGURE 4.36 The yolk sac (YS) early in gestation (a) is a flat membrane featuring widely dispersed knots of hemato-poietic precursors (*blood islands* [dotted circle]), which function as the principal sites for embryonic erythropoiesis during early organogenesis (approximately E8–E11). Later in gestation (b), the YS exhibits elongated folds lined by cuboidal epithelial cells along its outer surface, a thin core of loose connective tissue, and a smooth inner surface. (Republished from Bolon, B., Pathology analysis of the placenta, in *The Guide to Investigations of Mouse Pregnancy*, Croy, B.A., Yamada, A.T., DeMayo, F., Adamson, S.L., eds., Academic Press [Elsevier], San Diego, CA, pp. 175–188, Copyright 2014, with permission from Elsevier.)

a multi-layered discoid mass oriented in the mesometrial position (i.e., near the maternal arterial supply and maternal sinusoids) in the location formerly filled by the EPC. These layers are understood most readily by working outward from the flat base of the disc (i.e., closest to the embryo) to the outer, convex surface that contacts the uterine wall.

The base of the definitive placenta is the chorionic plate—a narrow, dense epithelial band that covers the central third of the disc bottom (Figure 4.39). This plate is the site where the allantois fuses, thereby forming a bridge between the placenta and embryo that ultimately becomes the umbilical cord. The remainder of the disc is partitioned into three distinct zones (Figure 4.40). The inner region—the labyrinth—is a dense mesh of intermingled embryonic capillaries and maternal sinusoids arranged in a parallel array to facilitate efficient countercurrent exchange. Embryonic capillaries, which are lined by an uninterrupted layer of pore-bearing endothelial cells, contain only large, immature, nucleated (i.e., primitive) erythroid cells at the beginning of organogenesis which over time are joined by increasing numbers of small, mature, nonnucleated (i.e., definitive) erythrocytes. Maternal sinusoids possess an intermittent lining of embryo-derived trophoblasts and contain only nonnucleated erythrocytes throughout the entire

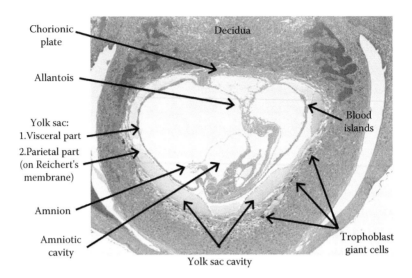

FIGURE 4.37   As shown in this GD 8.5 embryo, the early placenta consists chiefly of yolk sac (YS) and maternal decidua. Inversion of the mouse embryo at the egg cylinder stage (GD 5.0 to GD 5.5) results by this stage in apposition of the two YS walls (visceral [inner] and parietal [outer]). Reichert's membrane is a rodent-specific, nearly invisible but tough basement membrane formed between the trophoblast and the parietal endoderm cells. The allantois, a mesodermal protuberance from the caudal end of the embryo proper, has contacted and fused with the chorionic plate; the allantois forms the nascent umbilical cord, while the chorionic plate serves as the base upon which the labyrinthine portion of the definitive placenta is founded. The trophoblast giant cells arise from the mural trophectoderm and mark the boundary between the tissues of embryonic and maternal origin.

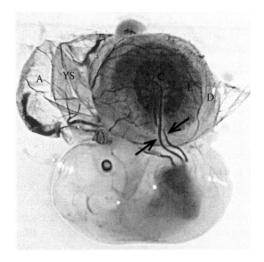

FIGURE 4.38   The definitive placenta, which forms fully by GD 12.5 (and is shown here at GD 13.5), appears as a discoid, highly vascularized mass that caps the mesometrial pole of the thin, translucent yolk sac (YS). The umbilical vessels (arrows) extend from the ventral abdomen of the embryo to the centrally located, red chorionic plate (C), which in turn is encompassed by concentric rings of pale red labyrinth (L) and tan decidua (D). The amnion (A) is the thin, translucent, avascular, inner membrane that envelops the embryo.

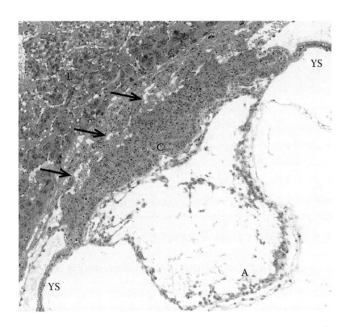

FIGURE 4.39 The chorionic plate (C) is a flat, broad strip of loosely woven epithelium attached along the central third of the inner surface of the labyrinth (L). Large vascular channels (arrows) scattered throughout this plate are derived from the allantois (A) and are connected to the umbilical blood vessels. YS, yolk sac. (Republished from Bolon, B., Pathology analysis of the placenta, in *The Guide to Investigations of Mouse Pregnancy*, Croy, B.A., Yamada, A.T., DeMayo, F., Adamson, S.L., eds., Academic Press [Elsevier], San Diego, CA, pp. 175–188, Copyright 2014, with permission from Elsevier.)

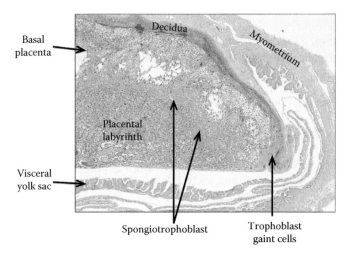

FIGURE 4.40 The definitive placenta (here shown at GD 13.5) consists of multiple distinct regions. The yolk sac appears as a convoluted membrane separating the remainder of the placenta from the embryo (not shown). This membrane is bordered by a mass of trophoblast subdivided into three distinct regions—an inner hemisphere of closely packed labyrinthine trophoblast, a middle vacuolated layer of spongiotrophoblast, and an outer thin band of trophoblast giant cells—which are intermingled with parallel arrays of embryonic capillaries and maternal sinusoids. The labyrinth rests atop the flat, broad chorionic plate (not shown), a thick vascular plate that serves as the terminus for the umbilical cord. The outer decidua is a maternal-derived zone of endometrial stromal cells.

GD 8.5   GD 13.5   GD 17.5

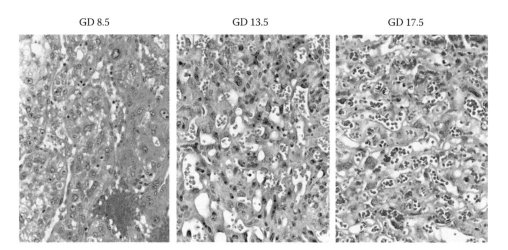

FIGURE 4.41   The labyrinth contains tissues of both embryonic and maternal origin intermingled in thin, interweaving columns separated by multiple tortuous vascular channels. The columns consist of embryonic labyrinthine trophoblast cells, which have small nuclei and narrow rims of eosinophilic cytoplasm. The labyrinthine trophoblasts encompass narrow maternal-origin sinusoids (indicated by a lack of endothelium). Adjacent embryo-derived capillaries are lined by flat endothelial cells. Early in gestation (GD 8.5), blood in this zone chiefly originates from the dam and so is characterized by many nonnucleated (adult) erythrocytes. In midgestation (GD 13.5), embryonic blood vessels are differentiated by their nucleated (embryonic) erythrocytes. Late in gestation (GD 17.5), most fetal erythrocytes are produced in the liver and (to a growing degree) bone marrow and therefore are nonnucleated (adult) cells.

in utero period. Both embryonic and maternal vasculature is remodeled all through pregnancy (Figure 4.41). The middle region—the junctional or basal zone—lies between the labyrinth and the uterine wall. Blood vessels in this area contain only maternal blood. A thin, interrupted layer of trophoblast giant cells occupies the margin between the junctional zone and the outer area—the decidua basalis (composed of modified endometrial stroma cells). In mice, retrograde trophoblast migration deep into maternal arteries that supply blood to the placenta is prominent.[1,34,147]

Specific embryonic cell types in the definitive placenta have distinct cytoarchitectural features and patterns of gene expression (Table 4.3), both of which are associated with their unique cell type–specific functions. These cell populations often exhibit spatial and/or temporal exclusivity. For example, in the labyrinth, all nucleated cells in the vessel walls represent either labyrinthine trophoblast or endothelium, both of which are derived from the embryo. Embryonic capillaries are encircled by two continuous layers of plump, multi-nucleated syncytiotrophoblasts, on top of which resides a single discontinuous layer of mononuclear cytotrophoblasts that project into the lumen of the maternal sinusoids (spaces filled with maternal blood); the sinusoids lack maternal endothelium in contrast to other adult organ-specific sinusoids, which always are lined by endothelium. The junctional zone contains two embryo-derived trophoblast lineages: an inner, vacuolated region containing myriad spongiotrophoblasts, and a thin outer rim of trophoblast giant cells (Figures 4.40 and 4.42). The spongiotrophoblasts are small, oblong cells with basophilic nuclei and slim rims of eosinophilic cytoplasm, while the giant cells have large, polyploid nuclei and abundant cytoplasm. As gestation advances, growing numbers of periodic acid Schiff (PAS)-positive glycogen cells[26] may be observed within the spongiotrophoblast layer and decidua (Figure 4.43).

Other than intravascular erythrocytes, decidual cells are the chief maternal contribution to the placenta (Figure 4.42). These elements are polygonal cells with oval, slightly open nuclei and indistinct cell membranes, which generally are arranged in dense sheets. The primary or antimesometrial decidual cells form at approximately GD 4.5 to GD 5.0 as a dense field of large multinucleate cells; these cells regress by apoptosis as the egg cylinder begins to expand, and their remnants are consumed by the adjacent trophoblast giant cells.[140,203] At about GD 7.0, decidual cells begin to form along the mesometrial boundary as small, mononuclear cells. The decidual cords are separated by blood vessels lined by thin, flat endothelial cells and filled with nonnucleated (maternal) erythrocytes.

TABLE 4.3

Selected Cell Type–Specific Markers for Principal Cell Populations in the Mouse Placenta

Site(s) of Expression	Marker	Marker Abbreviation	References
Decidua	Macrophage surface antigen-2	*Mac2 (Lgals3)*	[95]
Trophoblast			
Labyrinth trophoblast	Extraembryonic, spermatogenesis, homeobox 1	*Esx1*	[135]
	Glial cells missing-1	*Gcm1*	[3,193]
	Mammalian achaete–scute complex homolog 2	*Mash2 (Ascl2)*	[63]
	Transcription factor EB	*Tfeb*	[187]
Spongiotrophoblast	Mammalian achaete–scute complex homolog 2	*Mash2*	[63]
	Platelet-derived growth factor	*Pdgf*	[135]
	Proliferin	*Plf*	[135]
	Placental lactogen-II	*Prl3b1*	[79]
	Son of sevenless homolog 1	*Sos1*	[152]
	Trophoblast-specific protein α	*Tpbpa*	[104]
Trophoblast giant cells	Cathepsin Q	*Ctsq*	[79]
	Colony-stimulating factor 1 receptor	*Csf1r*	[135]
	Heart and neural crest derivatives expressed 1	*Hand1*	[159,179]
	MyoD family inhibitor domain containing	*I-mfa (Mdfi)*	[123]
	LIM domain kinase 1	*Limk1*	[23]
	Placental lactogen-1	*Prl3d1 (Pl-1)*	[128,135]
Glycogen-containing trophoblast	Prolactin-like protein N	*Prl7b1*	[79]
	Trophoblast-specific protein α	*Tpbpa*	[79]
Trophoblast stem cells	Caudal type homeobox 2	*Cdx2*	[135,183]
	Eomesodermin	*Eomes*	[167]
Natural killer (NK) lymphocytes	*Dolichos biflorus* agglutinin lectin	DBA lectin	[136]
Endothelium	Mesoderm-specific transcript	*Mest*	[79]
	Plasmalemma vesicle-associated protein	*Plvap* (MECA-32)	[96]
	Platelet/endothelial cell adhesion molecule 1	*Pecam1* (CD31)	[120]
Chorion	α4 integrin	*Itga4*	[226]
	Mammalian achaete–scute complex homolog 2	*Mash-2 (Ascl2)*	[63]
Allantois	Vascular cell adhesion molecule 1	*Vcam1*	[64,98]

Note: Adapted in part from Refs. [11,79].

Physiologic Characteristics of the Definitive Placenta

The expansion of the definitive placenta is a key feature throughout the postimplantation period. The placental volume of outbred ICR mice exhibits a threefold increase from GD 10.5 (0.01 mL) to GD 13.5 (0.03 mL) and an additional threefold rise nearing birth (0.09 mL at GD 18.5).[148] The bulk of the growth appears to reflect expansion of the labyrinth,[149] which represents approximately 50% of the placental size when it reaches its fully functional state (about GD 12–GD 12.5).[27] By comparison, the ratio of labyrinth to decidua evolves from 1:2 at GD 10.5 to almost 6:1 at GD 13.5 and nearly 15:1 at term.[149] Despite this precipitous shift in embryonic-to-maternal derivation, the mean placental weight peaks near the end of organogenesis (at GD 14) at about 100 g and remains relatively stable until birth—even while embryonic body weight increases by twofold during the fetal period.[78,108,160,229] Interestingly, the ratio of fetal-to-placental weights reported in C57BL/6 mice covers a broad range, from 7:1[160] to 14–15:1.[27] This degree of divergence suggests that continued viability of the conceptus is feasible even in cases

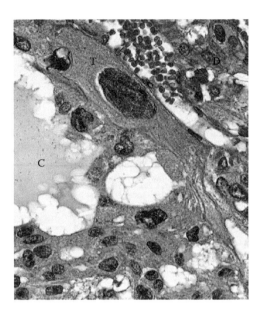

FIGURE 4.42 Trophoblast giant cells (upper middle) are enormous cells with a large nucleus and abundant eosinophilic cytoplasm. Spongiotrophoblasts (densely packed cells in lower middle)—comparatively small, oval cells with small basophilic nuclei, thin rims of pale eosinophilic cytoplasm, and a variable number of clear cytoplasmic vacuoles—occupy the *junctional zone* between the trophoblast giant cells (T) outside and the labyrinth inside (not shown). Large cavities (C) in the spongiotrophoblast layer give this area its distinctive *spongy* appearance. D, decidual cells.

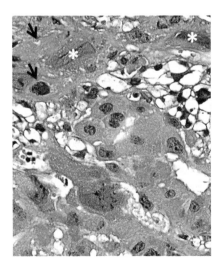

FIGURE 4.43 Trophoblast glycogen cells packed with eosinophilic, cytoplasmic granules grow in number in the decidual layer near the junctional zone as the definitive placenta is launched. Large nuclei (*) represent trophoblast giant cells forming a single row above the spongiotrophoblasts (small cells with clear, colorless cytoplasm). Mouse, GD 14.

where the placenta is comparatively small, an outcome made possible by increased placental efficiency. Nonetheless, placental size clearly correlates well with fetal size as larger placentas generally accompany bigger fetuses.[116,148] However, reductions in placental size due to maternal conditions like hypoxemia or malnutrition may be rectified by adjustments in placental morphology and/or gene expression to enhance metabolism and nutrient transport capacity.[209]

In addition to nutrient transfer, various placental cell lineages have important local and systemic regulatory roles during pregnancy. Both spongiotrophoblast cells and trophoblast giant cells produce hormones, such as luteotropic molecules to sustain the early and middle stages of pregnancy as well as lactogenic factors to kindle mammary gland development and milk generation in the dam later in gestation.[67,106] Trophoblast giant cells also create many angiogenic factors,[19,101,210] vasodilators,[57,227] and anticoagulants,[219] and a subset of them invade and remold the mural architecture of maternal spiral arteries.[1,34] Together, these actions increase the amount of maternal blood in proximity to embryonic capillaries, thereby improving the possibility of nutrient and waste exchange between the circulatory systems of the conceptus and the dam. The maternal decidua also secrete a plethora of cytokines, growth factors, and hormones that modulate placental function, and they can serve as phagocytes when the placenta is being refashioned.[234] The ability of decidual cells to store glycogen may represent a means for reliably supplying energy to the embryo until the definitive placenta can be built.

The structure of the placenta serves as an effective hurdle to inappropriate activation of an immune response. Decidual cells appear to form a barrier early in gestation that protects the uterus from excessive penetration by embryonic cells. This phenomenon results from both the dense nature of the decidual reaction, which provides a physical barrier, but also by the production of chemical inhibitors of tissue proteases.[2,157,196] In the mature definitive placenta (after GD 12.5), the decidual cells function as macrophage-like elements at the same time that gene expression patterns and tissue antigens are being revealed; they have abundant eosinophilic cytoplasm at this point but soon undergo regressive changes that lead to their degeneration after GD 14. The thick decidual layer also shields the embryo and its complement of paternal antigens from assault by the maternal immune system by limiting transfer of leukocytes and antibodies into close proximity with embryonic tissues until after the protective layers like Reichert's membrane and the YS have been generated.[140] The influx of uterine natural killer (uNK) cells into the placenta during early and mid-gestation represents a non-physical means of restraining the maternal immune response against the embryo. The uNK cells are bone marrow–derived, large granular leukocytes that collect in the decidua, especially in proximity to trophoblast giant cells and blood vessels, and also between the smooth muscle layers at implantation sites.[37,39,188,189] Early in gestation, they are thought to regulate the maintenance of implantation[5] as well as local angiogenic and immunological tasks,[37–39,205] while later they form a transient structure known as the mesometrial lymphoid aggregates of pregnancy (MLAp; alternate designations: metrial gland[38,145,188] and decidualized mesometrial triangle[146]) in the decidua basalis (Figure 4.44). The uNK cells are susceptible to infection by intracellular pathogens[170] and thus might represent a potential focus for microbial colonization of the conceptus. The uNK cells begin to regress by mid-gestation.

Several adaptations in placental organization are designed to increase the efficiency of gas, nutrient, and waste transfer. In many cases, the improved effectiveness depends on the continued structural evolution of the organ throughout gestation. For example, branching morphogenesis in the labyrinth involves extension of embryonic capillaries into trophoblast columns, thus increasing the amount of vascular surface available for exchange.[35,59] The outer layer of cytotrophoblasts that lines the maternal sinusoids becomes attenuated as gestation advances, while the two layers of syncytiotrophoblasts develop many surface microvilli[134] and invaginations[59] to increase the surface area in contact with maternal-derived fluid. The countercurrent direction of blood flow in maternal and embryonic vascular channels (Figure 4.45) ensures that the gradients for exchange between the two circulations always favor the transfer of oxygen and nutrients into embryonic capillaries and the removal of carbon dioxide and metabolic waste products into maternal sinusoids.[1]

Influence of Genetic Background on Placental Structure and Function

Placental anatomy and physiology may be impacted both by dissimilarities in genetic background and by the presence of gene mutations in one or more placental cell populations. Such variations have not been reviewed systematically.

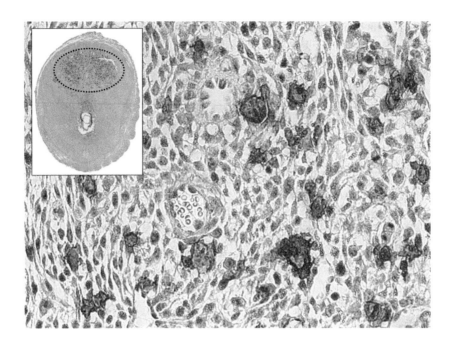

FIGURE 4.44 Uterine natural killer (uNK) cells (sometimes referred to as granulated metrial gland cells), shown here at GD 8.5, are large granular leukocytes that are recruited to form secondary lymphoid tissues to populate the uterus, where they cluster (inset, surrounded by dotted circle) near large maternal vessels coursing through the decidua basalis. On later gestational days, uNK cells also are enriched between the layers of the uterine wall musculature at decidual attachment sites, forming transient structures termed mesometrial lymphoid aggregates of pregnancy (MLAp). Activated uNK cells are thought to serve immunologic and angiogenic functions. Stain: Histochemistry for binding of *Dolichos biflorus* (DBA) lectin. (Image courtesy of Dr. Bruno Zavan, University of Alfenas, Alfenas, Brazil; reproduced from Bolon, B., Pathology analysis of the placenta, in *The Guide to Investigations of Mouse Pregnancy*, Croy, B.A., Yamada, A.T., DeMayo, F., Adamson, S.L., eds., Academic Press [Elsevier], San Diego, CA, pp. 175–188, Copyright 2014, with permission from Elsevier.)

Placental dimensions have been shown to be affected by many mutations carried in a placental cell lineage.[163] For example, enhanced placental size (i.e., placentomegaly) due to elevated numbers of spongiotrophoblasts is a recognized outcome of cloning[195] and epidermal growth factor receptor (*Egfr*) ablation.[41] The labyrinth is bigger in mice missing *Esx1* (extraembryonic/spermatogenesis/homeobox 1), which causes a pervasive decline in blood vessel numbers.[105] In contrast, labyrinth size often plunges if placental cells bear a null mutation that alters cell signaling (e.g., *Egfr* and fibroblast growth factor 2 receptor [*Fgf2r*]).[4,67] The extent to which a single-gene mutation modifies the development of a particular placental lesion will vary with the mouse strain.[40,41] Placental size also may be affected by the concurrent existence of mutations in multiple genes. For instance, female mice bearing only one X chromosome consistently undergo placental hyperplasia relative to females that carry two X chromosomes.[80] This phenomenon has been credited to the dampening effect of X-linked genes because embryos having an XY genotype possess demonstrably larger placentas than do littermates with an XX genotype.[80]

Placental mass varies among mouse strains. For instance, outbred ICR mice have placentas that are approximately 25% greater than those of age-matched inbred C57BL/6J (B6) mice.[149] This discrepancy is correlated with embryonic growth as B6 embryos are approximately 25% smaller than the corresponding ICR embryos.[149] A similar placental size increase of 20% has been reported in hybrid B6C3F1 mice relative to inbred C57BL/6 or C3H parent animals.[116] Strain differences in placental biology are not limited to structural effects as metabolic capacities and pathways in placental tissue also vary among mouse strains.[58]

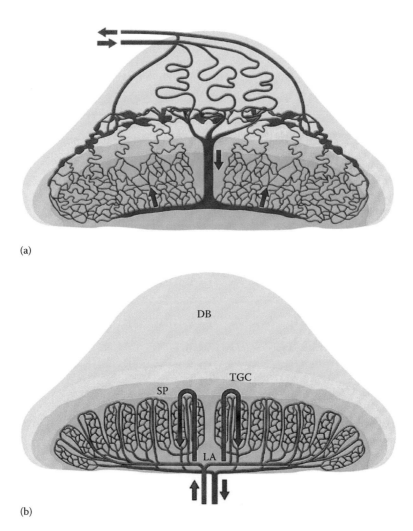

(a)

(b)

FIGURE 4.45 Schematic diagrams demonstrating the countercurrent relationship of the maternal (a) and embryonic (b) circulations in the mouse definitive placenta. Vessels in red bear the most highly oxygenated blood, while vessels in blue hold the least oxygenated blood. Arrows indicate the direction of blood flow. DB, decidua basalis; LA, labyrinth; SP, spongiotrophoblast; TGC, trophoblast giant cells. (Republished from *Dev. Biol.*, 250(2), Adamson, S.L., Lu, Y., Whiteley, K.J., Holmyard. D., Hemberger, M., Pfarrer, C., and Cross, J.C., Interactions between trophoblast cells and the maternal and fetal circulation in the mouse placenta, 358–373, Copyright 2002, with permission from Elsevier.)

Appendix: Timing of Major Milestones in Mouse Development

TABLE 4.A.1

Timing of Major Milestones in Embryonic and Fetal Development

Parameter	Age (Day)
Ovulation	0.0
Formation of one-celled embryo (zygote) by fertilization, in oviduct	0.5
Two-celled embryo	1.0
Four-celled embryo	2.0
Eight-celled embryo	2.5
Morula, in uterus	3.0
Blastocyst	3.5
Blastocyst shedding the zona pellucida (*hatching*)	4.0
Implantation	4.5
Differentiation of epiblast and hypoblast	4.75
Egg cylinder	5.0
Ectoplacental cone fills with maternal blood	6.0
Reichert's membrane forms	6.0
Gastrulation	6.0–6.5
Embryonic axis morphologically evident	6.5
Primitive streak initiated	6.5
Initial mesoderm formation (at caudal end of embryo)	6.75
Amnion forming	7.0
Neural plate	7.0
Head process	7.0
Allantois forming	7.25–7.5
Hematopoietic precursors cluster in yolk sac (*blood islands*)	7.5
Neural folds	7.5
Amniotic cavity	7.75
Heart primordium	7.75
Somite pair no. 1	7.75–8.0
One branchial arch	8.0
Allantois contacts chorionic plate	8.0
One pharyngeal pouch	8.3
Chorioallantoic fusion	8.5
Somite pair no. 10	8.5
Two branchial arches	8.5
Two pharyngeal pouches	8.5
Myocardial contractions begin	8.5
Dorsal flexure completely reversed	8.5–9.0
Rostral (anterior or cranial) neuropore closure	9.0
Caudal (posterior) neuropore closure	9.25–9.5
Three branchial arches	9.5
Three pharyngeal pouches	9[15]
Forelimb bud	9.5–9.75
Tail bud	9.5–10.0
Hind limb bud	10.0–10.3
Tail extends	10.5
End of the *human first trimester* equivalent	10.5–11.0
Umbilical hernia forms (containing intestinal loops)	11.0–12.0
	(Continued)

TABLE 4.A.1 *(CONTINUED)*

Timing of Major Milestones in Embryonic and Fetal Development

Parameter	Age (Day)
Pigmented eyes	11.0–12.0
External auditory meatus (ear canal)	12.0–13.0
Papillae rows on snout for vibrissae	12.0–13.0
Eyelids	12.0–13.0
Forepaw digital rays	12.3
Hind paw digital rays	12.8
Cloaca partitioned	13.0
Pinnae	13.0–14.0
Palatal shelves fuse	15.0
End of organogenesis (and the *human second trimester* equivalent)	15.0
Umbilical hernia recedes	15.5–16.5
External genitalia differentiated	17.0
Birth	19–20
End of the *human third trimester* equivalent	PND 10

Notes: Values represent gestational days for *in utero* events, or postnatal days (PND) for events occurring after birth.

Data adapted from values reviewed in Refs. [24,45,59,72,93,130,176,192,197,211].

TABLE 4.A.2

Timing of Major Milestones in Postnatal Development

Parameter	Postnatal Age (Days)
Cerebellum begins to differentiate	2
Hair begins to appear	2–4
Myelination (region-specific) starts in central nervous system	7–8
Nipples develop	8–10
Incisors erupt	10
Intake of solid food	10–11
External auditory canals open	11
Eyelids open	12
External nares open	19
Neural cell proliferation tapers	17–20
Testes descend	28–30
Vagina opens	35–36

Note: Data adapted from values reviewed in Refs. [24,45,130,182,197].

TABLE 4.A.3

Timing of Major Milestones in Circulatory System Development

Parameter	Age (Day)
Heart tubes fuse	7.0
S-shaped heart tube with no distinct chambers	8.0
First aortic arch	8.0[13]
Myocardial contractions begin	8.25
Aortic arch arteries begin forming	8.5
S-shaped heart with three distinct zones: common atrium, primitive left ventricle, bulbus cordis (anlage of the right ventricle)	8.5
Embryonic circulation to visceral yolk sac begins	8.5
Compact looped heart with clear paired atria, primitive left ventricle, bulbus cordis (right ventricle), and prominent aortic sac	9.0
Second aortic arch	9.0[4]
Dorsal aorta fuses	9.0[4]
Endocardial cushions form	9.0[18]
Third aortic arch	10.0
Cardiac septation begins in atria	10.0
Fully functional circulation is established in the embryo	~10.25
Heart rate ≈ 125 beats per minute (bpm)	10.5
Cardiac septation begins in ventricles	10.5
Pulmonary arteries form	10.5
Vitelline veins anastomose with liver vessels	10.5
Endocardial cushions fuse	11.0
Atrioventricular valves form	11.0
Atrial septum completed, common ventricle remains	11.5
Left atrium fuses with pulmonary veins	11.5
Foramen ovale	12.0
Heart rate ≈ 150 bpm	12.5
Caudal (inferior) vena cava enters heart	12.5
Right dorsal aorta regresses	13.0
Interventricular septum complete	13.0
Aorta and pulmonary arterial trunks separate	13.0
Cardiac valves initiated	13.5
Coronary arteries established	14.0
Heart rate ≈ 200 bpm	14.5
Heart rate ≈ 250 bpm	18.5

Notes: Values represent gestational days (lower-case numerals) and sometimes hours (superscript numerals). Data adapted from values reviewed in Refs. [45,72,92,115,130,173,176,197,211,228].

TABLE 4.A.4

Timing of Major Milestones in Digestive System Development

Parameter	Age (Day)
Foregut pocket and oral plate	7.8
One pharyngeal pouch	8^8
Hindgut pocket established	8.5
Liver diverticulum	8.8–9.0
Laryngotracheal groove	9.0
Oral membrane perforates	$9–9^2$
Gall bladder primordium	9.6
Liver epithelial cords established	9.5
Pancreas (dorsal lobe)	9.5
Pancreas (ventral lobe)	9.7–11.0
Esophagus partitions from trachea	11.0
Hepatopancreatic duct	11.0
Umbilical hernia initiated	11.0–12.3
Upper lip	11.0
Tongue	11.0
Stomach established	11.5
Pancreatic lobes fuse	11.5
Palatine shelves form	12.0
Cloacal membrane	13.0
Dental laminae	13.0
Anal membrane perforates	14.0
Salivary glands	14.0
Umbilical hernia at maximal size	14.5
Palatal folds fusing	15.0
Umbilical hernia begins to recede	14.0–15.0
Umbilical hernia completely reduced	15.5–16.5
Eruption of incisors	PND 10

Notes: Values represent gestational days (lower-case numerals) and sometimes hours (superscript numerals) for *in utero* events, or postnatal days (PND) for events occurring after birth.

Data adapted from values reviewed in Refs. [45,72,130,192,197].

TABLE 4.A.5

Timing of Major Milestones in Endocrine System Development

Parameter	Age (Day)
One pharyngeal pouch	8.3
Two pharyngeal pouches	8.5
Thyroid primordium	8.5
Pituitary gland (Rathke's pouch) forms from oral ectoderm	8.5
Thyroid gland (primordium) established	8.5
Pituitary gland (Rathke's pouch) contacts infundibulum	$9.5–9^9$
Three pharyngeal pouches	9^{15}
Adrenal gland (cortex) launched	11.0–11.5
Parathyroid gland begun	11.5
Pineal gland initiation	11.5
Pituitary gland (neurohypophysis [*pars nervosa*]) develops	11.5
Pituitary gland (Rathke's pouch) loses connection with oral ectoderm	12.0
Adrenal medulla	14.0
Pituitary gland (adenohypophysis [*pars distalis*]) matures	14.0
Pineal gland expands	14.0
Testicular interstitial cells (of Leydig)	14.5
Pancreatic islets (of Langerhans)	18

Notes: Values represent gestational days (lower-case numerals) and sometimes hours (superscript numerals).
Data adapted from values reviewed in Refs. [45,72,130,197].

TABLE 4.A.6

Timing of Major Milestones in Hemolymphatic System Development

Parameter	Age (Day)
Committed hematopoietic precursors reach yolk sac (YS)	7.0–7.25
Hemangioblasts form aggregates (*blood islands*) in YS	7.5
Differentiation of primitive erythroblasts in blood islands	8.0
Erythroblasts first reach embryo	8.25
Myocardial contractions begin	8.25
Hematopoietic stem cells (HSCs) reach aorta-gonad-mesonephros (AGM)	8.5–9.25
Liver colonized by waves of HSCs	10.0–12.0
Spleen primordium	11.0
Thymus primordium (from third pharyngeal pouch)	11.0–12.0
Lymph node primordia	11.5
Thymus colonized by HSCs (especially for lymphocytes)	12.0–13.0
YS hematopoiesis begins to decline	13.0
Spleen colonized by HSCs	15.0–15.5
Peyer's patch primordia	15.5
Bone marrow colonized by HSCs	16.0
Peyer's patches colonized by HSCs	16.5
Liver-centered hematopoiesis begins to decline	19.0

Notes: Values represent gestational days.
Data adapted from values reviewed in Refs. [45,72,115,130,180,197].

TABLE 4.A.7

Timing of Major Milestones in Integumentary System Development

Parameter	Age (Day)
Surface ectoderm becomes a continuous layer	8.0
First dermatomes initiated	8.0
Mammary hillocks form	10.5–11.0
Auricular hillocks form	11.5
Nasolacrimal groove	11.5
Milk line appears	12.0
Periderm (a transient outer layer of flattened epidermis, representing the initial epidermal covering of the embryo) present	12.0
Eyelids form	13.0
Pinnae develop	13.0
Hair follicles first form	13.0
Vibrissae (*whisker*) rudiments	13.0–13.5
Epidermal layers begin differentiating	13.5
Hair follicles on trunk	14.0–15.0
Keratin begins to form	15.0
Melanocytes (without processes) colonize epidermis	15.0
Epidermal layers have differentiated	16.0
Dendritic cells begin to appear in epidermis	16.0
Eyelids fuse	16.0–16.5
Hair follicles on head	16.0–17.0
Melanosomes proliferate in melanocytes	17.5–18.0
Hair begins to appear	PND 2–4
Nipples develop	PND 8–10

Notes: Values represent gestational days for *in utero* events, or postnatal days (PND) for events occurring after birth.

Data adapted from values reviewed in Refs. [45,72,130,175,182,192,197,220–222].

TABLE 4.A.8

Timing of Major Milestones in Musculoskeletal System Development

Parameter	Age (Day)
Somite pairs nos. 1–7	8.0
First myotomes initiated	8.0
First sclerotomes initiated	8.0
Somite pairs nos. 8–12	8.5
Somite pairs nos. 13–20	9.0
Somite pairs nos. 21–29	9.5
Forelimb bud (near somites 8–12)	9.5–9.75
Tail bud	9.5
Somite pairs nos. 30–34	10.0
Hind limb bud (near somites 23–28)	10.0–10.3
Somite pairs nos. 35–39	10.5
Forelimb bud develops subdivisions	10.5
Somite pairs nos. 40–45	11.0
Forelimb bud develops subdivisions	11.0
Tongue primordium	11.0
Somite pairs nos. 46–50	11.5
Somite pairs nos. 51–54	12.0
Palatal shelves form	12.0
Ribs begin to form	12.0
Forepaw digital rays	12.3
Somite pairs nos. 55–60	12.5
Dental laminae form	13.0
Ginglymi (*elbows*) can be distinguished	13.0
Ribs begin to develop chondrification centers	13.0
Carpi (*wrists*) can be distinguished	14.0
Distal phalanges of forepaw digits separate	14.0
First ossification centers are initiated in large bones of forelimb	14.5
Diaphragm completed	14.5
Chondrified ribs (first seven pairs) contact sternum	14.5
Palatal folds are fusing	15.0
Ossification centers in all large bones of forelimb	15.0
Ossification centers in pelvic bones	15.0
Ossification centers in all large bones of hind limb	15.0
Ossification centers in all ribs	15.5
End of somite formation	16
Ossification centers first seen in metacarpals III and IV	16
Ossification centers first seen in metatarsals II, III, and IV	16
Ossification center first seen in metacarpal V	17
Ossification center first seen in metatarsal V	17
Ossification center first seen in metacarpal I	17.5
Ossification centers first seen in fore paw: proximal phalanges II, III, and IV	17.5
Ossification centers first seen in fore paw: distal phalanges II, III, and IV	17.5
Ossification centers first seen in hind paw: proximal phalanges II, III, and IV	17.5

(*Continued*)

TABLE 4.A.8 (*CONTINUED*)

Timing of Major Milestones in Musculoskeletal System Development

Parameter	Age (Day)
Ossification centers first seen in hind paw: distal phalanges II, III, and IV	17.5
Ossification center first seen in fore paw: proximal phalanx V	18
Ossification centers first seen in fore paw: distal phalanges I and V	18
Ossification centers first seen in hind paw: distal phalanges I and V	18
Ossification center first seen in hind paw: distal phalanx V	18
Ossification centers first seen in fore paw: middle phalanges II, III, and IV	18.5
Ossification centers first seen in tarsus (calcaneus and talus)	18.5
Ossification centers first seen in hind paw: proximal phalanges IV and V	18.5
Ossification centers first seen in hind paw: middle phalanges II, III, and IV	18.5
Ossification center first seen in fore paw: middle phalanx V	PND 1
Ossification center first seen in hind paw: middle phalanx V	PND 1
Ossification centers first seen in carpus and metacarpal I	PND 7
Ossification centers first seen in fore paw: metacarpal I, proximal phalanx I	PND 7
Ossification centers first seen in hind paw: metatarsal I, proximal phalanx I	PND 7
Ossification centers in tarsus (remaining bones)	PND 7
Ossification center in patella	PND 14

Notes: Values represent gestational days for *in utero* events, or postnatal days (PND) for events occurring after birth.

Data adapted from values reviewed in Refs. [45,72,130,141,197].

TABLE 4.A.9

Timing of Major Milestones in Nervous System Development

Parameter	Age (Day)
Neural plate	7.0
Head process	7.0
Neural folds	7.5
Neural tube formation initiated at somite pairs nos. 4–5	8.5
Primordia form for three brain vesicles (prosencephalon [forebrain], mesencephalon [midbrain], rhombencephalon [hindbrain])	8.5
Rostral (anterior or cranial) neuropore closure	9.0
Caudal (posterior) neuropore closure	9.25–9.5
Cerebral hemispheres initiated	10.0
Pontine flexure forms	10.5
Olfactory bulbs	10.5
Basal nuclei (caudate/putamen) form[a]	11
Cerebrocortical neurons (layer VI) begin forming	11
Hippocampus (CA1, CA3, and dentate gyrus) neurons begin forming	11
Vomeronasal organ	11.5
Cortical subventricular zone (stem cell layer) forms[a]	12
Cerebellar primordia initiated	12.0
Colliculi (rostral [superior] and caudal [inferior]) are partitioned	12.0–12.5
Choroid plexus established	12.5–13.0
Cerebrocortical neurons (layer V) neurons begin forming	13
External capsule forms[a]	13
Rostral (anterior) commissure[a]	14
Cerebrocortical neurons (layers II, III) begin forming	16
Corpus callosum[a]	16–17
Corticothalamic connections established[a]	17
Corticospinal tracts reach cervical spinal cord segments[a]	PND 1
Synaptogenesis accelerates in the brain[a]	PND 3
Sensory barrels (for vibrissae) form in cerebral cortex[a]	PND 3
Corticospinal tracts reach lumbar spinal cord segments[a]	PND 7
Optic tract—onset of myelination[a]	PND 8
Hippocampus—onset of myelination[a]	PND 13
Corpus callosum—onset of myelination[a]	PND 15

Notes: Values represent gestational days for *in utero* events, or postnatal days (PND) for events occurring after birth.

Data adapted from values reviewed in Refs. [45,72,176,197].

[a] Predicted value based on statistical modeling.[24,25]

TABLE 4.A.10

Timing of Major Milestones in Pulmonary/Lower Respiratory Tract Development

Parameter	Age (Day)
Primary lung diverticula arise from ventral wall of foregut	9.5–9.75
Primary bronchi	10.0
Lung lobe asymmetry initiated	10.5
Trachea separates from esophagus	11.0
Secondary bronchi form	12.0
Bronchial tree branching	13.0
Laryngeal remodeling (narrowing of caudal portion)	13.0
Diaphragm completed	14.5
Bronchioles differentiate	16.5
Alveolar ducts and alveoli develop	GD 17.0 to PND 5
Surfactant	18.0
Alveolar septation, size, and number increase	PND 5 to PND 28

Notes: Values represent gestational days for *in utero* events, or postnatal days (PND) for events occurring after birth.
Data adapted from values reviewed in Refs. [45,70,72,130,197]·

TABLE 4.A.11

Timing of Major Milestones in Reproductive System Development

Parameter	Age (Day)
Germ cells in yolk sac epithelium	8.0
Germ cells migrating in mesentery	9.5
Gonadal ridge becomes distinct	11.0–11.5
Germinal epithelium in gonad	11.5–12.0
Differentiation (sex-specific) of gonad	12.0–12.5
Differentiation (sex-specific) of external genitalia	17.0
Regression of unneeded genital ducts	17.0–18.0
Male	
Mesonephric duct (Wolffian duct [progenitor of the epididymis, vas deferens, and accessory sex glands]) appears	9.5–10.0
Mesonephric duct reaches urogenital sinus/cloaca	11.0–11.5
Germinal epithelium (testis) appears	11.5
Testis differentiates	12.0–12.5
Interstitial cells (of Leydig) form	14.5
Prostate present as solid cords	18.0
Descent of testes	PND 28–30
Sexual maturity	PND 60
Female	
Paramesonephric duct (Müllerian duct [progenitor of the oviduct, uterus, cervix, and upper 65% of the vagina]) appears	10.0–11.0
Paramesonephric duct reaches urogenital sinus/cloaca	13.5
Primary ovarian follicles	18.0
Vagina opens	PND 35–36
Sexual maturity	PND 55

Notes: Values represent gestational days for *in utero* events, or postnatal days (PND) for events occurring after birth.
Data adapted from values reviewed in Refs. [45,72,130,197].

TABLE 4.A.12

Timing of Major Milestones in Sensory System Development

Parameter	Age (Day)
Eye (sight)	
Optic vesicles	8.0
Optic sulci	8.5
Lens placode forms	9.5
Lens placode (eye) begins to invaginate	10.0
Lens separated	10.5
Lens vesicle closed	11.0–11.5
Retinal pigmentation develops	11.5–12.0
Optic nerve axons reach optic chiasm[a]	12
Cornea differentiates	13.0
Eyelids form	13.0
Cones—peak generation[a]	14
Optic nerve axons reach visual cortex[a]	14
Anterior chamber differentiates	14.0
Retinal ganglion cells	14.0
Rods—onset of generation[a]	15
Eyelids fuse	16.0–16.5
Iris and ciliary body differentiate	18.0
Rods—peak generation[a]	PND 2
Optic tract—onset of myelination[a]	PND 8
Eyelids open	PND 12
Ear (hearing)	
Otic placodes form	8.0–8.5
Otic vesicle forms	8.5–8.75
Otic pits develop	9.0
Endolymphatic duct forms	10.0–10.5
Endolymphatic duct separates from otic vesicle	11.0
Saccules and utricle form	11.0
Cochlea and vestibular apparatus begin differentiating	11.0–11.5
Semicircular canals forming	11.5–12.0
Cochlea present	12.0
External auditory meatus	12.5–13.0
Pinnae form	13.0
Organ of Corti forms	14.0
Saccules and utricle are partitioned	14.5
Otic capsule forms (as cartilage)	14.5
Pinna overgrows and occludes external auditory meatus	16.5
External auditory meatus—opens	PND 11
Auditory tracts—onset of myelination	PND 11
	(*Continued*)

TABLE 4.A.12 *(CONTINUED)*

Timing of Major Milestones in Sensory System Development

Parameter	Age (Day)
Nose (smell)	
Olfactory placodes form	8.5–9.0
Olfactory pit develops	10.0–10.5
Olfactory bulb differentiates	11.0
Nasolacrimal duct	14.5
External nares open	18.5–19.0
Olfactory tract—onset of myelination[a]	PND 9

Notes: Values represent gestational days for *in utero* events, or postnatal days (PND) for events occurring after birth.

Data adapted from values reviewed in Refs. [45,72,130,192,197].

[a] Predicted value based on statistical modeling.[24,25]

TABLE 4.A.13

Timing of Major Milestones in Urinary System Development

Parameter	Age (Day)
Pronephros (kidney primordium)	8.0
Nephrogenic cord has tubules and associates with the mesonephric (or Wolffian) duct (the precursor of the epididymis, vas deferens, and accessory sex glands)	8.75–9.0
Mesonephros (embryonic kidney) appears	9.5
Paramesonephric duct (or Müllerian duct [precursor of the oviduct, uterus, cervix, and the upper 65% of the vagina]) appears	10.0
Metanephros (adult kidney in amniotes) appears	11.0
Ureteric bud develops	11.0–11.5
Mesonephric duct extends to cloaca	11.5–12.0
Glomeruli form	16.0

Notes: Values represent gestational days for *in utero* events, or postnatal days (PND) for events occurring after birth.

Data adapted from values reviewed in Refs. [45,72,130,197].

TABLE 4.A.14

Timing of Major Milestones in Placental Development

Parameter	Age (Day)
Trophectoderm forms the blastocyst wall	3.0–3.5
Implantation	4.5
Decidua (a pseudo-epithelioid derivative of the endometrial stroma) forms	4.5–5.0
Trophoblast giant cells first invade decidua	4.5
Extraembryonic endoderm (from hypoblast) begins lining the inner wall of the blastocoele and ventral surface of the inner cell mass (ICM [the precursor of the embryo])	4.5
Extraembryonic endoderm completely lines the surfaces of the blastocoele (now called the yolk sac [YS] cavity) and ICM	5.0
Creation of the ectoplacental cone (EPC) by polar trophoblast atop the ICM	5.0
Trophoblast giant cells begin forming	5.0–5.5
Uterine natural killer (uNK) cells begin collecting in the decidua, forming the mesometrial lymphoid aggregates of pregnancy (or *metrial gland*)	5.0–5.5
Amniotic cavity forms (as a space separating the EPC from the dorsal surface of the embryo)	6.0
Maternal blood enters the EPC	6.0
First appearance of Reichert's membrane (an acellular membrane that forms beneath the parietal wall of the YS in rodents)	6.0
Embryonic membranes expand to obliterate the lumen of the uterine gland crypt in which implantation has occurred	6.5
Allantois (precursor of the umbilical cord) forms at the caudal end of the embryo	7.0
Vascularization of the visceral YS (vitelline circulation) begins	7.0
Amnion is completely sealed	7.5
Allantois visibly elongates	7.5
Aggregates of hematopoietic stem cells (*blood islands*) develop in the YS	8.0
Allantois fuses with the chorionic plate	8.5
The yolk sac placenta begins to fully function	~8.5
Induction of the labyrinth	8.5–8.7
Amnion encircles the turned embryo	9.5
Assembly of the definitive placenta begins	9.5
Maternal blood enters the labyrinth	9.5–10.0
The chorioallantoic placenta commences to function	9.5–10.0
Reliance on hemotrophic nutrition (provided by the definitive placenta) begins	11.5–12
Definitive placenta is complete and fully functional	12.5
uNK cell numbers peak in decidua	12.5
Peak placental weight is achieved (approximately 100 mg)	14.0–14.5
uNK cell numbers in decidua start to decline	16.0

Notes: Values represent gestational days for *in utero* events, or postnatal days (PND) for events occurring after birth. Data adapted from values reviewed in Refs. [7,38,59,86,184,210,229].

REFERENCES

1. Adamson SL, Lu Y, Whiteley KJ, Holmyard D, Hemberger M, Pfarrer C, Cross JC (2002). Interactions between trophoblast cells and the maternal and fetal circulation in the mouse placenta. *Dev Biol* **250** (2): 358–373.

2. Alexander CM, Hansell EJ, Behrendtsen O, Flannery ML, Kishnani NS, Hawkes SP, Werb Z (1996). Expression and function of matrix metalloproteinases and their inhibitors at the maternal-embryonic boundary during mouse embryo implantation. *Development* **122** (6): 1723–1736.

3. Anson-Cartwright L, Dawson K, Holmyard D, Fisher S, Lazzarini RA, Cross JC (2000). The glial cells missing-1 protein is essential for branching morphogenesis in the chorioallantoic placenta. *Nat Genet* **25** (3): 311–314.

4. Arman E, Haffner-Krausz R, Chen Y, Heath JK, Lonai P (1998). Targeted disruption of fibroblast growth factor (FGF) receptor 2 suggests a role for FGF signaling in pregastrulation mammalian development. *Proc Natl Acad Sci USA* **95** (9): 5082–5087.

5. Ashkar AA, Croy BA (1999). Interferon-γ contributes to the normalcy of murine pregnancy. *Biol Reprod* **61** (2): 493–502.

6. Bany BM, Cross JC (2006). Post-implantation mouse conceptuses produce paracrine signals that regulate the uterine endometrium undergoing decidualization. *Dev Biol* **294** (2): 445–456.

7. Batten BE, Haar JL (1979). Fine structural differentiation of germ layers in the mouse at the time of mesoderm formation. *Anat Rec* **194** (1): 125–141.

8. Beckman DA, Feuston M (2003). Landmarks in the development of the female reproductive system. *Birth Defects Res B Dev Reprod Toxicol* **68** (2): 137–143.

9. Benirschke K (2007). Comparative placentation. http://placentation.ucsd.edu/ (last accessed December 1, 2014).

10. Blois SM, Klapp BF, Barrientos G (2011). Decidualization and angiogenesis in early pregnancy: Unravelling the functions of DC and NK cells. *J Reprod Immunol* **88** (2): 86–92.

11. Bolon B (2014). Pathology analysis of the placenta. In: *The Guide to Investigations of Mouse Pregnancy*, Croy BA, Yamada AT, DeMayo F, Adamson SL (eds.). San Diego, CA: Academic Press (Elsevier), pp. 175–188.

12. Bolon B, Couto S, Fiette L, La Perle KMD (2012). Internet and print resources to facilitate pathology analysis when phenotyping genetically engineered rodents. *Vet Pathol* **49** (1): 224–235.

13. Boyd KL, Bolon B (2010). Embryonic and fetal hematopoiesis. In: *Schalm's Veterinary Hematology*, 6th edn., Weiss DJ, Wardrop KJ (eds.). Ames, IA: Wiley-Blackwell, pp. 3–7.

14. Brown NA (1990). Routine assessment of morphology and growth: Scoring systems and measurements of size. In: *Postimplantation Mammalian Embryos: A Practical Approach*, Copp AJ, Cockroft DL (eds.). Oxford, U.K.: IRL Press, pp. 93–108.

15. Bruce NW, Wellstead JR (1992). Spacing of fetuses and local competition in strains of mice with large, medium and small litters. *J Reprod Fertil* **95** (3): 783–789.

16. Burgoyne PS (1993). A Y-chromosomal effect on blastocyst cell number in mice. *Development* **117** (1): 341–345.

17. Burkus J, Cikos S, Fabian D, Kubandova J, Czikkova S, Koppel J (2013). Maternal restraint stress negatively influences growth capacity of preimplantation mouse embryos. *Gen Physiol Biophys* **32** (1): 129–137.

18. Butler H, Juurlink BHJ (1987). *An Atlas for Staging Mammalian and Chick Embryos*. Boca Raton, FL: CRC Press.

19. Carney EW, Prideaux V, Lye SJ, Rossant J (1993). Progressive expression of trophoblast-specific genes during formation of mouse trophoblast giant cells *in vitro*. *Mol Reprod Dev* **34** (4): 357–368.

20. Chakraborty I, Das SK, Dey SK (1995). Differential expression of vascular endothelial growth factor and its receptor mRNAs in the mouse uterus around the time of implantation. *J Endocrinol* **147** (2): 339–352.

21. Chakraborty I, Das SK, Wang J, Dey SK (1996). Developmental expression of the cyclo-oxygenase-1 and cyclo-oxygenase-2 genes in the peri-implantation mouse uterus and their differential regulation by the blastocyst and ovarian steroids. *J Mol Endocrinol* **16** (2): 107–122.

22. Chen WS, Manova K, Weinstein DC, Duncan SA, Plump AS, Prezioso VR, Bachvarova RF, Darnell Jr JE (1994). Disruption of the HNF-4 gene, expressed in visceral endoderm, leads to cell death in embryonic ectoderm and impaired gastrulation of mouse embryos. *Genes Dev* **8** (20): 2466–2477.

23. Cheng AK, Robertson EJ (1995). The murine LIM-kinase gene (*limk*) encodes a novel serine threonine kinase expressed predominantly in trophoblast giant cells and the developing nervous system. *Mech Dev* **52** (2–3): 187–197.

24. Clancy B, Charvet CJ, Darlington RB, Finlay BL, Workman A (2013). Translating time (across developing mammalian brains). http://translatingtime.net/ (last accessed December 1, 2014).

25. Clancy B, Kersh B, Hyde J, Darlington RB, Anand KJ, Finlay BL (2007). Web-based method for translating neurodevelopment from laboratory species to humans. *Neuroinformatics* **5** (1): 79–94.

26. Coan PM, Conroy N, Burton GJ, Ferguson-Smith AC (2006). Origin and characteristics of glycogen cells in the developing murine placenta. *Dev Dyn* **235** (12): 3280–3294.

27. Coan PM, Fowden AL, Constância M, Ferguson-Smith AC, Burton GJ, Sibley CP (2008). Disproportional effects of *Igf2* knockout on placental morphology and diffusional exchange characteristics in the mouse. *J Physiol* **586** (Pt 20): 5023–5032.

28. Constantinescu GM (2011). *Comparative Anatomy of the Mouse and the Rat: A Color Atlas and Text*. Memphis, TN: American Association for Laboratory Animal Science.

29. Crawford LW, Foley JF, Elmore SA (2010). Histology atlas of the developing mouse hepatobiliary system with emphasis on embryonic days 9.5–18.5. *Toxicol Pathol* **38** (6): 872–906.

30. Cross JC (1998). Formation of the placenta and extraembryonic membranes. *Ann NY Acad Sci* **857**: 23–32.

31. Cross JC (2001). Genes regulating embryonic and fetal survival. *Theriogenology* **55** (1): 193–207.

32. Cross JC (2005). How to make a placenta: Mechanisms of trophoblast cell differentiation in mice—A review. *Placenta* **26** (Suppl. A): S3–S9.

33. Cross JC, Baczyk D, Dobric N, Hemberger M, Hughes M, Simmons DG, Yamamoto H, Kingdom JC (2003). Genes, development and evolution of the placenta. *Placenta* **24** (2–3): 123–130.

34. Cross JC, Hemberger M, Lu Y, Nozaki T, Whiteley K, Masutani M, Adamson SL (2002). Trophoblast functions, angiogenesis and remodeling of the maternal vasculature in the placenta. *Mol Cell Endocrinol* **187** (1–2): 207–212.

35. Cross JC, Nakano H, Natale DR, Simmons DG, Watson ED (2006). Branching morphogenesis during development of placental villi. *Differentiation* **74** (7): 393–401.

36. Croy BA, Yamada AT, DeMayo FJ, Adamson SL (2014). *The Guide to Investigation of Mouse Pregnancy*. London, U.K.: Academic Press (Elsevier).

37. Croy BA, He H, Esadeg S, Wei Q, McCartney D, Zhang J, Borzychowski A et al. (2003). Uterine natural killer cells: Insights into their cellular and molecular biology from mouse modelling. *Reproduction* **126** (2): 149–160.

38. Croy BA, van den Heuvel MJ, Borzychowski AM, Tayade C (2006). Uterine natural killer cells: A specialized differentiation regulated by ovarian hormones. *Immunol Rev* **214** (1): 161–185.

39. Croy BA, Zhang J, Tayade C, Colucci F, Yadi H, Yamada AT (2010). Analysis of uterine natural killer cells in mice. *Methods Mol Biol* **612**: 465–503.

40. Dackor J, Caron KM, Threadgill DW (2009). Placental and embryonic growth restriction in mice with reduced function epidermal growth factor receptor alleles. *Genetics* **183** (1): 207–218.

41. Dackor J, Li M, Threadgill DW (2009). Placental overgrowth and fertility defects in mice with a hypermorphic allele of epidermal growth factor receptor. *Mamm Genome* **20** (6): 339–349.

42. Das SK, Wang XN, Paria BC, Damm D, Abraham JA, Klagsbrun M, Andrews GK, Dey SK (1994). Heparin-binding EGF-like growth factor gene is induced in the mouse uterus temporally by the blastocyst solely at the site of its apposition: A possible ligand for interaction with blastocyst EGF-receptor in implantation. *Development* **120** (5): 1071–1083.

43. Daston GP, Manson JM (1995). Critical periods of exposure and developmental outcome. *Inhal Toxicol* **7** (6): 863–871.

44. Degenhardt K, Wright AC, Horng D, Padmanabhan A, Epstein JA (2010). Rapid 3D phenotyping of cardiovascular development in mouse embryos by micro-CT with iodine staining. *Circ Cardiovasc Imaging* **3** (3): 314–322.

45. DeSesso JM (2006). Comparative features of vertebrate embryology. In: *Developmental and Reproductive Toxicology: A Practical Approach*, 2nd edn., Hood RD (ed.). Boca Raton, FL: CRC Press, pp. 147–197.

46. Downs KM, Davies T (1993). Staging of gastrulating mouse embryos by morphological landmarks in the dissecting microscope. *Development* **118** (4): 1255–1266.

47. Drews U (1995). *Color Atlas of Embryology.* New York: Thieme Medical Publishers.
48. Duncan SA, Nagy A, Chan W (1997). Murine gastrulation requires *HNF-4* regulated gene expression in the visceral endoderm: Tetraploid rescue of *Hnf-4$^{-/-}$* embryos. *Development* **124** (2): 279–287.
49. Dunty Jr WC, Chen S-Y, Zucker RM, Dehart DB, Sulik KK (2001). Selective vulnerability of embryonic cell populations to ethanol-induced apoptosis: Implications for alcohol-related birth defects and neuro-developmental disorder. *Alcohol Clin Exp Res* **25** (10): 1523–1535.
50. Dzierzak E, Medvinsky A (1995). Mouse embryonic hematopoiesis. *Trends Genet* **11** (9): 359–366.
51. Edwards RG (2005). Genetics of polarity in mammalian embryos. *Reprod Biomed Online* **11** (1): 104–114.
52. EMAP (e-MOUSE Atlas) (2013). e-Mouse Atlas, v3.5. http://www.emouseatlas.org/emap/home.html (last accessed December 1, 2014).
53. Erb C (2006). Embryology and teratology. In: *The Laboratory Rat*, Suckow MA, Weisbroth SH, Franklin CA (eds.). San Diego, CA: Academic Press (Elsevier), pp. 817–846.
54. Fabian D, Bystriansky J, Cikos S, Bukovska A, Burkus J, Koppel J (2010). The effect on preimplantation embryo development of non-specific inflammation localized outside the reproductive tract. *Theriogenology* **74** (9): 1652–1660.
55. Fabian D, Makarevich AV, Chrenek P, Bukovska A, Koppel J (2007). Chronological appearance of spontaneous and induced apoptosis during preimplantation development of rabbit and mouse embryos. *Theriogenology* **68** (9): 1271–1281.
56. Findlater GS, McDougall RD, Kaufman MH (1993). Eyelid development, fusion and subsequent reopening in the mouse. *J Anat* **183** (1): 121–129.
57. Gagioti S, Scavone C, Bevilacqua E (2000). Participation of the mouse implanting trophoblast in nitric oxide production during pregnancy. *Biol Reprod* **62** (2): 260–268.
58. George JD, Manson JM (1986). Strain-dependent differences in the metabolism of 3-methylcholanthrene by maternal, placental, and fetal tissues of C57BL/6J and DBA/2J mice. *Cancer Res* **46** (11): 5671–5675.
59. Georgiades P, Ferguson-Smith AC, Burton GJ (2002). Comparative developmental anatomy of the murine and human definitive placentae. *Placenta* **23** (1): 3–19.
60. Goedbloed JF, Smits-van Prooije AE (1986). Quantitative analysis of the temporal pattern of somite formation in the mouse and rat. A simple and accurate method for age determination. *Acta Anat (Basel)* **125** (2): 76–82.
61. Goldman AS (1979). Critical periods of prenatal toxicity. *Clin Perinatol* **6** (2): 203–218.
62. Gossler A (1992). Early mouse development. In: *Early Embryonic Development of Animals*, Hennig W (ed.). Berlin, Germany: Springer-Verlag, pp. 151–201.
63. Guillemot F, Nagy A, Auerbach A, Rossant J, Joyner AL (1994). Essential role of *Mash-2* in extraembryonic development. *Nature* **371** (6495): 333–336.
64. Gurtner GC, Davis V, Li H, McCoy MJ, Sharpe A, Cybulsky MI (1995). Targeted disruption of the murine *VCAM1* gene: Essential role of VCAM-1 in chorioallantoic fusion and placentation. *Genes Dev* **9** (1): 1–14.
65. Harris SE, Gopichandran N, Picton HM, Leese HJ, Orsi NM (2005). Nutrient concentrations in murine follicular fluid and the female reproductive tract. *Theriogenology* **64** (4): 992–1006.
66. Hebel R, Stromberg MW (1986). *Anatomy and Embryology of the Rat.* Wörthsee, Germany: BioMed Verlag.
67. Hemberger M, Cross JC (2001). Genes governing placental development. *Trends Endocrinol Metab* **12** (4): 162–168.
68. Hew KW, Keller KA (2003). Postnatal anatomical and functional development of the heart: A species comparison. *Birth Defects Res B Dev Reprod Toxicol* **68** (4): 309–320.
69. Hill MA (2014). Embryology. https://embryology.med.unsw.edu.au/embryology/index.php?title=Main_Page (last accessed December 1, 2014).
70. Hill MA (2014). Mouse development. https://embryology.med.unsw.edu.au/embryology/index.php/Mouse_Development (last accessed December 1, 2014).
71. Hill MA (2014). Rat development stages (Witschi). https://embryology.med.unsw.edu.au/embryology/index.php?title=Rat_Development_Stages (last accessed December 1, 2014).
72. Hoar RM, Monie IW (1981). Comparative development of specific organ systems. In: *Developmental Toxicology*, Kimmel CA, Buelke-Sam J (eds.). New York: Raven Press, pp. 13–33.

73. Hogan B, Beddington R, Costantini F, Lacy E (1994). *Manipulating the Mouse Embryo: A Laboratory Manual*, 2nd edn. Plainview, NY: Cold Spring Harbor Laboratory Press.

74. Holsapple MP, West LJ, Landreth KS (2003). Species comparison of anatomical and functional immune system development. *Birth Defects Res B Dev Reprod Toxicol* **68** (4): 321–334.

75. Houghton FD, Thompson JG, Kennedy CJ, Leese HJ (1996). Oxygen consumption and energy metabolism of the early mouse embryo. *Mol Reprod Dev* **44** (4): 476–485.

76. Hu D, Cross JC (2010). Development and function of trophoblast giant cells in the rodent placenta. *Int J Dev Biol* **54** (2–3): 341–354.

77. Huet-Hudson YM, Dey SK (1990). Requirement for progesterone priming and its long-term effects on implantation in the mouse. *Proc Soc Exp Biol Med* **193** (4): 259–263.

78. Iguchi T, Tani N, Sato T, Fukatsu N, Ohta Y (1993). Developmental changes in mouse placental cells from several stages of pregnancy *in vivo* and *in vitro*. *Biol Reprod* **48** (1): 188–196.

79. Isaac SM, Langford MB, Simmons DG, Adamson SL (2014). Anatomy of the mouse placenta throughout gestation. In: *The Guide to Investigation of Mouse Pregnancy*, Croy BA, Yamada AT, DeMayo FJ, Adamson SL (eds.). London, U.K.: Academic Press (Elsevier), pp. 69–73.

80. Ishikawa H, Rattigan A, Fundele R, Burgoyne PS (2003). Effects of sex chromosome dosage on placental size in mice. *Biol Reprod* **69** (2): 483–488.

81. Jackson Laboratory (2001–2014). Mouse phenome database, v1.41. http://phenome.jax.org/ (last accessed December 1, 2014).

82. Jackson Laboratory (1996–2012). Mouse genome informatics, v5.20. http://www.informatics.jax.org (last accessed December 1, 2014).

83. Jacobwitz DM, Abbott LC (1998). *Chemoarchitectonic Atlas of the Developing Mouse Brain*. Boca Raton, FL: CRC Press.

84. Ji RP, Phoon CK, Aristizábal O, McGrath KE, Palis J, Turnbull DH (2003). Onset of cardiac function during early mouse embryogenesis coincides with entry of primitive erythroblasts into the embryo proper. *Circ Res* **92** (2): 133–135.

85. Johansson S, Wide M (1994). Changes in the pattern of expression of alkaline phosphatase in the mouse uterus and placenta during gestation. *Anat Embryol* (*Berl*) **190** (3): 287–296.

86. Jollie WP (1990). Development, morphology, and function of the yolk-sac placenta of laboratory rodents. *Teratology* **41** (4): 361–381.

87. Kaufman M, Nikitin AY, Sundberg JP (2010). *Histologic Basis of Mouse Endocrine System Development: A Comparative Analysis*. Boca Raton, FL: CRC Press (Taylor & Francis Group).

88. Kaufman MH (1992). *The Atlas of Mouse Development*, 2nd edn. San Diego, CA: Academic Press.

89. Kaufman MH (1997). Mouse and human embryonic development: A comparative overview. In: *Molecular Genetics of Early Human Development*, Strachan T, Lindsay S, Wilson DI (eds.). Oxford, U.K.: Bios Scientific Publishers, pp. 77–110.

90. Kaufman MH (2007). Mouse embryology: Research techniques and a comparison of embryonic development between mouse and man. In: *The Mouse in Biomedical Research*, 2nd edn., Vol. 1: *History, Wild Mice, and Genetics*, Fox JG, Barthold SW, Davisson MT, Newcomer CE, Quimby FW, Smith AL (eds.). San Diego, CA: Academic Press (Elsevier), pp. 165–209.

91. Kaufman MH, Bard JBL (1999). *The Anatomical Basis of Mouse Development*. San Diego, CA: Academic Press.

92. Keller BB, MacLennan MJ, Tinney JP, Yoshigi M (1996). *In vivo* assessment of embryonic cardiovascular dimensions and function in day-10.5 to -14.5 mouse embryos. *Circ Res* **79** (2): 247–255.

93. Kinder SJ, Tsang TE, Quinlan GA, Hadjantonakis AK, Nagy A, Tam PP (1999). The orderly allocation of mesodermal cells to the extraembryonic structures and the anteroposterior axis during gastrulation of the mouse embryo. *Development* **126** (21): 4691–4701.

94. Knight BS, Sunn N, Pennell CE, Adamson SL, Lye SJ (2009). Developmental regulation of cardiovascular function is dependent on both genotype and environment. *Am J Physiol Heart Circ Physiol* **297** (6): H2234–H2241.

95. Knisley KA, Weitlauf HM (1993). Compartmentalized reactivity of M3/38 (anti-Mac-2) and M3/84 (anti-Mac-3) in the uterus of pregnant mice. *J Reprod Fertil* **97** (2): 521–527.

96. Kruse A, Hallmann R, Butcher EC (1999). Specialized patterns of vascular differentiation antigens in the pregnant mouse uterus and the placenta. *Biol Reprod* **61** (6): 1393–1401.

97. Kuida K, Zheng TS, Na S, Kuan C, Yang D, Karasuyama H, Rakic P, Flavell RA (1996). Decreased apoptosis in the brain and premature lethality in CPP32-deficient mice. *Nature* **384** (6607): 368–372.
98. Kwee L, Baldwin HS, Shen HM, Stewart CL, Buck C, Buck CA, Labow MA (1995). Defective development of the embryonic and extraembryonic circulatory systems in vascular cell adhesion molecule (VCAM-1) deficient mice. *Development* **121** (2): 489–503.
99. Lauder JM (1988). Neurotransmitters as morphogens. *Prog Brain Res* **73**: 365–387.
100. Le Floc'h J, Cherin E, Zhang MY, Akirav C, Adamson SL, Vray D, Foster FS (2004). Developmental changes in integrated ultrasound backscatter from embryonic blood *in vivo* in mice at high US frequency. *Ultrasound Med Biol* **30** (10): 1307–1319.
101. Lee SJ, Talamantes F, Wilder E, Linzer DI, Nathans D (1988). Trophoblastic giant cells of the mouse placenta as the site of proliferin synthesis. *Endocrinology* **122** (5): 1761–1768.
102. Leese HJ (1995). Metabolic control during preimplantation mammalian development. *Hum Repro Update* **1** (1): 63–72.
103. Leese HJ, Baumann CG, Brison DR, McEvoy TG, Sturmey RG (2008). Metabolism of the viable mammalian embryo: Quietness revisited. *Mol Hum Reprod* **14** (12): 667–672.
104. Lescisin KR, Varmuza S, Rossant J (1988). Isolation and characterization of a novel trophoblast-specific cDNA in the mouse. *Genes Dev* **2** (12A): 1639–1646.
105. Li Y, Behringer RR (1998). *Esx1* is an X-chromosome-imprinted regulator of placental development and fetal growth. *Nat Genet* **20** (3): 309–311.
106. Lin J, Toft DJ, Bengtson NW, Linzer DI (2000). Placental prolactins and the physiology of pregnancy. *Recent Prog Horm Res* **55**: 37–51; discussion 52.
107. Lomaga MA, Henderson JT, Elia AJ, Robertson J, Noyce RS, Yeh W-C, Mak TW (2000). Tumor necrosis factor receptor-associated factor 6 (TRAF6) deficiency results in exencephaly and is required for apoptosis within the developing CNS. *J Neurosci* **20** (19): 7384–7393.
108. Louvi A, Accili D, Efstratiadis A (1997). Growth-promoting interaction of IGF-II with the insulin receptor during mouse embryonic development. *Dev Biol* **189** (1): 33–48.
109. MacLennan MJ, Keller BB (1999). Umbilical arterial blood flow in the mouse embryo during development and following acutely increased heart rate. *Ultrasound Med Biol* **25** (3): 361–370.
110. Malle D, Economou L, Sioga A, Toliou TH, Galaktidou G, Foroglou CH, Destouni CH, Nestoridis K, Toli A, Sparopoulou TH, Moula A (2004). Somitogenesis in different mouse strains. *Folia Anat* **32** (1): 5–10.
111. Maronpot RR (1999). *Pathology of the Mouse: Reference and Atlas*. Vienna, IL: Cache River Press.
112. Martin KL, Leese HJ (1999). Role of developmental factors in the switch from pyruvate to glucose as the major exogenous energy substrate in the preimplantation mouse embryo. *Reprod Fertil Dev* **11** (7–8): 425–433.
113. Marty MS, Chapin RE, Parks LG, Thorsrud BA (2003). Development and maturation of the male reproductive system. *Birth Defects Res B Dev Reprod Toxicol* **68** (2): 125–136.
114. McGeady TA, Quinn PJ, FitzPatrick ES, Ryan MT (2006). *Veterinary Embryology*. Oxford, U.K.: Blackwell Publishing.
115. McGrath KE, Koniski AD, Malik J, Palis J (2003). Circulation is established in a stepwise pattern in the mammalian embryo. *Blood* **101** (5): 1669–1676.
116. McLaren A (1965). Genetic and environmental effects on foetal and placental growth in mice. *J Reprod Fertil* **9**: 79–89.
117. Mitchell M, Bakos HW, Lane M (2011). Paternal diet-induced obesity impairs embryo development and implantation in the mouse. *Fertil Steril* **95** (4): 1349–1353.
118. Mittwoch U (2001). Genetics of mammalian sex determination: Some unloved exceptions. *J Exp Zool* **290** (5): 484–489.
119. Miyake T, Cameron AM, Hall BK (1996). Detailed staging of inbred C57BL/6 mice between Theiler's [1972] stages 18 and 21 (11–13 days of gestation) based on craniofacial development. *J Craniofacial Genet Dev Biol* **16** (1): 1–31.
120. Mo F-E, Muntean AG, Chen C-C, Stolz DB, Watkins SC, Lau LF (2002). CYR61 (CCN1) is essential for placental development and vascular integrity. *Mol Cell Biol* **22** (24): 8709–8720.
121. Mori N, Tsugane MH, Yamashita K, Ikuta Y, Yasuda M (2000). Pathogenesis of retinoic acid-induced abnormal pad patterns on mouse volar skin. *Teratology* **62** (4): 181–188.

122. Muntener M, Hsu Y-C (1977). Development of trophoblast and placenta of the mouse. A reinvestigation with regard to the *in vitro* culture of mouse trophoblast and placenta. *Acta Anat (Basel)* **98** (3): 241–252.

123. Nadra K, Anghel SI, Joye E, Tan NS, Basu-Modak S, Trono D, Wahli W, Desvergne B (2006). Differentiation of trophoblast giant cells and their metabolic functions are dependent on peroxisome proliferator-activated receptor β/δ. *Mol Cell Biol* **26** (8): 3266–3281.

124. Nagy A, Gertsenstein M, Vintersten K, Behringer R (2003). *Manipulating the Mouse Embryo*, 3rd edn. Cold Spring Harbor, NY: Cold Spring Harbor Laboratory Press.

125. Nagy A, Rossant J (2001). Chimaeras and mosaics for dissecting complex mutant phenotypes. *Int J Dev Biol* **45** (3): 577–582.

126. Natale DRC, Starovic M, Cross JC (2006). Phenotypic analysis of the mouse placenta. In: *Placenta and Trophoblast: Methods and Protocols*, Vol. 1, Soares MJ, Hunt JS (eds.). Totowa, NJ: Humana Press, pp. 275–293.

127. Nelson JF, Karelus K, Felicio LS, Johnson TE (1990). Genetic influences on the timing of puberty in mice. *Biol Reprod* **42** (4): 649–655.

128. Nieder GL, Jennes L (1990). Production of mouse placental lactogen-I by trophoblast giant cells *in utero* and *in vitro*. *Endocrinology* **126** (6): 2809–2814.

129. Nieman BJ, Turnbull DH (2010). Ultrasound and magnetic resonance microimaging of mouse development. *Methods Enzymol* **476**: 379–400.

130. Nishimura H, Shiota K (1977). Summary of comparative embryology and teratology. In: *Handbook of Teratology*, Vol. 3, Wilson JG, Fraser FC (eds.). New York: Plenum, pp. 119–154.

131. Noden DM, de LaHunta A (1985). *The Embryology of Domestic Animals: Developmental Mechanisms and Malformations*. Baltimore, MD: Williams & Wilkins.

132. Nonaka K, Sasaki Y, Yanagita K, Watanabe Y, Matsumoto T, Nakata M (1994). The effect of dam's strain on the intrauterine craniofacial growth of mouse fetuses. *J Assist Reprod Genet* **11** (7): 359–366.

133. O'Neill C, Li Y, Jin XL (2012). Survival signaling in the preimplantation embryo. *Theriogenology* **77** (4): 773–784.

134. Olovsson M, Nilsson BO (1993). Structural and functional properties of trophoblast cells of mouse egg-cylinders *in vitro*. *Anat Rec* **236** (2): 417–424.

135. Ouseph MM, Li J, Chen HZ, Pecot T, Wenzel P, Thompson JC, Comstock G et al. (2012). Atypical E2F repressors and activators coordinate placental development. *Dev Cell* **22** (4): 849–862.

136. Paffaro VA Jr., Bizinotto MC, Joazeiro PP, Yamada AT (2003). Subset classification of mouse uterine natural killer cells by DBA lectin reactivity. *Placenta* **24** (5): 479–488.

137. Pampfer S, Donnay I (1999). Apoptosis at the time of embryo implantation in mouse and rat. *Cell Death Differ* **6** (6): 533–545.

138. Paria BC, Huet-Hudson YM, Dey SK (1993). Blastocyst's state of activity determines the "window" of implantation in the receptive mouse uterus. *Proc Natl Acad Sci USA* **90** (21): 10159–10162.

139. Paria BC, Lim H, Das SK, Reese J, Dey SK (2000). Molecular signaling in uterine receptivity for implantation. *Semin Cell Dev Biol* **11** (2): 67–76.

140. Parr MB, Parr EL (1986). Permeability of the primary decidual zone in the rat uterus: Studies using fluorescein-labeled proteins and dextrans. *Biol Reprod* **34** (2): 393–403.

141. Patton JT, Kaufman MH (1995). The timing of ossification of the limb bones, and growth rates of various long bones of the fore and hind limbs of the prenatal and early postnatal laboratory mouse. *J Anat* **186** (1): 175–185.

142. Paxinos G, Halliday G, Watson C, Koutcherov Y, Wang H (2007). *Atlas of the Developing Mouse Brain at E17.5, P0, and P6*. San Diego, CA: Academic Press (Elsevier).

143. Pedersen RA, Meneses J, Spindle A, Wu K, Galloway SM (1985). Cytochrome P-450 metabolic activity in embryonic and extraembryonic tissue lineages of mouse embryos. *Proc Natl Acad Sci USA* **82** (10): 3311–3315.

144. Peterka M, Lesot H, Peterková R (2002). Body weight in mouse embryos specifies staging of tooth development. *Connect Tiss Res* **43** (2–3): 186–190.

145. Picut CA, Swanson CL, Parker RF, Scully KL, Parker GA (2009). The metrial gland in the rat and its similarities to granular cell tumors. *Toxical Pathol* **37** (4): 474–480.

146. Pijnenborg R (2000). The metrial gland is more than a mesometrial lymphoid aggregate of pregnancy. *J Reprod Immunol* **46** (1): 17–19.

147. Pijnenborg R, Robertson WB, Brosens I, Dixon G (1981). Review article: Trophoblast invasion and the establishment of haemochorial placentation in man and laboratory animals. *Placenta* **2** (1): 71–91.
148. Plaks V, Berkovitz E, Vandoorne K, Berkutzki T, Damari GM, Haffner R, Dekel N, Hemmings BA, Neeman M, Harmelin A (2011). Survival and size are differentially regulated by placental and fetal PKBalpha/AKT1 in mice. *Biol Reprod* **84** (3): 537–545.
149. Plaks V, Sapoznik S, Berkovitz E, Haffner-Krausz R, Dekel N, Harmelin A, Neeman M (2011). Functional phenotyping of the maternal albumin turnover in the mouse placenta by dynamic contrast-enhanced MRI. *Mol Imaging Biol* **13** (3): 481–492.
150. Poelmann RE, Mentink MM (1982). The maternal-embryonic barrier in the early post-implantation mouse embryo: A morphological and functional study. *Scan Electron Microsc* **3**: 1237–1247.
151. Pritchett KR, Taft RA (2007). Reproductive biology of the laboratory mouse. In: *The Mouse in Biomedical Research*, 2nd edn., Vol. 3: *Normative Biology, Husbandry, and Models*, Fox JG, Barthold SW, Davisson MT, Newcomer CE, Quimby FW, Smith AL (eds.). San Diego, CA: Academic Press (Elsevier), pp. 91–121.
152. Qian X, Esteban L, Vass WC, Upadhyaya C, Papageorge AG, Yienger K, Ward JM, Lowy DR, Santos E (2000). The Sos1 and Sos2 Ras-specific exchange factors: Differences in placental expression and signaling properties. *EMBO J* **19** (4): 642–654.
153. Raab G, Kover K, Paria BC, Dey SK, Ezzell RM, Klagsbrun M (1996). Mouse preimplantation blastocysts adhere to cells expressing the transmembrane form of heparin-binding EGF-like growth factor. *Development* **122** (2): 637–645.
154. Rai A, Cross JC (2014). Development of the hemochorial maternal vascular spaces in the placenta through endothelial and vasculogenic mimicry. *Dev Biol* **387** (2): 131–141.
155. Ramathal CY, Bagchi IC, Taylor RN, Bagchi MK (2010). Endometrial decidualization: Of mice and men. *Semin Reprod Med* **28** (1): 17–26.
156. Rendi MH, Muehlenbachs A, Garcia RL, Boyd KL (2012). Female reproductive system. In: *Comparative Anatomy and Histology: A Mouse and Human Atlas*, Treuting PM, Dintzis SM (eds.). San Diego, CA: Academic Press (Elsevier), pp. 253–284.
157. Reponen P, Leivo I, Sahlberg C, Apte SS, Olsen BR, Thesleff I, Tryggvason K (1995). 92-kDa type IV collagenase and TIMP-3, but not 72-kDa type IV collagenase or TIMP-1 or TIMP-2, are highly expressed during mouse embryo implantation. *Dev Dyn* **202** (4): 388–396.
158. Rice D, Barone SJ (2000). Critical periods of vulnerability for the developing nervous system: Evidence from humans and animal models. *Environ Health Perspect* **108** (Suppl. 3): 511–533.
159. Riley P, Anson-Cartwright L, Cross JC (1998). The Hand1 bHLH transcription factor is essential for placentation and cardiac morphogenesis. *Nat Genet* **18** (3): 271–275.
160. Robertson SA, Mau VJ, Young IG, Matthaei KI (2000). Uterine eosinophils and reproductive performance in interleukin 5-deficient mice. *J Reprod Fertil* **120** (2): 423–432.
161. Robertson SA, Prins JR, Sharkey DJ, Moldenhauer LM (2013). Seminal fluid and the generation of regulatory T cells for embryo implantation. *Am J Reprod Immunol* **69** (4): 315–330.
162. Rodier PM (1980). Chronology of neuron development: Animal studies and their clinical implications. *Dev Med Child Neurol* **22** (4): 525–545.
163. Rossant J, Cross JC (2001). Placental development: Lessons from mouse mutants. *Nat Rev Genet* **2** (7): 538–548.
164. Rossant J, Spence A (1998). Chimeras and mosaics in mouse mutant analysis. *Trends Genet* **14** (9): 358–363.
165. Rossant J, Tam PPL (2002). *Mouse Development: Patterning, Morphogenesis, Organogenesis.* San Diego, CA: Academic Press.
166. Rugh R (1990). *The Mouse: Its Reproduction and Development.* Oxford, U.K.: Oxford University Press.
167. Russ AP, Wattler S, Colledge WH, Aparicio SA, Carlton MB, Pearce JJ, Barton SC et al. (2000). Eomesodermin is required for mouse trophoblast development and mesoderm formation. *Nature* **404** (6773): 95–99.
168. Ruvinsky A (1999). Basics of gametic imprinting. *J Anim Sci* **77** (Suppl. 2): 228–237.
169. Ryan BC, Vandenbergh JG (2002). Intrauterine position effects. *Neurosci Biobehav Rev* **26** (6): 665–678.
170. Sanchez J, Buendia AJ, Salinas J, Bernabe A, Rodolakis A, Stewart IJ (1996). Murine granulated metrial gland cells are susceptible to *Chlamydia psittaci* infection *in vivo*. *Infect Immun* **64** (9): 3897–3900.
171. Sasaki Y, Nonaka K, Nakata M (1994). The effect of four strains of recipients on the intrauterine growth of the mandible in mouse fetuses. *J Craniofac Genet Dev Biol* **14** (2): 118–123.

172. Savolainen SM, Foley JF, Elmore SA (2009). Histology atlas of the developing mouse heart with emphasis on E11.5 to E18.5. *Toxicol Pathol* **37** (4): 395–414.
173. Sawyers CL (2009). Shifting paradigms: The seeds of oncogene addiction. *Nat Med* **15** (10): 1158–1161.
174. Schambra U (2008). *Prenatal Mouse Brain Atlas.* New York: Springer.
175. Schambra UB (2007). Electronic prenatal mouse brain atlas. http://www.epmba.org/ (last accessed December 1, 2014).
176. Schneider BF, Norton S (1979). Equivalent ages in rat, mouse and chick embryos. *Teratology* **19** (3): 273–278.
177. Schneider JE, Wysocki CJ, Nyby J, Whitney G (1978). Determining the sex of neonatal mice (*Mus musculus*). *Behav Res Meth Instrument* **10** (1): 105.
178. Scialli AR (1992). Embryology and principles of teratology. In: *A Clinical Guide to Reproductive and Developmental Toxicology.* Boca Raton, FL: CRC Press, pp. 1–27.
179. Scott IC, Anson-Cartwright L, Riley P, Reda D, Cross JC (2000). The HAND1 basic helix-loop-helix transcription factor regulates trophoblast differentiation via multiple mechanisms. *Mol Cell Biol* **20** (2): 530–541.
180. Seymour R, Sundberg JP, Hogenesch H (2006). Abnormal lymphoid organ development in immunodeficient mutant mice. *Vet Pathol* **43** (4): 401–423.
181. Sheng G, Foley AC (2012). Diversification and conservation of the extraembryonic tissues in mediating nutrient uptake during amniote development. *Ann NY Acad Sci* **1271**: 97–103.
182. Silver LM (2008). *Mouse Genetics: Concepts and Applications.* Oxford, U.K.: Oxford University Press. Available online in full at http://www.informatics.jax.org/silver/index.shtml (last accessed December 1, 2014).
183. Simmons DG, Cross JC (2005). Determinants of trophoblast lineage and cell subtype specification in the mouse placenta. *Dev Biol* **284** (1): 12–24.
184. Simmons DG, Fortier AL, Cross JC (2007). Diverse subtypes and developmental origins of trophoblast giant cells in the mouse placenta. *Dev Biol* **304** (2): 567–578.
185. Spencer TE, Bazer FW (2004). Conceptus signals for establishment and maintenance of pregnancy. *Reprod Biol Endocrinol* **2**: 49 (15 pp.).
186. Spencer TE, Bazer FW (2004). Uterine and placental factors regulating conceptus growth in domestic animals. *J Anim Sci* **82** (E-Suppl.): E4–E13.
187. Steingrímsson E, Tessarollo L, Reid SW, Jenkins NA, Copeland NG (1998). The bHLH-Zip transcription factor *Tfeb* is essential for placental vascularization. *Development* **125** (23): 4607–4616.
188. Stewart IJ (1990). Granulated metrial gland cells in the mouse placenta. *Placenta* **11** (3): 263–275.
189. Stewart IJ (1998). Granulated metrial gland cells in 'minor' species. *J Reprod Immunol* **40** (2): 129–146.
190. Sulik KK, Bream PRJ (Not given). Embryo images: Normal and abnormal mammalian development. http://www.med.unc.edu/embryo_images/ (last accessed December 1, 2014).
191. Sulik KK, Schoenwolf GC (1985). Highlights of craniofacial morphogenesis in mammalian embryos, as revealed by scanning electron microscopy. *Scan Electron Microsc* **4**: 1735–1752.
192. Szabo KT (1989). Comparative synopsis of major prenatal developmental events in laboratory and farm animals, Appendix I. In: *Congenital Malformations in Laboratory and Farm Animals.* San Diego, CA: Academic Press, pp. 287–290.
193. Takeda K, Ho VC, Takeda H, Duan LJ, Nagy A, Fong GH (2006). Placental but not heart defects are associated with elevated hypoxia-inducible factor α levels in mice lacking prolyl hydroxylase domain protein 2. *Mol Cell Biol* **26** (22): 8336–8346.
194. Tanaka N, Mao L, DeLano FA, Sentianin EM, Chien KR, Schmid-Schonbein GW, Ross JJ (1997). Left ventricular volumes and function in the embryonic mouse heart. *Am J Physiol* **273** (3 Pt. 2): H1368–H1376.
195. Tanaka S, Oda M, Toyoshima Y, Wakayama T, Tanaka M, Yoshida N, Hattori N, Ohgane J, Yanagimachi R, Shiota K (2001). Placentomegaly in cloned mouse concepti caused by expansion of the spongiotrophoblast layer. *Biol Reprod* **65** (6): 1813–1821.
196. Teesalu T, Blasi F, Talarico D (1998). Expression and function of the urokinase type plasminogen activator during mouse hemochorial placental development. *Dev Dyn* **213** (1): 27–38.
197. Theiler K (1972). *The House Mouse: Development and Normal Stages from Fertilization to 4 Weeks of Age.* Berlin, Germany: Springer-Verlag.

198. Theiler K (1983). Embryology. In: *The Mouse in Biomedical Research*, Vol. 3: *Normative Biology, Immunology and Husbandry*, Foster HL, Small JD, Fox JG (eds.). New York: Academic Press, pp. 121–135.
199. Theiler K (1989). *The House Mouse: Atlas of Embryonic Development*. New York: Springer-Verlag. Available online at http://www.emouseatlas.org/emap/ema/theiler_stages/house_mouse/book.html (last accessed December 1, 2014).
200. Thiel R, Chahoud I, Jürgens M, Neubert D (1993). Time-dependent differences in the development of somites of four different mouse strains. *Teratogen Carcinogen Mutagen* **13** (6): 247–257.
201. Thornhill AR, Burgoyne PS (1993). A paternally imprinted X chromosome retards the development of the early mouse embryo. *Development* **118** (1): 171–174.
202. Tsunoda Y, Tokunaga T, Sugie T (1985). Altered sex ratio of live young after transfer of fast- and slow-developing mouse embryos. *Gamete Res* **12** (3): 301–304.
203. Tung HN, Parr MB, Parr EL (1986). The permeability of the primary decidual zone in the rat uterus: An ultrastructural tracer and freeze-fracture study. *Biol Reprod* **35** (4): 1045–1058.
204. Turgeon B, Meloche S (2009). Interpreting neonatal lethal phenotypes in mouse mutants: Insights into gene function and human diseases. *Physiol Rev* **89** (1): 1–26.
205. van den Heuvel MJ, Xie X, Tayade C, Peralta C, Fang Y, Leonard S, Paffaro VA Jr., Sheikhi AK, Murrant C, Croy BA (2005). A review of trafficking and activation of uterine natural killer cells. *Am J Reprod Immunol* **54** (6): 322–331.
206. van Maele-Fabry G, Delhaise F, Picard JJ (1990). Morphogenesis and quantification of the development of post-implantation mouse embryos. *Toxicol In Vitro* **4** (2): 149–156.
207. Vandenbergh JG, Huggett CL (1995). The anogenital distance index, a predictor of the intrauterine position effects on reproduction in female house mice. *Lab Anim Sci* **45** (5): 567–573.
208. Varcoe TJ, Boden MJ, Voultsios A, Salkeld MD, Rattanatray L, Kennaway DJ (2013). Characterisation of the maternal response to chronic phase shifts during gestation in the rat: Implications for fetal metabolic programming. *PLoS One* **8** (1): e53800 (13 pages).
209. Vaughan OR, Sferruzzi-Perri AN, Coan PM, Fowden AL (2011). Environmental regulation of placental phenotype: Implications for fetal growth. *Reprod Fertil Dev* **24** (1): 80–96.
210. Voss AK, Thomas T, Gruss P (2000). Mice lacking HSP90β fail to develop a placental labyrinth. *Development* **127** (1): 1–11.
211. Waller BR 3rd, Wessels A (2000). Cardiac morphogenesis and dysmorphogenesis. An immunohistochemical approach. *Methods Mol Biol* **135**: 151–161.
212. Walthall K, Cappon GD, Hurtt ME, Zoetis T (2005). Postnatal development of the gastrointestinal system: A species comparison. *Birth Defects Res B Dev Reprod Toxicol* **74** (2): 132–156.
213. Wanek N, Muneoka K, Holler-Dinsmore G, Burton R, Bryant SV (1989). A staging system for mouse limb development. *J Exp Zool* **249** (1): 41–49.
214. Ward JM, Devor-Henneman DE (2000). Gestational mortality in genetically engineered mice: Evaluating the extraembryonal embryonic placenta and membranes. In: *Pathology of Genetically Engineered Mice*, Ward JM, Mahler JF, Maronpot RR, Sundberg JP, Frederickson RM (eds.). Ames, IA: Iowa State University Press, pp. 103–122.
215. Ward JM, Elmore SA, Foley JF (2012). Pathology methods for the evaluation of embryonic and perinatal developmental defects and lethality in genetically engineered mice. *Vet Pathol* **49** (1): 71–84.
216. Warkany J (1971). Sensitive or critical periods in teratogenesis: Uses and abuses of embryologic timetables. In: *Congenital Malformations*. Chicago, IL: Year Book Medical Publishing, Inc., pp. 49–52.
217. Wassarman PM, DePamphilis ML (1993). *Guide to Techniques in Mouse Development*. San Diego, CA: Academic Press.
218. Watson ED, Cross JC (2005). Development of structures and transport functions in the mouse placenta. *Physiology* **20**: 180–193.
219. Weiler-Guettler H, Aird WC, Rayburn H, Husain M, Rosenberg RD (1996). Developmentally regulated gene expression of thrombomodulin in postimplantation mouse embryos. *Development* **122** (7): 2271–2281.
220. Weiss LW, Zelickson AS (1975). Embryology of the epidermis: Ultrastructural aspects. I. Formation and early development in the mouse with mammalian comparisons. *Acta Derm Venereol* **55** (3): 161–168.
221. Weiss LW, Zelickson AS (1975). Embryology of the epidermis: Ultrastructural aspects. II. Period of differentiation in the mouse with mammalian comparisons. *Acta Derm Venereola* **55** (5): 321–329.

222. Weiss LW, Zelickson AS (1975). Embryology of the epidermis: Ultrastructural aspects. III. Maturation and primary appearance of dendritic cells in the mouse with mammalian comparisons. *Acta Derm Venereol* **55** (6): 431–442.

223. Welsh AO, Enders AC (1983). Occlusion and reformation of the rat uterine lumen during pregnancy. *Am J Anat* **167** (4): 463–477.

224. Wood SL, Beyer BK, Cappon GD (2003). Species comparison of postnatal CNS development: Functional measures. *Birth Defects Res B Dev Reprod Toxicol* **68** (5): 391–407.

225. Yamanaka Y, Ralston A, Stephenson RO, Rossant J (2006). Cell and molecular regulation of the mouse blastocyst. *Dev Dyn* **235** (9): 2301–2314.

226. Yang JT, Rayburn H, Hynes RO (1995). Cell adhesion events mediated by α4 integrins are essential in placental and cardiac development. *Development* **121** (2): 549–560.

227. Yotsumoto S, Shimada T, Cui CY, Nakashima H, Fujiwara H, Ko MS (1998). Expression of adreno-medullin, a hypotensive peptide, in the trophoblast giant cells at the embryo implantation site in mouse. *Dev Biol* **203** (2): 264–275.

228. Yu Q, Leatherbury L, Tian X, Lo CW (2008). Cardiovascular assessment of fetal mice by *in utero* echo-cardiography. *Ultrasound Med Biol* **34** (5): 741–752.

229. Zechner U, Hemberger M, Constância M, Orth A, Dragatsis I, Lüttges A, Hameister H, Fundele R (2002). Proliferation and growth factor expression in abnormally enlarged placentas of mouse interspe-cific hybrids. *Dev Dyn* **224** (2): 125–134.

230. Zoetis T, Hurtt ME (2003). Species comparison of anatomical and functional renal development. *Birth Defects Res B Dev Reprod Toxicol* **68** (2): 111–120.

231. Zoetis T, Hurtt ME (2003). Species comparison of lung development. *Birth Defects Res B Dev Reprod Toxicol* **68** (2): 121–124.

232. Zoetis T, Tassinari MS, Bagi C, Walthall K, Hurtt ME (2003). Species comparison of postnatal bone growth and development. *Birth Defects Res B Dev Reprod Toxicol* **68** (2): 86–110.

233. Zohn IE, Sarkar AA (2010). The visceral yolk sac endoderm provides for absorption of nutrients to the embryo during neurulation. *Birth Defects Res A Clin Mol Teratol* **88** (8): 593–600.

234. Zorn TM, Bijovsky AT, Bevilacqua EM, Abrahamsohn PA (1989). Phagocytosis of collagen by mouse decidual cells. *Anat Rec* **225** (2): 96–100.

235. Zucker RM, Hunter ES 3rd, Rogers JM (1999). Apoptosis and morphology in mouse embryos by confo-cal laser scanning microscopy. *Methods* **18** (4): 473–480.

236. Klinger S, Turgeon B, Lévesque K, Wood GA, Aagaard-Tillery KM, Meloche S (2009) Loss of Erk3 function in mice leads to intrauterine growth restriction, pulmonary immaturity, and neonatal lethality. *Proc Natl Acad Sci USA* **106** (39): 16710–16715.

5

Teratogens and Teratogenic Mechanisms in Mouse Developmental Pathology

Brad Bolon and Colin G. Rousseaux

CONTENTS

Well over a century of research in teratology—the evaluation of congenital defects, their causes, and the mechanisms responsible for their genesis (derived from the Greek *teras* [monster] and *logos* [study])—has shown that developing mouse embryos are vulnerable to both structural defects and functional alterations arising from many teratogenic influences. Regardless of the cause, these agents act by perturbing normal developmental processes during sensitive periods of development. For this reason, scientists engaged in mouse developmental pathology experiments must have a firm grasp of the processes responsible for normal development so that they may better understand the pathogenic consequences of disturbing biochemical reactions, molecular pathways, and cellular interactions. The major events of normal development have been recounted briefly in Chapter 4 and the references listed therein. This chapter concisely reviews the primary etiologies and mechanisms by which developmental malformations have been induced in mice. Additional insight regarding causes and mechanisms of abnormal development, and advanced methods for investigating them, of relevance to mice may be gained from books and book chapters devoted to developmental toxicology.[3,24,28,37,38,45,66]

Etiologic Agents That Cause Defective Development in Mice

Multiple influences have been shown to incite structural malformations in developing mice.[74] These may be broadly classified as genetic mutations, pathogens, physical agents, and toxicants. Genetic mutations commonly but not always arise from altered expression of elements that control gene transcription, cell growth, or cell stability, and may act to produce either increased or decreased quantities of the protein. In such cases, the miscommunication associated with disrupted protein cascades promotes malformations by preventing or warping normal developmental processes rather than causing damage to pre-existing cells and organs. In contrast, pathogens, physical agents, and toxicants often act by damaging cells (including stem cells), thereby resulting in death or inflammation within differentiating cell populations or organs. Examples of agents from each of these classes are showcased in Table 5.1. In some instances, damage to the conceptus is exacerbated by concurrent maternal illness.[2]

Considerable overlap exists among these broad classes of teratogens. For example, genetic mutations may arise spontaneously, be introduced deliberately by engineering of the nuclear (or rarely mitochondrial) genome, or induced by exposure to a mutagenic chemical or virus. Simultaneous exposure to multiple teratogens (of the same or different classes) may induce additive or synergistic damage whereby the combination of agents is more injurious than either teratogen alone.[47,75] In this regard, standard superovulation as practiced for genetic engineering experiments may potentiate the toxicity of concurrently administered teratogens.[49]

Even if not subject to deliberate exposures, pregnant wild-type mice typically will encounter one to several potential teratogens sometime during the course of gestation. Sources by which such agents may enter conventional colonies include adverse husbandry conditions (e.g., excessive heat or noise, disinfectant residues); contaminated consumables (e.g., food, water); medications; and newly introduced animals (e.g., carriers for novel microbial flora). The rarity of developmental malformations in the offspring born to normal mice under such conditions testifies to the abilities of maternal metabolic pathways and placental membranes to serve as barriers to reduce embryonic exposure as well as to the embryonic capacities to repair or transcend minor damage without lasting effect.

Fundamental Principles of Teratology

Successful development is dictated by many factors. Relevant parameters intrinsic to the conceptus (or neonate or juvenile) include the individual's genotype, the genotypes of nearby siblings, and the stage of development during which exposure to any external agent (e.g., a microbial pathogen or xenobiotic) or experimental manipulation takes place. The impact of such offspring-centered elements may be influenced by other factors, including the parental genotypes, the husbandry conditions in which the pregnant dam is kept, and exposure of the parents to any teratogenic agents.

Developmental pathologists require a clear comprehension of fundamental concepts that moderate the formation of developmental malformations in mice. The basic principles of teratology were defined nearly 50 years ago (Table 5.2),[85] and the list retains its relevance today. More importantly, while formulated as a means of evaluating the toxic potential of environmental chemicals and drugs, the concepts can be adapted readily to cover nontoxic causes of developmental lesions as well. For example, developmental defects arising from an engineered genetic mutation or exposure to a teratogenic chemical will be comparable if they affect the same tissue, organ, or system and strike the target during the same critical period of development. Taken together, these fundamental concepts indicate that one or more structures, functions, and/or molecular pathways will still be undergoing some degree of differentiation any time a genetic event is triggered or a xenobiotic exposure occurs during gestation or the early postnatal life of developing mice.

Developmental Stage

The *critical period* concept is a particularly important determinant of developmental vulnerability. In mice, critical developmental processes take place during narrow windows of time (encompassing hours or days) and at particular sites in all individuals (Figure 5.1). These periods may differ in timing among inbred strains,[27] but the chronological sequence of events and the sites at which they take place usually are conserved. Structural malformations in mice typically are engendered by damage that occurs

TABLE 5.1

Selected Etiologic Agents Capable of Inducing Structural Lesions or Growth Restriction in Developing Mice

Agent Class	Agent	Gene Symbol	Agent Function	References
Genetic mutations				
Engineered loss	Cytokeratin 19	*Ck19*	Intermediate filament	[80]
	Fibroblast growth factor 10	*Fgf10*	Growth factor	[51]
	Organic cation transporter 1	*Oct1*	Transcription factor	[71]
	Phosphatase and tensin homolog	*Pten*	Tumor suppressor gene	[6]
	Spectrin, alpha	*Spta*	Cytoskeletal protein	[84]
	Vascular cell adhesion molecule-1	*Vcam1*	Cell adhesion molecule	[33]
Engineered overexpression	Agrin	*Agrn*	Heparan sulfate proteoglycan	[31]
	Engrailed-1	*En1*	Transcription factor	[68]
	Myelocytomatosis, cellular variant	*c-Myc*	Oncogene	[40]
	Retinoid X receptor, alpha	*Rxra*	Nuclear receptor	[79]
	T-box 1	*Tbx1*	Transcription factor	[48]
	Tropomyosin, beta	*Tpm2*	Cytoskeletal protein	[53]
Spontaneous loss	Grainyhead-like 3 (formerly *curly tail*)	*Grhl3*	Transcription factor	[35]
	Paired box 3 (formerly *splotch*)	*Pax3*	Transcription factor	[7]
Pathogens				
Bacteria	*Brucella abortus*			[14]
	Chlamydia psittaci			[65]
	Fusobacterium nucleatum			[34]
	Listeria monocytogenes			[63]
	Salmonella enterica serovar *typhimurium*			[18]
Protozoa	*Plasmodium berghei*			[73]
	Toxoplasma gondii			[52]
Virus	Cytomegalovirus			[9]
	Enterovirus			[58]
	Influenza B			[19]
Physical agents				
Heat (hyperthermia)	Increased ambient temperature (35°C or higher)			[57]
Noise	High intensity (loud and long)			[56]
Radiation	X-ray			[5]
Toxicants				
Anesthetic	Ketamine			[1]
Endocrine disruptor	2,2',3,3',4-Pentachlorobiphenyl (PCB)			[50]
	2,3,7,8-Tetrachlorodibenzo-*p*-dioxin (TCDD)			[12]
Heavy metals	Lead			[22]
Hormone	Stress (excess corticosterone)			[22]
	Triiodothryonine (T$_3$)			[46]
Lifestyle drugs	Caffeine			[47]
	Cocaine			[1]
	Ethanol			[26]
	Nicotine			[21]

(Continued)

TABLE 5.1 (*CONTINUED*)

Selected Etiologic Agents Capable of Inducing Structural Lesions or Growth Restriction in Developing Mice

Agent Class	Agent	Gene Symbol	Agent Function	References
Medication	Colchicine		Antineoplastic agent	[77]
	Ethacrynic acid		Loop diuretic	[4]
	Fluconazole		Antifungal antibiotic	[82]
	Fluoxetine		Antidepressant	[76]
	Valproic acid		Anticonvulsant	[54]
Metabolic byproduct	Glucose		Energy substrate	[30]
	β-Hydroxybutyrate		Ketone	[39]
Solvent	2-Methoxyethanol			[81]
	1,1,1-Trichloroethane			[42]
Toxin	Nigericin (*Streptomyces hygroscopicus*)			[83]
	T-2 mycotoxin (*Fusarium* spp.)			[36]
Transplacental carcinogen	3-Methylcholanthrene			[41]
	N-Ethyl-*N*-nitrosourea (ENU)			[15]

TABLE 5.2

Fundamental Principles of Teratology

1. The severity and onset of teratogenic effects will be controlled by interactions between the genotype of an individual and numerous environmental factors.
2. Vulnerability to teratogens will be dictated by the stage of development during which an individual is exposed.
3. Teratogens cause developmental malformations by altering specific biochemical, cellular and/or molecular events.
4. Access of xenobiotic teratogens to an embryo or fetus (and neonate) depends on the agent's chemical and physical properties.
5. Manifestations of teratogenicity range from functional abnormalities (including after birth) to intrauterine growth retardation, overt malformations, and death.
6. Manifestations of developmental toxicity generally are related to the dose.

Source: Adapted from Wilson, J.G., *Environment and Birth Defects*, Academic Press, New York, 1973.

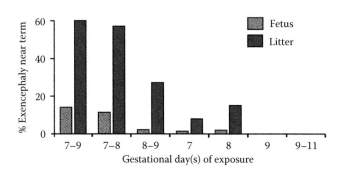

FIGURE 5.1 Critical periods of development that span narrow ranges of time (hours or days) define the times during the initial differentiation of each organ during which exposure to a teratogenic influence may cause overt malformations. In this example, maternal inhalation of methanol (15,000 ppm for 6 h/day during the gestational days [GD] noted on the x-axis) produced the highest incidence of exencephaly (i.e., exposure of the brain) at GD 17.5 when exposure occurred while the neural tube was forming and closing (GD 7 to GD 9), and specifically if the timing involved the developmental stage when or shortly after the head process (i.e., primordial brain) was first being specified (GD 7.0). The wide divergence between the numbers of affected fetuses (pale checkered bars) vs. litters (dark solid bars) demonstrates that only a few conceptuses per litter exhibited this major malformation. (Data adapted from Bolon, B. et al., *Fundam. Appl. Toxicol.* 21(4), 508, 1993.)

during gestation, although early postnatal damage may lead to gross defects in those organs that do not fully form until after birth (e.g., the cerebellum). In contrast, the continued evolution of cellular microanatomy and molecular pathways after birth (e.g., gonads, neural synapses) may lead to cell or organ dysfunction throughout the neonatal and juvenile periods, and often well into adulthood. Each organ will have its own critical period(s) of development, which will represent the cumulative vulnerabilities of individual cell populations and/or subregions that each follow their own developmental programs (Figure 5.2). Not all critical periods of development occur during gestation. Indeed, some brain domains (e.g., hippocampus and olfactory bulb) have distinct prenatal and postnatal critical periods for distinct cell types.

Developmental toxicants are likely to produce lesions in those organs and systems that have critical periods that overlap with the time of exposure. The existing knowledge regarding the timing of normal developmental processes (see Tables 4.A.1 through 4.A.16) permits the prediction of potential target organs and systems if the approximate time of exposure can be determined. The greatest sensitivity to anatomic lesions that might be detected using the routine developmental pathology techniques described in this book will occur when teratogenic agents or mutations are present during either organogenesis (i.e., gestational days [GD] 7.0 to GD 14.5), the fetal period (GD 15.0 to birth), or the neonatal stage (postnatal day [PND] 0–7).

Differential Sensitivity

The sensitivity of mice to teratogenic agents is established by many factors. In fully inbred mouse strains, susceptibility among individuals typically is equivalent as all animals bear the same genotype. However, different inbred strains may exhibit divergent responses to teratogenic agents, with those strains exhibiting a greater concordance of genomes tending to express similar effects. The variance among strains may stem from intrinsic biological differences, such as disparate timing for common developmental events[43] or divergent gene complements,[41] or they can arise from extrinsic differences related to dissimilar maternal metabolic abilities. For instance, A/J mice develop a greater incidence of cleft palate following exposure to 6-aminonicotinamide[10] or cortisone[11] than do C57BL/6J mice. This enhanced sensitivity seems to result mainly from two unique, genetically specified features intrinsic to embryos of the A/J strain: a higher number of glucocorticoid receptors in mesenchymal cells of the maxillary processes and a delay in initiating palatal shelf elevation.[60] However, the higher incidence of cleft palate also likely reflects the greater ability of A/J dams to process cortisone, which will boost the exposure of A/J embryos to potentially teratogenic steroids.

In outbred mouse stocks, vulnerability will differ to some degree among individuals, including littermates, depending on the complement of alleles carried by each conceptus. The significance of the embryonic genotype in defining developmental outcomes following exposure to exogenous teratogens is demonstrated clearly by the variation in responses for offspring within an exposed litter; in general, major malformations occur in only one or a few progeny (Figure 5.1), and the severity of the lesion differs among animals. Such discrepancies explain the choice of the litter, rather than the individual conceptus, as the experimental unit when designing developmental toxicology studies (see Chapter 18). Within-litter discordance may be magnified by other causes. Examples include subtle differences in maternal blood supply to different portions of the uterine horns and the influence afforded by molecules (e.g., hormones) secreted by neighboring conceptuses of the opposite gender or of another genotype.

Differential sensitivity may reflect divergence in the metabolic efficiency of mouse conceptuses from different strains.[32] In general, developing mice have limited metabolic capabilities during gestation, although most enzyme systems rise slowly as the time of birth approaches. *In utero* exposure to enzyme-inducing agents may boost the activity, although the extent of such increases differs among strains. For example, aryl hydrocarbon hydroxylase (AHH), which converts benzo[*a*]pyrene to more toxic metabolites, is induced with phenobarbital in C57BL/6J mice but not DBA mice. Accordingly, the teratogenic risk of benzo[*a*]pyrene in phenobarbital-treated, pregnant mice is greater in C57BL/6 animals.

Strain-dependent differences in maternal care may have long-lasting consequences on development.[61] The pathogenesis likely reflects repression of certain molecular pathways by an excess of stress-associated corticosteroids. Such changes typically present as biochemical or functional variations rather

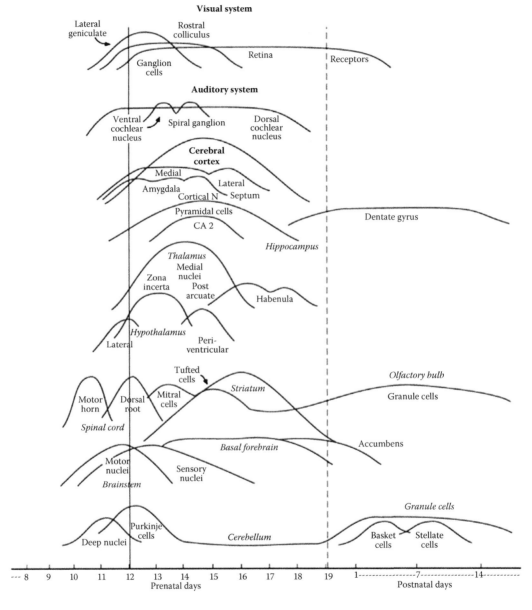

FIGURE 5.2 Within the central nervous system (CNS) as well as other organs, each region and cell population exhibits its own critical period of development. The highest point of each curve represents the developmental age (x-axis) for maximal neuron production (as defined by autoradiographic data) in each structure; the areas under the curves are not proportional to the final number of cells. The solid vertical line (at gestational day [GD] 12) denotes the last day on which teratogen-induced disruptions in neuron generation are likely to yield gross CNS malformations in mouse conceptuses, and the dashed vertical line defines the day of birth. The position of the regions/cell groups on the y-axis is arbitrary. (Data adapted from Rodier, P.M., *Dev. Med. Child Neurol.*, 22(4), 525, 1980, by permission of Oxford University Press.)

than structural changes, which would be expected since organogenesis is completed in nearly all organs and organ subregions prior to birth. However, specialized techniques of structural analysis may discern strain-specific divergence in microanatomic features.[69]

Dose

A foundational tenet of toxicology is that the dose determines whether or not a substance acts as a poison or as a remedy. The usual problem is too much of an entity, but cell damage and/or death also may result from too little of an essential molecule (e.g., micronutrients or some metals) (Figure 5.3). Toxicants with more generic mechanisms of action, such as cytotoxic or antiproliferative chemicals, target numerous tissues in an indiscriminant manner and therefore typically lead to retarded intrauterine growth at low doses and malformations or embryonic death at high doses. In contrast, agents with more specific modes of action, such as drugs that utilize receptor-mediated signaling cascades to modify cellular processes, cause dose-dependent effects in tissues expressing the receptor but generally do not alter the function and structure of tissues that lack the receptor. For some agents, toxicity requires an acute, high-dose exposure (i.e., maximal concentration [C_{max}]), while for others toxicity may represent a prolonged, low-level exposure (i.e., area under the curve [AUC]) leading to a gradual accumulation of damage over time. Structural malformations in mice that may be detected by routine developmental pathology methods usually result from acute, high-dose events.

A corollary concept to that of dose is the *threshold* effect. The threshold dose is one below which no significant increase is observed in the incidence of functional defects, structural malformations, or death. In general, teratogens do exhibit *no observed adverse effect level* (NOAEL) following exposure during the critical periods of organogenesis.[28,29] The threshold may be lower if genetic mutations are present that predispose the conceptus to develop a malformation. These predilections often are specific for a given organ or system. The presence of a NOAEL for gross malformations or death may yet be associated with more subtle microscopic lesions or functional deficits.

The dose concept may be adapted readily to defects induced by genetic mutations. The situation in which gene overexpression leads to excessive availability or activity of a protein mirrors the high-dose toxicology scenario. Similarly, the reduction of loss of a protein in gene-targeted (i.e., knockout) mice

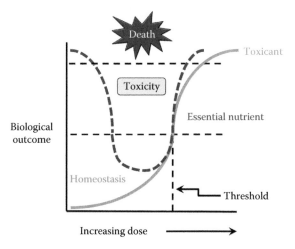

FIGURE 5.3 The dose determines the teratogenic potential of developmental toxicants. Low-level exposures to potential toxicants (solid orange line) have no effect until a threshold dose (lower dashed horizontal line) is reached. Below this exposure level, the individual remains in a state of homeostasis. As the exposure exceeds the threshold, toxic effects begin to accrue, which at high-level exposures will surpass the lethal threshold (upper dashed horizontal line) to result in death. For some essential nutrients (dashed blue parabola) like metals and vitamins, deficits (to the left side of the inverted curve) are as potentially injurious as excess levels (to the right of the inverted curve).

mimics the situation in which an essential constituent is missing. The resemblance to the toxicity setting becomes even closer if the gene engineering constructs are assembled to permit conditional (i.e., space- and/or time-dependent) control of the modified gene.

Tissue Specificity

Target tissues within the mouse conceptus are defined in large part by the nature of the teratogenic influence. Rapidly proliferating cell populations (e.g., limb bud, neural tube) are more vulnerable to cytotoxic and antiproliferative chemicals, likely because precipitous cell loss and/or lessened production suffered at such an active stage of development will impede normal formation of the structures dependent on expansion of the affected cell population. Toxicant exposure will still occur in neighboring populations of slowly proliferating cells, but the absence of the necessary target population of dividing cells will minimize the impact of exposure.

　Target tissues within the conceptus also depend on the differential sensitivities of various embryonic tissues. Divergence in responsiveness to toxicants usually is dictated by the presence or absence of specific receptor molecules (e.g., for hormone analogs), functional biotransformation machinery for activating or detoxifying parent compounds or their metabolites, and any physicochemical characteristics that will lead to accumulation or exclusion of the agent. In genetically engineered animals, tissue-specific vulnerability may be contrived via the use of conditional targeting constructs in which the gene of interest is controlled by a cell type- or tissue-specific promoter.

Maternal Toxicity

Developmental toxicity often occurs in tandem with maternal toxicity,[44] and factors that can disrupt homeostasis in the dam also may affect development of her progeny.[20,59] Standard protocols for testing potential developmental toxicants often utilize doses that induce either overt maternal toxicity (e.g., substantially reduced weight gain or absolute weight loss, clinical signs) or more subtle effects. In general, maternal toxicity is not considered to be responsible for inducing major fetal malformations, but it has been linked in rats with other manifestations of developmental toxicity such as decreased fetal weight and minor skeletal anomalies (e.g., wavy ribs).[23] Developmental toxicity cannot be confirmed definitively using the usual means for developmental toxicity investigations (i.e., weight measurements, gross observations of exterior and interior structures [see Chapter 7 for details]) if developmental defects occur only at doses where maternal toxicity also is observed. Instead, other methods (e.g., histopathological assessment) will be needed to reveal the existence of developmental damage at doses where maternal toxicity is not evident in the dams.

Pathogenic Mechanisms of Defective Development

Common responses to injury are detailed for the embryo, fetus, and neonate in Chapter 15, and for the placenta in Chapter 16. Many mechanisms have been proposed as potential explanations for the initiation and progression of developmental defects. The current section introduces the principal mechanisms thought to be of relevance to common categories of lesions that arise in the course of mouse developmental pathology studies.

Excessive Cell Death

Excessive cell loss is a known cause of abnormal embryonic development. Populations of cells that are rapidly dividing or starting to differentiate are the most sensitive to cytotoxic stimuli. This vulnerability likely reflects both the highly active genome, in which the open DNA conformation affords increased contact to many essential genes, and a reduced span in which to make necessary repairs to damaged or inappropriate nucleotides. Alternatively, rapidly proliferating cells may become starved for nutrients in

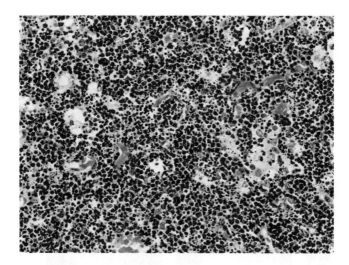

FIGURE 5.4 Significant cell loss due to a cytotoxic stimulus may remove large numbers of cells within an entire organ or specific subregion. In this case, the thymus of a neonatal mouse exhibits extensive programmed cell death (indicated by widespread nuclear fragmentation) due to corticosteroid exposure related to maternal stress. The organ would not have been malformed but might have been modestly reduced in size. Hematoxylin and eosin (H&E).

the presence of toxicants that alter energy metabolism. Cell death may affect an entire organ or a finite domain within an organ (Figure 5.4), either of which may lead to a structural malformation; importantly, damage to large numbers of cells at one time may release enough active cell enzymes to permit injury to nearby cells. The consequences of cell loss include not only reduction in the affected cell population but also the loss of inductive signals that may be critical for initiating the differentiation of adjacent cells and later events. In this way, minor cell deficits that occur early in gestation can propagate into a growing cascade of cell deficiencies during later development.

Interference with Programmed Cell Death (Apoptosis)

Programmed cell death (PCD) at carefully specified locations and times is a normal part of embryonic shaping and does not damage adjacent cells. The process of PCD occurs by apoptosis, which involves the deliberate entry into active self-destruction. Many embryonic organs rely on PCD, including the brain (to remove unnecessary neurons), digits (to properly separate the digital rays; Figure 5.5), and hollow organs (to initiate and mold cavities in solid anlagen). Molecules that can incite PCD usually are chemical mediators like hormones and growth factors. Interference with the induction of PCD may delay or prevent cell death and thus preclude organ remodeling.[25,86]

Reduced Cell Proliferation

Cell proliferation may decline following certain genetic mutations or as a consequence of exposure to developmental toxicants at doses below those that cause cell death. As with true cell loss, decreased cell production (i.e., hypoplasia) may either temporarily thwart or permanently prevent organ expansion during critical periods of organogenesis. Retarding growth in this fashion may have profound implications later in development for both the affected organ, which may not be able to generate enough cells to ensure proper function or an adequate functional reserve, and also for any secondary structures whose differentiation is dependent on factors released by the hypoplastic cell population. For example, the nephrons of the kidney will not differentiate if the metanephros (primitive kidney) is not penetrated by a sufficient mass of cells from the metanephric duct (primordium for the renal collecting ducts, renal pelves, and ureters; Figure 5.6). Divergence in the basal growth rates of different cell populations and tissues at any developmental stage means that the extent to which growth may be altered by toxicant-induced

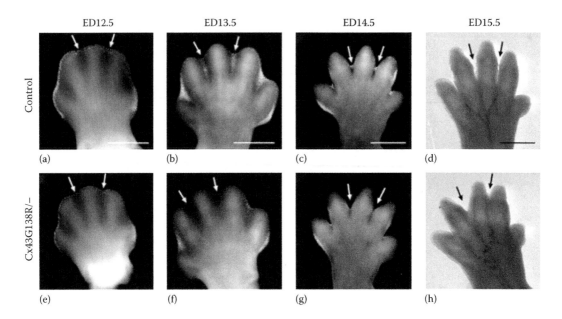

FIGURE 5.5 Properly localized and timed apoptosis (programmed cell death) is critical to the appropriate development of the mouse paw. In control animals (top row), a rise in apoptosis within the soft tissues of the interdigital rays (arrows) begins at gestational day (GD) 12.5 (panel a) and progresses gradually over time (panels b and c) to produce total separation of the digits by GD 15.5 (panel d). In stage-matched littermates with a point mutation in connexin-43 (a constituent of gap junctions), reduced interdigital apoptosis (panels e–g) leads to a variable degree of syndactyly (persistent fusion of digits, panel h). Scale bars: 500 μm. (Reproduced from Dobrowolski, R. et al., *Hum. Mol. Genet.*, 18(15), 2899, 2009, by permission of Oxford University Press.)

reductions in cell division will vary among organs and systems. In some cases, those cells that survive a lethal toxic insult and reach their final destination have the capacity to undergo normal albeit delayed differentiation, but they remain unable to replenish the decimated cell population.[8]

Failed Cellular Interactions

Developmental toxicants, genetic mutations, or both can disrupt information exchange among cell populations. Such effects may sever physical connections (e.g., axonal circuits); erode gradients or curtail delivery of necessary morphogens (e.g., hormones, growth factors); or mask extracellular proteins needed to specify position (e.g., cell adhesion molecules). For instance, aberrant cerebellar differentiation in the weaver mouse (gene symbol: *wv*) arises from a mutation that reduces the number of radial glia processes available to escort migrating external granule cells from the brain surface across the molecular layer into the granule cell layer.[62] Such interactions often are reciprocal, where factors derived from each of two cell populations serve to attract migrating cells or cell processes from the other. This mechanism works in synergy with other pathogenic mechanisms (described earlier) that are capable of altering the size of cell populations that might produce factors required to establish proper connectivity.

Inhibited Morphogenetic Movements

The migration of cells and cell processes may be hindered in many ways. Three major pathogenic mechanisms are diminished cell mobility, an extracellular matrix of deficient quantity or quality, and disruption of cytoskeletal microtubules or microfilaments. The main error embodied in these mechanisms is that defects are likely if the ability of cells to detect and interpret positional information is distorted. For example, altered production[72] or enhanced degradation[16] of certain glycosaminoglycans, which promote

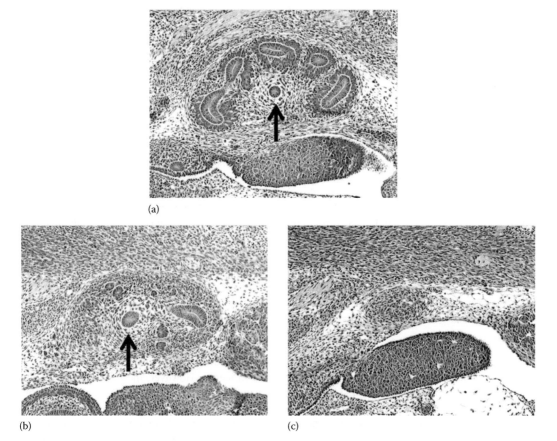

(a)

(b)   (c)

FIGURE 5.6 Diminished cell production leading to a metanephric duct (the primordium of the renal collecting ducts, renal pelves, and ureters [arrow]) of reduced size will thwart its appropriate entry into the metanephrogenic blastema (the primordium of the nephrons). In control animals (a), differentiating cortical tissue caps the primitive connecting ducts at the periphery of the organ, while in mice with null mutations for *glial cell-derived neurotrophic factor (GDNF) receptor-alpha 1 (Gfrα1)* the cortical tissue is reduced in mass (b) or the entire organ is missing (c). Hematoxylin and eosin (H&E). (The mice were the kind gift of Dr. Shuqian Jing, Amgen, Inc., Thousand Oaks, CA.)

cell migration and adhesion, have been proposed as an explanation for increased incidences of cleft palates following *in utero* exposure to chlorcyclizine or targeted removal of the transcription factor *single-minded, homolog 2 (Sim2)*. In a comparable fashion, toxicant-induced alterations in the polymerization, synthesis, or turnover of tubulin by various antineoplastic agents directly disrupt cytoskeletal interactions so severely that malformations develop in many organs.[77] Limb malformations also have been produced indirectly (in rats) by *in utero* administration of calcium chelators,[17,67] with one possible explanation being the roles of this cation in the normal assembly of the cytoskeletal framework as well as extension of cell processes during cell migration.

Reduced Biosynthesis of Essential Components

Decreased production of numerous cell constituents may have a significant impact on cell differentiation and organogenesis. Deficiencies in any of the main cellular building blocks needed to sustain and expand rapidly growing tissues—nucleic acids, proteins, and energy storage molecules—usually are sufficient to cause structural defects or death if they occur during a critical period of development. Many pathogenic mechanisms may be involved, alone or in combination, in disrupting biomolecular synthesis. Examples include inadequate supplies of precursor molecules, inability to assemble the synthetic organelles, inhibition of synthetic enzymes, insufficient time to repair damaged constituents, and errors in translating messenger RNA into protein or performing post-translational modifications to activate the final product. Abnormalities in cell processes associated with DNA and energy molecules typically result in substantial cell lethality, while aberrant protein generation usually leads to nonlethal cell injury. Nonetheless, disruptions in the supply of any of these constituents are effective means of producing developmental malformations.

Intracellular pH

Many rodent teratogens are weak organic acids.[55] Such toxicants accumulate inside embryonic cells and especially embryonic fluids (e.g., amniotic fluid), which have a higher pH than either maternal or embryonic blood.[78] Upon entry, the agents release a hydrogen (H^+) ion, thus forming charged molecules that cannot cross the lipid-rich plasma membrane to exit the cell (ion trapping) while simultaneously inducing a modest reduction (0.2–0.3 pH units) in intracellular pH (pHi).[70] The interaction of trapped anions with cell constituents and/or the drop in pHi can disrupt many cellular processes, including such essential functions as cytoskeletal assembly and stability, enzyme activity, and mitosis. The pHi of the rodent embryo decreases with advancing gestational age, so that buildup of weak acid teratogens in embryonic cells falls as *in utero* development proceeds.[55] Therefore, the susceptibility to the induction of malformations is limited chiefly to earlier postimplantation stages.

Conclusion

Mouse developmental pathology studies begin as an exercise in defining the presence and scope of phenotypic alterations that result from some teratogenic influence. However, the resulting description and interpretation of lesions is unsatisfying if a deeper understanding of the mechanisms responsible for induction of the changes cannot be obtained. Accordingly, individuals who engage in developmental pathology analyses in mice need to comprehend the major ways in which disruption of developmental processes are thought to impact cell differentiation and organogenesis. Such knowledge will be rewarding in its own right, but more importantly such appreciation will permit more productive and collegial interactions with fellow researchers.

REFERENCES

1. Abdel-Rahman MS, Ismail EE (2000). Teratogenic effect of ketamine and cocaine in CF-1 mice. *Teratology* **61**(4): 291–296.
2. Abzug MJ, Tyson RW (2000). Picornavirus infection in early murine gestation: Significance of maternal illness. *Placenta* **21**(8): 840–846.
3. Anderson DJ, Brinkworth MH (2007). *Male-Mediated Developmental Toxicology*. Cambridge, U.K.: Royal Society of Chemistry.
4. Anniko M, Nordemar H (1981). Ototoxicity or teratogenicity. An analysis of drug-induced effects on the early development of the mammalian otocyst. *Arch Otorhinolaryngol* **232**(1): 43–55.
5. Aolad HM, Inouye M, Hayasaka S, Darmanto W, Murata Y (1998). Congenital hydrocephalus caused by exposure to low level X-radiation at early gestational stage in mice. *Biol Sci Space* **12**(3): 256–257.
6. Backman SA, Stambolic V, Suzuki A, Haight J, Elia A, Pretorius J, Tsao MS, Shannon P, Bolon B, Ivy GO, Mak TW (2001). Deletion of *Pten* in mouse brain causes seizures, ataxia and defects in soma size resembling Lhermitte-Duclos disease. *Nat Genet* **29**(4): 396–403.
7. Baldwin CT, Hoth CF, Macina RA, Milunsky A (1995). Mutations in PAX3 that cause Waardenburg syndrome type I: Ten new mutations and review of the literature. *Am J Med Genet* **58**(2): 115–122.
8. Bannigan J, Langman J (1979). The cellular effect of 5-bromodeoxyuridine on the mammalian embryo. *J Embryol Exp Morphol* **50**: 123–135.
9. Baskar JF, Stanat SC, Huang ES (1985). Congenital defects due to reactivation of latent murine cytomegaloviral infection during pregnancy. *J Infect Dis* **152**(3): 621–624.

10. Biddle FG (1977). 6-Aminonicotinamide-induced cleft palate in the mouse: The nature of the difference between the A/J and C57BL/6J strains in frequency of response and its genetic basis. *Teratology* **16**(3): 301–312.

11. Biddle FG, Fraser RC (1977). Cortisone-induced cleft palate in the mouse. A search for the genetic control of the embryonic response trait. *Genetics* **85**(2): 289–302.

12. Birnbaum LS, Weber H, Harris MW, Lamb J 4th, McKinney JD (1985). Toxic interaction of specific polychlorinated biphenyls and 2,3,7,8-tetrachlorodibenzo-*p*-dioxin: Increased incidence of cleft palate in mice. *Toxicol Appl Pharmacol* **77**(2): 292–302.

13. Bolon B, Dorman DC, Janszen D, Morgan KT, Welsch F (1993). Phase-specific developmental toxicity in mice following maternal methanol inhalation. *Fundam Appl Toxicol* **21**(4): 508–516.

14. Bosseray N (1980). Colonization of mouse placentas by *Brucella abortus* inoculated during pregnancy. *Br J Exp Pathol* **61**(4): 361–368.

15. Branstetter DG, Stoner GD, Schut HA, Senitzer D, Conran PB, Goldblatt PJ (1987). Ethylnitrosourea-induced transplacental carcinogenesis in the mouse: Tumor response, DNA binding, and adduct formation. *Cancer Res* **47**(2): 348–352.

16. Brinkley LL, Morris-Wiman J (1987). Effects of chlorcyclizine-induced glycosaminoglycan alterations on patterns of hyaluronate distribution during morphogenesis of the mouse secondary palate. *Development* **100**(4): 637–640.

17. Brownie CF, Brownie C, Noden D, Krook L, Haluska M, Aronson AL (1986). Teratogenic effect of calcium edetate (CaEDTA) in rats and the protective effect of zinc. *Toxicol Appl Pharmacol* **82**(3): 426–443.

18. Chattopadhyay A, Robinson N, Sandhu JK, Finlay BB, Sad S, Krishnan L (2010). *Salmonella enterica* serovar *typhimurium*-induced placental inflammation and not bacterial burden correlates with pathology and fatal maternal disease. *Infect Immun* **78**(5): 2292–2301.

19. Chen B-Y, Chang H-H, Chen S-T, Tsao Z-J, Yeh S-M, Wu C-Y, Lin DP-C (2009). Congenital eye malformations associated with extensive periocular neural crest apoptosis after influenza B virus infection during early embryogenesis. *Mol Vis* **15**: 2821–2828.

20. Chernoff N, Rogers JM, Kavlock RJ (1989). An overview of maternal toxicity and prenatal development: Considerations for developmental toxicity hazard assessments. *Toxicology* **59**(2): 111–125.

21. Cohen G, Roux JC, Grailhe R, Malcolm G, Changeux JP, Lagercrantz H (2005). Perinatal exposure to nicotine causes deficits associated with a loss of nicotinic receptor function. *Proc Natl Acad Sci USA* **102**(10): 3817–3821.

22. Cory-Slechta DA, Stern S, Weston D, Allen JL, Liu S (2010). Enhanced learning deficits in female rats following lifetime Pb exposure combined with prenatal stress. *Toxicol Sci* **117**(2): 427–438.

23. Danielsson BR (2013). Maternal toxicity. In: *Teratogenicity Testing* (Barrow PC, ed.), *Methods in Molecular Biology*, vol. 947. Totowa, NJ: Human Press, pp. 311–325.

24. Daston P (1997). *Molecular and Cellular Methods in Developmental Toxicology*. Boca Raton, FL: CRC Press.

25. Dobrowolski R, Hertig G, Lechner H, Worsdorfer P, Wulf V, Dicke N, Eckert D, Bauer R, Schorle H, Willecke K (2009). Loss of connexin43-mediated gap junctional coupling in the mesenchyme of limb buds leads to altered expression of morphogens in mice. *Hum Mol Genet* **18**(15): 2899–2911.

26. Dunty WCJ, Chen S-Y, Zucker RM, Dehart DB, Sulik KK (2001). Selective vulnerability of embryonic cell populations to ethanol-induced apoptosis: Implications for alcohol-related birth defects and neuro-developmental disorder. *Alcohol Clin Exp Res* **25**(10): 1523–1535.

27. Epstein HT, Kaufman M, Saperstein A, Frank D, Huang S (1991). Strain differences in mouse brain weight gain and spatial-location scores during postnatal development. *Biol Neonate* **59**(3): 171–180.

28. Fawcett LB, Brent RL (2006). Pathogenesis of abnormal development. In: *Developmental and Reproductive Toxicology: A Practical Approach*, 2nd edn. (Hood RD, ed.). Boca Raton, FL: CRC Press (Taylor & Francis Group), pp. 61–92.

29. Fawcett LB, Buck SJ, Beckman DA, Brent RL (1996). Is there a no-effect dose for corticosteroid-induced cleft palate? The contribution of endogenous corticosterone to the incidence of cleft palate in mice. *Pediatr Res* **39**(5): 856–861.

30. Fine EL, Horal M, Chang TI, Fortin G, Loeken MR (1999). Evidence that elevated glucose causes altered gene expression, apoptosis, and neural tube defects in a mouse model of diabetic pregnancy. *Diabetes* **48**(12): 2454–2462.

31. Fuerst PG, Rauch SM, Burgess RW (2007). Defects in eye development in transgenic mice overexpressing the heparan sulfate proteoglycan agrin. *Dev Biol* **303**(1): 165–180.

32. George JD, Manson JM (1986). Strain-dependent differences in the metabolism of 3-methylcholanthrene by maternal, placental, and fetal tissues of C57BL/6J and DBA/2J mice. *Cancer Res* **46**(11): 5671–5675.

33. Gurtner GC, Davis V, Li H, McCoy MJ, Sharpe A, Cybulsky MI (1995). Targeted disruption of the murine *VCAM1* gene: Essential role of VCAM-1 in chorioallantoic fusion and placentation. *Genes Dev* **9**(1): 1–14.

34. Han YW, Redline RW, Li M, Yin L, Hill GB, McCormick TS (2004). *Fusobacterium nucleatum* induces premature and term stillbirths in pregnant mice: Implication of oral bacteria in preterm birth. *Infect Immun* **72**(4): 2272–2279.

35. Harris MJ, Juriloff DM (2010). An update to the list of mouse mutants with neural tube closure defects and advances toward a complete genetic perspective of neural tube closure. *Birth Defects Res A Clin Mol Teratol* **88**(8): 653–669.

36. Hood RD (1986). Effects of concurrent prenatal exposure to rubratoxin B and T-2 toxin in the mouse. *Drug Chem Toxicol* **9**(2): 185–190.

37. Hood RD (1997). *Handbook of Developmental Toxicology*. Boca Raton, FL: CRC Press.

38. Hood RD (2012). *Developmental and Reproductive Toxicology: A Practical Approach*, 3rd edn. London, U.K.: Informa Healthcare.

39. Hunter ES III, Sadler TW (1987). D-(-)-β-hydroxybutyrate-induced effects on mouse embryos *in vitro*. *Teratology* **36**(2): 259–264.

40. Ishibashi K, Yamamoto H, Hatano M, Koizumi T, Yamamoto M, Tokuhisa T (1999). Enlargement of the globe with ocular malformations in c-Myc transgenic mice. *Jpn J Ophthalmol* **43**(3): 201–208.

41. Jennings-Gee JE, Moore JE, Xu M, Dance ST, Kock ND, McCoy TP, Carr JJ, Miller MS (2006). Strain-specific induction of murine lung tumors following *in utero* exposure to 3-methylcholanthrene. *Mol Carcinog* **45**(9): 676–684.

42. Jones HE, Kunko PM, Robinson SE, Balster RL (1996). Developmental consequences of intermittent and continuous prenatal exposure to 1,1,1-trichloroethane in mice. *Pharmacol Biochem Behav* **55**(4): 635–646.

43. Juriloff DM, Harris MJ, Tom C, MacDonald KB (1991). Normal mouse strains differ in the site of initiation of closure of the cranial neural tube. *Teratology* **44**(2): 225–233.

44. Khera KS (1984). Maternal toxicity—A possible factor in fetal malformations in mice. *Teratology* **29**(3): 411–416.

45. Kimmel CA, Buelke-Sam J (1994). *Developmental Toxicology*, 2nd edn. In: Target Organ Toxicology Series (Hayes AW, Thomas JA, Gardner DE, series eds.). New York: Raven Press.

46. Lamb JC IV, Harris MW, McKinney JD, Birnbaum LS (1986). Effects of thyroid hormones on the induction of cleft palate by 2,3,7,8-tetrachlorodibenzo-*p*-dioxin (TCDD) in C57BL/6N mice. *Toxicol Appl Pharmacol* **84**(1): 115–124.

47. Leblebicioglu-Bekcioglu B, Paulson RB, Paulson JO, Sucheston ME, Shanfeld J, Bradway SD (1995). Effects of caffeine and nicotine administration on growth and ossification of the ICR mouse fetus. *J Craniofac Genet Dev Biol* **15**(3): 146–156.

48. Liao J, Kochilas L, Nowotschin S, Arnold JS, Aggarwal VS, Epstein JA, Brown MC, Adams J, Morrow BE (2004). Full spectrum of malformations in velo-cardio-facial syndrome/DiGeorge syndrome mouse models by altering *Tbx1* dosage. *Hum Mol Genet* **13**(15): 1577–1585.

49. Martinez F, Happa J, Arias F (1985). Biochemical and morphologic effects of ethanol on fetuses from normally ovulating and superovulated mice. *Am J Obstet Gynecol* **151**(4): 428–433.

50. Mayura K, Spainhour CB, Howie L, Safe S, Phillips TD (1993). Teratogenicity and immunotoxicity of 3,3′,4,4′,5-pentachlorobiphenyl in C57BL/6 mice. *Toxicology* **77**(1–2): 123–131.

51. Min H, Danilenko DM, Scully SA, Bolon B, Ring BD, Tarpley JE, DeRose M, Simonet WS (1998). *Fgf-10* is required for both limb and lung development and exhibits striking functional similarity to *Drosophila branchless*. *Genes Dev* **12**(20): 3156–3161.

52. Minamitani M, Tanaka J, Suzuki Y (1996). Pathomechanism of cerebral hypoplasia in experimental toxoplasmosis in murine fetuses. *Early Hum Dev* **44**(1): 37–50.

53. Muthuchamy M, Boivin GP, Grupp IL, Wieczorek DF (1998). β-Tropomyosin over-expression induces severe cardiac abnormalities. *J Mol Cell Cardiol* **30**(8): 1545–1557.

54. Nau H, Hauck RS, Ehlers K (1991). Valproic acid-induced neural tube defects in mouse and human: Aspects of chirality, alternative drug development, pharmacokinetics and possible mechanisms. *Pharmacol Toxicol* **69**(5): 310–321.

55. Nau H, Scott WJ Jr. (1986). Weak acids may act as teratogens by accumulating in the basic milieu of the early mammalian embryo. *Nature* **323**(6085): 276–278.

56. Nawrot PS, Cook RO, Staples RE (1980). Embryotoxicity of various noise stimuli in the mouse. *Teratology* **22**(3): 279–289.

57. Ozawa M, Yamasaki Y, Hirabayashi M, Kanai Y (2003). Viability of maternally heat-stressed mouse zygotes *in vivo* and *in vitro*. *Anim Sci J* **74**(3): 181–185.

58. Palmer AL, Rotbart HA, Tyson RW, Abzug MJ (1997). Adverse effects of maternal enterovirus infection on the fetus and placenta. *J Infect Dis* **176**(6): 1437–1444.

59. Paumgartten FJ (2010). Influence of maternal toxicity on the outcome of developmental toxicity studies. *J Toxicol Environ Health Pt A* **73**(13–14): 944–951.

60. Pratt RM (1985). Receptor-dependent mechanisms of glucocorticoid and dioxin-induced cleft palate. *Environ Health Perspect* **61**: 35–40.

61. Priebe K, Romeo RD, Francis DD, Sisti HM, Mueller A, McEwen BS, Brake WG (2005). Maternal influences on adult stress and anxiety-like behavior in C57BL/6J and BALB/cJ mice: A cross-fostering study. *Dev Psychobiol* **47**(4): 398–407.

62. Rakic P, Sidman RL (1973). Sequence of developmental abnormalities leading to granule cell deficit in cerebellar cortex of weaver mutant mice. *J Comp Neurol* **152**(2): 103–132.

63. Redline RW, Lu CY (1987). Role of local immunosuppression in murine fetoplacental listeriosis. *J Clin Invest* **79**(4): 1234–1241.

64. Rodier PM (1980). Chronology of neuron development: Animal studies and their clinical implications. *Dev Med Child Neurol* **22**(4): 525–545.

65. Rodolakis A, Bernard F, Lantier F (1989). Mouse models for evaluation of virulence of *Chlamydia psittaci* isolated from ruminants. *Res Vet Sci* **46**(1): 34–39.

66. Rousseaux CG, Bolon B (2013). Embryo and fetus. In: *Handbook of Toxicologic Pathology*, 3 edn., vol. 3 (Haschek WM, Rousseaux CG, Wallig MA, eds.). San Diego, CA: Academic Press (Elsevier), pp. 2695–2759.

67. Rousseaux CG, MacNabb LG (1992). Oral administration of D-penicillamine causes neonatal mortality without morphological defects in CD-1 mice. *J Appl Toxicol* **12**(1): 35–38.

68. Rowitch DH, Danielian PS, McMahon AP, Zec N (1999). Cystic malformation of the posterior cerebellar vermis in transgenic mice that ectopically express *Engrailed-1*, a homeodomain transcription factor. *Teratology* **60**(1): 22–28.

69. Schopke R, Wolfer DP, Lipp H, Leisinger-Trigona MC (1991). Swimming navigation and structural variations of the infrapyramidal mossy fibers in the hippocampus of the mouse. *Hippocampus* **1**(3): 315–328.

70. Scott WJ Jr, Schreiner CM, Nau H, Vorhees CV, Beliles RP, Colvin J, McCandless D (1997). Valproate-induced limb malformations in mice associated with reduction of intracellular pH. *Reprod Toxicol* **11**(4): 483–493.

71. Sebastiano V, Dalvai M, Gentile L, Schubart K, Sutter J, Wu GM, Tapia N et al. (2010). Oct1 regulates trophoblast development during early mouse embryogenesis. *Development* **137**(21): 3551–3560.

72. Shamblott MJ, Bugg EM, Lawler AM, Gearhart JD (2002). Craniofacial abnormalities resulting from targeted disruption of the murine *Sim2* gene. *Dev Dyn* **224**(4): 373–380.

73. Sharma L, Kaur J, Shukla G (2012). Role of oxidative stress and apoptosis in the placental pathology of *Plasmodium berghei* infected mice. *PLoS ONE* **7**(3): e32694.

74. Shepard TH, Lemire RJ (2010). *Catalog of Teratogenic Agents*, 13th edn. Baltimore, MD: Johns Hopkins University Press.

75. Shiota K, Shionoya Y, Ide M, Uenobe F, Kuwahara C, Fukui Y (1988). Teratogenic interaction of ethanol and hyperthermia in mice. *Proc Soc Exp Biol Med* **187**(2): 142–148.

76. Shuey DL, Sadler TW, Lauder JM (1992). Serotonin as a regulator of craniofacial morphogenesis: Site specific malformations following exposure to serotonin uptake inhibitors. *Teratology* **46**(4): 367–378.

77. Sieber SM, Whang-Peng J, Botkin C, Knutsen T (1978). Teratogenic and cytogenetic effects of some plant-derived antitumor agents (vincristine, colchicine, maytansine, VP-16-213 and VM-26) in mice. *Teratology* **18**(1): 31–47.

78. Srivastava M, Collins MD, Scott WJ Jr, Wittfoht W, Nau H (1991). Transplacental distribution of weak acids in mice: Accumulation in compartments of high pH. *Teratology* **43**(4): 325–329.
79. Subbarayan V, Mark M, Messadeq N, Rustin P, Chambon P, Kastner P (2000). RXRα overexpression in cardiomyocytes causes dilated cardiomyopathy but fails to rescue myocardial hypoplasia in RXRα-null fetuses. *J Clin Invest* **105**(3): 387–394.
80. Tamai Y, Ishikawa T, Bosl MR, Mori M, Nozaki M, Baribault H, Oshima RG, Taketo MM (2000). Cytokeratins 8 and 19 in the mouse placental development. *J Cell Biol* **151**(3): 563–572.
81. Terry KK, Stedman DB, Bolon B, Welsch F (1996). Effects of 2-methoxyethanol on mouse neurulation. *Teratology* **54**(5): 219–229.
82. Tiboni GM (1993). Second branchial arch anomalies induced by fluconazole, a bis-triazole antifungal agent, in cultured mouse embryos. *Res Commun Chem Pathol Pharmacol* **79**(3): 381–384.
83. Vedel-Macrander GC, Hood RD (1986). Teratogenic effects of nigericin, a carboxylic ionophore. *Teratology* **33**(1): 47–51.
84. Wandersee NJ, Roesch AN, Hamblen NR, de Moes J, van der Valk MA, Bronson RT, Gimm JA, Mohandas N, Demant P, Barker JE (2001). Defective spectrin integrity and neonatal thrombosis in the first mouse model for severe hereditary elliptocytosis. *Blood* **97**(2): 543–550.
85. Wilson JG (1973). *Environment and Birth Defects*. New York: Academic Press.
86. Yoshida H, Kong YY, Yoshida R, Elia AJ, Hakem A, Hakem R, Penninger JM, Mak TW (1998). Apaf1 is required for mitochondrial pathways of apoptosis and brain development. *Cell* **94**(6): 739–750.

Section III

Experimental Methods in Mouse Developmental Pathology

6

Principles of Experimental Design for Mouse Developmental Pathology Studies

Brad Bolon and Krista M.D. La Perle

CONTENTS

The proverb, "Failing to plan is planning to fail," is the most important consideration for every mouse developmental pathology study. Suitable advance preparation is the essential ingredient for ensuring that a given experiment has the best opportunity for success.[76] Whether the trouble lies in mistaken assumptions or suboptimal methodology or carelessness (or some combination of them), a badly designed experimental plan will make the acquisition, processing, and analysis of appropriate samples more difficult—assuming that the study can be completed and evaluated in a meaningful manner at all.

This chapter introduces several major issues that must be considered when designing mouse developmental pathology experiments as well as acceptable means for addressing them. In general, laboratories performing mouse developmental pathology investigations on a regular basis should devise a basic strategy to guide the initial design of new projects as well as the necessary standard operating procedures (SOPs) and training programs to ensure that the design may be followed with confidence. In the authors' experience, preplanning of this sort still permits sufficient experimental flexibility so that specimens may be gathered and examined even if an unexpected problem arises during the course of an ongoing study.

Principal Factors in Designing Mouse Developmental Pathology Studies

Multiple topics must be pondered during the planning phase that will impact the final experimental outcome. The simplest approach to avoiding errors in study design is to use a checklist (Table 6.1) to ensure that essential questions have been identified, considered, and answered before experimental activities begin. In our experience, checklists of this kind generally are more useful when available as an electronic or paper worksheet, but mental reflection alone is acceptable as long as all the topics

TABLE 6.1

Design Parameters for Mouse Developmental Pathology Studies

Objective(s)

- *Identify and/or describe in detail a developmental phenotype*
  - What specimen(s) is to be evaluated?
    - Embryo/fetus only
    - Placenta only
    - Embryo/fetus + placenta (i.e., the entire conceptus)
    - Neonate
    - Dam (typically uterus, mammary gland, and pituitary gland)
  - What outcome(s) is sought?
    - A list of affected cell populations, tissues, or organs
    - An understanding of lesion progression over time
    - A biochemical/molecular explanation for lesion induction
- *Test a hypothesis*
  - Does the stated hypothesis clearly distinguish between two distinct outcomes?
  - Can the two outcomes be related to the intervention (genetic manipulation, xenobiotic treatment, etc.)?
  - Will the study design and methods permit unambiguous distinction of the two outcomes?
  - Is a statistical test available to test the data set? (see Chapter 18)

Study Conditions

- *Animals*—is the model system defined in sufficient detail?
  - Genetic background (especially known sensitivity or resistance)
    - Strain/substrain (especially for genetic engineering studies)
    - Stock (particularly common for toxicity bioassays)
  - Source
  - Sex
  - Age—stage of development as defined by:
    - Morphologic landmarks (see Chapter 4)
      - Crown-rump length
      - Number of somite pairs
      - Theiler stages (TS)
    - Timing conventions (for the day that a vaginal plug is seen as evidence of recent mating)
      - Days *post coitum* (dpc)
      - Gestational day (GD) 0 (or embryonic day [E] 0)—where GD 0 is the plug-positive day
      - GD 0.5—where mating is assumed to have occurred at 12:01 AM on the plug-positive day
      - GD 1—where GD 1 is the plug-positive day
      - Portions of days are denoted by
        - Decimals (fractional days): GD 9.25 = 9¼ days
        - Superscripts (for hours): GD 9^6 = 9¼ days
- *Husbandry*—Is the protocol specified clearly and completely?
  - Environmental conditions
    - Air exchanges/hour[a] (within cage or within the room)
    - Light/dark cycle (typically 10–14 h of light)
    - Relative humidity
    - Room temperature
  - Management practices
    - Breeding practices[a]
      - Scheme (commonly one male and two females)
      - Schedule
        - Overnight—female introduced to male at the start of the dark cycle
        - Timed—female introduced to male for the last 1–2 h of the dark cycle

(Continued)

TABLE 6.1 (*CONTINUED*)

Design Parameters for Mouse Developmental Pathology Studies

- Caging[a]
  - Type of cage (usually filter-capped, static [non-ventilated], polystyrene [Micro-Isolator®])
  - Bedding (typically corn-cob fragments or wood shavings ± cotton fiber nesting material)
  - Population (usually one to five pregnant mice)
- Diet
  - Food (generally commercial rodent chow)
  - Water[a] (purified by acidification, distillation, or reverse osmosis)
- Microbial status

Experimental Design

- Is the manipulation (*treatment*) defined with exactitude?
  - Genetic experiments
    - *Treatment* = engineered (targeted or transgenic) or spontaneous mutation(s)
    - If known, the affected gene/gene product, tissue specificity, and spatial and temporal expression patterns should be stated
  - Toxicity studies
    - *Treatment* = toxic agent
    - The exposure regimen (route of administration, dose[s], frequency, and vehicle) should be specified
    - Lot numbers for reagents should be included, if available
- Are suitable control materials available?
  - Genetic experiments
    - Wild-type (WT or +/+)
    - Heterozygous (HET or +/−)
    - Control mice should share the same genetic background (inbred strain/substrain) and be exposed to the same husbandry as the experimental groups
  - Toxicity studies
    - Negative control
      - Vehicle—preferred as mice receive all manipulations experienced by a treatment group except for exposure to the toxic agent
      - Untreated
    - Positive control—a separate treatment group that has been exposed to an agent that causes a known developmental effect
      - Positive controls are done to ensure that the analytical methods are suited to detecting subtle defects
      - The exposure regimen of the positive control agent needs to be specified as it may differ from that of the experimental toxicant in some or all details
  - Multifactorial studies
    - Rationale—two factors are to be compared simultaneously:
      - Offspring carry two genetic mutations (A and B)
      - Dams are exposed to two chemicals (X and Y)
      - Offspring carry a mutation (A), and the dam is exposed to a chemical (X)
    - Negative controls—multiple are needed
      - Genetic study
        - Offspring carrying only mutation A
        - Offspring carrying only mutation B
        - Offspring carrying neither mutation
      - Toxicity study
        - Dams exposed only to chemical X
        - Dams exposed only to chemical Y
        - Dams exposed to neither chemical

(Continued)

TABLE 6.1 (*CONTINUED*)

Design Parameters for Mouse Developmental Pathology Studies

- • Mixed genetic and toxicity study
  - • Offspring carrying mutation A from dams not exposed to chemical
  - • Offspring carrying no mutation from dams exposed to chemical Y
  - • Offspring carrying no mutation from dams not exposed to chemical
  - • Untreated—inappropriate for multifactorial studies
- • Do manipulated and control groups differ only in the presence or absence of the treatment?
- • Are statistical considerations outlined in suitable detail? (see Chapter 18)
  - • Has the experimental unit been identified appropriately?
    - • Genetic experiments—the *individual* is the unit since the variable is whether or not the animal carries a mutation
    - • Toxicity studies—the *litter* is the unit since all conceptuses in a treated dam theoretically experienced an equivalent exposure
  - • Are numbers of experimental units sufficient for analysis?
    - • Are sample size calculations included?
    - • Is the sample size justified?
  - • If feasible, have experimental units been assigned randomly?
  - • Is the study design balanced?
  - • Is the analysis to be undertaken in coded (*blinded*) fashion?
  - • Are chosen statistical tests suited to the hypothesis and study design?
  - • Does the data set meet assumptions underlying the statistical tests?
  - • Are all experimental units included in the analysis, or are good reasons given for excluding some units?

Notes: Developmental biologists often prefer the E (embryonic day) designation as it avoids the need for separate nomenclature for the mouse F (fetal) period, which is quite short (4–5 days) relative to humans. The generic designation GD (gestational day) inherently covers both the E and F periods.

[a] These parameters are important but seldom reported explicitly, thereby bringing into question whether or nor such issues were considered in advance.

receive due consideration *in advance*. The sections of this chapter cover major design parameters in an order that we have found them to be addressed most conveniently when designing new developmental pathology studies.

Establishing Study Objectives

Mouse developmental pathology studies commonly are conducted to address one of two primary aims. The first is to identify the timing and catalog the lesions associated with prenatal or perinatal lethality, and eventually to understand the pathogenesis (i.e., sequence of events and mechanisms that produce a given lesion). The outcome of such investigations typically is a more or less detailed description of a lethal phenotype that arises during development. The second goal is to assess the validity of one or more hypotheses, typically using statistics (see Chapter 18) to test the likelihood that the experimental outcome resulted from the genetic event or toxicant exposure. In many cases, a good experimental design may provide specimens that are suitable for addressing both these objectives at the same time. Regardless of the goal(s), they should be defined clearly (ideally in writing) as the study is being designed to ensure that all elements are formulated in a fashion that actually will serve the stated objective(s).

Defining the Study Timing

For studies of novel mutations, the main consideration when devising an experimental plan to study an embryonic lethal phenotype is to decide when during gestation the examination should be conducted. The usual scenario leading to a presumption of embryonic lethality is the detection of no animals carrying a desired genetic mutation when genotyping is performed (commonly at or shortly after weaning). Similar issues will arise when conducting developmental toxicity studies. In both settings, further breeding is required to generate more pregnant dams so that follow-up studies may be done to examine the presence and viability of the litters during gestation or immediately after birth.

Prenatal Assessments

Investigations of embryonic lethality will proceed in a step-wise manner, with the number of steps dependent on how quickly the time of *in utero* lethality may be defined. In our experience (Figure 6.1), the most rapid way forward in defining a putative new embryonic lethal phenotype is to assess the genotypes of near-term fetuses (gestational day [GD] 16.0[71]–18.5[73]) to see whether or not individuals of the desired genotype are present within the litter. A common genotypic distribution for a study of animals bearing a single-gene mutation is the Mendelian 1:2:1 ratio for homozygous wild-type, heterozygous, and homozygous null mutant animals, respectively. The main rationale for this late timing is that fetuses are large enough to provide a large tissue sample (usually tail or distal limb) that is unambiguously of embryonic origin and not of maternal derivation. A crucial secondary consideration is that the appearance of the fetuses and placentae permit an immediate appraisal regarding the potential timing of a lethal *in utero* or perinatal event. Viable, near-term, wild-type mouse fetuses will be approximately 1.0–1.5 cm in

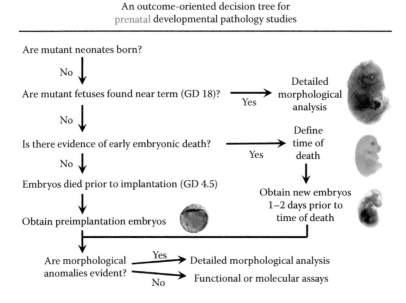

An outcome-oriented decision tree for
prenatal developmental pathology studies

FIGURE 6.1 An outcome-oriented decision tree for evaluating embryonic lethal phenotypes in developing mice. The initial analysis is undertaken near term (typically between gestational day [GD] 17.5 and GD 18.5) and involves the acquisition of qualitative changes (both gross lesions and histopathological findings) as well as quantitative measurements (crown-rump length and/or fetal body weight). If necessary, one or more follow-up experiments may be needed at earlier time points in gestation; the timing of these additional studies is established using external morphologic traits of involuted implantation sites (*resorptions*) or dead embryos. Common times for analyzing embryos earlier during gestation—given in the order dictated by lesions we usually observe near term—include GD 12.5 (when assembly of the placental labyrinth is finished), GD 10.5 (when abnormal embryonic or yolk sac circulation will halt growth of the conceptus), GD 8.5 (due to the absence of heart tube contractions and/or production of primitive erythrocytes in yolk sac blood islands), GD 7.5 (from abnormal gastrulation [i.e., differentiation of germ layers]), and GD 3.5 (due to embryo disintegration prior to implantation). The top-right image is a viable wild-type embryo at approximately GD 14.5 (shown by the adult-like body conformation and easily visualized, large-caliber blood vessels in the subcutis). The middle-right image is an age-matched littermate that died at about GD 13.5 (indicated by diffuse cutaneous pallor and incomplete digit separation on the limbs). The bottom-right image is a viable wild-type embryo at GD 9.5 (as demonstrated by the curve of the tail across the right side of the body and the visible branchial arches just under the mandible). The bottom-left image is a wild-type preimplantation embryo (specifically, a blastocyst with crescent-shaped inner cell mass capping a prominent blastocoele) at GD 3.5. (The blastocyst image was kindly provided by Mr. Joe Anderson, Amgen, Inc., Thousand Oaks, CA.)

length and weigh 1.10–1.65 g (depending on the stock, strain, and litter size)[20,54]; possess pale pink skin; and exhibit distinct torso and limb motility in response to external stimuli. In contrast, dead conceptuses usually will be discolored (i.e., tan or white skin, with or without foci of hemorrhage) and either friable (i.e., subject to fragmentation with minimal manipulation) or reduced in size relative to nearby viable littermates. A follow-up study to assess the lethal phenotype at the proper developmental stage may be designed once the genotypes are determined and matched with the phenotypes seen at necropsy (see Chapter 7). The decision regarding the timing to be used in the follow-up study is set based on such traits as body size, head contour, and viability of the fetus as well as the dimensions of the implantation site.

If fetuses with the mutant genotype are present in the near-term litter but are all dead, then the time of death *in utero* is predicted to be the developmental stage at which the structural evolution of the individual ceased. This decision is made by comparing the appearance of any dead conceptus to that of wild-type embryos at known developmental stages (see Chapter 4).[18,30] In such cases, the follow-up study should be scheduled so that conceptuses are harvested 1 or 2 days prior to the day on which development seems to have ended. In our experience, the appropriate timing for such follow-up studies often will fall between GD 10.0 and GD 12.5. The reason for collecting embryos prior to the stage where their morphological evolution ceased is that dead conceptuses typically are autolyzed so completely that histopathological evaluation will yield no useful insight regarding the cause of death.

Postnatal Assessments

If mutant fetuses are all viable near term, then the follow-up study will need to assess when during the postnatal period the pups actually expired. In most cases, the first time after birth during which to evaluate pup viability is the neonatal period (Figure 6.2). Lethal events during this time followed by maternal cannibalization are a very common explanation for the lack of mutants at weaning. In such cases, the design of the follow-up study will require hourly evaluation of dams at term (GD 19–20) to assess the viability of newborn pups before they can be eaten by the dam. Given the propensity of rodents to deliver litters during the dark cycle, this *death watch* often will necessitate an all-night vigil or video recording. Care must be taken to ensure that repeated intrusions at this stage are brief and very quiet so that stress is minimized that might incite the dam to scavenge the entire litter. The least objectionable means of accomplishing such viewings is to place dams in transparent cages on counter tops where the progress of delivery may be determined without the need to physically jostle the cage.

If visible mutant embryos are not detected near term but the number of resorptions (i.e., shrunken implantation sites) is increased, the follow-up study should be timed to examine the likelihood that embryos died shortly after they implanted in the uterus. In our experience, the appropriate scheduling of such peri-implantation follow-up studies often will fall between GD 5.5 and GD 7.0. This timing spans the period immediately after implantation (approximately GD 4.8–5.2) through the initial axis specification needed to begin forming organs. The small size of embryos within these new implantation sites coupled with the overwhelming preponderance of the maternal decidua nearby typically requires that genotyping at this stage be conducted by laser capture microdissection (LCM; see Chapter 10) from sections of embedded tissue to ensure that the embryonic cells are not contaminated by maternal elements. Another advantage of the LCM method is that the structures of the embryos to be genotyped also are available for histopathological examination, which permits linkages to be drawn between genotype and phenotype.

Finally, if mutant embryos cannot be identified near term and the litters contain a reduced number of mutant implantation sites, the follow-up study will need to explore the possibility that embryonic lethality is occurring prior to implantation. In this event, the follow-up experiment typically will be timed for GD 3.5–GD 4.0 so that blastocysts may be collected before they implant (see Chapter 7 for a description of the uterine flushing procedure). The collected embryos may be examined to determine whether or not some exhibit morphologic changes consistent with terminal damage, and one or more cells may be harvested for genotypic analysis.

Alternative timing schemes are possible when scheduling the initial study needed to define the timing of any follow-up experiments. The first option, which is best used if near-term mutant fetuses are viable, consists of moving forward during gestation at a defined interval (e.g., every 2 days[71]) rather than jumping straight to the GD 12.5 time point to ascertain when mutant individuals first begin to

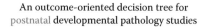

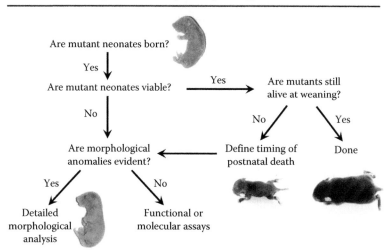

FIGURE 6.2 An outcome-oriented decision tree for evaluating lethal phenotypes in neonatal and juvenile mice, from birth through weaning. If mutant animals are not found at weaning—typically performed at postnatal day (PND) 21 or 22—but are seen to be viable near term, the next assessment generally is undertaken at PND 0 by monitoring the delivery to confirm that pups are not dying and being eaten by the dams soon after delivery. Common times for analyzing postnatal mice prior to weaning—given in the order dictated by lesions we usually observe near term—are PND 0–PND 1 (when traumatic delivery or functional deficits that limit breathing, energy metabolism, or systemic delivery of oxygen lead to death) and PND 7–PND 10 (when organ-specific dysfunction or systemic deficiencies in either excretory or metabolic pathways may lead to marked growth disruption). The top image is a wild-type neonate (PND 0) whose viability is demonstrated by its uniform pale pink skin and delicate web of red superficial blood vessels. The bottom-left image is an age-matched mutant littermate lacking expression of extracellular signal-regulated kinase 3 (Erk3); it died within hours of birth due to marked pulmonary hypoplasia, causing cyanosis (shown by diffuse, pale blue discoloration of the skin). The two haired PND 7 pups (portrayed at an identical magnification) are a severely runted knockout pup (left) lacking Mdgt, a long noncoding RNA (lncRNA), and a wild-type age-matched, littermate (right). (The two fetal images are reproduced from Klinger, S. et al., *Proc. Natl. Acad. Sci. USA*, 106(39), 16710, 2009; the two juvenile images are reproduced from Sauvageau, M. et al., *eLife*, 2, e01749, 2013.)

develop abnormally. This strategy has the advantage of providing fairly complete coverage of major gestational stages but will require more animals to complete. A second option is to time the first study so that the necropsy is undertaken at mid-gestation (between GD 9.5 and GD 12.5)[52] rather than near term; GD 12 is considered the appropriate starting point in one high-throughput mouse phenotyping laboratory (see Chapter 17). The reason for considering this scheme is that this early phase encompasses the portion of organogenesis (i.e., the period during which organs are first specified) for most major organs in the embryo and also brackets the evolution of the placenta from a nutrient exchange interface centered on the yolk sac to one dependent on appropriate formation of the definitive placenta. This second approach also is practical because the dam may be seen by eye or palpated by touch to be pregnant beginning at GD 10 (for large litters) or GD 11. If all embryos are viable and exhibit normal morphology at this point, the follow-up study often is scheduled to end at GD 14.5 or GD 15.0 (the end of organogenesis) or near term; the presence of abnormal embryos or more resorptions at either of these later times would dictate that any other follow-up studies be shifted back to an earlier time in gestation. In our experience, the principal disadvantage of starting at mid-gestation as a routine strategy in developmental pathology investigations is that neonatal lethality (i.e., an absence of dead mutant fetuses at birth due to maternal cannibalization) cannot be explored as readily as a potential explanation for an apparent embryonic lethal phenotype, but rather will require not only a study near term to confirm the viability of mutant fetuses just before birth but also another follow-up experiment to specifically examine neonatal survival.

Selecting the Study Conditions

The ability to replicate a mouse developmental pathology study depends primarily on the accurate and complete documentation of the model system. A full description of the animals and husbandry conditions are particularly vital points in this regard. These data should be fully explained in the "Materials" sections of study reports and papers rather than being scattered haphazardly in the text of "Results" sections, figure legends, or table footnotes.

Animals

Variables associated with the choice of animals include strain (or stock), source, age, and sex (Table 6.1). In general, mouse developmental pathology studies always account for the first three variables while sex is evaluated only when a simple test (e.g., visual estimation of the anogenital distance in fetuses or neonates [see Chapter 7], polymerase chain reaction [PCR] of tissue homogenates to measure the male-specific *Sry* gene[43]) may be incorporated into the battery of desired study endpoints.

The choice of strain or stock may influence the outcome of the entire developmental pathology study, especially if the protocol calls for evaluating the impact of a toxic agent. Inbred mouse strains typically have smaller litters due to both a reduction in the number of ova released during each estrous cycle and an increased incidence of preimplantation mortality associated with increased inbreeding.[41,64] Furthermore, many inbred mouse strains exhibit a higher incidence of strain-specific malformations after exposure to a toxicant. For instance, administration of acetazolamide to pregnant dams induces much more ectrodactyly (i.e., absence of all or part of at least one digit) in sensitive strains like the C57BL/6J, C57BL/10J, and CBA/J relative to insensitive strains such as BALB/cJ, C3H/HeJ, and DBA/2J.[3] Accordingly, experiments that compare the response of sensitive and resistant strains in a single study can provide a potent means for exploring the impact of genotype and phenotype (e.g., the presence, partial lack, or complete absence of a specific molecular mechanism) on the induction of certain developmental defects. In addition to inter-strain effects on outcome, care also should be taken to always include the correct nomenclature (see Chapter 2) when describing the test animal. This attention to detail is essential as seemingly subtle variations in the genetic background of the parent mice may invoke major differences in the constellation of developmental lesions encountered in the progeny. For example, targeted removal of epithelial growth factor receptor (*Egfr*) leads to embryonic lethality near implantation in inbred CF-1 embryos because of degeneration in the inner cell mass of the blastocyst and during mid-gestation in 129/Sv mouse conceptuses due to altered placental angiogenesis, while a similar mutation on the outbred CD-1 genetic background allows individuals to survive gestation.[65] Inbred mouse strains/substrains or their first-generation (F_1) hybrids usually are utilized for genetic studies so that the nearly complete genome homogeneity among individuals will emphasize the impact of the mutated gene, while outbred stocks commonly are used for toxicity assays since the extensive genetic variation among animals more closely resembles the wide genotypic heterogeneity that exists in the average human population.

A corollary of the overall concern with a standard genetic background is that the nature of the genetic mutation must be carefully defined and recorded. This attention to detail is required because mutations with the same overt effect in fact may cause distinct effects *in vivo*. For example, mice bearing a null mutation of the cytochrome P450 1a2 (*Cyp1a2*) gene induced by disruption but not deletion of exon 2 develop lethal respiratory distress as neonates,[55] while *Cyp1a2* knockout animals engineered by partial removal of exon 2 and complete ablation of exons 3, 4, and 5 exhibit no developmental phenotype.[35] Even after backcrossing to produce a congenic strain (defined in Chapter 2), modifier genes contained in the microsatellite regions of the transferred chromosome[24,36] or adjacent to the engineered locus[28,60] actually carry multiple linked genes that might be able to produce subtle differences in developmental outcomes.

The rate of embryonic development proceeds at different rates among various mouse strains.[48,64] The developmental age of the animals may be communicated in multiple ways. The first is to define conceptus age based on the presumed time of conception, which is set as sometime during the day on which a vaginal plug is observed (Table 6.1). If one of these conventions is employed to establish embryonic age, it is imperative that the chosen method be stated in the "Materials" section, and ideally in the study

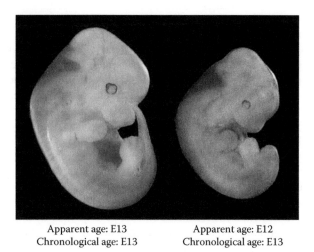

| Apparent age: E13 | Apparent age: E12 |
| Chronological age: E13 | Chronological age: E13 |

FIGURE 6.3 Mouse embryos from the same litter demonstrate the fact that apparent ages (i.e., the developmental stage) as indicated by external anatomic landmarks may differ by many hours from the chronological age (calculated as days *post coitum* [dpc]). The embryo on the left is at Theiler stage (TS) 21 (gestational day [GD] 13) as shown by distinct linear gray rays (the sites of future separation between the digits) on the paddle-shaped limb bud ends for both the forepaw and hind paw, while the sibling on the right is at TS 20 (GD 12) because rays are absent on both limb buds. Rays typically form on the forelimb bud at approximately GD 12.3 and on the hind limb bud at about GD 12.8.

abstract as well, of study reports and scientific publications. An alternative means of communicating developmental age is to concentrate on the stage as indicated by a certain set of structural traits rather than actual developmental chronology *per se*. The rationale for this choice is that the range of anatomic features observed for mouse conceptuses within a single multi-animal litter reveals that the time of conception usually spans a period of 12^{30}–$16\ h^{58}$ to as much as 24 h^{64} (Figure 6.3); this variation may arise from subtle differences in the time of conception or divergence in the maturity of male vs. female blastocysts. The most common morphology-based scheme for assigning conceptus age is the Theiler scale (TS; see Chapter 4),[25,62,63] which relies mainly on the presence of body and head characteristics. The evaluation of external morphological features permits differences in developmental progress to be compared among inbred mouse strains and among litters for a given strain; for example, CBA/J embryos tend to reach each stage later than do either C57BL/6 or C3H/He embryos.[48] Furthermore, additional understanding of mouse stages may be gained by comparing them to structural evolution in other species, particularly the closely related rat[26,74] and well-studied human[17,51,61] embryos. However, other staging schemes have been developed for mouse embryos that rely on the evolution of a particular structure, such as the face,[46] foot pads,[49] heart,[39] limb,[4,30,53,70] ossification centers in specific appendicular bones,[53] or teeth[54]; ratios of measured lengths[16]; numbers of somites[6,21,67,68]; or expression of various gene products.[47] Staging of mouse conceptuses may be performed *in vivo* using sonography.[8,11,22] All these methods are acceptable as long as the chosen option is described clearly whenever data are to be presented. The time-dependent differences that exist among strains and stocks for many events in development[48,64] require that the chosen staging scheme be confirmed for the particular genetic background on which the study is being performed. To promote technical proficiency among research personnel, we recommend that each laboratory choose only one method for staging, which will be employed in all developmental pathology studies.

Husbandry

A brief but detailed statement to outline major environmental conditions and management practices should be included in study documentation. Items that should be reported to facilitate repetition of the experiment (in either the original laboratory or by an independent institution) are listed in Table 6.1.

When performing developmental pathology studies in mice, the key critical factors to standardize are the breeding strategy, caging conditions, and the length of the mating period.

One factor regarding the animal model deserves special consideration when designing mouse developmental pathology studies, and that is the breeding strategy. In most cases, naturally bred mice will exhibit a strain-dependent litter size, which averages between 5 and 11 conceptuses.[56] This fact is important because the growth of a conceptus throughout gestation and during early life is dictated not only by its own genotype (or *nature*) but also by the amount of intra-uterine competition afforded by its littermates (*nurture*).[20,42,45] In general, the size of litters is reduced in highly inbred mouse strains relative to outbred stocks,[2,56,64] ostensibly because more complete genetic homogeneity reduces hybrid vigor and thereby hampers embryonic survival.[40] In our experience, the litters resulting from natural mating do not have any reproducible pattern of lesions associated with the chosen breeding paradigm. In contrast, litters resulting from superovulation (including transfer of 25–30 embryos into pseudo-pregnant females) often yield resorption rates of 40%–60%.[44] This attrition is to be expected since the high number of implanted embryos results in an excessive number of implantation sites relative to the usual strain-specific number that are accommodated during a normal pregnancy. The proposed reason for the large resorption rate in this scenario is a nutritional deficiency that leads to pruning of sites at suboptimal locations (i.e., located at a distance from major uterine vasculature branches). Even if the number of resorption sites is not increased, embryos in large superovulated litters tend to have chronological ages (based on the vaginal plug-positive day) that exceed the apparent developmental age (as assigned using external anatomic landmarks), and fetuses usually exhibit reductions in both length and weight. These indications of developmental delay based on the experimental design may mask genuine cases of tardy *in utero* progression that might be attributed to a faulty gene or toxicant exposure.

The degree of divergence among developmental events in littermates is influenced by the breeding schedule. For example, variability in body weight is reduced by limiting the contact between the breeding pair to the last 2 h of the dark cycle rather than using a longer (overnight or continuous) cohabitation period.[19,27,72] Timed mating does require more labor and so is not utilized for screening studies. However, restricted mating represents an important means of decreasing interlitter variability in development for experiments in which the important endpoints are confined to precise developmental stages.

Another aspect related to the breeding strategy is the requirement that maternal stress be minimized if a viable litter is to be obtained at all. Breeding conditions associated with cage crowding may yield significant changes in the sex ratio within their litters.[32] Pregnant mice that are shipped to another facility during pregnancy may be found at necropsy to have no or very few implantation sites, presumably because the spike in corticosterone associated with shipping stress kills blastocysts or interferes with their attachment. A related phenomenon attributed to a stress response is the tendency for some dams to kill and cannibalize their litter if the room holding their cage is entered too often in the perinatal period.[56] These preventable causes of lethality may be mitigated effectively by adopting management practices that ameliorate maternal stress during the breeding, incubation, and delivery phases. The breeding schedule must be arranged so that all mating pairs have an opportunity to pair during the same phase of the estrous cycle since early or delayed copulation also may alter the sex ratio in the litter.[33]

Choosing Appropriate Study Controls

Interpretations regarding the biological impact of a novel mutation or a xenobiotic exposure typically depend on comparisons of manipulated animals with their *normal* counterparts (i.e., *controls*). Such comparisons are essential to discriminating between an outcome induced by some treatment and an event that occurs by chance. Mouse developmental pathology studies may succeed or fail depending on whether or not the proper control animals are picked to minimize or cancel the influence of inter-individual variability. In general, control data for mouse developmental pathology studies should be derived from concurrently treated groups rather than by extrapolation from archived historical data. This approach will ensure that the genetic background (especially as related to genetic drift of substrains from different colonies over time) and husbandry conditions of treated and control mice actually do exhibit a tight correspondence.

One potential confounder unique to the developmental pathology setting is the stage of the conceptus. Ideally, the mice in all treatment groups should be at the same stage of development when treatment is begun. If the treatment is to be undertaken *ex vivo* in younger rodent embryos (up to GD 12.5), the simplest means to ensure that individuals in all groups share the same developmental stage is to segregate them into groups based on recognizable external features (e.g., body conformation, number of somites, or the presence and/or shapes of digital rays on limb buds)[7,10,13,18,63]; this strategy supports the identification of multiple developmental stages for each of these early gestational days. In older embryos (GD 13 and beyond), the lack of distinctive external features typically results in sorting based on chronological age, with divisions being made for a specific day or brief range of days during gestation.[18,58]

Other possible confounding factors are comparable to those in other biological tests. The most important of these is to control for the primary experimental manipulation (i.e., *treatment*). In genetic experiments, the optimal controls for mutant conceptuses—whether transgenic or nullizygous (i.e., homozygous null or *knockout* [KO or −/−])—are offspring that share an identical genetic background (i.e., strain) but lack the mutated allele (i.e., *wild-type* [WT or +/+]). These genotypes ideally are obtained using littermates from one or a few pregnancies. In some cases, the only available control animals with a similar genetic background are heterozygous individuals (HET or +/−) having one WT and one KO allele or even animals from a related background strain (and raised under similar conditions). However, these less precise control groups often exhibit greater variance in biological responsiveness than would a genetically identical cohort, which thus may necessitate an increase in the group sizes to discern a difference between the experimental and control groups. In toxicity assays, the best negative controls are progeny of vehicle-treated dams because these individuals encountered the identical treatment-related factors, like handling and vehicle exposure, which were experienced by litters carried in toxicant-treated dams; an untreated group is less suitable as a negative control since the dam was not subjected to all these treatment-linked stressors. In certain instances, the study design also may require inclusion of a positive control group (i.e., a cohort in which pregnant dams receive a known teratogen) to ensure that the test methods are sensitive enough to detect subtle changes in target organs and parameters of interest. Multiple control groups may be needed in experiments where combinations of factors are being assessed simultaneously. Common scenarios in which multiple factors are being evaluated together include studies in which the conceptuses of interest will carry two gene mutations, the dams will be exposed concurrently to two chemicals, or the combined effect of a mutation and a toxicant are to be assessed. In such multifactorial studies, additional control groups are required to examine the impact of each mutation and/or each toxicant independently (Table 6.1).

Other factors may need to be controlled as well. Examples include the sex of the animal, intrauterine position,[38,69] strain-specific differences in implantation between uterine horns,[48] the age[44] and nutritional status[2] of the dam, and the timing of mating relative to the stage of the estrus cycle.[33] For example, male (XY) embryos develop more rapidly than female (XX) embryos during the preimplantation period[9,66] and retain this advantage later in gestation.[72] These variables often are downplayed in study designs for mouse developmental pathology studies—if they are addressed at all—unless they represent all or part of a hypothesis that is being tested.

Defining the Group Size

Biological experiments are interpreted most easily when experimental and control groups have equivalent sample sizes.[15,29,75] A common practice for mouse developmental pathology studies is to submit only one or two conceptuses of each genotype for analysis. While an understandable adaptation to conserve individuals with rare genotypes, this approach does not yield sufficient statistical power to definitively distinguish between experimentally induced outcomes and random happenings. A modest increase in group size, to five individuals, will distinguish genuine phenotypes 95% of the time (i.e., $p \leq 0.05$) with only a 10% likelihood of producing a false-negative result, if the background incidence of the observed lesion is 1% or less. An expanded consideration of statistical factors that impact the design of mouse developmental pathology studies is presented in Chapter 18.

Setting the Study Methods

The analytical techniques to be employed for a given study will depend to a large extent on the information desired by the researchers who design the experiment. In some settings, studies are conducted chiefly to analyze in detail the spectrum of morphological lesions that are induced in all organs during a given stage of development. Other teams of investigators may be interested in elucidating the entire sequence of molecular changes that culminate in a specific lesion within one particular organ. These divergent interests define two strategies for analysis that may be chosen when designing a mouse developmental pathology study (Table 6.2).

The first tier is designed to screen for potential developmental phenotypes. This approach is appropriate for the examination (described earlier) of near-term fetuses to determine what time during gestation a follow-up study should be scheduled and also may serve to obtain the complete data set for regulatory-type toxicity studies that are undertaken to identify developmental toxicants. Surveys of this type generally are dedicated to cataloging such overt demonstrations of developmental disruption as increased incidences of implantation site resorption, macroscopic malformation, and sometimes microscopic abnormalities. The purpose of a tier I screen is to provide a basic description of the lesions induced in developing mice by a particular genetic alteration or toxicant exposure.

The second tier of developmental pathology studies is designed to deliver a more in-depth understanding of a previously identified phenotype. This approach constitutes an extensive investigation that ranges from confirming the basic lesion spectrum[12] to defining the pathogenesis for each lesion in terms of its biochemical/molecular basis.[1,23,34,37,50,57] Extended studies of this sort often also include functional assessments as well (e.g., behavioral testing[5,14]), which affords the opportunity for correlating the presence of the lesion to a particular developmental outcome. These molecular and functional tests may be undertaken on samples derived from the isolated embryo or fetus, but they also may utilize isolated organs or cells as substrates for analysis.

Requirements for tissue processing, and organ preservation in particular, will differ based on the desired study endpoints. In general, tier I experiments may be undertaken using specimens that have been prepared in a fixative that hardens tissues, which will allow even fragile embryos to withstand the modest amount of manipulation that must be done to complete their analysis. Bouin's solution or modified Davidson's solution are commonly used fixatives that serve this purpose. In contrast, tier II studies usually seek to evaluate both structural integrity and *in situ* molecular expression and so must utilize a gentler means of tissue preservation. Formaldehyde and paraformaldehyde represent the typical choices for this strategy. A more detailed description of fixative options and their relative advantages and disadvantages when used in mouse developmental pathology studies may be found in Chapter 9.

TABLE 6.2

Tiered Strategy for the Analysis of Developmental Phenotypes

- Tier I: Screening
  - *Purpose*: Basic description of the phenotype(s) elicited in a novel engineered construct
  - *Subjects*: Near-term fetuses (GD 17–GD 18.5) and placentae
  - *Endpoints*: Clinical observations (maternal and embryonic), macroscopic and microscopic anatomy, and (sometimes) clinical chemistry and/or hematology
- Tier II: Mechanistic studies
  - *Purpose*: Detailed characterization of the pathogenesis that produces a given phenotype
  - *Subjects*: Depend on the phenotype (likely will include both early and late embryos, with associated placentae)
  - *Endpoints*: Gross and microscopic anatomy, selected clinical chemistry and hematology tests, *in situ* molecular assays, functional tests *in vitro* (cells, isolated organs, whole mounts) and *in vivo* (heart rate, blood flow), etc.

Consequences of Poor Study Design

The study design and associated statistical analyses determine whether or not a mouse developmental pathology study has provided an adequate evaluation of the hypothesis for which the study was designed. Successful publication of a study does not automatically indicate that it was well designed or properly implemented, or that its conclusions might be considered valid. Mouse developmental pathology reports in the literature, including a number released in high-impact journals not specifically dedicated to developmental investigations, often include terse descriptions of the study design insufficient to permit replication of the experiment by independent laboratories and/or draw conclusions regarding the spectrum and pathogenesis of a potential phenotype based on a limited amount of evidence. It is equally unfortunate that articles with inadequacies in the study design are cited regularly by noncritical readers as justification for poorly conceived hypotheses or follow-on experiments. In this fashion, bad experimental designs and data sets may be perpetuated, thereby substantially retarding our collective understanding of important biological questions.

Properly designed mouse developmental pathology studies are constructed readily by appropriate attention to detail before initiating the experiment. The surest and best means of avoiding design flaws is to consult *in advance* with subject matter experts who have ample prior experience so that complex factors (e.g., statistical considerations, tissue acquisition, and processing protocols) may be optimized. A similar expert review, or better yet inclusion of comparative pathologists or mouse developmental pathologists as co-investigators/coauthors when appropriate, would quickly elevate the accuracy and utility of published articles in this field. Skill in designing studies is a crucial factor in learning to conduct an efficient and effective analysis, and as such should be a goal of all researchers engaged in mouse developmental pathology investigations.

REFERENCES

1. Asp J, Abramsson A, Betsholtz C (2006). Nonradioactive *in situ* hybridization on frozen sections and whole mounts. *Methods Mol Biol* **326**: 89–102.
2. Berry ML, Linder CC (2007). Breeding systems: Considerations, genetic fundamentals, genetic background, and strain types. In: *The Mouse in Biomedical Research*, 2nd edn. Vol. 1: History, Wild Mice, and Genetics (Fox JG, Barthold SW, Davisson MT, Newcomer CE, Quimby FW, Smith AL, eds.). San Diego, CA: Academic Press (Elsevier), pp. 53–78.
3. Biddle FG (1988). Genetic differences in the frequency of acetazolamide-induced ectrodactyly in the mouse exhibit directional dominance of relative embryonic resistance. *Teratology* **37** (4): 375–388.
4. Boehm B, Rautschka M, Quintana L, Raspopovic J, Jan Z, Sharpe J (2011). A landmark-free morphometric staging system for the mouse limb bud. *Development* **138** (6): 1227–1234.
5. Bolon B, St. Omer VE (1989). Behavioral and developmental effects in suckling mice following maternal exposure to the mycotoxin secalonic acid D. *Pharmacol Biochem Behav* **34** (2): 229–236.
6. Brown NA (1990). Routine assessment of morphology and growth: Scoring systems and measurements of size. In: *Postimplantation Mammalian Embryos: A Practical Approach* (Copp AJ, Cockroft DL, eds.). Oxford, U.K.: IRL Press, pp. 93–108.
7. Brown NA, Fabro S (1981). Quantitation of rat embryonic development *in vitro*: A morphological scoring system. *Teratology* **24** (1): 65–78.
8. Brown SD, Zurakowski D, Rodriguez DP, Dunning PS, Hurley RJ, Taylor GA (2006). Ultrasound diagnosis of mouse pregnancy and gestational staging. *Comp Med* **56** (4): 262–271.
9. Burgoyne PS (1993). A Y-chromosomal effect on blastocyst cell number in mice. *Development* **117** (1): 341–345.
10. Butler H, Juurlink BHJ (1987). *An Atlas for Staging Mammalian and Chick Embryos*. Boca Raton, FL: CRC Press.
11. Chang CP, Chen L, Crabtree GR (2003). Sonographic staging of the developmental status of mouse embryos *in utero*. *Genesis* **36** (1): 7–11.

12. Cleary JO, Modat M, Norris FC, Price AN, Jayakody SA, Martinez-Barbera JP, Greene ND et al. (2011). Magnetic resonance virtual histology for embryos: 3D atlases for automated high-throughput phenotyping. *NeuroImage* **54** (2): 769–778.
13. Copp AJ, Cockroft DL (1990). *Postimplantation Mammalian Embryos: A Practical Approach.* Oxford, U.K.: IRL Press.
14. Crawley JN (2007). *What's Wrong with My Mouse? Behavioral Phenotyping of Transgenic and Knockout Mice*, 2nd edn. New York: Wiley-Liss.
15. Daniel WW (2009). *Biostatistics: A Foundation for Analysis in the Health Sciences*, 9th edn. Hoboken, NJ: John Wiley & Sons.
16. Downs KM, Davies T (1993). Staging of gastrulating mouse embryos by morphological landmarks in the dissecting microscope. *Development* **118** (4): 1255–1266.
17. Drews U (1995). *Color Atlas of Embryology.* New York: Thieme Medical Publishers.
18. EMAP (Electronic Mouse Atlas Project) (2013). e-Mouse Atlas, v3.5. http://www.emouseatlas.org/emap/home.html (last accessed November 1, 2014).
19. Endo A, Watanabe T (1988). Interlitter variability in fetal body weight in mouse offspring from continuous, overnight, and short-period matings. *Teratology* **37** (1): 63–67.
20. Fekete E (1954). Gain in weight of pregnant mice in relation to litter size. *J Hered* **45** (2): 88–89.
21. Goedbloed JF, Smits-van Prooije AE (1986). Quantitative analysis of the temporal pattern of somite formation in the mouse and rat. A simple and accurate method for age determination. *Acta Anat (Basel)* **125** (2): 76–82.
22. Greco A, Ragucci M, Coda AR, Rosa A, Gargiulo S, Liuzzi R, Gramanzini M et al. (2013). High frequency ultrasound for *in vivo* pregnancy diagnosis and staging of placental and fetal development in mice. *PLoS ONE* **8** (10): e77205 (7 pages).
23. Hargrave M, Bowles J, Koopman P (2006). *In situ* hybridization of whole-mount embryos. *Methods Mol Biol* **326**: 103–113.
24. Hide T, Hatakeyama J, Kimura-Yoshida C, Tian E, Takeda N, Ushio Y, Shiroishi T, Aizawa S, Matsuo I (2002). Genetic modifiers of otocephalic phenotypes in *Otx2* heterozygous mutant mice. *Development* **129** (18): 4347–4357.
25. Hill MA (2014). Mouse development. http://php.med.unsw.edu.au/embryology/index.php?title=Mouse_Development (last accessed November 1, 2014).
26. Hill MA (2014). Rat development stages (Witschi). http://embryology.med.unsw.edu.au/embryology/index.php?title=Rat_Development_Stages (last accessed November 1, 2014).
27. Holson JF, Scott WJ, Gaylor DW, Wilson JG (1976). Reduced interlitter variability in rats resulting from a restricted mating period, and reassessment of the "litter effect". *Teratology* **14** (2): 135–141.
28. Johnen H, Gonzalez-Silva L, Carramolino L, Flores JM, Torres M, Salvador JM (2013). Gadd45g is essential for primary sex determination, male fertility and testis development. *PLoS ONE* **8** (3): e58751 (7 pages).
29. Julien E, Willhite CC, Richard AM, Desesso JM (2004). Challenges in constructing statistically based structure-activity relationship models for developmental toxicity. *Birth Defects Res A Clin Mol Teratol* **70** (12): 902–911.
30. Kaufman MH (1992). *The Atlas of Mouse Development*, 2nd edn. San Diego, CA: Academic Press.
31. Klinger S, Turgeon B, Levesque K, Wood GA, Aagaard-Tillery KM, Meloche S (2009). Loss of Erk3 function in mice leads to intrauterine growth restriction, pulmonary immaturity, and neonatal lethality. *Proc Natl Acad Sci USA* **106** (39): 16710–16715.
32. Krackow S (1997). Effects of mating dynamics and crowding on sex ratio variance in mice. *J Reprod Fertil* **110** (1): 87–90.
33. Krackow S, Burgoyne PS (1997). Timing of mating, developmental asynchrony and the sex ratio in mice. *Physiol Behav* **63** (1): 81–84.
34. Lazik A, Liu Y, Bringas P, Sangiorgi F, Maxson R (1996). A sensitive method for analyzing β-galactosidase reporter gene expression in tissue sections of mouse embryos. *Trends Genet* **12** (11): 445–447.
35. Liang HC, Li H, McKinnon RA, Duffy JJ, Potter SS, Puga A, Nebert DW (1996). *Cyp1a2*(−/−) null mutant mice develop normally but show deficient drug metabolism. *Proc Natl Acad Sci USA* **93** (4): 1671–1676.
36. Liljander M, Andersson A, Holmdahl R, Mattsson R (2009). Increased litter size and super-ovulation rate in congenic C57BL mice carrying a polymorphic fragment of NFR/N origin at the *Fecq4* locus of chromosome 9. *Genet Res* **91** (4): 259–265.

37. Linask KK, Tsuda T (2000). Application of plastic embedding for sectioning whole-mount immuno-stained early vertebrate embryos. *Methods Mol Biol* **135**: 165–173.
38. Louton T, Domarus H, Hartmann P (1988). The position effect in mice on day 19. *Teratology* **38** (1): 67–74.
39. Marcela SG, Cristina RM, Angel PG, Manuel AM, Sofia DC, Patricia de LR, Bladimir RR, Concepcion SG (2012). Chronological and morphological study of heart development in the rat. *Anat Rec (Hoboken)* **295** (8): 1267–1290.
40. McCarthy JC (1965). The effect on litter size of crossing inbred strains of mice. *Genetics* **51** (2): 217–222.
41. McCarthy JC (1967). The effects of inbreeding on the components of litter size in mice. *Genet Res* **10** (1): 73–80.
42. McCarthy JC (1967). Effects of litter size and maternal weight on foetal and placental weight in mice. *J Reprod Fertil* **14** (3): 507–510.
43. McClive PJ, Sinclair AH (2001). Rapid DNA extraction and PCR-sexing of mouse embryos. *Mol Reprod Dev* **60** (2): 225–226.
44. McLaren A, Michie D (1959). Superpregnancy in the mouse. 1. Implantation and foetal mortality after induced superovulation in females of various ages. *J Exp Biol* **36** (2): 281–300.
45. McLaren A, Michie D (1959). Superpregnancy in the mouse. 2. Weight gain during pregnancy. *J Exp Biol* **36** (2): 301–316.
46. Miyake T, Cameron AM, Hall BK (1996). Detailed staging of inbred C57BL/6 mice between Theiler's [1972] stages 18 and 21 (11–13 days of gestation) based on craniofacial development. *J Craniofacial Genet Dev Biol* **16** (1): 1–31.
47. Miyake T, Cameron AM, Hall BK (1997). Stage-specific expression patterns of alkaline phosphatase during development of the first arch skeleton in inbred C57BL/6 mouse embryos. *J Anat* **190** (Pt 2): 239–260.
48. Miyake T, Cameron AM, Hall BK (1997). Variability of embryonic development among three inbred strains of mice. *Growth Dev Aging* **61** (3–4): 141–155.
49. Mori N, Tsugane MH, Yamashita K, Ikuta Y, Yasuda M (2000). Pathogenesis of retinoic acid-induced abnormal pad patterns on mouse volar skin. *Teratology* **62** (4): 181–188.
50. Neidhardt L, Gasca S, Wertz K, Obermayr F, Worpenberg S, Lehrach H, Herrmann BG (2000). Large-scale screen for genes controlling mammalian embryogenesis, using high-throughput gene expression analysis in mouse embryos. *Mech Dev* **98** (1–2): 77–94.
51. O'Rahilly R, Müller F (1987). *Developmental Stages in Human Embryos*. Washington, DC: Carnegie Institute.
52. Papaioannou VE, Behringer RR (2012). Early embryonic lethality in genetically engineered mice: Diagnosis and phenotypic analysis. *Vet Pathol* **49** (1): 64–70.
53. Patton JT, Kaufman MH (1995). The timing of ossification of the limb bones, and growth rates of various long bones of the fore and hind limbs of the prenatal and early postnatal laboratory mouse. *J Anat* **186** (1): 175–185.
54. Peterka M, Lesot H, Peterková R (2002). Body weight in mouse embryos specifies staging of tooth development. *Connect Tiss Res* **43** (2–3): 186–190.
55. Pineau T, Fernandez Salgueuro P, Lee SST, McPhail T, Ward JM, Gonzalez FJ (1995). Neonatal lethality associated with respiratory distress in mice lacking cytochrome P450 1A2. *Proc Natl Acad Sci USA* **92** (11): 5134–5138.
56. Pritchett KR, Taft RA (2007). Reproductive biology of the laboratory mouse. In: *The Mouse in Biomedical Research*, 2nd edn. Vol. 3: *Normative Biology, Husbandry, and Models* (Fox JG, Barthold SW, Davisson MT, Newcomer CE, Quimby FW, Smith AL, eds.). San Diego, CA: Academic Press (Elsevier), pp. 91–121.
57. Richardson L, Venkataraman S, Stevenson P, Yang Y, Moss J, Graham L, Burton N, Hill B, Rao J, Baldock RA, Armit C (2014). EMAGE mouse embryo spatial gene expression database: 2014 Update. *Nucleic Acids Res* **42** (D1): D835–D844.
58. Rugh R (1990). *The Mouse: Its Reproduction and Development*. Oxford, U.K.: Oxford University Press.
59. Sauvageau M, Goff LA, Lodato S, Bonev B, Groff AF, Gerhardinger C, Sanchez-Gomez DB et al. (2013). Multiple knockout mouse models reveal lincRNAs are required for life and brain development. *eLife* **2**: e01749 (24 pages).
60. Simpson EM, Linder CC, Sargent EE, Davisson MT, Mobraaten LE, Sharp JJ (1997). Genetic variation among 129 substrains and its importance for targeted mutagenesis in mice. *Nature Genet* **16** (1): 19–27.

61. Steding G (2009). *The Anatomy of the Human Embryo: A Scanning Electron-Microscopic Atlas.* Basel, Switzerland: Karger.

62. Theiler K (1972). *The House Mouse: Development and Normal Stages from Fertilization to 4 Weeks of Age.* Berlin, Germany: Springer-Verlag.

63. Theiler K (1989). *The House Mouse: Atlas of Embryonic Development.* New York: Springer-Verlag.

64. Thiel R, Chahoud I, Jürgens M, Neubert D (1993). Time-dependent differences in the development of somites of four different mouse strains. *Teratogen Carcinogen Mutagen* **13** (6): 247–257.

65. Threadgill DW, Dlugosz AA, Hansen LA, Tennenbaum T, Lichti U, Yee D, LaMantia C et al. (1995). Targeted disruption of mouse EGF receptor: Effect of genetic background on mutant phenotype. *Science* **269** (5221): 230–234.

66. Tsunoda Y, Tokunaga T, Sugie T (1985). Altered sex ratio of live young after transfer of fast- and slow-developing mouse embryos. *Gamete Res* **12** (3): 301–304.

67. van Maele-Fabry G, Delhaise F, Picard JJ (1990). Morphogenesis and quantification of the development of post-implantation mouse embryos. *Toxic In Vitro* **4** (2): 149–156.

68. van Maele-Fabry G, Delhaise F, Picard JJ (1992). Evolution of the developmental scores of sixteen morphological features in mouse embryos displaying 0 to 30 somites. *Int J Dev Biol* **36** (1): 161–167.

69. von Domarus H, Louton T, Lange-Wuhlisch F (1986). The position effect in mice on day 14. *Teratology* **34** (1): 73–80.

70. Wanek N, Muneoka K, Holler-Dinsmore G, Burton R, Bryant SV (1989). A staging system for mouse limb development. *J Exp Zool* **249** (1): 41–49.

71. Ward JM, Elmore SA, Foley JF (2012). Pathology methods for the evaluation of embryonic and perinatal developmental defects and lethality in genetically engineered mice. *Vet Pathol* **49** (1): 71–84.

72. Watanabe T, Endo A (1988). Digit development and embryonic weight in mice: Analysis of sex-related time difference and mating period-related interlitter variability. *Teratology* **38** (2): 157–163.

73. Wendling O, Teletin M, Ghyselinck NB, Mark M (2011). Une procédure dédiée au phénotypage de souris porteuses de mutations ciblées, létales in utero ou à la naissance (A procedure dedicated to the phenotyping of mice carrying targeted mutations, lethal *in utero* or at birth) [original in French]. *Rev Fr Histotechnol* **24** (1): 47–57.

74. Witschi E (1962). Development: Rat. In: *Growth Including Reproduction and Morphological Development* (ltman PL, Dittmer DS, eds.). Washington, DC: Federation of American Societies for Experimental Biology, pp. 304–314.

75. Zar JH (2010). *Biostatistical Analysis.* Englewood Cliffs, NJ: Prentice-Hall.

76. Zeiss CJ, Ward JM, Allore HG (2012). Designing phenotyping studies for genetically engineered mice. *Vet Pathol* **49** (1): 24–31.

7

Necropsy Sampling and Data Collection for Studying the Anatomy, Histology, and Pathology of Mouse Development

Susan Newbigging, Jerrold M. Ward, and Brad Bolon

CONTENTS

The analysis of lethal phenotypes in developing mice usually begins with the evaluation of readily identifiable features related to the individual animal, its littermates, and the dam. The nature of the parameters to be assessed and the procedures used during the examination will vary depending on both the developmental age of the presumed lethal event and the chronological age of the animals. In our experience, an optimal macroscopic analysis should involve the acquisition of qualitative and selected quantitative data. Certain endpoints and investigational methods are common to all developmental stages, while others are best suited to animals of a particular age. Techniques for specimen collection and macroscopic analysis are flexible and, thus, may be adapted to suit the needs of individual laboratories and specific research questions. That said, due care must be taken to plan the examination in advance so that samples and quantitative measurements are acquired in an appropriate order and using methods designed to minimize the time between tissue harvesting and tissue preservation. Such planning is necessary because any gross defects in structure and/or size often may be explained only by a subsequent microscopic assessment or molecular analysis.

This chapter will introduce standard techniques for isolating tissue samples of sufficient quality to obtain usable macroscopic data sets from mouse developmental pathology studies while ensuring that microscopic and molecular endpoints may be obtained after additional processing of the specimens. Further details regarding necropsy (animal autopsy[26]) methods for embryonic, fetal, and neonatal mice may be found in Chapter 17 of this volume as well as selected review articles[7,8,36,45] and reference books[14,33,35,46] that discuss the gross analysis of developing animals with lesions induced by a mutation or known toxicant exposure.

Sample Collection Practices during the Prenatal Period

The procurement of specimens for evaluation of embryonic lethality requires a flexible stepwise approach. The first stage, removal of the female reproductive tract, is undertaken in a standard fashion for all stages of gestation. Procedures for this first stage are given in the next subsection. Methods used to examine young animals will depend on the developmental stage during which the individuals are to be assessed. Therefore, a menu of options needed to remove and evaluate developing mice at different stages follows in other subsections. The exact protocol used for any given mouse developmental pathology study will depend on the specific parameters of interest. For larger organs, an initial macroscopic evaluation may be useful, conducted using either the unaided eye or with the advantage of magnification (often a stereomicroscope or illuminated binocular headband magnifier, both of which are available commercially from many vendors). For smaller organs and younger animals, magnification at the macroscopic level is insufficient for analysis, so the better method is to process and embed the intact specimen and then conduct the examination on one or more histologic sections.

The most critical point to remember when specimens are to be obtained from developing mice is that the degree of advance preparation can make or break an entire experiment. All necessary instruments and reagents should be available at the necropsy station before the procedure is initiated. These materials need to be arranged in a fashion that provides a convenient work flow for the prosector (Figure 7.1). Inattention to detail when preparing the theater of operations may result in delays at one or more steps during the isolation and tissue sampling portions of the experiment. Short delays (a minute or two) usually are of little consequence to the quality of the final specimens, but extended interruptions (to search for missing instruments, prewarm fluids, etc.) may impact the preservation of more delicate organs and cell populations, especially when dealing with embryos younger than gestational day (GD) 9.5. The immediate availability of fixative solutions is particularly vital as poor tissue preservation (especially if accompanied by a large degree of trauma related to tissue removal) is a common reason for the failure to identify the mechanism of embryonic lethal phenotypes.

Removal of the Maternal Reproductive Tract

The first step in harvesting mouse conceptuses during gestation is to isolate the dam's reproductive tract. This process typically is undertaken after the mother has been humanely euthanized using a method approved by one's institutional animal care and use committee (ACUC). Commonly employed, acceptable means of euthanizing pregnant mice include carbon dioxide (CO_2) inhalation (supplied by gas canister and not sublimation from dry ice bricks), cervical dislocation, decapitation, and parenteral injection of an anesthetic agent.[2] The first three methods are used when acquisition of specimens uncontaminated by exogenous chemicals is necessary. For CO_2 inhalation, the dam should be lowered into a closed chamber that is gradually filled with CO_2; placement into a prefilled chamber (i.e., 100% CO_2) is extremely stressful and thus considered to be an unethical practice.[2] The key point for both cervical dislocation and decapitation is that their use should be limited to personnel who have been trained to be proficient in the techniques. The choice of parenteral anesthesia (e.g., 2,2-tribromoethanol [Avertin®] given by intraperitoneal injection as two doses, separated by 2 min, of 250 mg/kg [1.0 mL for a 50 g near-term dam]) is

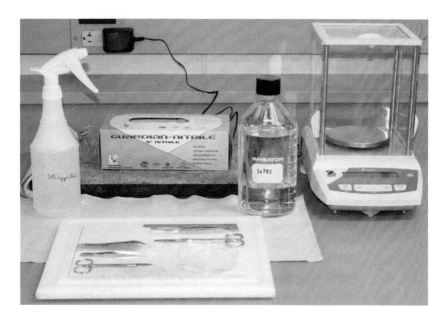

FIGURE 7.1 Arrangement of a necropsy station suitable for extracting the gravid uterus from pregnant mice. Starting at the foreground and progressing in clockwise fashion, the station is equipped with a hard plastic cutting board holding two sets of surgical instruments (scissors, rat-toothed forceps, and serrated forceps), a plastic Petri dish, and a digital caliper; a spray bottle filled with 70% ethanol (to wet the fur); another board with cork surface (under the box of gloves); a bottle of PBS (pH 7.4); and an analytical balance (for weighing the conceptuses). (Image courtesy of Dr. Kelli Boyd, Vanderbilt University.)

a good choice to assuage both pain in the mouse and distress to research staff. Isoflurane inhalation is an acceptable alternative if administered at a level (typically 2% or higher) sufficient to achieve a surgical plane of anesthesia. Reproductive tract removal may start once the female is dead (approximately 5 min after the second parenteral anesthetic dose and within 15 s for the other three methods). Death (or a surgical plane of unconsciousness if isoflurane is being used) should be confirmed prior to exposing the uterus. This corroboration is obtained easily by pinching the skin web between two hind paw digits (where failure to retract the limb shows that pain is not being perceived). Some investigators prefer to use two euthanasia procedures in series (e.g., injected anesthetic followed by cervical dislocation) to ensure that the dam has expired.

For prenatal investigations, the reproductive tract must be exposed to gain access to the developing conceptuses (Figure 7.2). The dam is placed in dorsal recumbency (i.e., on its back). Pins may be pushed through all four paws into a cork-faced dissection board to secure the body during the necropsy, but immobilization of this kind is optional. The abdominal fur is wet liberally with 70% ethanol to minimize hair contamination of the exposed organs. A forceps is used to grasp and elevate the ventral abdominal wall so that a pair of scissors may be used to introduce a longitudinal incision along the ventral midline, starting from the pelvis and extending to the sternum. Two additional incisions are made, one on each side, starting at the midline opening and proceeding laterally and dorsally along the diaphragm. These latter two incisions will form two flaps of abdominal wall that may be reflected to either side to expose the abdominal organs. The diaphragm should be punctured at this time to allow air to enter the thoracic cavity as this breech will ensure that the dam cannot revive. The gravid uterus may be identified readily in the caudal abdomen by its bilateral position, multiple large swellings (i.e., implantation sites), and bright red wall (due to active engorgement [i.e., *physiologic congestion*] of the blood vessels). For very early studies in which embryos have not yet implanted in the uterus (i.e., GD 4.5 and earlier), the uterus typically will be appreciated as two small-caliber, tan or pale red horns that are anchored caudally where they converge on a single cervix but are restrained cranially and dorsolaterally by their individual connections to one ovary.

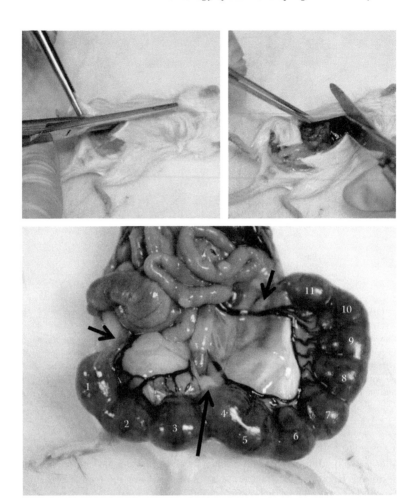

FIGURE 7.2   To expose the uterus, the euthanized dam is placed in dorsal recumbency on a cutting board and sprayed with 70% ethanol to wet the fur. A pair of forceps (rat-toothed or serrated) held in the nondominant hand is used to grasp and raise (*tent*) the skin of the ventral abdomen so that a midline incision may be made (left panel) toward the thorax using a large pair of sharp–sharp scissors. The incision should be extended forward into the thoracic cavity to ensure that the animal cannot revive. Two additional cuts are made along the diaphragm, reaching from the midline incision toward the vertebral column, so that the abdominal walls may be reflected (right panel). The intestines may be shifted forward to reveal the uterus, which appears as a large, red, bifurcated organ with multiple saccular swellings (each of which represents an implantation site). The white cervix (long arrow) functions as a convenient boundary between the left and right uterine horns. The uterus is anchored in the abdomen by the broad mesometrium (recognized as the translucent membrane harboring the very large, dark red, uterine vessels) and the oviduct (short arrows). The implantation sites (all viable) are numbered consecutively beginning near the right ovary. Gravid uterus at GD 13. (Images courtesy of Dr. Kelli Boyd, Vanderbilt University.)

The remaining steps in the necropsy process will be dictated by the developmental stage during which the investigation is being conducted. The options for different life phases are given in some detail below.

Isolation of Preimplantation Embryos (GD 0–GD 4.5)

In our experience, embryonic lethal phenotypes that occur prior to implantation usually are explored in laboratories with a special focus on developmental biology rather than as a service provided by a shared comparative pathology resource. This division of labor reflects the fact that developmental biologists working on early embryos are often experts in using the specialized methods needed to evaluate the undifferentiated cells that comprise embryos during these early stages of gestation; such techniques often are not available in conventional pathology service laboratories, while subtle dissimilarities among stem cells in early embryos often are completely indistinguishable or exhibit only minimal differences when viewed by routine pathology techniques. Nonetheless, scientists who participate in phenotypic analyses of developing mice at later stages of gestation also should be familiar with basic means for obtaining and evaluating embryos during the preimplantation period.

Preimplantation embryos are isolated by flushing them from the reproductive tract. This process is undertaken using a stereomicroscope. The portion of the tract to be processed will depend on the stage at which the embryos are to be examined. The isolated embryos may be evaluated for structural changes immediately after collection (i.e., while immersed in buffer), after fixation and processing into paraffin for routine histologic sectioning (see Chapter 9 for processing details), or serially over time by growing them in a suitable nutrient medium for several days. The procedures used for sterile collection and culture of mouse embryos are beyond the scope of this brief chapter, but comprehensive protocols for acquiring them and evaluating their structure are available in the literature.[33,35] Such assessments are centered on analyzing cell numbers, sizes, and cytoarchitectural features. Common morphological traits indicative of early embryonic injury include fewer cells than anticipated for a given time after conception (i.e., a developmental delay), reduced cell size, cytoplasmic vacuolation (i.e., degeneration), or cell fragmentation (i.e., death) as well as decreased embryonic size as a whole.[36] These characteristics are examined using a microscope, most commonly with either specially equipped bright-field (Figure 7.3) or differential interference contrast (DIC) instrumentation.

To retrieve zygotes (one-celled embryos; about GD 0.5) up to early morulae (compact multicelled embryos; about GD 2.5), both oviducts and the first 1 cm of their associated uterine horns are detached and placed in a Petri dish containing phosphate-buffered saline (PBS, pH 7.4). The outer surface of each oviduct is grasped with a pair of fine pointed forceps (e.g., Dumont #5 jeweler's forceps, available

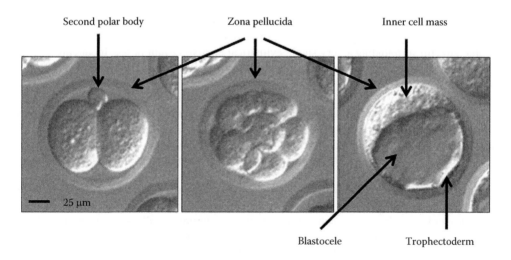

FIGURE 7.3 Photomicrographs of a preimplantation wild-type mouse embryo during the first 3.5 days after fertilization demonstrating the normal morphological appearance at different developmental stages. The two-cell embryo (left panel) develops by GD 1.0 via completion of the first cleavage and extrusion of the second polar body (a nucleus with little cytoplasm, arising from the oocyte during meiosis). The *zona pellucida* is a glycoprotein shell secreted by the oocyte and follicular cells that serves as a binding site for spermatozoa. The eight-cell embryo (middle panel) is found by GD 2.0 after two additional rounds of cleavage, and the blastocyst (right panel) is formed by GD 3.5 following another two to three divisions. The earliest cell lineage specification results in the formation of the inner cell mass (primordial animal) and trophectoderm (nascent placenta). This embryo was cultured *in vitro* for 3 days, and images were captured by time-lapse video microscopy using a bright-field light microscope equipped with a Hoffman objective. (From Marikawa, Y. and Alarcón, V.: Establishment of trophectoderm and inner cell mass lineages in the mouse embryo. *Mol. Reprod. Dev.* 2009. 76. 1019–1032. Copyright Wiley-VCH Verlag GmbH & Co. KGaA. Reproduced with permission.)

commercially from many vendors) so that the proximal end of the oviduct slides over a hypodermic needle; a 30-gauge or 32-gauge hypodermic needle that has had the tip polished by grinding it against an abrasive stone or fine sandpaper is suitable for this purpose. Gently press the tip of the needle (inside the oviduct) against the floor of the Petri dish (to prevent the oviduct from slipping off) and then flush the oviduct with 0.1–0.2 mL of PBS. The embryos may be collected from the PBS using a pipette and then transferred onto a glass slide and air-dried to make a cytological preparation. Alternatively, the embryos may be placed in a microfuge tube and concentrated using a brief period of low-speed centrifugation. The cell-free supernatant is removed by gentle pipetting, after which the pellet of embryos is moved to one chamber of a 24-well tissue culture plate (Multiwell™, available from many vendors) and covered with 1% agar (as described in detail in Chapter 9). The agar-covered embryos may be fixed by immersion in a suitable solution (usually neutral buffered 10% formalin [NBF, i.e., 3.7% formaldehyde, usually combined with a stabilizing agent like 1%–5% methanol] or 4% paraformaldehyde [PFA, i.e., 4% formaldehyde without any stabilizer]) and then embedded in paraffin to allow conventional histological sectioning.

A similar procedure is used to obtain blastocysts (multicelled, cystic embryos; about GD 3.5–GD 4.0).[33] For these larger preimplantation embryos, both uterine horns (with their attached oviducts) are removed in their entirety and placed in a PBS-filled Petri dish. The caudal end of each uterine horn is cut transversely just adjacent to the cervix to provide an exit portal for fluid. The cranial end of each horn is grasped with jeweler's forceps near the uterotubal junction, after which the tip of a polished 25-gauge or 26-gauge needle is inserted into the uterine lumen at an angle, with the tip directed toward the former cervical attachment. The uterotubal junction functions as a one-way valve, so the flushing solution and embryos will only move toward the cut end of the horn and not pass back into the oviduct. The horn is flushed with 0.2–0.25 mL of PBS, after which the blastocysts are processed for morphological evaluation as described earlier.

Isolation of Newly Implanted Embryos (GD 5.0–GD 7.0)

Developmental biologists and comparative pathologists often participate cooperatively in evaluations of embryonic lethal phenotypes that arise soon after implantation. In such cases, the implanted conceptuses consist chiefly of a nodule comprised of decidua (a modification of endometrial stroma that supports embryonic growth) surrounding a very small, cylindrical embryo (2 mm or less in length). This small size may preclude simple removal and analysis of the conceptuses, although very skilled practitioners can remove intact conceptuses from the uterus starting as early as GD 5.5.[33] This task may be undertaken using jeweler's forceps or very fine straight scissors to open the uterus and jeweler's forceps to sever the decidual attachments that secure the conceptuses to the uterine wall. However, in our experience, such delicate dissections require very fine motor control (and a willingness to forego caffeine and other stimulants!), so they typically should be performed only by very proficient technicians, and only to provide a more detailed characterization of a phenotype that was explored previously by means of an *in situ* microscopic analysis.

For screening purposes, the easiest approach to identifying and analyzing new embryonic lethal phenotypes during this period of development is to remove the entire reproductive tract as a single unit that includes the ovaries, oviducts, uterine horns, and cervix, and to process all the organs as a single specimen, the structures of which may be examined after suitable handling by evaluating stained histologic sections (Figure 7.4). The whole tract may be placed in a tissue cassette for fixation; inclusion of the ovaries permits the number of corpora lutea to be counted (Figure 7.5), the sum of which provides an estimate of how many ova were released. To maintain the optimal Y-shaped orientation, the specimen may be applied to a piece of index card; the stickiness of the serosal (outer) surface allows the tract to adhere to the card stock and retain its position during fixation. Certain positions in the uterine horns are prone to harbor conceptuses that exhibit lower intrauterine growth[27,43] or enhanced vulnerability to toxic agents,[4,17] so discrimination of the actual implantation sites with respect to both the horn (i.e., side of the dam) and the position within it often is appropriate. In such cases, left and right sides may be differentiated by removing one ovary, by tying a small ligature around one horn (near the cervix), or by placing a small mark on one oviduct or uterine horn using indelible histological ink. In our experience, use of a ligature is the least desirable scheme as it requires additional time and motor skill to deploy.

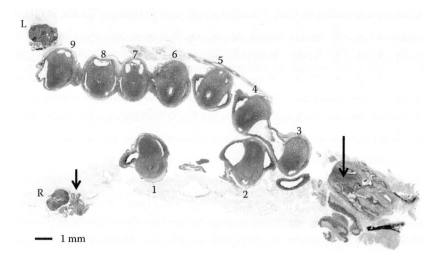

FIGURE 7.4 Photomicrograph demonstrating a histologic preparation that permits simultaneous assessment of many implantation sites *in situ* by embedding the entire reproductive tract in a single block. Each conceptus consists of a central, elongate, gastrulating embryo separated by a thin layer of trophectoderm (an embryonic derivative) from a thick basophilic cushion of decidua (derived from the endometrial stroma). The cervix (long arrow) functions as a convenient boundary between the left and right uterine horns. The implantation sites (all viable) are numbered consecutively beginning near the right (R) ovary. Additional annotation: L, left ovary; short arrow, oviduct. Gravid uterus at GD 7.0. Histologic processing: fixation in Bouin's solution, embedding in paraffin, staining with H&E.

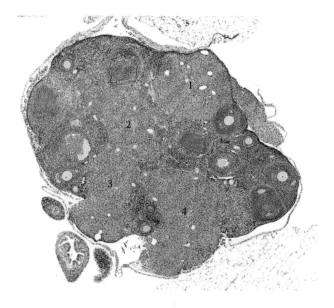

FIGURE 7.5 Photomicrograph from a pregnant mouse fetus (GD 8.5) showing four corpora lutea within the ovary. The number of corpora lutea in both ovaries should match the number of implantation sites if no preimplantation embryonic lethal phenotype is present. Processing: Bouin's, paraffin, H&E.

Isolation of Evolving Embryos (GD 7.5–GD 15.0) and Fetuses (GD 15.5–GD 19)

Skilled practitioners can remove intact embryos and yolk sacs from the uterus with reasonable facility between GD 7.5 and GD 9.5 using well-established techniques generally utilized for mouse whole embryo culture.[9,11,16,21,47] For researchers who perform embryo isolations infrequently, our experience has been that this activity is easier for conceptuses at a developmental age of GD 10.0 or older because the accumulation of fluid within the amniotic cavity improves the likelihood that the combination of elevated fluid pressure and relatively thick extra-embryonic membranes will protect the embryo during the isolation process.

Euthanasia

Considerations for humane handling of postimplantation embryos and fetuses vary with the gestational age. Embryos at GD 14.5 or younger usually may be chilled first and then isolated and immersed while intact in solutions needed to fix or extract tissue constituents. The reasons for this recommendation are that maternal euthanasia using routine methods (particularly CO_2 inhalation[34]) induces rapid systemic anoxia, including in tissues of the conceptus, and that embryos are incapable of registering pain due to their incompletely formed nociceptive circuits and the absence of myelin in the central nervous system (reviewed in Ref. 12). In contrast, older embryos and fetuses (GD 15.0 and older) appear capable of responding to some noxious stimuli with behaviors (e.g., limb withdrawal) consistent with perception of pain (reviewed in Refs. 12, 34), so they should be euthanized humanely before any invasive procedures are started. Acceptable methods for fetal termination include cervical dislocation, decapitation with a razor blade or scissors, or injectable anesthetic (including large doses given to the dam, as long as the fetus can actually be shown to be dead).[2,34] If desired, the animals may be rendered insensate using hypothermia, which can be achieved easily by placing the conceptus (i.e., fetus inside the intact amniotic sac) in ice-cold (4°C–8°C) PBS until it stops moving. A further means of assuring that death has occurred in hypothermia-anesthetized conceptuses, which may be used if it will not impact endpoints of interest, is to place a short linear incision in the lateral thoracic wall between two ribs to open the thoracic cavity; the resulting pneumothorax will lead rapidly to unconsciousness and death. If physical means of euthanasia are selected, personnel should be reminded that excessive force induced by violent compression (for cervical dislocation) or cutting (for decapitation) will traumatize small organs in the region (e.g., autonomic ganglia, larynx, and parathyroid and thyroid glands) to some degree, thereby reducing the likelihood or entirely preventing the possibility of any meaningful microscopic analysis.

Removal Using Scissors

Where feasible, the simplest and fastest means for retrieving the conceptuses is to use scissors. If necessary for the experimental design, care must be used at this point to obtain uncontaminated embryonic tissue (e.g., yolk sac or a portion of the body) for genotyping (as described in the following text). This technique may be performed without additional magnification and works readily for embryos at GD 10.0 and over. However, with care scissors may be used reliably from the start of organogenesis (approximately GD 8.0). The uterus is permitted to remain in the abdominal cavity, anchored over its entire length by the mesometrium as well as at its ends by its connections to the cervix and the two oviducts (Figure 7.2). A pair of forceps with fine serrated tips (typically held in the nondominant hand) is used to grasp the uterine wall at a point along the anti-mesometrial border; a useful starting point often is the approximate center, immediately across from the cervix. The forceps are used to pull the uterus caudally, thereby applying tension to the wall. One point of a pair of sharp/sharp scissors (held in the dominant hand) is inserted through the uterine wall at a location adjacent to an implantation site (Figure 7.6). The scissors point is angled out toward the endometrial wall rather than in toward the lumen and conceptuses, and the scissors arm is moved in a complete motion all the way along one uterine horn until the tip has moved past the last implantation site. The scissors need not be *worked* (i.e., no cutting motion is needed) since the stretched uterine wall parts easily when in contact with the blade. An identical incision is made

FIGURE 7.6 To remove older conceptuses (GD 10.0 or older) from the uterus, a pair of forceps (serrated) held in the nondominant hand is used to grasp and raise a portion of the anti-mesometrial uterine wall between two swellings (implantation sites). A pair of sharp–sharp scissors is used to cut the wall, taking care that the scissor tip inside the lumen is angled slightly outward so that the conceptus is not damaged. The incision should be extended to the proximal ends of both horns to expose all offspring. A blunt pair of scissors, flat pair of forceps, or metal spatula is used to tease the placenta away from the uterine wall, thereby detaching the conceptus. The isolated specimens are transferred to a Petri dish filled with PBS (if they are to be further dissected or weighed) or placed intact into fixative. Gravid uterus at GD 13. (Image courtesy of Dr. Kelli Boyd, Vanderbilt University.)

to open the second uterine horn. At this point, the scissors or another pair of forceps may be inserted between the uterine wall on the mesometrial side and the thickest portion of each conceptus's placenta so that it will pull away from its attachment site. The isolated conceptus then is moved to a separate Petri dish where it can be removed easily from the placenta, most of which is translucent and expanded by fluid at GD 10.0 and later. For convenience in matching the various parts of the conceptus, the embryo or fetus commonly is permitted to remain attached to the placenta by the umbilical cord (Figure 7.7) until the time comes for the external examination and tissue collection.

Removal Using Forceps

For greater control in extracting tiny embryos during early organogenesis (GD 8.0–GD 9.75), the dissection may be done with a stereomicroscope after the intact uterus has been removed from the abdominal cavity of the dam (Figure 7.8) and placed in a PBS-filled Petri dish. To prevent contamination of the PBS with maternal blood and tissue (mainly floating decidual fragments and adipose tissue), the uterus is dissected in stages using additional PBS-filled Petri dishes and multiple sets of clean instruments (Figure 7.9). Each Petri dish is placed on the microscope stage and illuminated from above, ideally with the bulb oriented overhead (i.e., a *ring* format) or at both sides (i.e., a dual fiber optic *goose-neck* arrangement). Two pairs of jeweler's forceps are used to tease apart the uterine wall (Figure 7.10). One pair (typically that held in the nondominant hand) is used to pinch the uterine wall between two swellings (i.e., implantation sites). Once closed, the two tips are grounded on the Petri dish floor to keep the muscular contractions in the uterus from making it writhe as the organ is being manipulated. The second pair (grasped in the dominant hand) is used to tease apart the smooth muscle layers immediately adjacent to the other forceps, producing a hole in the uterine wall next to a conceptus. This hole may be placed anywhere in the wall, but it is easiest to visualize if positioned on the anti-mesometrial surface since the engorged blood vessels within the mesometrium are prone to tearing (which releases copious quantities of blood) and are surrounded by fat (which often obscures the work area). Once a small hole has been made in the wall, one arm of each forceps is introduced into the uterine lumen, taking care that the tips are pointed out toward the endometrium and not in toward the conceptus. The forceps are closed, and then the one

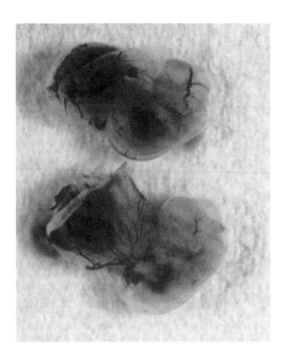

FIGURE 7.7 Isolated conceptuses may be dissected to free the embryo or fetus from the placenta, but the umbilical vessels connecting the two elements as well as the yolk sac and other membranes should be left intact so that the specimens from the same implantation site are easily matched. Embryos and placentas at GD 13 (suspended in PBS). (Image courtesy of Dr. Kelli Boyd, Vanderbilt University.)

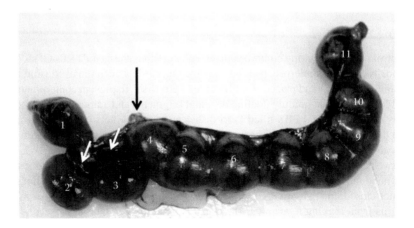

FIGURE 7.8 An alternative means for removing conceptuses is to remove the uterus from the abdominal cavity for subsequent dissection in a Petri dish filled with PBS. The implantation sites (all viable) are numbered consecutively beginning near the former attachment to the right ovary. Two offspring (Nos. 2 and 3) have been freed from the uterus to demonstrate that conceptuses appear as fluid-filled translucent sacs (the combined amnion and yolk sac) surrounding the embryo and capped by the dark red, discoid, definitive placenta (short arrows). The proximal tips of the horn are twisted so that the cut mesentery has rolled approximately 90° from its normal position, thus demonstrating that *postmortem* muscular contractions in the uterine wall tend to make the organ writhe during dissection. Long arrow = cervix. Gravid uterus at GD 13. (Image courtesy of Dr. Kelli Boyd, Vanderbilt University.)

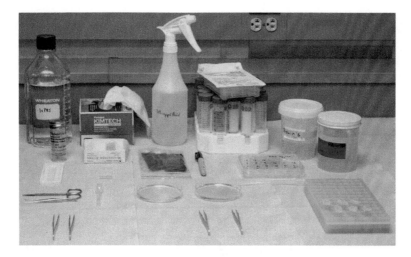

FIGURE 7.9 Arrangement of a necropsy station suitable for extracting younger conceptuses (GD 11.5 or earlier) from an isolated uterus. Starting at the foreground and progressing in clockwise fashion, the station is equipped with two pairs of Dumont No. 5 jeweler's forceps (used to isolate and subdivide the conceptuses); serrated forceps and sharp–sharp scissors (to remove the uterus); a tray of putty and vial of capillary hematocrit tubes (for collecting embryonic blood samples); a bottle of PBS (pH 7.4, to fill Petri dishes); a box of absorbent paper (KimWipes®) to gently wick fluid away from small organs that are to be measured or weighed; a box of clean glass microscope slides (to prepare blood smears); a spray bottle filled with 70% ethanol (to clean instruments); a bag filled with scalpel blades; a black marker (to prelabel specimen containers); a rack of 50 mL conical centrifuge tubes (to house embryos older than GD 13.5 and also fetuses); a multiwell culture plate (to hold embryos up to GD 13.0); two short, round containers of fixative (Bouin's solution [yellow] and NBF [colorless]); and a green rack containing microfuge tubes (to hold tissue samples for genotyping). Two PBS-filled Petri dishes are located in the center of the station. Equipment is oriented for optimal work flow by a right-handed person, with items needed for sample collection during nearly all studies (e.g., fixatives, genotyping tubes) located to the right side of the work area. (Image courtesy of Dr. Kelli Boyd, Vanderbilt University.)

in the dominant hand is used to slowly tear the wall and extend the diameter of the hole. Once the hole is wide enough, the forceps are used to detach the conceptuses from the uterine wall and move them into another Petri dish filled with fresh PBS (Figure 7.10).

Early conceptuses (at or before GD 9.0 to GD 9.5) usually appear as decidua-encased, asymmetric, oblong masses with a roughened, pale pink to pale purple exterior. At this point, each conceptus may be fixed in its intact state by immersion in either NBF or PFA. Alternatively, the embryo may be isolated from these units by inserting the tips of the two jeweler's forceps into a small depression located at the center of the broad end; this site is a weak spot near the nascent definitive placenta, which forms at the mesenteric pole (Figure 7.10). The nondominant forceps is used to immobilize the conceptus. One point of the dominant forceps is pushed through the tissue, after which the two arms of the forceps are closed and the dominant forceps are used to pull the combined decidua (outer maternal layer) and trophoblast (inner embryonic derivative) away from the yolk sac. This tissue typically will pull away as a single thick sheet, leaving the yolk sac and embryo that it encases in a central position in the narrower pole. The isolated embryos and placentas may be evaluated prior to fixation for gross structural abnormalities.

Fixation of Embryos and Fetuses

The choice of fixative and the means for ensuring that fluids needed to preserve and stabilize tissue structures in delicate embryos, fetuses, and placentas are the most important considerations in procuring samples of suitable quality for subsequent analysis of lethal phenotypes that arise during middle to late gestation. Complete details of fixation and processing options are given in Chapter 9. The current paragraphs reiterate major principles that might impact the choice of fixative, and they are included here for the convenience of readers.

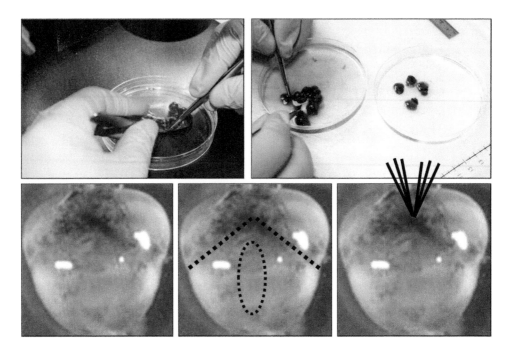

FIGURE 7.10   To remove younger conceptuses (GD 11.5 or older) from the uterus with minimal trauma, the detached uterus is submerged in a PBS-filled Petri dish, and two pairs of Dumont No. 5 jeweler's forceps are used to gently tear the anti-mesometrial uterine wall between two swellings (implantation sites) (upper left). Next, the conceptuses are teased away from the uterine wall and transferred to a Petri dish filled with fresh PBS (upper right). Decidua-encased specimens (GD 9.5 or younger) are slightly wider along their mesometrial (top) surface, which represents the eventual location of the definitive placenta (lower left). The embryo occupies an elongated, slightly bulbous chamber in the center (dotted oval), while the nascent placenta forms a slightly raised cap (above the dotted lines), which covers the mesometrial end (lower middle). The embryo is exposed by piercing the decidua with both forceps (represented by solid intersecting lines) at a small, depressed, typically dark focus (here a deep red [at the junction of the dotted lines in the lower middle]) near the margin of the nascent definitive placenta (lower right). Gravid uterus at GD 11.5 (upper images; courtesy of Dr. Kelli Boyd, Vanderbilt University), and an isolated GD 8.0 conceptus (lower images).

Fixation is undertaken to preserve cell and tissue architecture and to harden specimens so that they can withstand the manipulation required to orient them for examination, trimming, and embedding. Embryos (up to GD 14.5) may be fixed acceptably by immersion in NBF or PFA for up to 48 h because their small size and thin skin permit the ready entry of fixative and processing fluids. Bouin's solution and modified Davidson's solution are two other good alternatives as their ability to coagulate proteins greatly increases the hardness of embryonic tissues. Another advantage of Bouin's solution when used as a fixative for smaller embryos (GD 11 and earlier) is that the picric acid imparts a bright yellow color to specimens, which makes finding them inside a cassette much easier. Furthermore, Bouin's solution often is preferred by histotechnologists because the tissues are easier to section and the contrast among structures in tissues stained with hematoxylin and eosin is intensified substantially. Fetuses (GD 15.5 and older) usually should be fixed by immersion in either Bouin's solution or modified Davidson's solution for up to 48 h since the mixture of acids, alcohols, and aldehydes in these agents produces more effective fixation than aldehyde-only options like NBF and PFA. In addition, Bouin's solution also acts to demineralize bone from the fetal skeleton. However, if *in situ* gene expression or antigen localization may be main endpoints for the experiment, NBF and PFA likely will prove to be better choices for fixation as the aldehyde-only formulation is gentler on delicate tissue antigens. In such cases, the thoracic and abdominal cavities should be injected with fixative or opened to ensure that the tissues are given early access to the preservative fluid. In our experience, if a port is made to permit fluid entry, the usual practice is to incise the body wall approximately 2 mm to the left of the ventral midline. The incision should extend from the pelvis up to the mid-thoracic region. Care should be taken not to cut or displace organs within the body cavities. In general, an internal examination of unfixed tissues should not be attempted prior to fixation. Finally, for certain purposes (e.g., evaluating the regional distribution of functioning enzymes as markers for engineered genes [e.g., expression of bacterial lacZ]), the tissues should not be fixed at all, but rather should be preserved by flash-freezing the tissue. A technique for cryopreservation is described in Chapter 9.

Histological Processing of Embryos and Fetuses

Preparation of postimplantation conceptuses for histological sectioning requires advanced planning to ensure that collected specimens are handled in such a fashion that decent tissue sections may be obtained for histopathologic analysis. The most critical issue is that early embryos (up to approximately GD 10.5) are not lost during processing. Embryos at these developmental stages consist mainly of water (about 80% of their mass) and thus will shrink remarkably as specimens undergo repeated agitation when transferring among the many chemical vats needed to replace tissue water with wax. Loss of very small embryos is prevented readily by using special tissue cassettes with smaller openings or built-in chambers (Figure 7.11). To avoid the added expense of specialty products, mini-inserts that fit inside cheaper standard cassettes (Figure 7.11) may be utilized on occasions where extremely tiny conceptuses (typically GD 7.5 or below) must be protected. Other options include processing smaller specimens (e.g., by leaving embryos younger than GD 8.5 within the decidua for sectioning) or wrapping all isolated embryos (e.g., GD 9.5 to GD 11.5) in lens paper before placing them into the cassette. Lens paper should be used with caution as embryos may adhere to it, and thus be broken as the specimen is being manipulated after processing.

A related but less critical issue is processing of intact fetuses, which typically are too wide to fit into standard cassettes. One option is to trim fixed fetuses into blocks (two longitudinal halves, or multiple cross sections) that are thin enough (about 0.4 cm) to fit inside a standard cassette. An advantage of this approach is that a macroscopic analysis of internal organs may be completed during the trimming procedure. However, a considerable disadvantage is that positions of internal organs may shift once the adjacent structures and body wall no longer confine them to a single location; indeed, small organs that lack a firm anchor to adjacent structures may be lost completely. Displacement of structures may

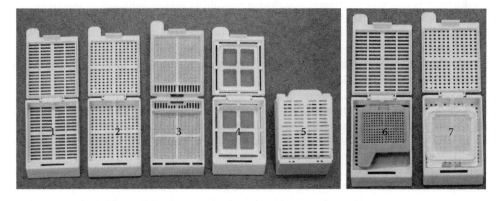

FIGURE 7.11 Tissue cassettes having potential utility for mouse developmental pathology studies. The pores in the cassette sides should be selected to accommodate the final size of the conceptuses after histological processing; specimen volume may be reduced by 30%–60%, especially in young water-rich embryos (GD 11.5 and earlier). Units with large pores are best suited to larger conceptuses (GD 13.0–GD 16.0 for cassette variant 1, GD 16.5 and older for variant 5), while specialty cassettes (variants 2, 3, and 4) with smaller pores are adapted to younger embryos. If cost is an option, standard cassettes (i.e., variant no. 1) may be fitted to hold small specimens by using miniature inserts equipped with small pores and hard (variant 6) or mesh (variant 7) covers. Embryos also may be placed within the fold of a permeable paper (e.g., KimWipe®) cut to fit the cassette, but generally, they should not be restrained with sponges as upon shrinkage the friable tissues may get ensnared in the holes and tear. Cassette types: 1, Tissue-Tek® Uni-Cassette® (Sakura Finetek, Torrance, CA); 2 and 3, Biopsy Uni-Cassette® (Sakura Finetek); 4, Tissue-Loc HistoScreen Cassette (Cancer Diagnostics, Inc. Durham, NC); 5, Mega-Cassette® (Sakura Finetek); 6, CellSafe Biopsy Cassette Insert (Cancer Diagnostics, Inc.); 7, Activflo Mini Biopsy Cassette (Leica Biosystems, Buffalo Grove, IL). (Images courtesy of Dr. Kelli Boyd, Vanderbilt University.)

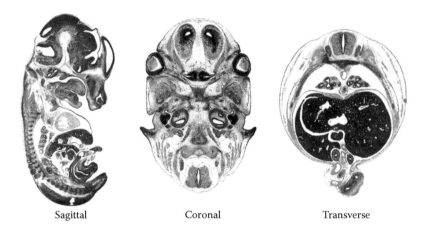

| Sagittal | Coronal | Transverse |

FIGURE 7.12   Representative sections of mouse embryos at GD 13.0 demonstrating the possible orientations in which specimens may be sectioned. The sagittal (longitudinal) plane is obtained by positioning the embryo in the embedding mold on its side (usually the left, since the tail curls forward and laterally on the right side in most animals). The coronal (*frontal*) view is obtained by positioning the embryo in the mold with either the tip of its nose or the highest curve of its back touching the mold floor. The transverse orientation (*cross section*) is acquired by trimming the torso into multiple blocks using incisions perpendicular to the long axis of the body and then resting one cut surface of each block on the floor of the mold. Processing: Bouin's, paraffin, H&E. (Images provided to Dr. J.M. Ward, courtesy of Dr. M.H. Kaufman.)

be prevented by wrapping the tissue blocks in lens paper to maintain the position of the organs. The other option is to employ extra-deep ("mega") cassettes that can accept wider tissue blocks (up to about 0.9 cm; Figure 7.11). The advantages to this approach are that fetuses remain intact, thereby preventing loss of organs, and yet are not crushed by trying to jam the cassette lid shut on a large sample. Tissue compression caused by a tightly wedged lid often promotes uneven wax penetration, which in turn makes procurement of complete tissue sections difficult or even impossible. A disadvantage of processing intact fetuses is that wax infiltration into thicker specimens may be impeded by both their thicker skin and greater tissue mass. Such constraints typically are addressed by performing histological processing under vacuum to enhance entry of fluids needed for tissue dehydration and wax infiltration (see Chapter 9 for details).

Embryos and fetuses may be embedded in any orientation (Figure 7.12). Each option offers advantages and disadvantages, depending on the question to be answered. Prior to GD 13.5 or so, intact embryos may be laid flat in the cassette on their left side (i.e., with the tip of the curled tail pointing *up*) to reproducibly obtain high-quality sagittal sections. At later stages of gestation, embryos and fetuses often are bisected at or just to the left of the midline, and both halves are placed with the cut surface oriented *down* in the cassette. The sagittal orientation is a good choice for evaluating axial structures, longitudinal fiber tracts in the brain, and viewing major organs in the body in one or a few sections. Coronal sections are well suited for assessing the brain, and specifically its bilateral symmetry, and offer a unique view of the vessels entering and leaving the heart. Transverse sections provide an optimal means for comparing paired organs all along the body axis, including the mirrored sides of the spinal cord.

Sample Collection Practices during the Postnatal Period

After birth, developing mice have their own independent existence and thus, in general, must be treated in the same fashion as adult animals. Accordingly, neonates and juveniles need to be euthanized humanely before specimens may be acquired. The methods described earlier for the dam and fetus may be used for developing mice after birth. In our experience, decapitation with scissors is acceptable for small individuals (e.g., postnatal day [PND] 4 or younger) as long as the stroke is rapid and the major blood vessels are completely cut, while older animals may be euthanized by cervical dislocation or parenteral administration of an injectable anesthetic at a lethal dose. As noted earlier for fetuses, the cervical dislocation and decapitation methods are likely to damage regional tissues in the neck. Certain traditional methods of euthanasia should no longer be used as stand-alone means of terminating young mice during the postnatal period. These practices include evisceration of unanesthetized animals (which is an essential prerequisite for *in situ* skeletal staining) and freezing, although hypothermia (4°C–8°C) is acceptable for anesthesia of animals up to PND 7 as long as the skin does not come into direct contact with ice or precooled metal surfaces.[2]

Very young pups (up to approximately PND 3) may be processed for evaluation in the same fashion as near-term fetuses due to their small size. In such cases, each individual receives a brief physical examination to assess functional capabilities followed by an external examination to define any structural abnormalities. Evidence of functional deficits includes such features as aberrant posture, altered skin coloration, labored breathing, and rapid death. Structural defects typically involve distortions in the shape and position of body contours or alternatively variations in the size or even presence of one or more appendages. Examples of developmental lesions that may be seen readily by a quick but thorough external examination are shown in Chapter 15.

In contrast, older developing mice should not be processed as intact specimens once they achieve a certain cross-sectional diameter (1 cm, or approximately PND 4 [with modest variations in timing depending on the litter size and strain]). At this point, young animals instead should be necropsied so that fixatives and fluids used in histological processing may penetrate sufficiently to properly preserve cellular and tissue constituents. The necropsy procedure for developing mice is comparable to that utilized for adult animals, with some minor variations. Once the mouse has been euthanized, the body is positioned on a cutting board in dorsal recumbency with the four limbs stretched outward to apply tension to the body wall. The limbs may be anchored in place using a thick rubber band (Figure 7.13); pins pushed through the paws and into the cutting board (see Figure 17.4); or small cords (i.e., thin strings looped around the distal limbs and then tied to nearby pins). The advantage to the rubber bands and the pins is that these methods of attachment may be done quickly, while the rationale for using rubber bands or cords is that these techniques do not damage tissue and thus permit histopathologic examination of the distal limbs. Next, the ventral body wall is opened (Figure 7.13) using a pair of serrated forceps (held in the nondominant hand to grasp and raise the body wall) and small pair of straight sharp–sharp scissors (gripped in the dominant hand) to open the ventral body wall from the pelvis to the neck. Major organs are examined for overt malformations, after which either major organs are removed or the entire carcass (with organs retained *in situ*) are immersed in fixative (generally NBF since it is available commercially and allows developmental specimens to be processed on the same instruments that handle samples from adult mice). An advantage to organ removal at this stage is that

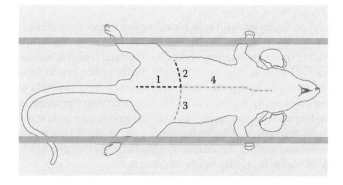

FIGURE 7.13 Diagram showing how to nontraumatically restrain neonatal and juvenile mice for necropsy using elastic bands. The body cavities are opened with four cuts (made in the order in which they are numbered, starting with "1") to permit an internal evaluation and excellent penetration of fixative solution. (The diagram was produced by Mr. Tim Vojt, The Ohio State University.)

those structures prone to early autolysis (self-digestion), like bone marrow and the gastrointestinal tract (including gall bladder and pancreas), may be removed first to ensure that their morphology is as pristine as possible when histopathologic evaluation is to be performed. A detailed protocol for harvesting tissues of fetal, neonatal, and juvenile mice, including procedures for perfusion fixation as a means of enhancing the preservation of dense tissues (e.g., central nervous system) and deep organs, is given in Chapter 17 and, thus, is not repeated here.

Data Collection during Mouse Developmental Studies

In principle, standard necropsy techniques developed for use in adult mice are applicable for use in developing mice of all ages, with some minor adjustments necessitated by stage-specific differences in animal size and tissue consistency. Mouse developmental pathology studies can incorporate a number of endpoints depending on the nature of the project and the interests of the research team. The collective health of the litter may be probed, particularly in developmental toxicology studies, by analyzing many maternal characteristics. Numerous kinds of data have direct relevance to assessing the fitness of a given individual. This section will consider the types of parameters that commonly are evaluated during developmental pathology studies, communicating them in the order in which they typically are gathered.

Data Collection for Maternal Endpoints in Developmental Pathology Studies

The well-being of an individual conceptus depends in part on the support it receives from the dam. For this reason, initial parameters to be assessed during any mouse developmental pathology study should be those related to the mother. The gradual changes in maternal traits that occur over the course of gestation serve as an important gauge of the litter's well-being.

In Vivo *Maternal Parameters Indicative of Litter Health*

Several parameters that illustrate the health of the litter may be evaluated effectively by examination of the pregnant mother during life (Table 7.1). In our experience, the principal endpoint that is defined during the life of the dam is the rise in total body weight that accompanies the cumulative growth of the entire litter as gestation progresses. Careful abdominal palpation to assess the increase in uterine volume is another fast way to estimate litter size, although this qualitative method is less precise than quantifying the change in maternal body weight over time. Noninvasive imaging methods,[38] in particular ultrasound biomicroscopy (UBM) (described in some detail in Chapter 14), that allow individual conceptuses to be visualized *in vivo* over time often afford a more concrete evaluation of litter size, with the added advantage that they also provide an initial indication regarding whether or not major malformations are present in one or more of the conceptuses.

The weight of a typical nonpregnant, wild-type female mouse ranges from 20 to 25 g. Weights for dams carrying a usual litter of multiple conceptuses (approximately 5–10 viable implantation sites, depending on the stock or strain) rise above this range starting about GD 10[15,29] and continue increasing in a linear fashion throughout the remainder of gestation. This rate of weight change results in a doubling or more of maternal weight by the end of gestation, and as expected litter size impacts the extent of this weight gain; near term, a dam with a litter of ten gains nearly 1.5 times as much as one carrying five offspring and nearly 3.5 times as much as a mouse holding a single conceptus.[18] Most of the gain results from transient addition of tissue (embryo and placenta) and fluid associated with the many conceptuses.[29] However, expansion of maternal fat depots (as a source of stored energy) and mammary gland (mainly the acini needed to make milk) does contribute mildly to the elevation in carcass weight of the dam.

Gentle manipulation of the abdomen may be used during later stages of pregnancy as a rapid survey of litter health. A demonstrable increase in uterine size is evident as a series of gradually expanding, palpable swellings along the sides of the abdomen. An experienced technician can detect the first enlargement of the uterine horns beginning at about GD 10–GD 11. Discrimination of implantation sites is easier in

TABLE 7.1

Litter-Based Endpoints of Developmental Events Seen in the Dam

---

Parameters for Individual Mothers

In vivo endpoints

    Body weight—changes over time

    Litter size

        Abdominal palpation

        Noninvasive imaging

Postmortem endpoints

    Implantation sites per dam

        Total

        Viable (containing live offspring)

    Deaths per litter

        Resorptions—placental remnants with no visible embryo

        Embryo lethality—implantation sites with nonviable embryos

    Malformations per litter

        Total

        Lethal

        Organ specific—often reflecting the research focus of the laboratory

Parameters for Entire Treatment Groups

Number of affected litters

Recurring lesions among affected litters

---

larger litters since more conceptuses are present to be detected. If palpation is chosen as a means of following the progression of pregnancy, care should be taken not to place too much pressure on the uterus. In addition, we recommend that this procedure be undertaken no more frequently than every second or third day in order to avoid inflicting potential trauma to the embryos.

UBM permits serial *in utero* evaluation of developing organs in mouse conceptuses. The anatomic position of structures within UBM images may be matched reliably with organ position in routine tissue sections,[19] thus allowing the onset and evolution of normal and abnormal morphologic traits to be followed in time. This modality is most effective when utilized in conceptuses at GD 10.5 or older as embryos at later stages possess more distinct body and organ contours, but useful information may be garnered from embryos as young as GD 7.5 (Chapter 14). The low equipment cost (relative to instruments needed for more invasive high-resolution techniques, like magnetic resonance microscopy and microscopic computer-assisted tomography) should enable many laboratories to exploit UBM technology to address developmental pathology questions in mice.

Postmortem *Maternal Parameters Indicative of Litter Health*

Once a decision has been made regarding when during gestation to assess the viability and lesion spectrum in conceptuses (see Chapter 6 for details concerning how this choice is made), pregnant female mice are terminated on the appropriate GD so that the gravid reproductive tract may be removed and the conceptuses isolated. The techniques to be followed for this purpose to some degree vary across gestation depending on the developmental age of the litter; specific steps to be followed when harvesting the litter are given in the next section. Isolation of preimplantation embryos (i.e., GD 4.5 or younger) from the oviducts or uterine horns generally does not induce maternal reproductive tract changes of interest to the final developmental pathology data set. However, all developmental stages that take place after implantation has occurred (i.e., at or shortly after GD 4.5) typically will produce several maternal changes that can be amassed as additional evidence regarding the health of the litter (Table 7.1).

The most informative *postmortem* endpoints indicative of a potential embryonic lethal phenotype are gained by examining the number and structure of the implantation sites and/or offspring within the maternal uterus. Common measurements address both the quantity of conceptuses (i.e., total number) as well as their quality (numbers of living vs. malformed vs. dead embryos, numbers of viable vs. resorbed implantation sites). Studies conducted with genetically engineered mice typically gather data for individual conceptuses because the genotype of each embryo is the variable (*treatment*). Experiments performed to assess development toxicity also amass information for individual progeny, but these assays also gather cumulative data for the entire litter since a treatment applied to the dam is the variable that may affect one or several or even all individuals in the litter. As with the individual animal findings, measurements of uterine changes made on a *whole litter* basis analyze the quantity of any toxic effects (i.e., number of affected litters) as well as their quality (number of litters with malformed or dead conceptuses and/or resorbed implantation sites). Findings for a given dam (i.e., litter) should be listed in detail on a data collection form (Figure 7.14), for which the maternal identification number was entered

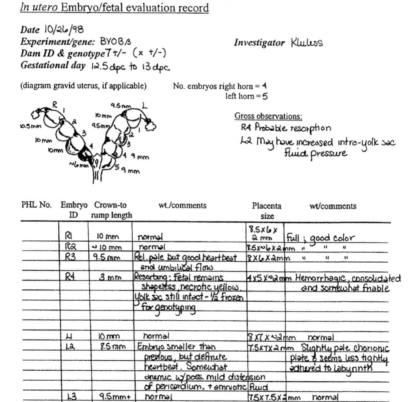

FIGURE 7.14 Diagram showing a data collection form suitable for entry of both litter-based and individual animal data at necropsy. Other formats are acceptable, but the best formats include a predrawn diagram of the uterus suitable for annotation as well as premade columns for entering particular types of data. (The original form was prepared by Ms. Deborah E. Devor-Henneman and is reprinted from Ward, J.M. and Devor-Henneman, D.E., Gestational mortality in genetically engineered mice: Evaluating the extraembryonic embryonic placenta and membranes, in: Ward, J.M. et al. (eds.), *Pathology of Genetically Engineered Mice*, Iowa State University Press, Ames, IA, 2000, pp. 103–122, through courtesy of Iowa State University Press.)

in advance. The same form also serves as the device for capturing information for individual progeny within the litter (see below). For greatest convenience, the form should be arranged so that all information for the litter and the fetuses within it can be entered on one side.

Data Collection for Individual Offspring in Developmental Pathology Studies

The developmental fitness of an individual conceptus depends in great part on its unique genetic heritage and *in utero* environment. For this reason, data sets acquired during mouse developmental pathology studies must be built from detailed descriptions of a wide spectrum of endpoints pertaining to the individual's well-being. The time spent in this endeavor will not impact the quality of tissue preservation in the conceptuses as their tolerance for the low-oxygen intrauterine environment will permit them to remain viable for at least an hour after uterine blood flow is stopped.[2]

The benchmark against which the appropriateness of an animal's phenotype is judged represents a composite impression of *normal* gained from both concurrent control (i.e., littermates with different genotypes, progeny from control litters) and historical control cohorts. The control data set should incorporate data only from individuals that share the same genetic background (e.g., strain/substrain/stock) and developmental stage since mouse strain and line differences are often apparent.

A critical aspect of the assessment is to confirm the identity of the individual progeny so that specimens from different individuals will not be mingled, and thus lost to any further analysis. After birth, such segregation is relatively simple since only one mouse likely will be undergoing necropsy at a given necropsy station at any point in time. In this setting, the key need is to verify the individual animal number, which typically exists as a pattern of intradermal ink spots (introduced shortly after birth by subcutaneous injection using a 26–30-gauge needle) or clipped toes for older animals or black marker stripes for short-term studies (up to 24 h). Greater care must be taken when dissecting conceptuses from the uterus. In our experience, mistakes during gestational analyses are prevented readily by leaving the embryo attached to the placenta when performing the evaluation, and by isolating and sampling one conceptus at a time.

Recording Individual Animal Data

The most convenient means for collecting a high-quality data set is to place entries for each individual in a litter directly onto one form (Figure 7.14) or directly into a semiautomated computer database (see Chapter 17 for details). The first entries should assess the position and appearance of each implantation site before the uterus has been opened. The next entries should describe the morphology of each implantation site once it has been exposed and confirm whether or not special sample types were collected, such as fluid or tissue for genotypic analysis or quantification of accumulated metabolite or toxicant levels. The bulk of the entries likely will be dedicated to a detailed accounting of structural traits for each animal. Common types of data gathered on an individual animal basis during developmental pathology studies include descriptions of the macroscopic appearance of the body (external examination) and sometimes selected major organs (internal examination); photographic documentation of detectable malformations; procurement of selected body measurements (e.g., crown-rump length, total body weight, and sometimes placental weight); acquisition of a tissue sample for genotypic analysis; and the microscopic features of multiple organs in one (or more likely several) tissue sections. An individual's functional abilities may be assessed by a brief clinical evaluation centered on such traits as breathing, movement (baseline and in response to simple nonpainful stimuli), and posture.

Positions and Appearances of Implantation Sites

Regardless of the gestational timing, the morphology of each implantation site should be described in full before any conceptus is harvested. This task includes a depiction of the location, shape, and size of the site before the uterus is opened as well as a delineation of the morphologic appearance of each site once the conceptus has come into view. Implantation site morphology is defined in terms of features like color (of tissues and fluids), shape, and size of each site as well as any anomalies within them (e.g., dead embryos/fetuses, partial or complete resorption [i.e., sites containing a little or no embryo, respectively]). Implantation sites having the same color, shape, and size as the majority of sites in the litter (or in stage-matched control litters that were not exposed to a toxicant) likely contain viable embryos at the proper developmental stage, while sites that are reduced in size usually hold a developmentally delayed embryo (i.e., exhibiting extensive focal or multifocal tissue degeneration/necrosis) or an autolyzed embryo (i.e., a *death* event) or consist chiefly of decidua with no remaining embryo (i.e., a *resorption*) (Figure 7.15). In some cases, resorption sites also may exhibit an altered color—commonly dark red or purple (since auto-lyzing decidua often contains lakes of blood) but sometimes dark green (which seems to represent large masses of cell debris filled with pigments derived from prior degradation of blood pools) (Figures 7.15 and 7.16). Cloudiness of the amniotic fluid, which may obscure the appearance of the embryo (Figure 7.16),

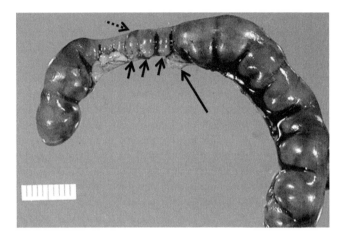

FIGURE 7.15   Gravid uterus at GD 13.0 in which three early resorptions (short arrows) have the size normally observed for implantation sites at GD 7.5. The resorption site at the left has a small, dark green discoloration (dotted arrow), which was demonstrated to be cell debris (including many erythrocytes) by histopathologic examination. The stump of the cervix is evident at the tip of the long arrow. (Image courtesy of Mr. Stephen Kaufman, Amgen, Inc., Thousand Oaks, CA.)

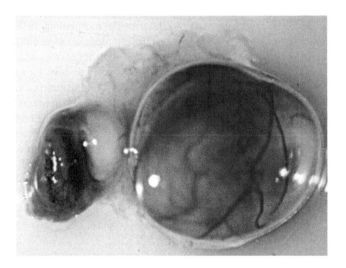

FIGURE 7.16   Implantation site isolated at GD 15.0 containing an embryo that died earlier during gestation. Evidence of death includes the murky amniotic fluid, which is discolored both red (indicating blood leakage) and white (suggesting protein or cell effusion); the uniform pallor of the embryo (which is most evident at the left side of the yolk sac); and the large amorphous plug of green (likely necrotic) cell debris at the left of the specimen.

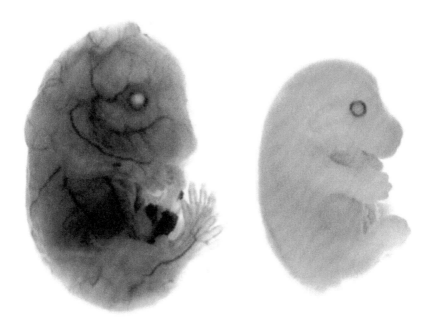

FIGURE 7.17 Photograph comparing viable (left) and dead (right) littermates evaluated at GD 14.5. The viable embryo exhibits a distinct brown liver spot (indicative of abundant hematopoiesis), a robustly branching superficial vasculature, and long digits. The dead embryo is pale tan uniformly except for the retina, demonstrating that circulation and hematopoiesis ceased some time back. The developmental stage at which death occurred can be estimated as GD 13.5 based on the presence of obvious albeit short digits on the forepaw and digital rays (which arise about GD 12.8) but very minimal digits on the hind paw. Histopathologic examination of the dead animal revealed diffuse and severe autolysis (i.e., suggesting that the lethal insult had occurred at least 12–18 h prior to death).

may serve as additional evidence that the conceptus has suffered a lethal or at least a very severe injury. In such instances, red discoloration of the fluid is indicative of discharged blood, while fluid with a white tinge to the murkiness reflects accumulation of other cell types and/or proteins. Once the extra-embryonic membranes have been reflected, dead or dying embryos usually exhibit a uniform pallor in which all or nearly all color is absent, and/or they are likely to be smaller relative to still-viable, developmental stage-matched littermates (Figure 7.17).

The precise position of each implantation site within a particular uterine horn, especially those that hold an abnormal conceptus, should be specified in detail. A convenient way of documenting this information is to photograph the entire uterus, either *in situ* (Figure 7.2) or after its removal (Figure 7.8); advantages to acquiring the image in its native intra-abdominal position are that the horns are attached, and thus fully available for viewing, and the clearly visible cervix provides a clear demarcation between the left and right horns. Another acceptable means of recording such data is a detailed written record of the uterine appearance. The description may be qualitative (i.e., the location of an aberrant site relative to other sites in the uterus) or quantitative (i.e., the exact distance of an anomalous site from a set reference point, like the uterotubal junction or cervix). In our experience, a qualitative assessment typically is sufficient unless the accumulation of data over multiple litters shows that distance (rather than position) might have an impact on the final phenotype.[27,43] Position locations for sites usually are numbered most easily starting with "1" for the site nearest the right ovary and then increasing the integer sequentially while moving distally down the right horn (1, 2, 3, …). The sequence typically is continued into the left horn (4, 5, 6, …), starting with the implantation site nearest the cervix and ending with the site closest to the left ovary. Other labeling schemes may be utilized at the preference of the laboratory (e.g., starting in the left horn or labeling sites in each horn with a letter ["L" for "left" and "R" for "right"] followed by an integer that is not repeated within that horn). Any of these schemes should prove suitable as long as the same system is used consistently within the laboratory.

Sample Collection for Genotyping or Chemical Analysis

Specimens dedicated to any chemical or molecular analysis typically are acquired after the uterus has been opened to expose the conceptuses but before the detailed examination of the individual units has begun. Samples may be obtained while the conceptuses are still attached to the implantation site or after they have been removed to a bench-top location that is not contaminated by maternal fluids or tissues.

Tissue for genotyping typically is acquired by careful removal of a small portion of tissue from any site that can be guaranteed to represent cell lineages that are solely of embryonic origin (Figure 7.18). Options for animals *in utero* include the amnion or yolk sac (for embryos up to GD 14.5) or the tail tip (for fetuses at GD 15 or older). Selection of a distal limb also may serve this purpose for conceptuses over GD 12. Possible areas to sample after animals have been born include amputation of the tail tip, one or more digits, or an ear. Samples are harvested using jeweler's forceps or a small pair of scissors and then placed in prelabeled microfuge tubes for retention until they may be analyzed. These tiny tissue pieces may be retrieved after the external evaluation of the animal as a prelude to fixing the conceptus, performing an internal evaluation, or undertaking microdissection to obtain samples for further biochemical or functional assays.

Material for chemical extraction or molecular studies may be acquired using similar techniques. Individual organs or tissues may be removed using a stereomicroscope to magnify smaller regions, especially in embryos undergoing organogenesis (GD 8.0–GD 14.5) and younger fetuses (GD 15.0–GD 17.0). The site to be sampled generally depends on the hypothesis to be addressed. The tissue-harvesting protocol will require sterile media (usually warmed to 37°C) and sterile instruments if isolated cells or tissue fragments are to be grown in culture. These specimens typically should be acquired after the examination of the animal (and, if warranted, its placenta) has been completed.

Categories of Individual Animal Data

Detailed evaluation of the embryo, fetus, neonate, or juvenile represents the raison d'être for undertaking a mouse developmental pathology study. For embryos and fetuses, the examination should include the placenta as well. The best practice for phenotypic evaluation will include both a clinical assessment and morphologic analysis.

A simple and brief clinical examination should precede the detailed necropsy. In our experience, this practice provides important clues regarding the animal's functional status. Clinical examinations in developing mice typically rely on visual observations of readily apparent morphological features that can

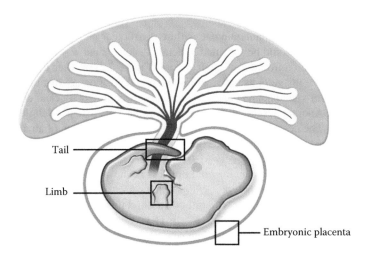

FIGURE 7.18   Diagram showing potential sites at which samples derived solely from the embryo may be taken for genotyping analysis. The embryonic placenta (yolk sac or amnion) is a preferred site because it is readily accessible, and its acquisition does not damage the embryo. (The diagram was produced by Mr. Tim Vojt, The Ohio State University.)

be tied to specific physiological processes rather than the more detailed evaluation performed in adult animals, which may include physical manipulation of body parts and/or behavioral testing. Examples of parameters to be assessed in developing mice include such traits as breathing, color, excretion, motion, and posture. Functionally normal mice wriggle and extend their heads up and back as they first begin to breathe, soon develop a diffuse pink skin color as a regular breathing pattern takes hold, should move their limbs and lips if these parts are stroked, will develop visible white spots on their left abdomen (a *milk spot* showing the location of the engorged stomach), and on occasion may be seen to exhibit fecal staining (typically pale tan to yellow/brown in suckling mice) in the perineal regions adjacent to the anus. Abnormalities in any clinical attributes are recorded easily on the data collection form (Figure 7.14). In most cases, laboratories that routinely perform mouse developmental pathology work follow the convention that the absence of an entry implies that the animal exhibited no overt functional deficits. Should the need arise for more focused functional assessment, simple behavioral tests have been devised to explore the integrity of certain organs and systems. In many cases, these screens may be undertaken with either no or only simple homemade apparatuses (see Refs 1, 5, 20 for review).

The necropsy protocol for developing mice typically is short but nonetheless thorough. The primary essential observations to record for each animal at necropsy are the macroscopic (gross) features of the body upon external examination (Table 7.2). Such evaluations serve mainly as a screen for overt structural malformations. In addition, the sex of older animals (typically fetuses at GD 16 or older and all postnatal individuals) should be determined using the anogenital distance (Figure 7.19), which is twice as great in males.[39] In most cases, one or more measurements of body size (e.g., crown-rump length, total body weight) will be desirable as a quantitative means of assessing whether or not growth has proceeded in a normal fashion (Table 7.2); placental weight also may be obtained although this practice often is followed only if placental biology is the focus of the research program or a readily distinguishable difference in placenta size is evident among the treatment groups. These simple measurements are rapidly acquired using a digital caliper (available from many vendors) for length and analytical scale for weight. In general, measurements should be taken at necropsy, especially for embryos (i.e., GD 14.5 or earlier), since their high water content will lead to a considerable degree of shrinkage (in terms of length and weight) as water is removed during histological processing. To prevent collapse of small specimens during the manipulation required to obtain the measurements, delicate embryos (especially GD 11.5 and earlier) may be placed into weigh boats that contain PBS, measured, and removed. The weight of the animal is calculated by subtracting the weight of the fluid-filled boat alone from that of the fluid-filled boat plus the embryo. In some instances, lengths of specific body parts (e.g., limbs) may be obtained as an additional means of evaluating subtle perturbations, which may be present at that site. If an unusually small embryo is demonstrably dead or dying (Figures 7.17 and 7.20), measurements may be eschewed in favor of using unique external body features as a means of estimating when during gestation the lethal event took place.

For many developmental pathology studies, intact animals are placed into fixative once external features have been evaluated and measurements are completed. Convenient vessels for holding conceptuses during fixation include standard centrifuge tubes (usually 20 or 50 mL), other small tubes or vials, and multiwell tissue culture plates (especially the 24-well variety as it has enough chambers to include

TABLE 7.2

Individual Gross Endpoints of Pathologic Events Seen in Developing Mice

Malformations per conceptus
Total
Lethal
Organ specific—often reflecting the research focus of the laboratory
Offspring metrics
Body size
Crown-rump length
Total weight
Regional dimensions[25,32]—typically lengths or volumes of key structures like brain vesicles, cardiac chambers, limb buds, and optic or otic vesicles

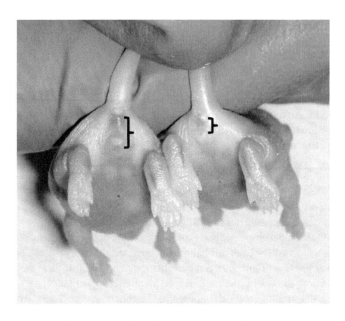

FIGURE 7.19 Photograph comparing the anogenital distances of male (left) and female (right) mouse pups at PND 1. The center of the anus (dark spot at top) is approximately twice as far from the center of the genital tubercle of the male as is the case in the female. The anus of the male has a large dark zone between the anus and the genital tubercle[39] due to the large size of the bulbourethral gland[13] in the superficial subcutis. (Image courtesy of Dr. Cynthia Besch-Williford, IDEXX BioResearch, Columbia, MO.)

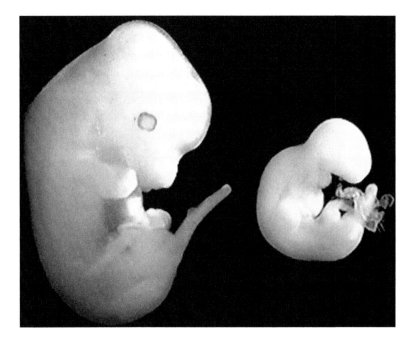

FIGURE 7.20 Photograph comparing viable (left) and dead (right) littermates evaluated at GD 13.0. The viable embryo has distinct digital rays on both forepaws and hind paws; these rays first occur at GD 12.3 on fore paws and at GD 12.8 on hind paws (see Tables 4.A.3 and 4.A.10). The dead embryo is pale white uniformly and exhibits superficial contours, including distinct branchial arches, indicative of developmental cessation at approximately GD 9.5.

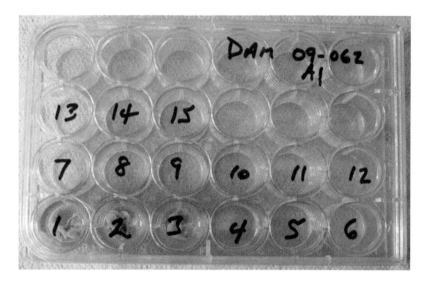

FIGURE 7.21 A 24-well tissue culture plate outfitted to hold an entire litter of embryos during fixation. The lid was prelabeled with the study no. (09-062), the dam identification number (A1), and the maximum possible number of conceptuses that might be harvested from the mouse strain used in the study. Once the exact litter size was defined ($N = 3$), the appropriate number of wells was filled with Bouin's solution. The chambers hold both the embryo and the placenta.

all embryos from even the largest litter) (Figure 7.9). The cover of the culture plate may be prelabeled with the identification numbers for the dam and multiple progeny (Figure 7.21), but chambers need not be filled with fixative until the number of implantation sites has been counted at necropsy. After addition of the embryos to the wells, the four edges of the plate may be sealed with an impermeable wrap (e.g., Parafilm®, which is available commercially from many vendors) to prevent inadvertent leakage of the fixative. Many laboratories do not retain the placenta in fixative, but in our experience, this omission is short-sighted because many embryonic lethal phenotypes result from placental rather than embryonic defects (see placental lesions in Chapter 16) or from a combination of embryonic and placental damages. Therefore, the placenta should be fixed whole and cut along with its embryo/fetus even if the initial assessment is centered only on the animal. The placenta typically is fixed in NBF and not in Bouin's solution because the latter fixative may rupture erythrocytes (causing them to appear as *ghost cells* or be invisible altogether).

The examination of external structures continues and assessment of internal organs usually begins during the histopathological examination. For example, the shape and size of viscera for smaller individuals (e.g., embryos, fetuses, and neonates up to PND 4) generally are first available for viewing in histologic sections, where they may be investigated readily at very low magnifications (5×–10×) prior to the detailed analysis of fine structural details of cells, tissues, and organs at higher magnifications (usually 40×–400×). Internal organs in larger animals (e.g., juveniles at PND 5 or older) may be assessed as they are removed at necropsy, but in many cases, the organs are given only a cursory screen of surface features and then left *in situ* to minimize the possibility of their displacement during fixation. Therefore, analysis of organ color, contours, and relationships at low magnifications remains a vital component of the histopathologic assessment for developing mice of all ages. Internal features often are screened in a single section per organ for juveniles in which specimens were removed at necropsy and subsequently trimmed and positioned in a particular orientation. Multiple step sections or even serial histologic sections often are utilized for individuals that are processed in their intact state to make sure that at least one section will possess a sufficient portion of each major organ.

In some instances, special procedures are utilized to gain a greater appreciation for the evolution of particular organs or systems over time. For example, a macroscopic evaluation of internal organs may be undertaken using a sharp blade to place several parallel incisions into the torso (Figure 7.22) or head (Figure 7.23) of fixed animals, thereby exposing internal organs for examination *in situ* or after

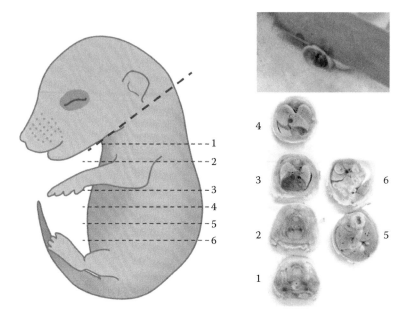

FIGURE 7.22   Schematic diagram of a near-term (GD 17) mouse fetus (left panel) showing locations for making a series of parallel, freehand transverse incisions (i.e., Wilson's technique, shown in upper right panel) to permit a macroscopic evaluation of major internal organs in fetal and neonatal mice. The numerical labels in the schematic diagram demonstrate the placement of a cut (dashed lines) designed to demonstrate the internal structures in the torso block labeled with the corresponding number (right panel). The subject was a near-term mouse fetus that had been fixed for 72 h in Bouin's solution to harden soft tissues and decalcify the bony skeleton. (The diagram was produced by Mr. Tim Vojt, The Ohio State University. The photographs were taken by Dr. Elizabeth R. Magden, Colorado State University.)

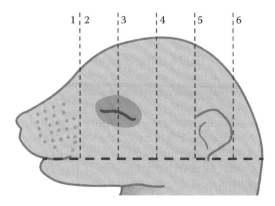

FIGURE 7.23   Schematic diagram showing locations for making a series of parallel, freehand coronal incisions (i.e., Wilson's technique) to permit a macroscopic evaluation of major brain regions in fetal and neonatal mice. The numbers identify the location of the tissue block located between two adjacent incisions (dashed lines). Should a microscopic examination of the blocks be desired, the cut surface closest to the number identifying the block should be placed *down* in the cassette to ensure that each section will demonstrate unique neuroanatomical features (as illustrated in Figure 7.24). (The diagram was produced by Mr. Tim Vojt, The Ohio State University.)

removal; this method is known as Wilson's technique after its inventor, J.G. Wilson (the father of modern teratology).[3,40] The incisions typically are made in free-hand fashion using external landmarks and generally are made in the transverse orientation (termed a *cross section* for the torso or a *coronal section* for the head). Considerable practice is required to gain proficiency in identifying organs since the technique often is performed in specimens fixed in Bouin's solution, which hardens tissues efficiently for

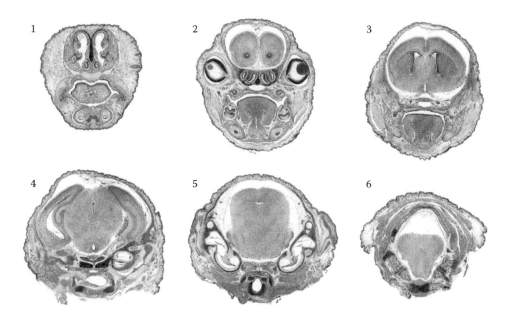

FIGURE 7.24 Coronal sections of the head from a near-term mouse fetus (GD 18.0) showing the features available for analysis if blocks are produced by freehand trimming near the levels shown in Figure 7.23. Processing: Bouin's, paraffin, H&E.

sectioning but obscures the contrast between organs. The blocks produced by Wilson's technique may be processed for histopathological examination in order to obtain a more detailed view of large organs' regions and the cell populations within them (Figure 7.24).

An alternative approach can be used to expose axial structures for macroscopic and microscopic evaluation. This method employs two freehand, parallel, parasagittal incisions placed on either side of the midline (Figure 7.25). The cutting planes are aligned using the base of the tail, the umbilicus, and center of the neck (or the mandibular symphysis [center point at the apex of the lower jaw] if the head remains attached to the body). If the blocks are sectioned, organs located specifically on the left side of

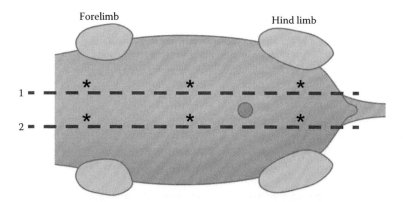

FIGURE 7.25 Schematic diagram showing positions to place two parallel, freehand longitudinal incisions on the ventral midline, located to either side of an imaginary line connecting the base of the tail, the umbilicus (small central circle), and the center of the neck. The numbers identify the preferred position of the incisions (dashed lines). Should a microscopic examination of the blocks be desired, the cut surface denoted by the asterisks (*) should be placed *down* in the cassette to ensure that each section will demonstrate unique neuroanatomical features (as illustrated in Figures 7.26 through 7.28). (The diagram was produced by Mr. Tim Vojt, The Ohio State University.)

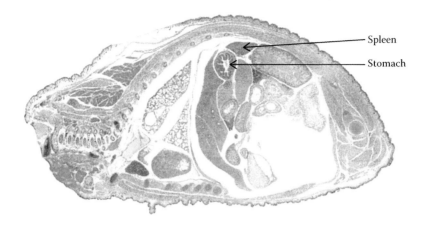

FIGURE 7.26 Parasagittal section of the torso from a near-term mouse fetus (GD 18.0) showing the features available for analysis if a block is produced by freehand trimming near level 1 shown in Figure 7.25. The presence of the spleen and stomach demonstrate that this section is from the left side of the animal. Processing: Bouin's, paraffin, H&E.

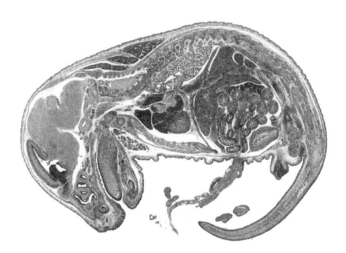

FIGURE 7.27 Parasagittal section of the torso from a near-term mouse fetus (GD 18.0) showing the features available for analysis if a block is produced by freehand trimming midway between levels 1 and 2 shown in Figure 7.25. Processing: Bouin's, paraffin, H&E.

the abdomen (e.g., spleen, stomach; Figure 7.26) and just off (Figure 7.27) or exactly on (Figure 7.28) the midline structures will be visible in two large sections. This presentation permits examination of over 30 major organs and tissues (depending on the exact trimming plane) in the thoracic and abdominal regions, and it often is preferred as a single image for demonstrating molecular expression throughout the animal. The main drawback to this longitudinal approach is that bilateral symmetry cannot be evaluated in paired organs.

Skeletal morphology may be examined using stains to highlight mineralized bone and/or cartilage.[24,33] A typical technique for this purpose is to render soft tissues of the animal translucent and then emphasize skeletal components using a double staining method (applied in sequence or simultaneously), with Alcian blue to show cartilage and Alizarin red S to reveal bone; a detailed procedure for this technique is given in Protocol 7.1, but any of the many published variants (e.g., Refs. 23, 24, 33, 41, 42) should prove suitable. If desired, the dyes may be used singly to focus the assessment on either cartilage[13,14] or bone.[33] In general, cartilage predominates from GD 12.5 to 15.5, while both bone and cartilage are evident albeit

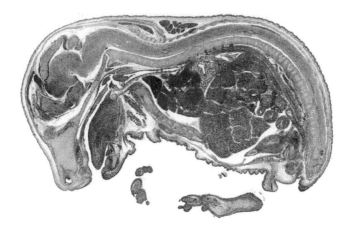

FIGURE 7.28 Mid-sagittal section of the torso from a near-term mouse fetus (GD 18.0) showing the features available for analysis if a block is produced by freehand trimming near level 2 shown in Figure 7.25. The mid-axial view is achieved by facing into the block until the spinal cord is visible throughout the entire length of the vertebral column. Processing: Bouin's, paraffin, H&E.

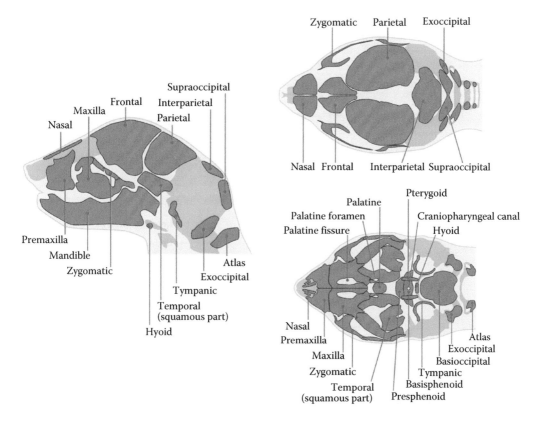

FIGURE 7.29 Schematic diagrams demonstrating the position of major bones within the normal craniofacial skeleton in a cleared near-term (GD 18.5) mouse fetus. Ossified bone is depicted in pale red, while cartilage primordia are portrayed in pale blue. The images shown are left lateral (side of face [left]), dorsal (top of head [upper right]), and ventral (bottom of skull [lower right]) views. (The diagram was produced by Mr. Tim Vojt, The Ohio State University.)

in shifting proportions from GD 16.0 to PND 21.[37] Thus, this skeletal staining approach may be used on developing mice of any age. In general, skeletal double staining is a standard component of developmental toxicology studies (i.e., near-term fetuses) and is routine in evaluations of animals with an engineered or spontaneous mutation that appears to induce a conformational change in skeletal organization. Changes in skeletal arrangement relative to the normal colors and contours in developmental stage-matched controls (Figure 7.29) are scored qualitatively by pinpointing deformed or missing elements; this requires a reasonable familiarity with normal skeletal anatomy (Figure 7.30). An exceptional reference to help identify individual bones in the developing skeleton has been developed using near-term rat fetuses,[30] while animal age may be estimated using the skeletal staining patterns of various appendicular (limb, paw, and digit) bones.[37] If bone contours, the degree of ossification, or malformations at a given time near term are the chief endpoints of interest, the skeleton may be examined by radiography (see Chapter 14 for more detail) with less labor and greater speed.[10] Other common special methods yield a targeted evaluation of gene expression (e.g., localization of the gene localization marker β-galactosidase[31,33] or cell type–specific molecules using either immunohistochemistry[33] or *in situ* hybridization[33] in mouse embryo whole mount preparations) or vascular system integrity (e.g., injection of Evan's blue[22] dye or India ink[33]). In our experience, many procedures of utility in evaluating lesions in adult mice may be adapted for examination of developing mice should the need arise.

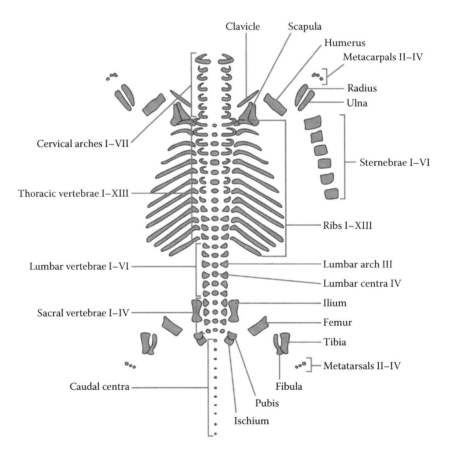

FIGURE 7.30 Schematic diagram demonstrating the position of major bones within the normal axial and appendicular skeletons of a near-term rodent fetus. The view shows the dorsal presentation of the major ossification centers and their approximate relationships in a cleared near-term (GD 18.5) mouse fetus. (The diagram was produced by Mr. Tim Vojt, The Ohio State University.)

PROTOCOL 7.1 SKELETAL STAINING IN THE DEVELOPING MOUSE

Bone and cartilage may be stained simultaneously (Option 1) or individually (Option 2).

For both options, Alcian blue stains cartilage, while Alizarin red S stains bone. The method may be performed conveniently in multiwell tissue culture plates (for smaller specimens like embryos and near-term fetuses) or screw-capped vials (specimens of any age).

Option 1: Simultaneous Staining of Bone and Cartilage (adapted from Ref. 24 as applied in Ref. 6)

Reagent recipes

Stock staining solutions

Alcian blue	0.14% (w/v)	
Alcian blue 8GX	(Sigma-Aldrich, St. Louis, MO)	140 mg
Ethanol, 70%		100 mL
Alizarin red S	0.12% (w/v)	
Alizarin red S	(Sigma-Aldrich, St. Louis, MO)	120 mg
95% ethanol		100 mL

For both stocks:
Mix to dissolve solute.
Filter to remove undissolved particulates.
Store in the dark at 4°C (for up to 3 months).

Stock clearing solution

Potassium hydroxide (KOH)	5% (w/v)	
KOH	(Sigma-Aldrich, St. Louis, MO)	25 g
Distilled water		500 mL

For stock:
Place beaker containing water on ice (the reaction generates heat!).
Slowly add KOH pellets to water while stirring.
Store at 4°C (for up to 6 months).

Working solution for double staining
Reconstitute before each use, as follows, for 250 mL of working solution:

Alcian blue stock	2 mL	(Final concentration: 0.001%)
Alizarin red S stock	4 mL	(Final concentration: 0.002%)
Glacial acetic acid	35 mL	(Final concentration: 14%)
70% ethanol	209 mL	

Approximate pH: 2.7–2.8

Working solutions for tissue clearing
Reconstitute as follows for 250 mL of working solution:

KOH, 2% (v/v)	
KOH, 5% stock	40 mL
Distilled water	60 mL
KOH, 1% (v/v)	
KOH, 5% stock	20 mL
Distilled water	80 mL

KOH, 1%/glycerol (1:1 mixture)

KOH, 5% stock	10 mL
Distilled water	40 mL
Glycerol (Sigma-Aldrich, St. Louis, MO)	50 mL

Procedure

1. Euthanize animals (leaving the head attached to the body).
2. Open the thorax and abdomen and remove all viscera.
3. Immerse fetus in hot water (approximately 65°C–75°C) for 30–60 s.
4. Use jeweler's forceps to remove all skin from the carcass, taking care that delicate appendages (e.g., digits and tail) are not detached.
5. Double stain for 48–96 h at room temperature (RT)—using less time for younger animals, which have less skeleton to stain.

   Note: Be sure to remove all air bubbles from recesses in the torso to ensure uniform access to staining solution.

6. Clear soft tissues of stained animals by serial immersions in KOH-based working solutions at RT, as follows:
   a. 2% KOH, 12 h.
   b. 1% KOH, 12 h—repeat as needed (usually once or twice) until all soft tissues are nearly translucent.

      Note: Appendages have less soft tissue, and so will clear faster than the torso.

      Note: Very rapid clearing (i.e., 2–3 h) may lead to specimen fragmentation due to loss of soft tissues needed to maintain the positions of various skeletal elements. If proceeding too quickly, the reaction may be slowed by chilling the vials to 4°C using ice or a refrigerator.

   c. 1% KOH/glycerin (1:1 mixture)—repeat as needed until soft tissues are clear.

      Note: Cleared specimens will be fully translucent except for stained skeletal elements. (Incomplete clearing usually presents as pale tan or white opacity of the muscles located along the vertebral column.)

      Note: Used solutions may be removed by decanting or by using a pipette attached to a vacuum line. If a vacuum is used, keep the pipette tip angled away from the specimen because it will soften enough over time to be drawn into the mouth of the pipette (which typically will disrupt specimen integrity to the point that a meaningful analysis is no longer possible).

7. Store stained and cleared animals indefinitely at RT in 100% glycerin.

Option 2: Sequential Staining of Bone and Cartilage (adapted from Ref. 33)

Reagent recipes

Working staining solutions

Alcian blue	0.015% (w/v)	
Alcian blue 8GX		15 mg
95% ethanol		95 mL
Glacial acetic acid	5% (v/v)	5 mL
Alizarin red S	0.005% (w/v)	
Alizarin red S		25 mg
KOH	1% (w/v)	5 g
Distilled water		500 mL

In making this solution, add Alizarin red S after the KOH has been dissolved in water.

Working solutions for tissue clearing (generate as described previously for Option 1)
Potassium hydroxide, 1% (v/v)
Potassium hydroxide, 1%/glycerol (1:1 mixture)

Procedure
1. Euthanize animals (leaving the head attached to the body).
2. Open the thorax and abdomen and remove all viscera.
3. *Optional*: Immerse the specimen in tap water at RT overnight to loosen the skin.
4. Immerse fetus in hot water (65°C–75°C) for 20–30 s.
5. Use jeweler's forceps to remove all skin from the carcass, taking care that delicate appendages (e.g., digits and tail) are not detached.
6. *Optional:* Fix the specimen overnight in 95% ethanol at RT to coagulate proteins.
7. *Optional:* Immerse the specimen overnight in acetone at RT to eliminate fats.
8. Rinse in deionized water for 15 s.
9. Incubate in Alcian blue staining solution (just enough to cover the specimen) for 24 h at RT, making sure to remove all bubbles from recesses in the opened thoracic and abdominal cavities.

   Note: This step and the next one may be omitted if the examination is to be limited to bony structures.
10. Wash Alcian blue–stained specimens in 70% ethanol for 6–8 h at RT, changing the fluid every 1–2 h.
11. *Optional:* Clear the specimen at RT in 1% KOH until tissues have visibly cleared (or overnight).
12. Counterstain in Alizarin red S solution overnight at RT.

    Note: This step may be omitted if the examination is to be limited to cartilaginous structures.
13. Clear soft tissues of stained animals by serial immersions in KOH-based working solutions at RT, as follows:
    a. 1% KOH, 12 h—repeat as needed (usually once or twice) until the soft tissues are nearly translucent.

       Note: Appendages have less soft tissue, and so will clear faster than the torso.

       Note: Very rapid clearing (i.e., 2–3 h) may lead to specimen fragmentation due to loss of soft tissues needed to maintain the positions of various skeletal elements. If proceeding too quickly, the reaction may be slowed by chilling the vials to 4°C.
    b. 1% KOH/glycerin (1:1 mixture)—repeat as needed until soft tissues are clear.

       Note: Cleared specimens will be fully translucent except for stained skeletal elements. Incomplete clearing usually presents as pale tan or white opacity of the muscles located along the vertebral column.

       Note: Used solutions may be removed by decanting or by using a pipette attached to a vacuum line. If a vacuum is used, keep the pipette tip angled away from the specimen because it will soften enough over time to be drawn into the mouth of the pipette (which typically will disrupt specimen integrity to the point that a meaningful analysis is no longer possible).
14. Store stained and cleared animals indefinitely at RT in 100% glycerin

Conclusion

The necropsy, whether of a pregnant female carrying an entire litter or of an individual developing mouse, often provides the first direct evidence regarding the true nature of an embryonic lethal phenotype. Accordingly, it is imperative that dissections be undertaken with care at all times. The identification and characterization of new phenotypes is aided substantially by adoption of a standard operating procedure for evaluating maternal and offspring endpoints that might provide clues regarding the constellation of lesions and the mechanism(s) responsible for their induction. Standardization of sample and data collection methods also greatly speeds the acquisition of developmental pathology skill in research staff while avoiding mistakes that arise when personnel are performing malleable tasks.

A well-executed macroscopic examination culminates in the detailed documentation (ideally both in images and words) of all lesions that reliably occur for a phenotype. Such representations provide a major portion of the data set needed to produce a compelling grant application, credible scientific publication, and complete product development package. Proper execution is also essential to obtain appropriately preserved tissue samples suitable for later histopathologic and molecular analyses. Accordingly, individuals who wish to begin building acumen as developmental pathologists will be well rewarded by beginning their embryological education in honing their ability to recognize and characterize gross lesions.

Appendix (Figures 7.A.1 through 7.A.10)

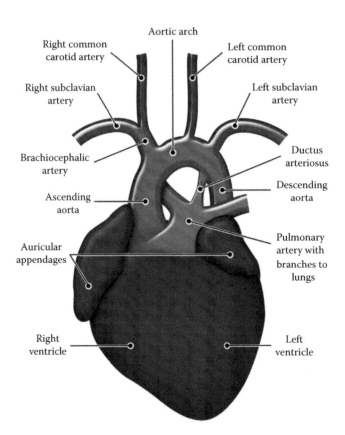

FIGURE 7.A.1   Diagram showing the heart profile and major intrathoracic arteries in a near-term (GD 18.5) mouse fetus, as viewed from the cranial face. (The diagram was produced by Mr. Tim Vojt, The Ohio State University.)

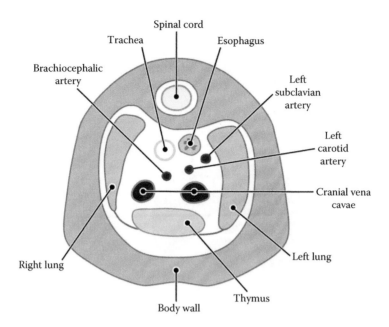

FIGURE 7.A.2 Diagram showing major structures located in the cranial thorax (near the thoracic inlet) in a near-term (GD 18.5) mouse fetus. This plane is consistent with level 1 as drawn in Figure 7.22. (The diagram was produced by Mr. Tim Vojt, The Ohio State University.)

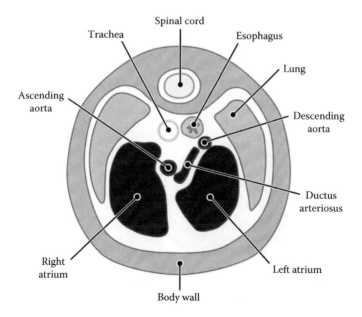

FIGURE 7.A.3 Diagram showing major structures located in the cranial thorax (near the carina, the point where the trachea branches to form the bronchi) in a near-term (GD 18.5) mouse fetus. This plane is consistent with level 2 as drawn in Figure 7.22. (The diagram was produced by Mr. Tim Vojt, The Ohio State University.)

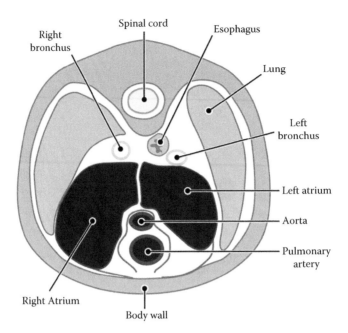

FIGURE 7.A.4 Diagram showing major structures located in the middle thorax in a near-term (GD 18.5) mouse fetus. This plane would fall midway between levels 2 and 3 as drawn in Figure 7.22. (The diagram was produced by Mr. Tim Vojt, The Ohio State University.)

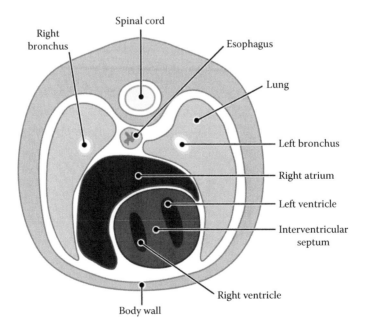

FIGURE 7.A.5 Diagram showing major structures located in the middle thorax in a near-term (GD 18.5) mouse fetus. This plane is consistent with level 3 as drawn in Figure 7.22. (The diagram was produced by Mr. Tim Vojt, The Ohio State University.)

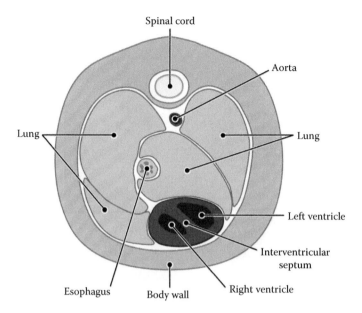

FIGURE 7.A.6 Diagram showing major structures located in the caudal thorax in a near-term (GD 18.5) mouse fetus. This plane is consistent with level 4 as drawn in Figure 7.22. (The diagram was produced by Mr. Tim Vojt, The Ohio State University.)

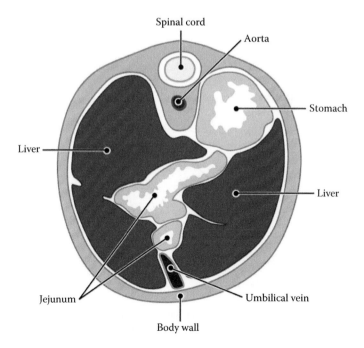

FIGURE 7.A.7 Diagram showing major structures located in the cranial abdomen in a near-term (GD 18.5) mouse fetus. This plane is near level 5 as drawn in Figure 7.22. (The diagram was produced by Mr. Tim Vojt, The Ohio State University.)

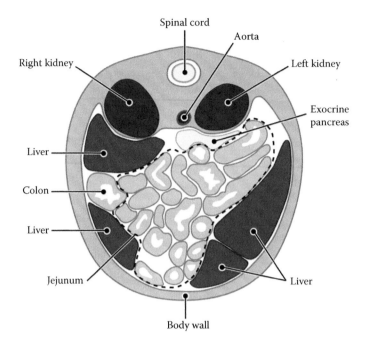

FIGURE 7.A.8   Diagram showing major structures located in the middle abdomen in a near-term (GD 18.5) mouse fetus. This plane is near level 5 as drawn in Figure 7.22 but is located more caudally. (The diagram was produced by Mr. Tim Vojt, The Ohio State University.)

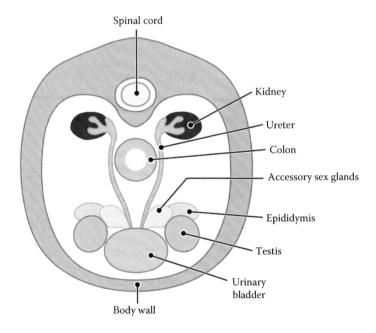

FIGURE 7.A.9   Diagram showing major structures located in the caudal abdomen of a male near-term (GD 18.5) mouse fetus. This plane is caudal to level 6 as drawn in Figure 7.22. (The diagram was produced by Mr. Tim Vojt, The Ohio State University.)

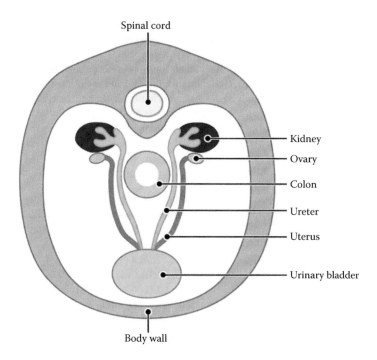

FIGURE 7.A.10 Diagram showing major structures located in the caudal abdomen of a female near-term (GD 18.5) mouse fetus. This plane is caudal to level 6 as drawn in Figure 7.22. (The diagram was produced by Mr. Tim Vojt, The Ohio State University.)

REFERENCES

1. Adams J, Buelke-Sam J (1981). Behavioral assessment of the postnatal animal: Testing and methods. In: *Developmental Toxicology*, Kimmel CA, Buelke-Sam J (eds.). New York: Raven Press, pp. 233–258.
2. AVMA (American Veterinary Medical Association) (2013). *AVMA Guidelines for the Euthanasia of Animals: 2013 Edition.* https://www.avma.org/kb/policies/documents/euthanasia.pdf (last accessed May 1, 2014).
3. Barrow MV, Taylor WJ (1969). A rapid method for detecting malformations in rat fetuses. *J Morphol* **127** (3): 291–305.
4. Beck SL (1967). Effects of position in the uterus on fetal mortality and on response to trypan blue. *J Embryol Exp Morphol* **17** (3): 617–624.
5. Bolon B, St. Omer VE (1989). Behavioral and developmental effects in suckling mice following maternal exposure to the mycotoxin secalonic acid D. *Pharmacol Biochem Behav* **34** (2): 229–236.
6. Bolon B, Welsch F, Morgan KT (1994). Methanol-induced neural tube defects in mice: Pathogenesis during neurulation. *Teratology* **49** (6): 497–517.
7. Brayton C, Justice M, Montgomery CA (2001). Evaluating mutant mice: Anatomic pathology. *Vet Pathol* **38** (1): 1–19. [Erratum in *Vet. Pathol* 38 (**5**): 568, 2001.]
8. Bronson RT (2001). How to study pathologic phenotypes of knockout mice. In: *Methods in Molecular Biology: Gene Knockout Protocols*, Tymms MJ, Kola I (eds.). Totowa, NJ: Humana, pp. 155–180.
9. Brown NA (1990). Routine assessment of morphology and growth: Scoring systems and measurements of size. In: *Postimplantation Mammalian Embryos: A Practical Approach*, Copp AJ, Cockroft DL (eds.). Oxford, U.K.: IRL Press, pp. 93–108.
10. Burdan F, Rozylo-Kalinowska I, Katarzyna Rozylo T, Chahoud I (2002). A new rapid radiological procedure for routine teratological use in bone ossification assessment: A supplement for staining methods. *Teratology* **66** (6): 315–325.

11. Calegari F, Marzesco AM, Kittler R, Buchholz F, Huttner WB (2004). Tissue-specific RNA interference in post-implantation mouse embryos using directional electroporation and whole embryo culture. *Differentiation* **72** (2–3): 92–102.

12. Committee on Guidelines for the Use of Animals in Neuroscience and Behavioral Research, U.S. National Academies of Science (2003). Guidelines for the care and use of mammals in neuroscience and behavioral research. http://oacu.od.nih.gov/GdeMammNeuro.pdf (last accessed May 1, 2014).

13. Constantinescu GM (2011). *Comparative Anatomy of the Mouse and the Rat: A Color Atlas and Text.* Memphis, TN: American Association for Laboratory Animal Science.

14. Croy A, Yamada AT, DeMayo FJ, Adamson SL (2014). *The Guide to Investigation of Mouse Pregnancy.* London, U.K.: Academic Press.

15. Davis FC (1989). Daily variation in maternal and fetal weight gain in mice and hamsters. *J Exp Zool* **250** (3): 273–282.

16. Dorman DC, Bolon B, Struve MF, LaPerle KMD, Wong BA, Elswick B, Welsch F (1995). Role of formate in methanol-induced exencephaly in CD-1 mice. *Teratology* **52** (1): 30–40.

17. Eluma FO, Sucheston ME, Hayes TG, Paulson RB (1984). Teratogenic effects of dosage levels and time of administration of carbamazepine, sodium valproate, and diphenylhydantoin on craniofacial development in the CD-1 mouse fetus. *J Craniofac Genet Dev Biol* **4** (3): 191–210.

18. Fekete E (1954). Gain in weight of pregnant mice in relation to litter size. *J Hered* **45** (2): 88–89.

19. Foster FS, Zhang M, Duckett AS, Cucevic V, Pavlin CJ (2003). *In vivo* imaging of embryonic development in the mouse eye by ultrasound biomicroscopy. *Invest Ophthalmol Vis Sci* **44** (6): 2361–2366.

20. Fox WM (1965). Reflex ontogeny and behavioural development of the mouse. *Anim Behav* **13** (2): 234–241.

21. Fujinaga M, Brown NA, Baden JM (1992). Comparison of staging systems for the gastrulation and early neurulation period in rodents: A proposed new system. *Teratology* **46** (2): 183–190.

22. Haghighi Poodeh S, Salonurmi T, Nagy I, Koivunen P, Vuoristo J, Räsänen J, Sormunen R, Vainio S, Savolainen MJ (2012). Alcohol-induced premature permeability in mouse placenta-yolk sac barriers *in vivo*. *Placenta* **33** (10): 866–873.

23. Kaufman MH (1992). *The Atlas of Mouse Development*, 2nd edn. San Diego, CA: Academic Press.

24. Kimmel CA, Trammell C (1981). A rapid procedure for routine double staining of cartilage and bone in fetal and adult animals. *Stain Technol* **56** (5): 271–273.

25. Lanning JC (1987). Whole embryo morphometry in teratogen screening. *Teratology* **36**: 265–270.

26. Law M, Stromberg P, Meuten D, Cullen J (2012). Necropsy or autopsy? It's all about communication! *Vet Pathol* **49** (2): 271–272.

27. Louton T, Domarus H, Hartmann P (1988). The position effect in mice on day 19. *Teratology* **38** (1): 67–74.

28. Marikawa Y, Alarcón VB (2009). Establishment of trophectoderm and inner cell mass lineages in the mouse embryo. *Mol Reprod Dev* **76** (11): 1019–1032.

29. McLaren A, Michie D (1959). Superpregnancy in the mouse. 2. Weight gain during pregnancy. *J Exp Biol* **36** (2): 301–316.

30. Menegola E, Broccia ML, Giavini E (2001). Atlas of rat fetal skeleton double stained for bone and cartilage. *Teratology* **64** (3): 125–133.

31. Mercer E (1995). Bacterial beta-galactosidase histochemistry bible. In: *The Whole Mouse Catalog*. http://www.rodentia.com/wmc/ (under the subject area "Laboratory" and subheading "Technical Guides and Protocols," in an entry entitled "Relatively complete guide to staining for lacZ with X-gal (E. Mercer)") (last accessed May 1, 2014).

32. Mu J, Slevin JC, Qu D, McCormick S, Adamson SL (2008). *In vivo* quantification of embryonic and placental growth during gestation in mice using micro-ultrasound. *Reprod Biol Endocrinol* **6**: 34.

33. Nagy A, Gertsenstein M, Vintersten K, Behringer R (2003). *Manipulating the Mouse Embryo*, 3rd edn. Cold Spring Harbor, NY: Cold Spring Harbor Laboratory Press.

34. Office of Animal Care and Use, Animal Research Advisory Committee, U.S. National Institutes of Health (2013). Guidelines for the euthanasia of rodent fetuses and neonates. http://oacu.od.nih.gov/ARAC/documents/Rodent_Euthanasia_Pup.pdf (last accessed April 1, 2014).

35. Papaioannou VE, Behringer RR (2005). *Mouse Phenotypes: A Handbook of Mutation Analysis.* Cold Spring Harbor, NY: Cold Spring Harbor Laboratory Press.

36. Papaioannou VE, Behringer RR (2012). Early embryonic lethality in genetically engineered mice: Diagnosis and phenotypic analysis. *Vet Pathol* **49** (1): 64–70.
37. Patton JT, Kaufman MH (1995). The timing of ossification of the limb bones, and growth rates of various long bones of the fore and hind limbs of the prenatal and early postnatal laboratory mouse. *J Anat* **186** (1): 175–185.
38. Powell KA, Wilson D (2012). 3-Dimensional imaging modalities for phenotyping genetically engineered mice. *Vet Pathol* **49** (1): 106–115.
39. Schneider JE, Wysocki CJ, Nyby J, Whitney G (1978). Determining the sex of neonatal mice (*Mus musculus*). *Behav Res Meth Instrument* **10** (1): 105.
40. Seegmiller RE, Cook N, Goodwin K, Leishman T (2012). Assessment of gross fetal malformations: The modernized Wilson technique and skeletal staining. *Methods Mol Biol* **889**: 451–463.
41. Staples RE, Schnell VL (1964). Refinements in rapid clearing technic in the KOH-Alizarin red S method for fetal bone. *Stain Technol* **39** (1): 61–63.
42. Trueman D, Stewart J (2014). An automated technique for double staining mouse fetal and neonatal skeletal specimens to differentiate bone and cartilage. *Biotech Histochem* **89** (4): 315–319.
43. von Domarus H, Louton T, Lange-Wuhlisch F (1986). The position effect in mice on day 14. *Teratology* **34** (1): 73–80.
44. Ward JM, Devor-Henneman DE (2000). Gestational mortality in genetically engineered mice: Evaluating the extraembryonal embryonic placenta and membranes. In: *Pathology of Genetically Engineered Mice*, Ward JM, Mahler JF, Maronpot RR, Sundberg JP, Frederickson RM (eds.). Ames, IA: Iowa State University Press, pp. 103–122.
45. Ward JM, Elmore SA, Foley JF (2012). Pathology methods for the evaluation of embryonic and perinatal developmental defects and lethality in genetically engineered mice. *Vet Pathol* **49** (1): 71–84.
46. Ward JM, Mahler JF, Maronpot RR, Sundberg JP (2000). *Pathology of Genetically Engineered Mice.* Ames, IA: Iowa State University Press.
47. Webster WS, Brown-Woodman PD, Ritchie HE (1997). A review of the contribution of whole embryo culture to the determination of hazard and risk in teratogenicity testing. *Int J Dev Biol* **41** (2): 329–335.

8

Clinical Pathology Analysis in Developing Mice

Kelli L. Boyd, Brad Bolon, and Denise I. Bounous

CONTENTS

Gestational and perinatal lethality is a common occurrence in mice following deliberate genetic engineering or exposure to toxicants. In many research programs, pathology evaluation of lethal phenotypes is confined to morphologic assessment for gross and microscopic defects. Unfortunately, investigations limited to structural evaluations often are unrewarding because many lethal phenotypes result from complicated physiologic abnormalities that do not manifest as anatomic lesions. As a result, a cause for any deaths that occur during the embryonic, fetal, or perinatal stages of development often remains undetermined.

Additional information may be gained when investigating the mechanisms responsible for lethal phenotypes by using routine clinical pathology methods. This approach typically utilizes automated instruments (though these may misclassify the unique cell forms found during certain stages of development) and standard diagnostic techniques to assess the cellular and/or chemical compositions of body fluids or organs. The strategy is comparable to that used in diagnosing disease in medical and veterinary practice except that control values specific to mice of the affected developmental age are required when interpreting the potential significance of analytical values obtained from unborn, neonatal, or juvenile subjects.

Clinical pathology analytes in mouse developmental pathology studies typically consist of biochemical constituents in fluids (e.g., clinical chemistry) or of cells in either fluids (e.g., hematology) or organs (e.g., cytology). Common specimens for assessing these elements are plasma or serum (for biochemical components), touch preparations of organs (for parenchymal cells and, in some cases, hematopoietic precursors), and whole blood (for more mature blood cells). Because abnormal hematopoiesis is a frequent cause of lethality in developing mice,[5,22,30,31,46,47] most literature reports incorporating clinical pathology endpoints as a component of developmental pathology studies in mice emphasize the assessment of blood cells.

The current chapter provides a foundation for clinical pathology analysis of lethal phenotypes in developing mice. Essential strategies and protocols will be detailed for clinical chemistry and hematology assessment in prenatal and early postnatal animals. In particular, information will be provided to help researchers select an appropriate specimen source, determine a suitable sample volume, and interpret a clinical pathology data set obtained from developing mice.

Sample Collection

Whole blood is the most common clinical pathology sample in prenatal and neonatal mice, and is a standard specimen in juvenile animals. The next most frequent material is a touch preparation from certain viscera (especially liver), which play a prominent role as sites of early hematopoiesis. These specimens are reported more frequently due to the critical importance of cardiovascular function—including production of sufficient erythrocyte numbers—in sustaining viability during the prenatal and early postnatal periods of development.

Biochemical constituents may be evaluated in plasma, serum, or other fluids (e.g., amniotic fluid, cerebrospinal fluid [CSF], or urine). However, in our experience, such alternative samples are not acquired as a routine testing practice in mouse developmental pathology studies. Instead, such specimens are collected and analyzed to answer specific experimental questions.

Preparation for Sample Collection

The most important requirement for clinical pathology assessment of developing mice is to have all materials prepared in advance of sample collection. Attention to detail in this regard will greatly reduce the number of nondiagnostic samples—a critical concern in developmental pathology studies of genetically engineered mice (GEM) where low numbers of animals are available for testing—and may provide less divergence among individuals because less time will be needed to sample a given cohort of animals.

Preparations for sample collection are straightforward. Appropriate tubes (for blood, plasma, serum, or another fluid) or slides (for blood smears or organ touch preparations) should be prelabeled. In our experience, the number of tubes or slides per animal that are to be prelabeled can equal the exact number needed for the analysis, but additional unlabeled tubes and slides should be kept in the collection area should an extra sample be necessary from one or more individuals. Furthermore, documentation as far as possible should be completed in advance to reduce the amount of time required for data entry during the collection. Finally, if samples are to be processed by a clinical pathology facility rather than within the researcher's own laboratory, the clinical pathology facility should be appraised of the date and approximate time at which samples will be delivered as well as the number and kind of specimens that will be submitted so that materials can be promptly processed; in our experience, at least 48 h of advance notice is warranted. Failure to notify the clinical pathology laboratory in advance may delay processing, which may increase the inter-individual variability and/or reduce the quality of the resulting data.

An obvious but sometimes neglected aspect of developmental pathology studies is that the ability to quantify any given clinical pathology analyte should be validated in advance by the clinical pathology facility. This precaution is mandated because many analytes in developing mice vary greatly across the gestational, neonatal, and juvenile periods and are often distinct from the values measured in adult mice. Therefore, the instrument settings and assays used for clinical pathology analysis in adult mice cannot be applied automatically to developing animals.

Blood Collection from Developing Mice

Withdrawal of blood during development generally is undertaken as a terminal procedure in fetuses (gestational day [GD] 15.0–18.5), neonates (postnatal day [PND] 0–5), or juveniles (PND 7–21). Blood may be harvested in several fashions, with the choice depending on the endpoints to be assessed and the speed with which the sample is to be obtained. For all approaches, the mice to be sampled are isolated (either by necropsy of the dam and removal of the conceptuses from the uterus, or by transfer of neonates or juveniles out of the home cage to a holding container) before the actual collection is initiated.

Blood volume generally is considered to represent about 7%–8% of an individual's total body weight. Therefore, body size will influence the sample size, with bigger individuals carrying a larger quantity of blood. Accordingly, the expected weight of the subjects to be sampled should be determined in advance from pilot studies or historical control databases so that collection tubes of the proper size are available. For example, the mean blood volume available for collection rises from approximately 40 µL at GD 15 to 135 µL at GD 18.[8] These sample volumes will partially to completely fill standard capillary tubes and are suitable for processing in standard microsized collection tubes (e.g., Microtainer®; BD Diagnostics, Franklin Lakes, NJ). Greater blood volumes may be obtained at any given developmental age from outbred mouse stocks, which tend to be larger in size and thus have a correspondingly higher blood volume compared to inbred mouse strains. Invasive methods like decapitation or vascular penetration[8] typically will yield more blood relative to less aggressive approaches like incising the skin over a superficial blood vessel[21] or severing the umbilical cord and its blood vessels.

Techniques for Blood Collection

The most straightforward strategy for blood withdrawal is acquisition from the exposed neck vessels following decapitation.[31,37,44,45] This procedure is quite successful in fetuses and neonates (up to PND 5) because it provides rapid blood flow combined with instantaneous euthanasia. The most reliable technique is to restrain the subject between the thumb and the forefinger (index finger) of the nondominant hand. The other hand is used to hold a clean pair of sharp scissors or a new razor blade. Position the animal over a clean surface, like a piece of plastic wrap or a table top (laminate [e.g., Formica®] or stainless steel). The individual's head is removed rapidly, and the collection apparatus—usually a glass capillary tube (e.g., Microvette®; Sarstedt, Inc., Newton, NC) or plastic collection tube with integrated funnel (e.g., Microtainer; BD Diagnostics) (Table 8.1)—is positioned in the center of the blood pool accumulating about the severed neck vessels. The carcass should be held at a slight downward angle to maximize the sample size. Another means of increasing sample volume is to gently compress (*milk*) the thorax to

TABLE 8.1

Selected Disposables Used in Collecting Common Types of Blood Samples from Developing Mice

Item	Common Brand	Catalog No.	Vendor
Capillary tube (EDTA- or heparin-coated)	Multiple	Multiple	Multiple
Pipette bulb for capillary tube	—	51675	Globe Scientific, Inc., Paramus, NJ
Collection tube with integrated EDTA-coated capillary tube	Microvette	Multiple	Multiple
Collection tubes with integrated funnel	Microtainer		BD Diagnostics, Franklin Lakes, NJ
Chemistry: serum			
No anticoagulant or clot activator (red top)		365957	
Clot activator, no anticoagulant (gold top)		365956	
Hematology: whole blood			
EDTA as anticoagulant (lavender top)		365973	
Heparin as anticoagulant (green top)		365958	

Note: EDTA, ethylenediaminetetraacetic acid.

mimic normal circulation, but applying such pressure may distort visceral structure and therefore should be avoided if blood collection is to be followed by histopathological evaluation. The main disadvantage of the decapitation method is that the sample will contain a mixture of arterial and venous blood along with a variable quantity of tissue fluid (though the latter component should be small relative to the quantity of blood). In our experience, this intermingling does not impact hematologic analysis and interpretation to a significant degree even though the tissue interstitial fluid may activate the clotting cascade.

Other approaches to whole blood collection have been reported in the mouse developmental pathology literature. Such strategies involve severing a specific arterial trunk (e.g., carotid, umbilical) or vascular bed (e.g., brachial plexus) or alternatively by opening a cardiac ventricle or intrathoracic artery followed by blood collection with capillary tubes, micropipettes, or short, blunt, 20-gauge needles.[8,21,35] A possible advantage to these strategies is that tissue fluid contamination may be reduced by limiting tissue trauma to the immediate perivascular tissues. However, in the authors' experience, several disadvantages often offset this potential benefit. The most important is that dissection to open the heart or a large, deeply located artery requires that the animal be anesthetized (see Chapter 7) prior to collection. From a technical perspective, developing mice are anesthetized readily *in utero* by keeping the dam at a surgical plane of anesthesia for approximately 10 min. However, profound sedation will reduce the heart rate and thus the pace at which blood circulates, which in turn may extend the time required for blood collection and also reduce the final sample size. This effect may be compounded by selective entry into only one or two arteries, which usually offers more sluggish blood flow relative to the rate offered by the massive opening of multiple vessels attained during decapitation. In our experience, slower blood collection also is more likely to produce a sample of lesser quality. Finally, the need to sequentially introduce a cutting instrument (to transect vessels) and the collection apparatus into the thorax likely will induce mild or greater trauma to the heart and/or lungs. Given the importance of histopathological examination of these organs in conventional mouse phenotyping protocols, we typically perform such invasive collections only in individuals not slated for later microscopic evaluation, or we utilize a peripheral site (e.g., carotid or umbilical artery) for blood collection.

Once the sample is obtained, the collection tube should be capped. If necessary, the blood should be well mixed with anticoagulant by several inversions (typically 5–10) or gentle vortexing (3–5 s).

Timing of Blood Collection

The key to obtaining high-quality clinical pathology data from blood samples is to properly match the collection schedule with the experimental design. In mouse developmental pathology studies, this balance may require a trade-off between the collection times for anatomic pathology and clinical pathology specimens as both sample types are harvested *postmortem*.

In general, mouse developmental pathology studies include multiple littermates (usually 4–12, depending on the strain and the extent of gestational lethality) or pups (typically 2–5 per genotype). The inherent inter-individual variability of clinical pathology values may be reduced to some degree by collecting blood samples from all littermates or pups at the beginning of the necropsy (i.e., immediately after a brief assessment ± photodocumentation of any gross external malformations) while delaying collection of tissue samples for histological examination until all clinical pathology specimens have been gathered. This strategy is possible because embryonic, fetal, and neonatal mice are resistant to hypoxia[20,32,33]; therefore, gross external observations and blood may be taken from a series of developing animals at once without a substantial impact on the structural integrity of tissue samples slated for subsequent histopathological evaluation. The order in which blood is collected should be randomized across all treatment groups when cohorts of similar mice can be defined reliably in advance based on the experimental design (e.g., control and dosed groups in a developmental toxicology study).

The brief delay in tissue fixation that results from acquiring the clinical pathology specimens from all animals at the start of the necropsy typically is not an issue for subsequent anatomic pathology analysis. In the usual case, embryos, fetuses, and neonates (up to PND 5 or so) may be immersed in fixative without an examination for gross internal malformations, so the delay between blood sampling and tissue preservation is limited to the time required to collect the blood sample and open the thoracic and abdominal cavities. Juvenile mice (PND 7 or older) typically receive a gross external examination comparable to that accorded to adult animals, which may lead to a longer delay between death/blood collection and tissue fixation. In our experience, necropsies in juvenile mice often consist of opening the ventral thorax and abdomen to permit viewing of organ colors, locations, shapes, and sizes *in situ* rather than the typical practice in adult animals of extraction followed by a detailed analysis of individual organs; this juvenile assessment can be performed in 2–3 min by an experienced researcher and so should not greatly delay the time between death and tissue fixation. Autolysis may be slowed by placing juveniles from which blood has been taken on ice covered with saline-moistened gauze until they may be examined in their turn. Should the decision be made to further reduce this brief delay, a reasonable option is to have one technician collect all blood samples and then have a rotation of additional technicians (usually one or two) available to undertake the gross examinations.

Blood Sample Types in Mouse Developmental Pathology

The kind of blood specimen to be collected is dictated by the requirements of the experiment and the equipment. Modern clinical pathology laboratories perform as many processing and analytical methods as feasible using automated systems. The three main categories of clinical pathology specimens derived from blood are whole blood (used for hematological evaluation), and plasma and serum (used for clinical chemistry measurements). The battery of analytes slated for measurement will depend on both the sample type and the sample size (Table 8.2).

TABLE 8.2

Common Clinical Pathology Tests Suitable for Use in Mouse Developmental Pathology Studies

Test Category	Sample	Test	Organ/System Assessed by Test	Assay Volume[a]
Hematology	Blood	CBC[b]	Hematopoiesis (all lineages)	20 µL
		Blood cell differential (manual)	Hematopoiesis (all lineages)	1–2 µL
		Hematocrit	Hematopoiesis (RBCs, WBCs)	1–2 µL
Cytology	Tissue[c]	Blood smear	Hematopoiesis (all lineages)	1–3 µL
		Squash preparation	Hematopoiesis (all lineages)	1–2 mm³
		Touch preparation	Hematopoiesis (all lineages)	N/A
Clinical chemistry	Fluid[d]	Albumin	Kidney—glomerular integrity Liver—hepatocyte function	3 µL
		Amylase	Pancreas—exocrine cell integrity	3 µL
		Aspartate aminotransferase (AST)	Liver—hepatocyte integrity	14 µL
		Bilirubin (total)	Liver—biliary function	20 µL
		Blood urea nitrogen (BUN)	Kidney—excretion ± Liver—hepatocyte function	3 µL
		Calcium	Metabolism—bone	3 µL
		Cholesterol	Metabolism—fat and liver	3 µL
		Globulins	Immune system	Calculated
		Glucose	Metabolism—kidney and liver	3 µL
		Phosphorus	Metabolism—bone and kidney	3 µL
		Protein (total)	Kidney—glomerular integrity Liver—hepatocyte function	3 µL
		Triglycerides	Metabolism—fat and liver	3 µL

Notes: CBC, complete blood count; RBC, red blood cells (erythrocytes); WBC, white blood cells (leukocytes).

[a] Sample sizes as shown are the actual volumes required for analysis. In general, when calculating sample needs for automated analysis, double the sample size to accommodate *dead space* within tubing as well as waste during pipetting.

[b] Standard CBC includes blood cell differential and hematocrit determinations.

[c] Common cytology specimens are from blood-forming organs (chiefly liver, but also spleen and thymus in fetuses or bone marrow and/or spleen in neonates and juveniles).

[d] Common clinical chemistry samples include plasma (fluid from blood containing an anticoagulant), serum (fluid from coagulated blood), and occasionally urine.

Hematology

Immediately after acquisition, whole blood must be mixed thoroughly with an anticoagulant to prevent clotting. The main anticoagulants are ethylenediaminetetraacetic acid (EDTA; lavender [*purple*]-topped blood collection tubes) and heparin (green-topped tubes). The choice of anticoagulant should be confirmed with the clinical pathology laboratory prior to sample harvesting because some automated hematology analyzers preferentially employ one anticoagulant solution. The usual choice anticoagulant for mouse developmental pathology studies will be EDTA.

Blood samples treated with EDTA are suitable for automated measurements of standard hematology analytes (e.g., complete blood count [CBC], cell differential, hematocrit) as well as preparation of blood smears for manual assessment of blood cell morphology. The whole blood volume required for hematologic analysis varies with both the parameters to be measured (Table 8.2) and the analytical instruments to be used. The least amount of blood required for automated hematology analysis is 20 µL, that needed to determine a microhematocrit is 1–2 µL,[15] while a smear may be made using a single drop of blood (approximately 1–3 µL[15]). These sample size restrictions indicate that analysis of all major blood analytes from individual mice will be limited to fetuses (GD 15 or older), neonates, and juveniles. Assessment of blood analytes in younger embryos (GD 14 and earlier), as well as for smaller fetuses, typically will require a more limited battery of tests[15] or combining (*pooling*) samples from multiple subjects (see the section on *Sample Extension and Pooling*). Individual values may be acquired from younger conceptuses using EDTA-coated capillary tubes, but the limited sample volume accommodated by such tubes (1–2 µL) usually will limit the menu of hematological measurements to either the hematocrit and acquisition of a very small plasma sample (approximately 0.5–1 µL) or to preparation of a blood smear.

Blood samples mixed with EDTA may be retained at 4°C for up to 96 h. The most likely reason for delaying the clinical pathology portion of a mouse developmental pathology study is to await the outcome of genotyping data needed to define the sample pooling strategy. If the automated analysis is to be delayed, blood smears should be made within 6 h of sample collection to avoid artifactual changes to leukocyte (white blood cell [WBC]) morphology.

Clinical Chemistry

The plasma (fluid) obtained when an anticoagulated (EDTA or heparin) blood sample is centrifuged is an acceptable substrate for common clinical chemistry analytes. The exception to this is calcium (Ca^{++}) measurements in EDTA-treated specimens as this cation is chelated and removed by the EDTA. Plasma (up to 1 µL) may be collected from capillary tubes after the hematocrit has been measured. The tube is scored just above the leukocyte layer (the *buffy coat*, a thin white band of cells separating the red blood cell [RBC] column from the fluid plasma) so that the tube may be broken near the border of the fluid with the buffy coat. The plasma now is expelled from the capillary tube into another container using a pipette bulb fitted to the intact end. Alternatively, plasma may be withdrawn from blood collection tubes (following centrifugation) by using a narrow-tipped glass pipette. The fluid may be held at 4°C for up to 24 h or for up to 4 weeks at −20°C if analysis must be delayed. Plasma analytes that have been evaluated during mouse development include dye binding by plasma proteins[28] as well as levels of glucose[14] and various hormones.[49] Plasma from embryos and fetuses also serves a good substrate for analyzing the pharmacokinetic properties of chemicals and drugs.[39,41,43]

Serum (the fluid produced when whole blood is allowed to coagulate) may also be used as a substrate in mouse developmental pathology studies. This fluid is collected by harvesting blood into tubes that do not contain an anticoagulant (red-topped tubes), although some tubes may have a clotting activator (gold-topped tubes). After approximately 30 min, specimens are centrifuged to express serum from the clot, after which the serum is removed by pipette. This fluid may be stored under similar conditions as those used for plasma. Examples of clinical chemistry analysis performed in serum of developing mice are measurements of glucose and hormone levels in near-term fetuses[37] and circulating iron levels at PND 7.[40] In our experience, serum is chosen much less frequently as a substrate for mouse developmental pathology studies because the blood volumes of embryos, fetuses, and neonates are too small to permit splitting the sample between collection vessels containing and lacking anticoagulant.

TABLE 8.3

Tiered Strategy for Low-Volume Clinical Pathology Analysis during Mouse
Developmental Pathology Studies

	Test Category	Sample	Test	Assay Volume (μL)[a]
Tier I	Hematology	Blood	Complete blood count (CBC)	20
			Blood cell differential (manual)	1–3
Tier II	Hematology	Blood	Complete blood count (CBC)	20
			Blood cell differential (manual)	1–2
	Clinical chemistry	Fluid[b]	Blood urea nitrogen (BUN)	3
			Calcium	3
			Cholesterol	3
			Glucose	3
			Phosphorus	3
			Protein (total)	3

[a] Sample sizes as shown are the actual volumes required for analysis. In general, when calculating sample needs for automated analysis, double the sample size to accommodate dead space within tubing as well as waste during pipetting.

[b] Common clinical chemistry samples include plasma (fluid from blood containing an anticoagulant) or serum (fluid from coagulated blood).

Many different enzymes and molecules are routinely measured in clinical chemistry panels for adult mice, but the small plasma volume available from developing animals will limit the number of analytes that can be examined in a single specimen. For screening purposes, we recommend a small battery of tests that assess the function of major organ systems (Table 8.3). The exact menu of constituents to be measured can be varied depending on the experimental question and the amount of plasma or serum available for analysis.

When warranted by the clinical presentation or predicted phenotype, other fluids may be subjected to clinical chemistry analysis. Common choices in this regard are amniotic fluid, CSF, and urine. All these fluids may be collected at necropsy by inserting small-gauge needles or drawn glass micropipettes into fluid-filled cavities; urine also may be harvested by holding live neonatal or juvenile mice over a clean metal or plastic surface (e.g., stainless steel table top or plastic wrap) to catch voluntarily voided urine. Amniotic fluid typically has been examined to measure the levels of chemical mediators,[10,25] metabolic substrates,[7,16,21,38] and enzyme activities.[17,19] Analysis of embryonic CSF has evaluated similar types of endpoints.[13,25,51] Urine generally is used to measure analytes indicative of renal damage (e.g., altered osmolarity,[50] proteinuria[9]).

Other Considerations in Fluid Analysis for Mouse Developmental Pathology Studies

Unique features of developmental pathology studies often impose additional constraints on the experimental design when fluids from developing mice are to be included in the evaluation. Particular characteristics that may require compromises when establishing the sampling scheme include the small size of embryos and early fetuses, which limits the fluid volume that may be collected, and the difficulty in obtaining adequate numbers of conceptuses that are matched for both developmental age/stage and treatment (e.g., genotype, xenobiotic exposure). The impact of these factors may be minimized by using one or more pragmatic adjustments to the sample and/or the sampling paradigm so that information may be acquired from individual animals, thereby allowing correlation of the clinical pathology data with anatomic pathology findings (e.g., body and organ weights, gross and microscopic findings, noninvasive imaging).

Deliberate Selection of a Limited Test Battery

The simplest approach to ensuring that clinical pathology results may be tied to a single individual is to reduce the number of tests performed on small-volume samples. Many essential clinical pathology tests may be performed on the small specimens that are commonly derived from a single mouse embryo or fetus (Table 8.3). Routine hematologic procedures like a blood smear or microhematocrit may be made easily with 3 μL of whole blood on suitable analyzers (see below). In like manner, given appropriate instrumentation (see the following text), multiple chemistry analytes may be measured using only 3 μL of plasma, serum, or other fluids; common preferences include albumin, BUN, calcium, cholesterol, glucose, phosphorus, total protein, and triglycerides. Taken together, these tests would require an initial blood volume of 6 μL for the two hematologic endpoints and 24 μL of plasma or serum (i.e., 48 μL of whole blood) for these eight blood chemistry assays. In practice, however, we have found that the assay volumes used to calculate the desired sample volume should be increased by 50%–100% to compensate for the dead space in instrument tubing and to counteract any potential sample wastage during processing.

Additional parameters may be evaluated if larger sample volumes are available for analysis (Table 8.2). The most useful option for routine mouse developmental pathology investigations is the automated CBC, with an assay volume of 20 μL (in whole blood). Other obvious choices to evaluate are AST and total bilirubin (which necessitate assay volumes of 14 and 20 μL of plasma or serum, respectively).

In practice, we often employ a two-tiered testing strategy when designing the clinical pathology analysis for mouse developmental pathology studies (Table 8.3). Tier I utilizes an automated CBC, and often a manual examination of a blood smear, to examine hematopoietic system integrity. The necessary sample is whole blood (approximate sample volume, 25 μL). This approach reflects the importance of abnormal oxygen delivery as a cause of decreased prenatal and perinatal viability, and the possibility that defective erythropoiesis plays a role in deficient circulation. Tier II encompasses selected hematologic parameters (automated CBC and blood smear) as well as major clinical chemistry analytes (e.g., BUN, calcium, cholesterol, glucose, phosphorus, and total protein). Taken together, this expanded battery evaluates multiple organs and systems as well as general metabolic function. The fluid volume needed for this chemistry analysis is 18 μL (harvested from approximately 36 μL of whole blood).

Sample Extension and Pooling

In our experience, a good guideline when planning a mouse developmental pathology study is that enough blood may be gotten from a single developing mouse (i.e., neonates or juveniles that are 1.5 g or larger in weight) to permit either a comprehensive hematology analysis *or* a fairly complete clinical chemistry analysis, but not both. Small subjects (i.e., fetuses younger than GD 16) usually provide a sample volume sufficient for only a few tests unless some means of increasing the quantity is employed.

A common means of extending sample volumes is to mix an individual sample with an equal volume of physiologic saline (pH 7.4). This strategy should permit the automated assessment of blood and plasma specimens from fetuses (GD 15 and older) to be assessed as individuals. The main drawbacks to this approach are that the results likely will fall below the validated range of linearity for the analytical instrument, thereby rendering the data invalid, and that values for some clinical chemistry analytes (mainly electrolytes) likely will be affected by the composition of the diluent.

Another alternative for acquiring a sample of adequate size is to pool blood or plasma from equivalent animals (i.e., those that have an identical genotype and/or that received the same treatment). The usual practice is to combine samples from as few animals as possible so that the largest possible number of values will be obtained for analysis. In our experience, pooling blood from two (at GD 18.5) to four (at GD 15.0) fetuses will provide enough sample for conventional hematology analysis *or* clinical chemistry analysis. We have not pooled samples from embryos during the course of our mouse developmental pathology projects. Pooling might be possible for studies where large numbers of similarly treated embryos can be obtained at a given time (e.g., for a developmental toxicity experiment) but is unlikely to be feasible for most GEM projects in which a few litters—and even fewer embryos of a given genotype—are available for assessment at any given time.

Timing of Analytical Runs

The inter-individual variability for clinical pathology analytes in age-matched mice of a particular geno-type or given a specific dose of toxicant typically is greater than that observed for structural changes. Accordingly, care must be taken to reduce biases resulting from skewing of the experimental design.

In the ideal case, all treatment groups will be necropsied on the same day over a short period. This strategy will allow a randomized collection order, intermingling control and treated animals over time. Similarly, samples can be processed quickly and in random order, often in a single analytical run. Taken together, these tactics will substantially lower the likelihood that bias will have occurred during sample collection and analysis, which improves the confidence accorded to the statistical analysis of the data.

In practice, however, mouse developmental pathology studies often require many necropsies scattered over several days, especially in lethal phenotypes for GEM lines where the affected genotype occurs at a low frequency. Such scenarios usually are dealt with by collecting samples and storing them until a single large analytical run may be conducted. However, when specimen pooling is not required, we recommend that clinical pathology samples be analyzed immediately (along with samples from appropriate control animals) to reduce any artifactual changes that might develop during extended storage.

Cytological Preparations

Assessments of cell morphology in clinical pathology usually are directed toward evaluation of the hematopoietic system. This strategy may prove especially useful for the analysis of lethal phenotypes in developing mice, which often result from alterations in primitive (*embryonic*) or definitive (*adult*) hematopoiesis or both.[4,12] Cell populations that might need to be examined in this regard include the cir-culating blood cells as well as the stem cells within blood-forming organs (liver mainly, but also spleen, thymus, and yolk sac during gestation as well as bone marrow in neonates and juveniles).

Smear Preparations of Blood

A drop of blood is placed near one end of a clean glass microscope slide, usually within 5–10 mm of the label's edge. A *spreader* slide is then placed on the longer (unlabeled) end of the slide, with the drop located between the label and the *spreader*, positioned at approximately a 45° angle. The *spreader* slide then is pulled back into the blood drop, the drop is allowed to spread along the edge of the angled slide by capillary action, and the *spreader* is rapidly advanced along the slide in the direction away from the label to produce a tapering oval smear. The smear naturally will be thicker near the site of the original drop and become less dense (*feathered*) at the distal edges.

Touch and Squash Preparations of Blood-Forming Organs

Organ cytology during mouse developmental pathology studies usually is confined to the blood cell lin-eages in the mid-gestational liver. This preference is dictated by the rapid evolution of hematopoietic cell populations within this organ from approximately GD 10 to term, with peak production between GD 12 to GD 16,[4] as well as by the relatively large size of this organ, which permits the specimen manipulations required to obtain a good-quality touch preparation.

The liver is isolated from the subject and grasped at one margin using a fine forceps. A new razor blade is used to create a flat, unencapsulated surface through the middle of one or more large lobes. Next, this flat surface is touched to a clean glass slide several times (usually four to eight), starting near the label and progressively moving away.

An alternative approach is to produce a squash preparation. In this method, a small drop of dilu-ent is placed at the center of a clean glass slide. Suitable diluents include 5% bovine serum albumin (BSA), allogenic mouse serum, or fetal calf serum (FCS) mixed in the ratio 2:1 with 7.5% EDTA (in PBS, pH 7.4); small (e.g., 0.5–1 mL) aliquots of diluent can be stored at −20°C in individual vials for months and thawed as needed for studies. Next, a drop of diluent is placed in the center of the slide, and a tiny (several mm^3) piece of tissue (typically removed from the core of a liver lobule, but also including spleen and thymus) is placed within the center of the drop. A second *squasher* slide is placed on the drop; the long axes of the slides should be perpendicular (at right angles) to each other. The weight of the *squasher* slide should be the only pressure applied to the tissue. The two slides are then pulled apart, taking care not to crush the sample and thereby disrupt the cells.

Technical Considerations for Cytological Preparations

Several technical issues should be regarded when undertaking cytological preparations in mouse developmental studies. The two most common are subsequent processing steps and the potential for inducing artifacts.

Blood smears as well as squash and touch preparations are processed in a comparable way. Typically, freshly prepared specimens are air-dried for 10 min to overnight in an enclosed container (to prevent dust from settling on the cells). The samples may be stained without fixation if cytological examination will be performed within 24 h. If a longer delay is anticipated, dried specimens should be fixed by immersion in absolute methanol (1–2 min at room temperature [RT]) and then air-dried again. Fixed and dried slides generally are stored in air- and moisture-tight boxes at −20°C until analysis.

Common stains for cytomorphology are conventional Romanowsky stains such as Giemsa and Wright-Giemsa. More consistent staining is usually possible if a high-quality brand of stain is acquired from a reputable vendor and the specimens are processed using an automated stainer. Distinct staining times are recommended by the vendor for blood smears, while staining times for tissue preparations are adjusted empirically to be longer due to the greater thickness of the latter samples. An alternative to validating an extended staining protocol for tissue preparations is to process them twice on the automated instrument using the vendor-supplied method for staining smears. Special procedures may be required to evaluate particular cell lineages (e.g., supravital dyes like new methylene blue for reticulocytes, myeloperoxidase for granulocytes[6]).

The primary artifacts in cytological preparations of blood-forming organs are cell lysis and excessive thickness. The main cause of the former defect is a traumatic preparatory technique, while the major source of the latter artifact is prolonged drying, leading to cell condensation and aggregation. Both anomalies are to be avoided as they may compromise subsequent efforts to identify lineage-specific cytoarchitectural features.

Clinical Pathology Analysis

The key to reliable assessment and interpretation of clinical pathology data acquired in mouse developmental pathology studies is to devise and maintain optimized analytical protocols. This prerequisite is fairly simple in practice as most clinical pathology laboratories possess automated instruments that are pre-equipped with mouse-specific analytical software.

Instrumentation

Two hematology instruments that are commonly used for mouse developmental pathology studies are the Forcyte™ Hematology Analyzer (Oxford Science, Inc., Oxford, CT) and the Hemavet® Hematology System (Model 950 LV [low volume]; Drew Scientific, Waterbury, CT). Both instruments reliably provide a CBC and blood cell differential using only 20 μL of anticoagulated blood. A needed ancillary instrument is a capillary tube centrifuge for processing hematocrit samples; these are available from many vendors. All this equipment is used according to the manufacturer's protocols.

Many different clinical chemistry analyzers are available in diagnostic laboratories that do clinical pathology analyses of laboratory animal specimens. Sample size requirements vary with the instrument and thus with the facility, so researchers will need to consult their laboratory prior to designing experiments with clinical chemistry endpoints. In our experience, standard diagnostic instruments such as the VetACE™ Clinical Chemistry System (Alfa Wasserman Diagnostic Technologies,

West Caldwell, NJ) are suitable for small-volume mouse samples. The small sample size generally will limit the number of analytes that can be examined in a single specimen. For screening purposes, we recommend a small battery of clinical chemistry tests that assess the function of major organ systems, to be selected from a larger menu (Table 8.2) based on the observed phenotype for the mouse line being investigated and/or the research interests of the primary investigator.

Interpretation of Hematology Data from Developing Mice

A primary role for clinical pathology analysis in mouse developmental pathology studies is to assess aberrant hematopoiesis as a potential explanation for early lethality during the prenatal, perinatal, or pre-weaning periods. This emphasis stems from the critical requirement for a fully functional cardio-vascular system beginning early in embryogenesis. Successful interpretation of clinical pathology data from developing mice necessitates a good preliminary understanding of previously documented hema-topoietic events.

Erythrocytes (RBCs) produced during different stages of development have distinct anatomic and functional traits. Mouse erythropoiesis occurs in two overlapping waves, a primitive (*fetal*) phase from early embryogenesis to term and a definitive (*adult*) phase that begins in the middle of embryogenesis and continues through the remainder of the animal's life. The main distinguishing feature of primitive erythrocytes (Figure 8.1) is their large size (average volume, 465–530 fL)—which is approximately six times greater than that of definitive erythrocytes (Figure 8.1).[29] This attribute can be reliably evaluated using both automated hematology analyzers and cytological examination of blood smears and liver touch preparations. Both primitive and definitive RBCs are released during most of embryogenesis (E10–E18), although the ratio gradually shifts over time to include a larger percentage of definitive cells. Primitive erythrocytes remain in circulation until as late as PND 5.[18]

Primitive erythrocytes are formed in the *blood islands* of the yolk sac (Figure 8.2). These collections of hematopoietic precursors are active from about GD 7.0 to GD 11.5 (reviewed in Ref. [4]). These RBCs circulate as nucleated cells with abundant cytoplasm until approximately GD 12.5. Subsequently, their nuclei become condensed and are gradually lost between GD 14.5 and GD 16.5.[18,29] Recently formed primitive erythrocytes have pale basophilic cytoplasm as a consequence of their extensive ribosomal

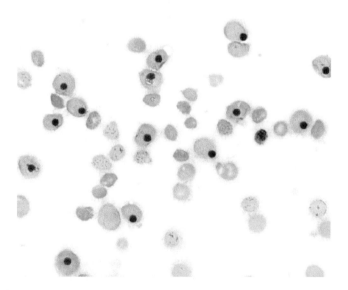

FIGURE 8.1 The main distinguishing features of primitive erythrocytes are their prominent nuclei and their large size (approximately six times greater than definitive erythrocytes). The blue cytoplasmic granules in the reticulocytes (smaller, enucleated erythrocytes) represent ribonucleic acid (RNA). Peripheral blood smear from an embryonic mouse at GD 13. Modified Romanowsky stain, 1000×.

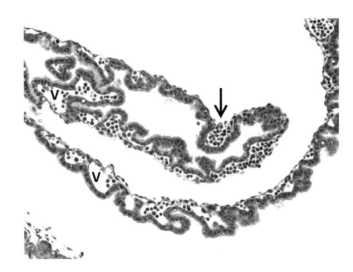

FIGURE 8.2   Hematopoiesis in the developing mouse embryo (here at GD 9.5) is initiated in densely clustered aggregates of small, dark precursors cells (*blood islands* [arrow]) within the visceral yolk sac. Circulating erythrocytes are visible in yolk sac blood vessels (V). H&E, 200×.

capacity for protein production. The color fades to its normal pale eosinophilic shade as the maximal hemoglobin content (80–100 pg/cell[29]) is attained.

As noted earlier, definitive erythrocytes (Figure 8.1) can be readily distinguished from their primitive counterparts by their smaller size. These RBCs are produced primarily by the liver during gestation (GD 10.0 to term, with peak production between GD 12 and GD 16), followed later by the spleen (starting at GD 15; Figure 8.4) and bone marrow (beginning about GD 16) (reviewed in Ref. [4]). An ancillary diagnostic feature is that definitive RBCs do not circulate as nucleated cells under normal physiological conditions. As with primitive cells, newly formed definitive erythrocytes have pale basophilic cytoplasm that gradually reaches its normal pale eosinophilic tint as hemoglobin content peaks. Consistent with their smaller size, definitive RBCs have a maximal hemoglobin amount that is about sixfold less than that of primitive erythrocytes.[29]

A significant component of the hematological assessment in developing mice, especially prior to birth, is routine histopathologic examination of the liver. Major characteristics of interest will be the number of hematopoietic cell lineages that may be discerned, as well as the cell numbers and patterns in which the various lineages are distributed within the organ (Figure 8.3). In general, the mass of all hemato-poietic cells dwarfs that of the hepatic parenchyma during mid-gestation, decreasing gradually until extramedullary hematopoiesis within the liver after birth dwindles to scattered cell islands. Erythrocytic precursors are the most abundant blood cell type, and are visible as aggregates of dark basophilic cells within hepatic sinusoids. Myeloid progenitors also are quite common, and typically tend to congregate near larger blood vessels; most of these large, pale cells have features consistent with those of neutrophils (e.g., twisted nuclear profiles) and macrophages. Megakaryocytes appear as individual large, multinucle-ated cells within sinusoids. Histopathologic evidence of altered hematopoiesis generally includes the reduced production of one or sometimes multiple cell lineages (Figures 8.4 and 8.5), the extent of which generally correlates well with the degree of hematopoietic disruption suggested by routine hematologic analysis of whole blood (Figure 8.5).

Anatomic and functional features of other hematopoietic lineages also evolve substantially during the course of development. For instance, primitive megakaryocytes from the yolk sac contain fewer nuclei of lower ploidy, have smaller diameters (about half as large), and respond differently to cytokines relative to definitive megakaryocytes from adult mouse bone marrow.[42] Similarly, primitive macrophages arise in the yolk sac starting at approximately E8.0.[2] A key to understanding hematopoietic evolution during development is that many hemangioblasts are only partially committed, and thus can give rise to more than one blood cell lineage (reviewed in Ref. [4]).

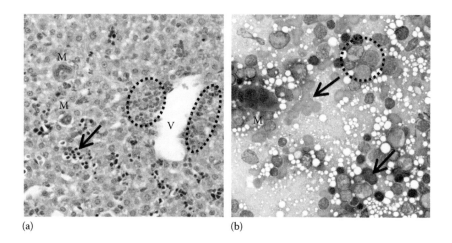

(a)   (b)

FIGURE 8.3   Definitive hematopoiesis (shown here at postnatal day [PND] 1) is extensive in the liver of developing mice from mid-gestation (GD 10) until well after birth. Multiple hematopoietic lineages may be seen in both tissue sections (a) and impression smears (b). Clusters of small, dark erythrocytic precursors (arrows) and giant, multinucleated megakaryo-cytes (adjacent to M) are dispersed randomly in sinusoids. Aggregates of large, pale blue myeloid cells (broken circles) encircle larger vessels (V). Mature erythrocytes (arrowheads) form stacks of small, pale pink round cells within the impres-sion smear. (a) H&E, 600×; (b) Wright's, 1000×.

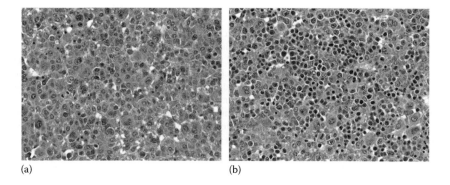

(a)   (b)

FIGURE 8.4   Abnormal erythropoiesis in an anemic knock-out mouse fetus (GD 15.5) is characterized by a profound reduction in the number of small, dark precursor cells within sinusoids (a) relative to the myriad progenitors visible in the age-matched, wild-type counterpart (b). H&E, 400×.

Numbers of circulating blood cells in young animals typically diverge substantially from correspond-ing cell numbers in adults (Table 8.4). For instance, the hematocrit[35] and erythrocyte counts[23] (as reported in Refs. [27] and [35]) more than double between birth and young adulthood. In contrast, young mice have at least 10-fold more circulating reticulocytes than adults, with numbers reported to range from 40%[27] to 90%.[1] Total WBC numbers just before or at birth are approximately 20% of adult levels[35] and undergo a further transient decline during the early postnatal period before increasing to adult numbers by 6–7 weeks of age[36]; this phenomenon appears to reflect a temporary deficit of lymphocytes, and is reflected in the rudimentary features of lymphocyte-rich structures such as the splenic white pulp (Figure 8.6) and thymus (Figure 8.7). However, counts of circulating monocytes are threefold to fourfold higher in near-term fetal mice relative to 9-week-old adults.[35] Circulating platelet counts in neonates are approximately one-third of those found in adults, rising to reach mature numbers shortly after weaning.[35] Young mammals also exhibit substantial variance in cell size and protein profile. The most obvious example is the broad range of cytoarchitectural features in erythrocytes of normal young animals[1] rela-tive to the fairly uniform morphologic appearance characteristic of healthy adults.

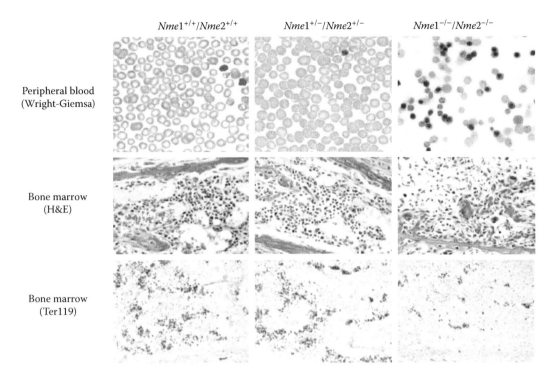

FIGURE 8.5 Altered hematopoiesis typically impacts both circulating cells (assessed by conventional hematology) and hematopoietic centers (evaluated by cytology and/or histopathology). In this example, near-term (GD 18.5) mouse embryos lacking both copies of the related kinases *Nme1* and *Nme2* (i.e., *Nme1*$^{-/-}$/*Nme2*$^{-/-}$ [right column]) had peripheral blood smears (top row) with fewer blood cells overall—most of which were large, basophilic (*blue*), nucleated erythroblasts rather than mature, acidophilic (*red*), enucleated erythrocytes—and smaller and less cellular hematopoietic cell colonies within the bone marrow (middle and bottom rows). In contrast, littermates with two copies (*Nme1*$^{+/+}$/*Nme2*$^{+/+}$ [left column]) or only one copy (*Nme1*$^{+/-}$/*Nme2*$^{+/-}$ [middle column]) of these two genes had normal peripheral blood smears composed of mature erythrocytes as well as larger, well-defined clusters of hematopoietic cells within the bone marrow (especially in sections labeled with the mouse erythroid marker Ter119. (The images were adapted from Postel, E.H. et al., *Dev. Dyn.*, 238(3), 775, 2009; Copyright 2009 by Wiley-Liss, Inc.)

Interpretation of Clinical Chemistry Data from Developing Mice

A number of treatises on clinical chemistry analysis of various animal species have been written in recent years. Several references emphasize rodent-based research and include substantial sections on adult mice.[11,24,34] However, in our experience, clinical chemistry analysis appears to be a less common undertaking in mouse developmental pathology studies than is hematological assessment. Instead, baseline values for clinical chemistry analytes generally must be gleaned from isolated literature reports. Accordingly, inclusion of appropriate controls (matched as necessary by age, sex, genotype, and/or treatment) is an essential precondition when designing clinical pathology experiments in developing mice.

Technical Considerations for Interpreting Clinical Pathology Data from Developing Mice

A primary issue encountered in evaluating lethal phenotypes in GEM lines is the trouble in obtaining sufficient numbers of animals of each genotype to populate treatment groups and support an effective statistical analysis. This difficulty is particularly acute for mouse lines bearing multiple engineered mutations, where some genotypes may be conceived at frequencies as low as one conceptus per one or two litters.

TABLE 8.4

Published Reference Values for Selected Clinical Pathology Analytes in Developing Mice

Test Category	Analyte	Age[a]	Gender	N	Mean Value	Value Relative to Adult[b]	Reference
Hematology	Hemoglobin	GD 17	Male	15	8.9 g/dL	63%	[35]
		GD 17	Female	15	8.3 g/dL	55%	[35]
	Erythrocytes (RBC)	GD 17	Male	15	3.44×10^6 cells/μL	38%	[35]
		GD 17	Female	15	3.24×10^6 cells/μL	33%	[35]
		NS	NS	NS	$3.2–3.7 \times 10^6$ cells/μL	33%–41%	[23]
	Erythrocytes: nucleated	GD 17	Male	15	48.6%	NOA	[35]
		GD 17	Female	15	53.7%	NOA	[35]
	Erythrocytes: reticulocytes	NS	NS	NS	40%	NOA	[27]
		NS	NS	NS	90%	NOA	[1]
	Leukocytes (WBC)	GD 17	Male	15	1.72×10^3 cells/μL	16%	[35]
		GD 17	Female	15	1.53×10^3 cells/μL	15%	[35]
	Platelets	GD 17	Male	15	4.53×10^5 cells/μL	33%	[35]
		GD 17	Female	15	4.54×10^5 cells/μL	34%	[35]
	Neutrophils: segmented	GD 17	Male	15	21.9%	81%	[35]
		GD 17	Female	15	25.1%	109%	[35]
	Neutrophils: metamyelocytes	GD 17	Male	15	1.5%	NOA	[35]
		GD 17	Female	15	0.8%	NOA	[35]
	Neutrophils: myelocytes	GD 17	Male	15	0.1%	NOA	[35]
		GD 17	Female	15	0.1%	NOA	[35]
	Eosinophils	GD 17	Male	15	2.5%	71%	[35]
		GD 17	Female	15	1.0%	24%	[35]
	Lymphocytes	GD 17	Male	15	31.8%	52%	[35]
		GD 17	Female	15	30.4%	45%	[35]
	Lymphocytes: immature	GD 17	Male	15	2.1%	1050%	[35]
		GD 17	Female	15	2.4%	2400%	[35]
	Monocytes	GD 17	Male	15	18.7%	322%	[35]
		GD 17	Female	15	19.7%	458%	[35]
	Monocytes: immature	GD 17	Male	15	0.9%	NOA	[35]
		GD 17	Female	15	1.1%	NOA	[35]
	Blasts	GD 17	Male	15	0.5%	NOA	[35]
		GD 17	Female	15	0.8%	NOA	[35]
Clinical chemistry	Calcium	GD 18	NS	15	5.8 mg/dL (1.45 mM/L)	NS	[3]
	Cholesterol	GD 19.5	NS	6	77 mg/dL (2.0 mM/L)[c]	NS	[45]
		PND 1	NS	6	151 mg/dL (3.9 mM/L)[c]	NS	[45]
	Glucose	GD 17.5	NS	22	21 mg/dL (1.14 mM/L) (fasted)	40%[c]	[26]
		GD 17.5	NS	18	84 mg/dL (4.65 mM/L) (fed)	85%[c,d]	[26]
		PND 0	NS	NS	~60 mg/dL[c]	NS	[48]
		PND 7	NS	NS	~100 mg/dL[c]	NS	[48]
	Insulin	GD 17.5	NS	22	0.52 ng/L (fasted)	500%[c]	[26]
		GD 17.5	NS	18	0.85 ng/L (fed)	135%[c,d]	[26]
	Protein (total)	PND 1–21	NS	22	<30 mg/dL	Similar	[9]

Notes: GD, gestational day; NOA, not observed in adults; NS, not stated; PND, postnatal day; RBC, red blood cells (erythrocytes); WBC, white blood cells (leukocytes).

[a] Age denotes the developmental age relative to the date of conception.

[b] *Adult* is defined variously as ranging in age from 8 weeks[23] (as reported in Ref. [27]) to 25 weeks.[35]

[c] Values estimated from data presented in graphic form.

[d] Nonfasted sample.

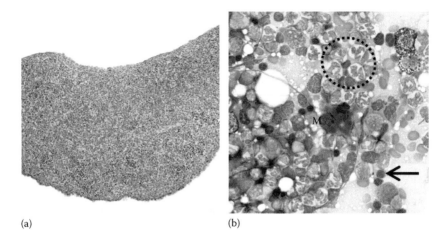

(a)   (b)

FIGURE 8.6 Definitive hematopoiesis (shown here at PND 1) is widespread in the spleen of developing mice from the end of organogenesis (GD 15) until well after birth. Tissue sections (a) are comprised entirely of red pulp, indicating that lymphocyte production is minimal prior to birth. In impression smears (b), hematopoietic cell lineages include small, dark erythrocytic precursors (arrows); large, pale myeloid progenitors (primarily neutrophils [broken circles]); and multinucleated megakaryocytes (adjacent to M). (a) H&E, 200×; (b) Wright's, 1000×.

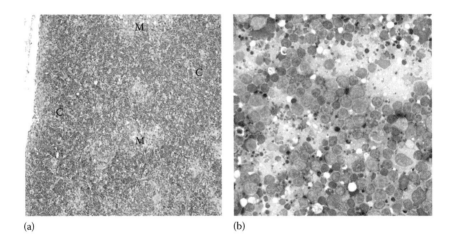

(a)   (b)

FIGURE 8.7 Lymphopoiesis (shown here at PND 1) in the thymus increases slowly after birth as demonstrated in tissue sections (a) by reduced cellularity of the cortex (dark peripheral zone [C], the location of the early events in thymocyte differentiation) relative to the medulla (pale central region [M]). Impression smears (b) demonstrate modest numbers of small lymphocytes and larger lymphoblasts (precursor cells) along with numerous tiny, condensed nuclear fragments consistent with apoptotic bodies. (a) H&E, 200×; (b) Wright's, 1000×.

Several options exist for mitigating this situation. For clinical chemistry, the most common approach is to acquire samples across multiple litters and experimental days, storing them at −20°C until all specimens can be gathered. This delay will permit any necessary pooling and generally also allows the analysis of all samples in a single run on an automated instrument. The usual strategy for ensuring a reliable data set is to pair up multiple breeders at the same time so that enough pregnancies will be ongoing to produce several animals of the low-frequency genotype within a short time frame. An alternative is to collect blood without pooling samples, in which case samples may be analyzed immediately and the data set may be built gradually. In both cases, it is imperative that the analysis include appropriate control materials, ideally concurrently collected samples from wild-type littermates or vehicle-treated animals. These specimens should be obtained from age-matched, and if necessary sex-matched, individuals to minimize the analytical bias.

REFERENCES

1. Bannerman RM (1983). Hematology in the mouse in biomedical research. In: *The Mouse in Biomedical Research*, Vol III: Normative Biology, Immunology, and Husbandry (Foster HL, Small JD, Fox JG, eds.). San Diego, CA: Academic Press; pp. 293–312.
2. Bertrand JY, Jalil A, Klaine M, Jung S, Cumano A, Godin I (2005). Three pathways to mature macrophages in the early mouse yolk sac. *Blood* **106** (9): 3004–3011.
3. Bond H, Dilworth MR, Baker B, Cowley E, Requena Jimenez A, Boyd RD, Husain SM, Ward BS, Sibley CP, Glazier JD (2008). Increased maternofetal calcium flux in parathyroid hormone-related protein-null mice. *J Physiol* **586** (7): 2015–2025.
4. Boyd KL, Bolon B (2010). Embryonic and fetal hematopoiesis. In: *Schalm's Veterinary Hematology*, 6th edn. (Weiss DJ, Wardrop KJ, eds.). Ames, IA: Wiley-Blackwell; pp. 3–7.
5. Chan JY, Kwong M, Lu R, Chang J, Wang B, Yen TS, Kan YW (1998). Targeted disruption of the ubiquitous CNC-bZIP transcription factor, Nrf-1, results in anemia and embryonic lethality in mice. *EMBO J* **17** (6): 1779–1787.
6. Crawford LW, Foley JF, Elmore SA (2010). Histology atlas of the developing mouse hepatobiliary system with emphasis on embryonic days 9.5–18.5. *Toxicol Pathol* **38** (6): 872–906.
7. Dawson PA, Sim P, Simmons DG, Markovich D (2011). Fetal loss and hyposulfataemia in pregnant NaS1 transporter null mice. *J Reprod Dev* **57** (4): 444–449.
8. de Lurdes Pinto M, Rodrigues P, Coelho AC, Antunes L, Gonçalves C, Bairos V (2008). Technical report: Mouse fetal blood collection—Taking the best out of the old needle-syringe method. *Scand J Lab Anim Sci* **35** (1): 5–8.
9. Donoviel DB, Freed DD, Vogel H, Potter DG, Hawkins E, Barrish JP, Mathur BN et al. (2001). Proteinuria and perinatal lethality in mice lacking NEPH1, a novel protein with homology to NEPHRIN. *Mol Cell Biol* **21** (14): 4829–4836.
10. Elovitz MA, Brown AG, Breen K, Anton L, Maubert M, Burd I (2011). Intrauterine inflammation, insufficient to induce parturition, still evokes fetal and neonatal brain injury. *Int J Dev Neurosci* **29** (6): 663–671.
11. Evans GO, ed. (2009). *Animal Clinical Chemistry: A Practical Handbook for Toxicologists and Biomedical Researchers*, 2nd edn. Boca Raton, FL: CRC Press.
12. Everds NE (2007). Hematology of the laboratory mouse. In: *The Mouse in Biomedical Research*, 2nd edn., Vol. III: *Normative Biology, Husbandry, and Models* (Fox JG, Barthold SW, Davisson MT, Newcomer CE, Quimby FW, Smith AL, eds.). Boston, MA: Elsevier; pp. 133–170.
13. Hatta T, Matsumoto A, Ono A, Udagawa J, Nimura M, Hashimoto R, Otani H (2006). Quantitative analyses of leukemia inhibitory factor in the cerebrospinal fluid in mouse embryos. *NeuroReport* **17** (18): 1863–1866.
14. He Y, Hakvoort TB, Vermeulen JL, Labruyère WT, De Waart DR, Van Der Hel WS, Ruijter JM, Uylings HB, Lamers WH (2010). Glutamine synthetase deficiency in murine astrocytes results in neonatal death. *Glia* **58** (6): 741–754.
15. Humbert PO, Rogers C, Ganiatsas S, Landsberg RL, Trimarchi JM, Dandapani S, Brugnara C, Erdman S, Schrenzel M, Bronson RT, Lees JA (2000). E2F4 is essential for normal erythrocyte maturation and neonatal viability. *Mol Cell* **6** (2): 281–291.
16. Hundertmark S, Dill A, Ebert A, Zimmermann B, Kotelevtsev YV, Mullins JJ, Seckl JR (2002). Foetal lung maturation in 11β-hydroxysteroid dehydrogenase type 1 knockout mice. *Horm Metab Res* **34** (10): 545–549.
17. Jacobsen J, Poulsen K (1984). Characterization of inactive renin in mouse amniotic fluid. *J Hypertens* **2** (5): 523–527.
18. Kingsley PD, Malik J, Fantauzzo KA, Palis J (2004). Yolk sac-derived primitive erythroblasts enucleate during mammalian embryogenesis. *Blood* **104** (1): 19–25.
19. Kingsley PD, Whitin JC, Cohen HJ, Palis J (1998). Developmental expression of extracellular glutathione peroxidase suggests antioxidant roles in deciduum, visceral yolk sac, and skin. *Mol Reprod Dev* **49** (4): 343–355.

20. Klaunberg BA, O'Malley J, Clark T, Davis JA (2004). Euthanasia of mouse fetuses and neonates. *Contemp Top Lab Anim Sci* **43** (5): 29–34.
21. Kovacs CS, Manley NR, Moseley JM, Martin TJ, Kronenberg HM (2001). Fetal parathyroids are not required to maintain placental calcium transport. *J Clin Invest* **107** (8): 1007–1015.
22. Kruger I, Vollmer M, Simmons DG, Elsasser HP, Philipsen S, Suske G (2007). *Sp1/Sp3* compound heterozygous mice are not viable: Impaired erythropoiesis and severe placental defects. *Dev Dyn* **236** (8): 2235–2244.
23. Kunze H (1954). Die Erythropoese bei einer erblichen Anämie römtgenmutierte Mäuse. *Folia Haematol* **72**: 392–436.
24. Loeb WF, Quimby FW (1999). *The Clinical Chemistry of Laboratory Animals*, 2nd edn. Philadelphia, PA: Taylor & Francis Group.
25. Matsumoto A, Hatta T, Ono A, Hashimoto R, Otani H (2011). Stage-specific changes in the levels of granulocyte-macrophage colony-stimulating factor and its receptor in the biological fluid and organ of mouse fetuses. *Congenit Anom* (*Kyoto*) **51** (4): 183–186.
26. Mogami H, Yura S, Tatsumi K, Fujii T, Fujita K, Kakui K, Kondoh E, Inoue T, Fujii S, Yodoi J, Konishi I (2010). Thioredoxin binding protein-2 inhibits excessive fetal hypoglycemia during maternal starvation by suppressing insulin secretion in mice. *Pediatr Res* **67** (2): 138–143.
27. Moore DM (2000). Hematology of the mouse (*Mus musculus*). In: *Schalm's Veterinary Hematology*, 5th edn. (Feldman BF, Zinkl JG, Jain NC, eds.). Boston, MA: Lippincott, Williams & Wilkins; pp. 1219–1224.
28. Moos T, Møllgård K (1993). Cerebrovascular permeability to azo dyes and plasma proteins in rodents of different ages. *Neuropathol Appl Neurobiol* **19** (2): 120–127.
29. Palis J, Yoder MC (2001). Yolk-sac hematopoiesis: The first blood cells of mouse and man. *Exp Hematol* **29** (8): 927–936.
30. Perkins AC, Sharpe AH, Orkin SH (1995). Lethal β-thalassaemia in mice lacking the erythroid CACCC-transcription factor EKLF. *Nature* **375** (6529): 318–322.
31. Postel EH, Wohlman I, Zou X, Juan T, Sun N, D'Agostin D, Cuellar M, Choi T, Notterman DA, La Perle KMD (2009). Targeted deletion of Nm23/nucleoside diphosphate kinase A and B reveals their requirement for definitive erythropoiesis in the mouse embryo. *Dev Dyn* **238** (3): 775–787.
32. Pritchett-Corning KR (2009). Euthanasia of neonatal rats with carbon dioxide. *J Am Assoc Lab Anim Sci* **48** (1): 23–27.
33. Pritchett K, Corrow D, Stockwell J, Smith A (2005). Euthanasia of neonatal mice with carbon dioxide. *Comp Med* **55** (3): 275–281.
34. Quimby FW, Luong RH (2007). Clinical chemistry of the laboratory mouse. In: *The Mouse in Biomedical Research*, 2nd edn., Vol. III: *Normative Biology, Husbandry, and Models* (Fox JG, Barthold SW, Davisson MT, Newcomer CE, Quimby FW, Smith AL, eds.). Boston, MA: Elsevier; pp. 171–216.
35. Rugh R (1990). *The Mouse. Its Reproduction and Development*. Oxford, U.K.: Oxford University Press.
36. Rugh R, Somogyi C (1968). Pre- and postnatal normal mouse blood cell counts. *Proc Soc Exp Biol Med* **127** (4): 1267–1271.
37. Saedler K, Hochgeschwender U (2011). Impaired neonatal survival of pro-opiomelanocortin null mutants. *Mol Cell Endocrinol* **336** (1–2): 6–13.
38. Simmonds CS, Karsenty G, Karaplis AC, Kovacs CS (2010). Parathyroid hormone regulates fetal-placental mineral homeostasis. *J Bone Miner Res* **25** (3): 594–605.
39. Sinjari T, Darnerud PO (1998). Hydroxylated polychlorinated biphenyls: Placental transfer and effects on thyroxine in the foetal mouse. *Xenobiotica* **28** (1): 21–30.
40. Takeda A, Takatsuka K, Connor JR, Oku N (2001). Abnormal iron accumulation in the brain of neonatal hypotransferrinemic mice. *Brain Res* **912** (2): 154–161.
41. Terry KK, Elswick BA, Welsch F, Conolly RB (1995). Development of a physiologically based pharmacokinetic model describing 2-methoxyacetic acid disposition in the pregnant mouse. *Toxicol Appl Pharmacol* **132** (1): 103–114.
42. Tober J, Koniski A, McGrath KE, Vemishetti R, Emerson R, de Mesy-Bentley KK, Waugh R, Palis J (2007). The megakaryocyte lineage originates from hemangioblast precursors and is an integral component both of primitive and of definitive hematopoiesis. *Blood* **109** (4): 1433–1441.
43. Van Calsteren K, Verbesselt R, Van Bree R, Heyns L, de Bruijn E, de Hoon J, Amant F (2011). Substantial variation in transplacental transfer of chemotherapeutic agents in a mouse model. *Reprod Sci* **18** (1): 57–63.

44. van Meer H, van Straten EM, Baller JF, van Dijk TH, Plosch T, Kuipers F, Verkade HJ (2010). The effects of intrauterine malnutrition on maternal-fetal cholesterol transport and fetal lipid synthesis in mice. *Pediatr Res* **68** (1): 10–15.

45. van Straten EM, Huijkman NC, Baller JF, Kuipers F, Plosch T (2008). Pharmacological activation of LXR *in utero* directly influences ABC transporter expression and function in mice but does not affect adult cholesterol metabolism. *Am J Physiol Endocrinol Metab* **295** (6): E1341–E1348.

46. Wang VE, Schmidt T, Chen J, Sharp PA, Tantin D (2004). Embryonic lethality, decreased erythropoiesis, and defective octamer-dependent promoter activation in Oct-1-deficient mice. *Mol Cell Biol* **24** (3): 1022–1032.

47. Warren AJ, Colledge WH, Carlton MB, Evans MJ, Smith AJ, Rabbitts TH (1994). The oncogenic cysteine-rich LIM domain protein Rbtn2 is essential for erythroid development. *Cell* **78** (1): 45–57.

48. Watanabe N, Hiramatsu K, Miyamoto R, Yasuda K, Suzuki N, Oshima N, Kiyonari H et al. (2009). A murine model of neonatal diabetes mellitus in *Glis3*-deficient mice. *FEBS Lett* **583** (12): 2108–2113.

49. Yano M, Watanabe K, Yamamoto T, Ikeda K, Senokuchi T, Lu M, Kadomatsu T et al. (2011). Mitochondrial dysfunction and increased reactive oxygen species impair insulin secretion in sphingomyelin synthase 1-null mice. *J Biol Chem* **286** (5): 3992–4002.

50. Yoshioka W, Akagi T, Nishimura N, Shimizu H, Watanabe C, Tohyama C (2009). Severe toxicity and cyclooxygenase (COX)-2 mRNA increase by lithium in the neonatal mouse kidney. *J Toxicol Sci* **34** (5): 519–525.

51. Zappaterra MD, Lisgo SN, Lindsay S, Gygi SP, Walsh CA, Ballif BA (2007). A comparative proteomic analysis of human and rat embryonic cerebrospinal fluid. *J Proteome Res* **6** (9): 3537–3548.

9

Histotechnological Processing of Developing Mice

Brad Bolon, Diane Duryea, and Julie F. Foley

CONTENTS

A firm grasp of basic histotechnological practices is required before meaningful information may be obtained from the microscopic analysis of tissue structure. This principle is especially true for specimens harvested from developing mice. Relative to adult mice, samples from developing animals are much more liable to collapse or fragment when manipulated during histotechnological processing because of their small size and relatively high water content. Nonetheless, with regular practice researchers should be able to obtain suitable data (and publication-quality images) via microscopic examination of tissues from developing mice. The length of this practice period may be shortened considerably if the individual directly responsible for histotechnological processing is either a trained histotechnologist or is able to observe someone who is already experienced in the processing of specimens from developing mice.

This chapter is designed to serve as both a basic histotechnology primer and a thorough review of the advanced methods that will be necessary for proper mouse developmental pathology analyses. Standard protocols based on the authors' experience are presented for several common histological methods. However, these protocols will need to be optimized anew in your laboratory as variables such as reagent purity and environmental conditions (e.g., relative humidity, room temperature [RT]) may affect the end product.[5] Assistance in modifying these sample protocols for your own setting may be obtained from standard histotechnology texts.[1,3,8,11]

Sequence for Processing Mouse Developmental Pathology Specimens

Developmental pathology studies in mice commonly deal with rare events that must be discerned and characterized in small, fragile specimens. Accordingly, considerable care must be taken when planning the order of histotechnological processing to ensure that artifactual damage is not introduced selectively into a specific group of samples. Care in this regard will prevent bias when interpreting the final analytical results.

The two major factors during tissue preparation for microscopic examination that can negatively influence the study outcome are (1) the treatment of the specimens before and during histotechnological processing and (2) the sequence in which study samples are processed. All specimens—both experimental groups (e.g., genetically engineered animals, toxicant-treated animals) and corresponding controls—should be fixed for an identical period of time, preferably using fixative from the same lot. In the ideal study, experimental and control tissues should be processed, embedded, and stained together at the same time as well. Batch processing of all specimens from a single experiment in this fashion quells one major source of variation that can ruin a study. However, in practice, such simultaneity may be challenging due to the large number of consecutive sections produced when sectioning mouse embryos (especially very large near-term fetuses). If the study is so large that all samples cannot be processed in a single batch, minor adjustments to the processing sequence can reduce the likelihood that artifactual damage associated with tissue preparation will confound the final outcome. In particular, each processing batch of a large study must contain a random assortment of experimental and treated tissues. Under no circumstances should all the controls be processed in a single batch or on a single day if experimental specimens will remain to be processed at another time.

Stages of Histotechnological Processing

Morphologic analysis at the tissue and cell levels requires specific preparation to preserve and stabilize structural and molecular components. Standard histotechnological processing procedures entail an ordered series of stages to achieve this purpose (Table 9.1). These stages are discussed in the following text in the order in which they are undertaken when processing tissues.

Fixation

Basic principles of tissue fixation have been detailed elsewhere.[1,3,4,8,12] Fixation is initiated at necropsy (i.e., before specimens are given to the histotechnology facility for histotechnological processing) in most developmental biology laboratories. Nevertheless, unless the research staff is highly experienced in preparing tissue for microscopic analysis, a trained histotechnologist and/or comparative pathologist should be consulted when designing the experiment to ensure that the most appropriate fixative is selected for the question being investigated. At many institutions, suitable fixatives may be supplied to individual research laboratories as a service by the histotechnology facility.

Two means of fixation may be considered for mouse developmental pathology specimens. Chemical fixatives typically act either by forming covalent bonds among molecules (i.e., *cross-linking* agents, such as aldehydes and osmium tetroxide) or by precipitating molecules into condensed aggregates (i.e., *coagulating/denaturing* agents, like acetic acid and alcohol). Chemical fixatives used for developmental pathology studies often mix both cross-linking and denaturing agents (Table 9.2) as a compromise to optimize morphological preservation (by reducing the extent of denaturation) while retaining molecular reactivity (by reducing the number of cross-links that mask antigens). Physical means of fixation (e.g., heat, irradiation) may be utilized for isolated cells spread on a flat surface (e.g., blood smears and tissue squash preparations on a glass slide [see Chapter 8]) but generally are unsuitable for whole mouse conceptuses and isolated but intact organs.

Common fixatives for mouse developmental pathology studies (listed in the order in which they often are considered) are Bouin's solution (Figure 9.1); 4% paraformaldehyde (PFA), pH 7.4; neutral buffered 10% formalin (NBF [research grade, which does not include methanol as a stabilizer]), modified Davidson's solution (mDF) (Figure 9.1),[10] and alcohol (Table 9.2). In our experience, the first four fixatives, but not alcohol, provide good structural resolution at the tissue and cellular levels. Bouin's solution and modified Davidson's solution stiffen the delicate tissues to a greater degree, which generally reduces the degree of iatrogenic damage caused by tissue distortion during subsequent processing steps. Bouin's solution typically is not used as a fixative for placenta as this agent may disrupt erythrocytes, which obscures the morphology of intravascular contents. Simultaneous preservation of both anatomic

TABLE 9.1

Automated Processing to Prepare Paraffin-Embedded Tissues

		Developmental Stage				
Age in gestational days (GD)		≤6	6.5–9.5	9.5–11.5	12–15	>15.5[a]
Theiler stage (TS)		≤8	9–14	15–19	20–23	>23
Size of isolated conceptus (mm)		<2	3–5	6–15	16–25	>25
Process	**Solution**	**Time**				
Fixation	Aldehydes	1–2 h	3–6 h	6–12 h	12–24 h	48–96 h
	Bouin's washout	1–2 h	2–4 h	4–8 h	8–24 h	24–48 h
		Serial 70% ethanol washes until fluid does not turn yellow				
	Modified Davidson's	1–2 h	2–4 h	4–8 h	8–24 h	24–48 h
Dehydration[b]	70% Ethanol[c]	30 min	1–2 h	1–2 h	1–2 h	2 h
	80% Ethanol[c]	30 min	1–2 h	1–2 h	1–2 h	2 h
	95% Ethanol[c,d]	30 min	1–2 h	1–2 h	1–2 h	2 h
	95% Ethanol[c]	30 min	1–2 h	1–2 h	1–2 h	2 h
	100% Ethanol[c]	30 min	1–2 h	1–2 h	1–2 h	2 h
	100% Ethanol[c]	30 min	1–2 h	1–2 h	1–2 h	2 h
Clearing	Xylene[c]	30 min	1–2 h	1–2 h	1–2 h	2 h
	Xylene[c]	30 min	1–2 h	1–2 h	1–2 h	2 h
Infiltration	Paraffin (60°C)[c]	30 min	1–2 h	1–2 h	1–2 h	2 h
	Paraffin (60°C)[c]	30 min	1–2 h	1–2 h	1–2 h	2 h
	Paraffin (60°C)[c]	30 min	1–2 h	1–2 h	1–2 h	2 h

Note: Instrument is a Tissue-Tek® VIP-5 Automated Processor (Sakura Finetek USA, Torrance, CA).

[a] Processing may be inadequate unless the specimens are manipulated to improve penetration of the fixative and paraffin (e.g., blanching to remove the thick skin, injecting into or opening the thoracic and/or abdominal cavities).

[b] Embryos, especially younger ones (GD 15 or earlier), should not be left for extended periods (e.g., over the weekend) in ethanol baths.

[c] Step performed under pressure (setting: 35 kPa) and vacuum (setting: −67 kPa).

[d] Contains eosin to add color to small translucent specimens.

features and fragile protein epitopes (for immunohistochemistry or enzyme histochemistry) is usually best achieved using PFA or NBF; these agents are also preferred when lipids and lipid-soluble materials must be demonstrated as the other fluids contain alcohol, which can remove lipids. Alcohol is used as the fixative for certain antigens that are destroyed by exposure to aldehydes (i.e., a main ingredient in both Bouin's and modified Davidson's solutions as well as NBF and PFA).

Certain research questions require special fixative chemicals. The most common developmental pathology endpoint of this kind is ultrastructural analysis, specifically transmission electron microscopy (TEM) to evaluate subcellular structures. Specimens slated to undergo TEM must be fixed rapidly and extensively to conserve fine detail. The usual fixative for this purpose is glutaraldehyde or glutaraldehyde mixed with PFA (Karnovsky's solution; Table 9.2). PFA followed by osmium tetroxide is a preferred choice when specimens contain abundant lipid or are to be embedded in plastic. Additional details regarding fixation options and procedures for electron microscopy analysis of developing mice are given in Chapter 13.

For all these chemicals (Table 9.2), the length of fixation is generally proportional to the size of the specimen (Table 9.1) and the quantity of fixative. In general, the size of the specimen must be small relative to the volume of fixative solution. In the mouse developmental pathology setting, 1 volume of tissue is placed into 10–20 volumes of fixative (or approximately 10–20 mL of solution for every 1 cm of length). Furthermore, since fixative solutions infiltrate slowly and to a limited depth (2–3 mm past the surface), especially in dense parenchymal organs, care must be taken to ensure that specimens are not so thick (i.e., >5 mm) that solutions cannot penetrate to their cores. Embryos at gestational day

TABLE 9.2

Composition of Common Fixatives Used in Mouse Developmental Pathology

Fixative	Constituent	Quantity
Bouin's solution	Picric acid (saturated in H_2O)	75 mL
	Formaldehyde (37%, w/w in H_2O)	25 mL
	Glacial acetic acid	5 mL
4% Paraformaldehyde (PFA)	PFA	4.0 g
	Phosphate-buffered saline (~pH 7.4)	100 mL
10% Formalin (neutral buffered) = NBF	Formaldehyde (37%, w/w in H_2O)	10 mL
	Distilled water	90 mL
	Disodium diphosphate (Na_2HPO_4)	0.65 g
	Monosodium phosphate (NaH_2PO_4)	0.40 g
Modified Davidson's solution	Ethanol (100%)	15 mL
	Formaldehyde (37%, w/w in H_2O)	30 mL
	Glacial acetic acid	5 mL
	Distilled water	50 mL
Alcohol (ethanol or methanol)	Absolute (100%)	
Modified Karnovsky's fixative (final: 2.5% glut/2.0% PFA)	Glutaraldehyde (25% in H_2O)	10 mL
	PFA (8% in H_2O)	25 mL
	Phosphate buffer (0.2 M, pH 7.2)	50 mL
	Distilled water	15 mL

Note: Recipes are for preparing fresh batches of ~100 mL in the laboratory. Reagents and pre-formulated fixative solutions may be purchased premade from many commercial vendors.

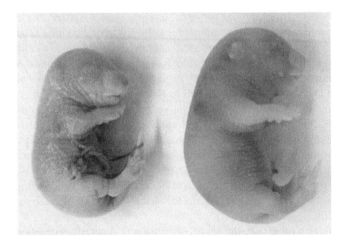

FIGURE 9.1 Common fixatives for near-term (GD 17.5) mouse fetuses include Bouin's solution (left) and modified Davidson's (mD) solution (right). Both these fixatives penetrate well to provide rapid tissue preservation and harden soft tissues so that they trim easily. An added advantage of Bouin's solution is that it does decalcify mineralized bone. Disadvantages of Bouin's include reduced contrast (due to diffuse, bright yellow coloration by picric acid) and moderate shrinkage (indicated by the smaller body profile) due to extensive soft tissue dehydration. Fetuses fixed in NBF have an appearance that resembles the mD-fixed fetus, but the slow penetration of NBF yields very soft tissues and typically promotes autolysis of deeply buried abdominal viscera.

(GD) 14.5 or younger as well as placentas may be fixed by immersion due to their small size and thin external covering. For larger specimens (e.g., fetuses [GD 15 or older] and neonates up to postnatal day [PND] 5), their thickness and the barrier posed by the keratinizing skin require that fixative be introduced directly into the thorax and abdomen in some fashion. We have tried four basic strategies for ensuring adequate internal fixation in such thick individuals. One means is by injecting a small quantity of fluid into each cavity (approximately 0.05 mL into the thorax and 0.1 mL into the abdomen) using a short, thin hypodermic needle (e.g., 26 gauge or higher, 3/8 in. long; D. Devor-Henneman, personal communication). The needle should be entered slowly at a shallow angle (essentially parallel to the body wall) to avoid damaging the organs. A second option for introducing fixative is to remove the head just caudal to the mandible and then placing a narrow slit into the cavity walls using a sharp scalpel; again, care should be taken to avoid piercing the viscera. A third aid is to blanch very large fetuses (e.g., GD 18.5) in boiling tap water for 20 s, followed by immersion in an ice water bath to permit removal of the outer skin (epidermis). Fixative penetration using these three methods can be enhanced by agitating the specimen on a shaker overnight (18–24 h). A fourth possibility is to immerse the intact subject in fixative for 24 h, trim it in a parasagittal (longitudinal) plane (typically cutting 3–5 mm to the left of the midline), and then immerse both halves in fresh fixative for another 24–48 h. The position of cut organs can be maintained during processing by placing the two halves within a cassette (cut-side down) on a piece of cardstock or biodegradable paper (e.g., Bio-Paper™ or Bio-Wraps® biopsy wraps, available from several vendors). Possible disadvantages to this latter approach are that internal organs will undergo some degree of autolysis during the first 24 h and that unfixed or incompletely fixed tissues will be damaged during trimming; these issues are especially likely to arise when using aldehyde-only preservatives such as NBF and PFA. In general, therefore, we do not advocate this fourth strategy when processing large specimens.

Longer fixation times are permissible when the primary thrust is to evaluate structural features, while shorter periods are required when attempting to evaluate functional activity (e.g., enzyme histochemistry for proteins) and many molecular endpoints (e.g., immunohistochemistry). If a short length of fixation is necessary, the exchange from fixative to transport solution (e.g., 70% ethanol for Bouin's solution and NBF) should occur at the necropsy facility or within the primary research laboratory. However, the time spent in the transport solution is also critical, as 70% ethanol will rapidly dehydrate small embryos due to their very high water content. If an extended fixation period is acceptable, removal of tissues from fixative may be delegated to the staff of the histotechnology facility.

Optimal histotechnological processing is more likely if the task of transporting specimens to the histotechnology facility is undertaken in a stereotypical manner. All submission procedures of the histotechnology service should be followed exactly. In general, these practices will include proper labeling of all containers (Table 9.3); transport of specimens within tightly sealed vessels (to prevent fluid leakage) carried within a large bucket or deep-sided tray (to contain any fixatives that might inadvertently leak); and completion of the appropriate documentation for the study (Table 9.3). An excellent, mutually beneficial habit is to contact the histotechnology service well in advance (ideally 7–14 days before) to reserve a place in their tissue-processing queue. This courtesy is crucial if the services of a particular histotechnologist with expertise in processing tissues from developing mice must be scheduled. Early warning is also imperative if specimens being submitted were flash-frozen at necropsy without prior fixation as freezer space for tissue storage will have to be cleared to receive the samples.

Decalcification

Hard tissues like bone can destroy the edge of regular microtomy blades. In doing so, the combination of bone and mineral fragments coupled with the jagged grooves that are etched into the cutting surface will shred the specimen that is being cut. This concern is avoided by including a processing step to chemically remove mineral from bone before embedding.

Three factors dictate whether or not a given mouse developmental pathology specimen will require decalcification. The first is the developmental age of the animal. The skeleton is first laid down during gestation as a cartilaginous model, which over time is converted to bone. Mineral deposits develop at specific sites and in an ordered sequence for each bone; for example, in the limbs the first depots

TABLE 9.3

Information Recommended for Specimen Submission

	Location of Information	
Identifying Datum	**Submission Form**	**Container Label**
Contact information for researcher		
Name	X	X
Telephone number	X	X
E-mail	X	
Alternate contact		
Name	X	
Telephone number	X	
E-mail	X	
Study-specific information		
Study no.	X	X
Animal no.	X	X[a]
Species	X	X
Strain or stock	X	
Developmental age	X	
Fixative used	X	X
Length of fixation[b]	X	
Special instructions	X	

[a] The animal number is required if all tissues within the container are from a single individual, but optional if tissues from multiple animals have been included.

[b] The preferred practice is to give the actual length of time in fixative, but an alternative is to state the date on which the specimen was first exposed to fixative.

form in the diaphysis (middle of the shaft) beginning about GD 15.5 (see Chapter 4, Table 4.A.10). Clearly, demineralization is not required for younger mouse conceptuses, and it cannot be performed if examination of skeletal mineralization is the desired endpoint for the study. It may be necessary for near-term fetuses (GD 17 or older) and will be required for neonates and juveniles. The second factor is the nature of the sample to be evaluated. Mineral is concentrated in bone rather than soft tissues, so decalcification is not necessary if bone will not be evaluated. The final factor is the choice of fixative. Bouin's and modified Davidson's solution both exhibit demineralizing activity, and are routinely used for this purpose to facilitate gross evaluation of internal organs in near-term fetuses and neonates. Thus, decalcification often will not be necessary for these subjects. In our experience, Bouin's solution (which contains both acetic acid and picric acid) produces more rapid demineralization than does modified Davidson's solution (which contains only acetic acid). Other common developmental pathology fixatives (NBF, PFA, alcohol) usually will require a decalcification step if bony specimens are to undergo histotechnological preparation.

Three chemical decalcification processes may be used to remove mineral from bony tissues in developing mice. The usual approach is to employ a fixative that doubles as a demineralization agent (e.g., Bouin's solution, modified Davidson's solution), where the weak organic acid components of the solution separate the inorganic calcium from the organic matrix. Additional decalcification typically is not required if these agents are employed. A second alternative for aldehydes like NBF or PFA or for coagulating agents like 70% ethanol is to immerse fixed tissues in an acidic solution. The advantage of acidic decalcification is its rapid action. However, great care must be taken to remove the sample from the solution as soon as possible after the calcium is eliminated because prolonged exposure to acids will reduce the intensity of staining in some structures (particularly hematoxylin uptake by nuclei). The third option is to employ a calcium-chelating agent such as ethylene diamine tetraacetic acid (EDTA), which preferentially binds calcium (and magnesium) with high affinity under physiologic conditions (pH 7.0–7.4). The main benefit of chelators is that they do not damage tissue structure or staining, while their main disadvantage is that their rate of decalcification is much slower. Decalcification is typically performed at RT, but the rate of calcium removal can be slowed by performing this step at 4°C.

In our experience, specimens from developing mice require no decalcification unless an intact near-term fetus or neonate or an isolated bone is to be analyzed and the fixative solution used for tissue preservation was NBF, PFA, or alcohol. In this case, we use formate (the weak organic anion derived from formic acid; see Table 9.4 for protocol) in preference to a strong (inorganic [e.g., hydrochloric]) acid or a chelating agent. The rationale for this choice is a trade-off between an acceptable degree of tissue damage (provided by the more moderate pH of formic acid relative to the low pH of a strong acid) and a faster rate of decalcification (since acidic decalcification is faster than chelation). Most mouse developmental pathology studies will involve multiple specimens, which ideally are processed as a group. The need to test for decalcification coupled with the inability to predict the length of this step usually warrants that this procedure be performed manually.

In our experience, the completeness of decalcification need not be checked in fetuses or neonates fixed in Bouin's or modified Davidson's solution. However, several means are available to determine whether or not decalcification is complete should the need arise. The simplest and most direct way is to cut through a deeply buried bony structure to see whether or not the blade catches or crunches an incompletely demineralized structure; we usually use the proximal femur or the upper cervical region for this test. An alternative is to radiograph the animal to see whether or not the normal opacity of the skeleton has been lost. This method is generally limited to juvenile mice (PND 5 or older) due to the low opacity of the bony skeleton in younger animals and the labor and time required to perform this method but fetuses as young as GD 16 do exhibit ossified bones when radiographed. A radiographic screen using a bench-top instrument (Faxitron® Model No. 43855A; Faxitron Bioptics, Tucson, AZ) set at 0.3 mA and 55 KVP for 49 s is suitable for this purpose in near-term (GD 18.0) mouse fetuses. A third approach is to use a chemical test to determine whether or not calcium is present in the decalcifying solution (see Table 9.5 for protocol). If calcium cannot be precipitated, then decalcification may be assumed to be complete. As with radiography, the labor required for this chemical test coupled with the small amount of calcium to be removed from developing mice does not warrant this assay for routine purposes.

TABLE 9.4

Kristensen's Protocol for Acid Decalcification of Bony Tissue

---

1. Stock solutions
   a. Fluid A: 8 N formic acid (1000 mL)
      i. Formic acid, concentrated (410 mL)
      ii. Deionized water (590 mL)
   b. Fluid B: 1 N sodium formate (1000 mL)
      i. Sodium formate (68 g)
      ii. Deionized water (1000 mL)
2. Working solution
   a. Composition
      i. Fluid A (200 mL)
      ii. Fluid B (800 mL)
   b. pH: 3.01–3.02
3. Procedure
   a. Specimens must be fixed in NBF or 70% ethanol for 24 or more hours. If fixed in NBF, specimens may be transferred to 70% ethanol for storage until decalcification.
   b. The volume of working solution used to decalcify fetuses (GD 15 to term) or neonates should be approximately 15–20 times the volume of the tissue.
   c. Complete decalcification is confirmed by physical properties of the specimen (e.g., tissue flexibility, softness, and ease of cutting) and/or chemical testing (see Table 9.5), at the discretion of the histotechnologist.
   d. Decalcification of intact near-term fetuses (GD 17.5 to term) or neonates typically requires 1–2 days of decalcification.
   e. Rinse for 3–5 min in deionized water before placing into 70% ethanol to begin processing.

Source: Adapted from Kristensen, H.K., *Biotech. Histochem.*, 23(3), 151, 1948.

TABLE 9.5

Calcium Oxalate Assay for Completion of Decalcification

1. Reagents
   a. Ammonium hydroxide, concentrated
   b. Ammonium oxalate, aqueous (saturated)
2. Procedure
   a. Place 2 mL of used decalcifying solution in a test tube.
   b. Add a strip of litmus paper.
   c. At room temperature (RT), add ammonium hydroxide drop by drop (shaking after each drop) to the used decalcifying solution until a neutral pH (7.0) is achieved (i.e., when the litmus paper first turns blue).

      Note: If the litmus paper becomes extremely blue due to overaggressive titration, add more used decalcifying solution and retitrate with a fresh litmus paper.
   d. Interim interpretation.
      i. Immediate formation of a white precipitate (calcium hydroxide) indicates the presence of a large quantity of calcium in the decalcifying solution, thus demonstrating that the tissue needs further decalcification.
      ii. If no precipitate forms, continue the assay using the same sample.
   e. Mix 2 mL of ammonium oxalate with sample.
   f. Incubate at RT for 15–30 min.
      i. Eventual formation of a white precipitate (calcium oxalate) indicates the presence of a minute quantity of calcium in the used decalcifying solution, thus suggesting that decalcification is almost complete.
      ii. If no precipitate forms after 30 min, decalcification is complete.

      Note: This test is unsuitable for solutions with an acid content exceeding 10%.

Note: This protocol is derived from the Clayden test.[2]

Processing for Paraffin Sectioning

Most mouse developmental pathology studies are performed in specimens that have been embedded in paraffin wax (Figure 9.2). Several additional steps are needed to prepare tissues for paraffin infiltration. Collectively, these steps involve the progressive replacement of tissue water with organic chemicals that are miscible with paraffin. The entire process may be performed manually or on an automated station, but in either event, the small size and increased water content of developing tissues typically requires that they be processed alone rather than mixed with specimens harvested from adult animals.

Dehydration is the first step: Specimens are gradually transferred through a graded series of ethanol to remove all water (Table 9.1). The timing of each step is of vital importance because excessive dehydration may shrink and harden tissues so much that they will fragment during sectioning. Extended soaking of the block face may partially restore tissue water, but this rehydrating process will substantially slow histological processing and also may reduce the quality of the resulting cut sections. The tissue is then moved from the 100% ethanol bath into a 1:1 mixture of 100% ethanol and an organic solvent (usually benzene [mainly of historical interest due to its toxicity],[5] toluene, or xylene). The length of time that the tissue must spend in each fluid is proportional to the size of the specimen (Table 9.1). To aid identification and orientation during embedding, small embryonic specimens may be colored at this point by adding a few drops of eosin into one of the alcohol baths (typically 80% or 90%).

Clearing follows dehydration. In this step, dehydrated specimens are moved into pure organic solvent. Benzene has been advocated as the preferred agent for embryos with a crown-rump length (CRL) below 6 mm, while xylene has been deemed to be more suitable for embryos 6 mm in length or longer.[5] However, in our experience, either toluene or xylene also are acceptable for all postimplantation stages of development. Using these latter two solvents avoids staff exposure to benzene, which is highly toxic, and also permits specimens collected from developing mice to be handled using standard chemical baths on automated histological processing systems.

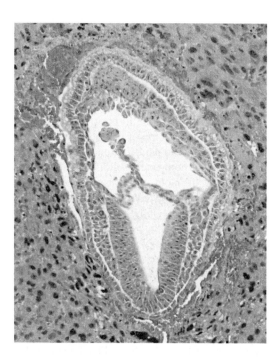

FIGURE 9.2　Morphological features are distinct at the cellular level but less so at the subcellular level in routine paraffin-embedded (4- to 8-μm-thick) sections. This gastrulating mouse embryo (GD 7.5) was fixed *in situ* by immersion in NBF for 48 h, dehydrated in graded alcohols and xylene, infiltrated overnight in paraffin, and then embedded in paraffin. Stain: hematoxylin and eosin (H&E). (This photomicrograph was kindly provided by Dr. Dorothy French, Genentech, Inc., South San Francisco, CA.)

Paraffin infiltration is the next step before embedding. Samples are passed through several changes of hot liquid paraffin (usually two or three) so that the wax will permeate the tissue. Heat may be applied using a heat lamp, an oven (typically set at about 60°C), or on an automated processing station. Again, the time required for infiltration is proportional to the size of the specimen (Table 9.1). Intact larger embryos and fetuses will have to be infiltrated under vacuum to ensure that penetration of the paraffin is complete.

Paraffin embedding is the final step required for routine histological processing. In modern histotechnology laboratories, embedding stations are flat platforms with three working areas. The first zone is a hot paraffin bath into which cassettes are placed until they can be processed, the adjacent central zone is a heated metal plate, and the final zone is a cold metal plate. During embedding, one cassette at a time is removed from the hot paraffin bath, placed on the hot plate, and opened. A metal mold is selected such that the volume of the mold is between 5- and 10-fold larger than the specimen that will be placed within it. The mold is placed on the hot plate and filled with liquid paraffin. Next, the tissue is moved from the cassette to the mold using a pair of heated forceps; the heat may be supplied by placing the forceps in ports located on the hot plate or by tethering the electrified tool to an adjustable control module. We prefer the former option because it allows more mobility and removes the power cord as a source of clutter near the station. However, a potential advantage of the electronic forceps is that paraffin will not accumulate on the tips, whereas cooling wax will build up on the unpowered instrument once it is removed from the heated port. This latter difficulty is overcome by rotating among three or four nonelectrified forceps.

The primary factor in maintaining good structural integrity at the embedding step rests with the manner in which the specimen is handled as it is moved into the mold and positioned for embedding. Obviously, all motions should be gentle. However, they should also be relatively rapid to avoid fusion of the paraffin-infiltrated tissue to the end of the nonelectrified forceps. In addition, the specimen cannot be grasped using the standard forceps *pincer* motion as the pressure will distort or fracture it. Instead, the tips of the open forceps should be gently nudged under the edge of the sample so that it can be lifted from the cassette into the mold. Similarly, the specimen should be repositioned within the liquid paraffin by gentle pushing and rolling rather than by direct lifting. Once the proper orientation has been achieved, the mold is moved from the hot plate to the cold plate. The paraffin at the bottom of the mold will begin congealing immediately, which will hold the tissue in position. Once the position is confirmed to be correct, the bottom of the cassette is placed on top of the mold (both to hold the paraffin block and to identify the sample) and then filled with liquid paraffin (to anchor the block to the cassette). Molds are left on the cold plate until the paraffin has completely solidified. If the orientation of the sample shifted as the mold was moved to the cold plate, the mold can always be moved back to the hot plate to re-liquefy the paraffin. In general, care must be taken to avoid too many such *freeze–thaw* cycles as the repeated manipulation may damage the specimen (especially any small protuberances, such as limb buds and tails of early embryos).

The key to an accurate developmental pathology analysis is to select and maintain a sample orientation appropriate to the research question of interest. This decision should be made in advance to avoid hasty or repeated manipulations of the infiltrated specimens, because excessive handling often fractures smaller embryos (especially GD 12.5) and placentas. An option to prevent such trauma is to pre-embed small embryos (GD 6.5–GD 12.5) in agar, so that the fragile tissue is not subject to direct manipulation. Using a stereomicroscope, fixed embryos are embedded in 1% agar (technical or microbiological grade) in distilled water. For this purpose, melted agar (prepared in a microwave) is cooled to 50°C and then maintained in liquid state in a heated water bath (46°C). Fixed embryos less than GD 12.5 are placed one at a time into wells of a 24-well tissue culture plate (Multiwell™, available commercially from multiple vendors). A narrow-bore pipette is used to remove excess fixative from the embryo, taking care not to touch the specimen. Agar is added to the well drop by drop until the embryo is covered. As the agar cools, disposable pipette tips (200 μL) are used to manipulate the embryo toward the center of the well and into the correction orientation; repositioning is accomplished by moving the adjacent agar rather than by directly touching the embryo. When the specimen is correctly oriented, the next embryo is placed into the adjacent well and embedded similarly. When all embryos have been embedded, the Multiwell plate is cooled in the refrigerator for 10 min to help solidify the agar; elimination of this step will result in agar that is friable, which may lead to cracks in the embedding medium and embryo upon removal from the well. Agar-embedded embryos are removed from wells by inserting a narrow, flat weighing spatula between the agar and side of the well, thereby freeing the agar block. Excess agar should be removed prior to placement in the cassette. The bottom surface of the block will be flat, while the upper surface will be concave; the flat surface (i.e., where the embryo rests) should be placed *down* in the cassette. In general, deeper cassettes (e.g., Mega-cassette®) should be used to ensure that the agar block and embryo are not squashed. The cassettes holding agar-embedded samples are returned to the fixative solution for submission to the histology laboratory. The routine tissue processing schedule may be followed for agar-embedded specimens. However, microwaves should not be used in the processing scheme to avoid melting of the agar and loss of the embryo.

Multiple specimens may be embedded in the same mold. Possible combinations usually include an embryo and its placenta, or multiple segments taken from a single embryo or fetus. In principle, several small embryos (GD 11.5 or younger) from the same treatment group of a toxicity study may also be placed in a single mold. For near-term fetuses and neonates, deeper molds are needed to avoid crushing tissues and to guarantee that sufficient paraffin can be added to fully support these large specimens.

Paraffin Sectioning

Paraffin blocks may be sectioned once they have solidified, but in our experience, additional preparation of the block can substantially improve the quality of the final sections. The first step is to remove excess paraffin at the margins of the block using a sharp blade. The paraffin border on each side can be reduced to a width that is equal to the shortest dimension of the sample. This pruning permits long ribbons of serial sections to be produced with minimal intervening paraffin. An alternative to trimming the blocks is to maintain molds of many sizes so that the specimen may be placed in a receptacle of appropriate size; however, in our experience, excess paraffin often still needs to be removed. Next, the block is mounted on the microtome, with the longest dimension of the tissue oriented vertically. This positioning is required so that the microtome cuts along the long axis of the specimen; attempts to cut through the short axis may result in fractures along the specimen's longitudinal axis at major stress points (i.e., the boundaries between tissues of differing firmness, such as the lungs and ribs in the thorax or the intestines and the liver in the abdomen). Finally, the face of the block may be covered with a wet paper towel to rehydrate the surface of the specimen. In our experience, this practice is useful for all developing mouse specimens, especially older (larger) ones, and can be accomplished using a tap water application of 5–10 min; this procedure may need to be repeated intermittently when extremely large tissue blocks are being serially sectioned. Rehydration in this manner may help prevent tearing of the sections during microtomy.

Routine mouse developmental pathology experiments are typically performed with paraffin sections cut at between 4 μm to 8 μm. Any thickness in this range is suitable for larger specimens (embryos at GD 10 or older), while younger embryos are usually sectioned at 4 μm due to their small size. The microtome used to cut sections may have either an automated or manual action as long as the cutting strokes are made with a relatively constant speed. The ribbons of paraffin sections are transferred to a warm-water bath (40°C–42°C) to spread, after which they are separated and applied to glass slides. Sections adhere more firmly if the slides have previously been coated with an adhesive gelatin, poly-L-lysine, or silane.[7] The first sections to be acquired from a specimen are placed closest to the slide label, while later sections are located farther from the label. In our experience, a single standard microscope slide (25 mm × 75 mm) may hold up to 40 serial longitudinal sections from a GD 8 embryo or two longitudinal sections from a GD 17 fetus. Slides with adhered sections typically are dried overnight at 60°C unless a procedure designed to detect protein function (e.g., enzyme histochemistry) is envisioned, in which case a physiologic drying temperature (37°C) is used. Dried sections may be kept at RT for weeks before staining. Temporary slide storage should be within a closed container (e.g., slide box) or sealed cabinet to keep dust from contaminating the surface.

In most instances, blocks should be completely sectioned at the original sitting. Otherwise, at least a few sections will be lost when the block has to be faced again. The choices of sectioning interval and capture scheme will depend on several factors. The primary drivers will be the research question under consideration and the cost that can be borne in performing the analysis (see Chapter 6 for experimental design details). If resources permit, developing mice should be serially sectioned so that organs from all specimens may be evaluated in an identical orientation. However, very large specimens (near-term fetuses, neonates, and isolated organs with relatively homogeneous structural features [kidney, liver, lung, placenta, etc.]) may be subjected to a more limited sectioning protocol in which one or a few sections are acquired from the face of the block. In such cases, the block bearing the residual tissue will be returned to the archive for long-term storage. Blocks with rare or critical specimens may be resealed with a thin layer of paraffin to protect the exposed surface of the specimen, while blocks that hold less unique or less important tissues may be returned to storage without any further preparation.

Plastic Sectioning

Efforts to section early postimplantation embryos (GD 8.5 or earlier) may benefit from plastic embedding (Figure 9.3). The reason for this choice is that plastic is much harder than paraffin, and thus provides more support when such small, fragile specimens are being sectioned. Advance preparation is required if plastic is to be employed due to the special fixatives (namely, osmium tetroxide) and embedding reagents that are required. The osmium tetroxide is concentrated in lipid-rich cell membranes. The resulting black color facilitates orientation of the specimen during embedding.

Two forms of plastic embedding media are available commercially. The first is a hard plastic (epoxy resin [e.g., Epon™]), which is the conventional choice for TEM evaluation. Specimens embedded in hard plastic may be cut readily as 1-μm-thick sections for routine light microscopic analysis and as 200- to 500-nm-thick sections for ultrastructural assessment. Hard plastic affords superior resolution of fine

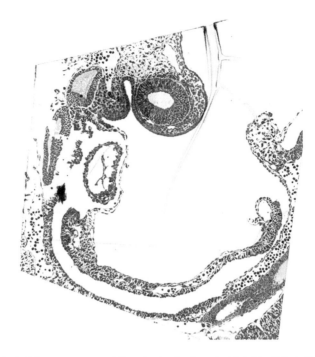

FIGURE 9.3 Morphological features are especially crisp in plastic-embedded thin (0.6- to 2-μm thick) sections. This mouse embryo (GD 9.5) was immersed in modified Karnovsky's fixative for 48 h, embedded in 1% agar, postfixed in 2% osmium tetroxide for 1 h, dehydrated in graded alcohols and polypropylene oxide, infiltrated overnight in serial changes of Poly/Bed® resin (a mixture of Araldite and Epon-like resins, which vary in their hardness and penetration ability), and then embedded in Poly/Bed. Stain: toluidine blue.

subcellular detail relative to that available in paraffin sections (typically cut at thicknesses of 4–8 μm). Further details regarding hard plastic sectioning may be found in Chapter 13. The second option is a soft plastic (e.g., glycol methacrylate [GMA]), which is usually cut at 2 μm. The resolution of soft plastic is comparable to that offered by paraffin, so we recommend that paraffin be used where feasible since it utilizes standard processing systems and less toxic reagents. A major advantage of soft plastics relative to hard plastics is that soft plastic sections can be processed using many special molecular techniques, while hard plastics can only be stained using chemical dyes. Once the decision to use a plastic medium has been made and the kind of plastic has been selected, appropriate commercial embedding kits may be purchased. These kits are used according to the manufacturer's instructions for mouse developmental pathology studies.

Soft plastic sections may be cut using a standard microtome, while sectioning hard plastic requires a specialized ultramicrotome. Sections slated for light microscopic analysis can be captured on regular microscope slides and retained at RT for weeks. Small specimens in soft plastic should be completely sectioned. Tissues embedded in hard plastic are rarely sectioned in their entirety, so the residual tissue and surrounding block are returned to the archive. No special preparation is required to prepare such samples for long-term archiving.

Cryosectioning

Frozen specimens (see Chapter 7 for tissue collection protocols) are removed from the tissue bank freezer (−70°C to −80°C) and placed in the cryostat (−20°C) to equilibrate for 15–30 min. One of several metal or plastic chucks within the cryostat chamber is positioned so that a mound of clear, viscous embedding medium (e.g., Tissue-Tek® O.C.T.™ [Optimal Cutting Temperature] Compound; Sakura Finetek USA, Torrance, CA) may be applied to the broad flat specimen surface. One sample at a time is removed from its wrappings. Previously embedded samples are applied to the viscous medium and surrounded to half of their height with more embedding compound; unembedded samples are pushed into the compound on the chuck and then entirely covered with additional medium. The medium is hardened by either submersing the chuck for 20 s in liquid nitrogen-cooled isopentane or using a freezing spray (available commercially from many vendors). The chuck is then mounted on the cryomicrotome within the cryostat chamber, with the long axis oriented vertically so that the section will be cut along its long axis. Sections are acquired using a slow consistent stroke and are generally cut at thicknesses of 6–10 µm. A fine-haired paintbrush (stored inside the cryostat to maintain thermal equilibrium with its surroundings) may be used to tease sections away from the surface of the block.

Sections are captured by applying a warm glass slide to one surface. Slides bearing tissues are commonly allowed to air-dry in the cryostat chamber in wire racks until a sufficient number have been acquired to transfer the full rack into a freezer (ideally at −70°C or lower, although −20°C is suitable for brief periods [up to 72 h]) until further processing can occur. Racks are placed within sealable plastic bags (e.g., Ziploc®) to prevent frost from forming on the tissues, as ice crystals are a common source of tissue damage in frozen sections. Alternatively, slides may be fixed immediately by immersion in a suitable fixative (e.g., acetone, 70% ethanol, NBF, or PFA) for up to 5 min and then air-dried for immediate use. When sectioning is completed, the chuck is removed from the cryomicrotome. Specimens are typically removed from metal chucks (which are expensive, and therefore subject to reuse) for return to the tissue bank, but samples mounted on plastic chucks are left attached to the chuck, and both tissue and chuck are achieved together. The face of the specimen is typically covered with O.C.T medium, and the sample is then wrapped in fresh aluminum foil (precooled squares [5–6 cm²] equilibrated in the cryostat) that in turn is enveloped in a flexible thermoplastic (e.g., Parafilm®) or placed in a sealable plastic bag. All these storage products are available commercially from multiple vendors.

Considerations for Orienting Specimens

Developmental pathology studies in mice require accurate positioning within the block to ensure that histotechnologic sections will be oriented so that a meaningful analysis may be performed. Large specimens (generally intact embryos at or older than GD 13), which have been isolated from the extra-embryonic membranes, may be oriented readily using the unassisted eye. In contrast, a magnifying device (either mounted on a headset or on a stand) or stereomicroscope will greatly assist efforts to position younger isolated embryos and small detached organs. Specimens naturally arrange themselves in lateral recumbency (resting on their sides) because this surface is the largest and relatively flattest plane. Magnification is particularly useful when trying to position small (GD 8) to mid-sized (GD 12) embryos in any orientation other than lateral recumbency. In general, specimens should be positioned head-down when transverse (cross) sections are desired, left side down when sagittal (longitudinal [from the side]) sections are necessary, and nose-down when coronal (longitudinal [from the front]) sections are required.

Extremely early postimplantation embryos (GD 7.5 or earlier) require special handling during the embedding process. They are quite difficult to isolate without damage, so they generally should be retained within the isolated decidual swelling or intact uterine horn to facilitate handling (see Figure 7.5). An alternative is to remove the egg cylinder (embryo and surrounding yolk sac), which permits partial visualization of the embryo. Added advantages of these *in situ* preparations are that the extra-embryonic membranes are available for evaluation and that their relationship with the embryo is preserved. The decidua is pear shaped and large enough to be easily manipulated without traumatizing the embryo within (see Figure 7.10). If it is positioned on its end, serial transverse sections of the embryo should be obtained in a reproducible manner. In contrast, longitudinal positioning of the decidua will yield sections unpredictably in one of three orientations: sagittal, coronal, or at some oblique angle. Thus, mouse developmental pathology studies in very early embryos in which an endpoint cannot be collected in the transverse plane often require that multiple decidual swellings be processed and sectioned to ensure that enough specimens are obtained with the proper orientation for analysis and acquiring relatively homologous photomicrographs.

Staining

Paraffin and plastic sections are rehydrated progressively by reverse passage through organic solvents and a series of graded alcohols (Table 9.6). Next, sections are placed in running tap water for at least 10 min to complete the restoration of tissue water. Care must be taken not to place the sections directly in the high-velocity stream of water, or small tissue fragments or entire sections may detach from slides and float away. A solution to this issue is to use a low flow rate and to funnel the water stream to the bottom of the water bath so that the greatest force is dissipated on the sides of the container rather than on the slides themselves. Rehydrated slides are ready for immediate staining.

Sections may be stained manually[5] or using an automated histostainer (Table 9.6). The standard stain for routine screening studies and detailed histopathologic analyses is hematoxylin and eosin (H&E) (Figure 9.2). Nucleic acid–rich structures such as nuclei and the cytoplasm of cells that are actively engaged in protein production bind hematoxylin and thus exhibit some shade of blue, while tissues with fewer nuclei and/or more extracellular matrix or cytoplasm bind eosin and appear pink in section. The staining times for both manual and automated procedures should be tested and, if necessary, modified in your laboratory to suit your needs. Once set, however, times should remain constant over time regardless of apparent differences in staining intensity. Such discrepancies usually stem from physiologic variations in tissue structure (early embryos bind little eosin due to the large numbers of actively proliferating cells and relative scarcity of extracellular stroma) or deviations earlier in the processing protocol (e.g., excessive time in acidic fixative or decalcifying solutions). Standardizing the staining protocol is a reliable means of compensating for these differences, and can be readily accomplished through automation.

TABLE 9.6

Automated Staining of Paraffin-Embedded Embryo Sections

Process	Step	Solution	Time
Dewax	1	Xylene	10 min
	2	Xylene	10 min
Rehydration	3	100% Ethanol	40 s
	4	100% Ethanol	40 s
	5	95% Ethanol	40 s
	6	80% Ethanol	40 s
	7	Water	1 min
Staining	8	Hematoxylin[a] 560®	3 min (exact)
Differentiation	9	Water	1 min
	10	Define[a]	20 s (exact)
	11	Water	1 min
	12	Blue buffer 8[a]	40 s (exact)
	13	Water	1 min
	14	95% Ethanol	40 s (exact)
Counterstain	15	Alcoholic eosin[a] Y 515®	12 s (exact)
Dehydration	16	95% Ethanol	40 s (exact)
	17	95% Ethanol	40 s (exact)
	18	100% Ethanol	2 min
	19	100% Ethanol	2 min
	20	100% Ethanol	3 min
Clearing	21	Xylene	3 min
	22	Xylene	3 min
Mounting			—

Note: Instrument is a Tissue-Tek® Prisma Automated Processor (Sakura Finetek USA, Torrance, CA).

[a] Selectech® reagents (Leica Microsystems Inc., Bannockburn, IL).

A few other stains are used on a regular basis to dye tissues of developing mice. Toluidine blue is perhaps the most common, especially for plastic-embedded tissues (Figure 9.3). Sections stained with this agent exhibit dark blue nuclei and pale blue cytoplasm and extracellular matrix. Various protocols for evaluating cell death and proliferation (see Chapter 12) or localizing the distribution of expressed genes (see Chapter 10) or proteins (see Chapter 11) are critical components of mouse developmental pathology studies, but they require specialized *in situ* molecular pathology methods that are beyond the scope of this chapter. Histochemical techniques (e.g., alkaline phosphatase[5,6]) may be undertaken to demonstrate sites of intracellular enzyme activity (i.e., functional proteins). Some histochemical pathways remain active in paraffin-embedded tissues, while others may only be evaluated in frozen sections. Finally, region-specific differences in the brain and spinal cord are poorly delineated by H&E, so special neurohistotechnological stains are necessary to effectively evaluate changes in these organs; common procedures for this purpose are cresyl violet (for nuclei), silver stains (for axons), and Luxol fast blue (for myelin). Details of these functional and neural-specific methods have been discussed elsewhere.[1,3,4,8,11]

Mounting

Following a final water wash to remove excess eosin, routine sections are then dehydrated and cleared using graded ethanols and organic solvents (Table 9.6). Care must be taken that eosin is not entirely removed by passage through the alcohol solutions. Once sections have been transferred into the organic solvent, a drop of a suitable mounting medium (e.g., Permount™, a toluene-based synthetic resin available commercially from multiple vendors) is applied to the section and a coverslip is applied over the tissues. We prefer conventional glass coverslips to plastic coverslips or plastic tape because glass does not scratch during routine handling and permits better light transmission for photomicrography. Further, glass is more readily removed should access to the section be required in the future.

Aqueous mounting media (e.g., Crystal Mount™, also available for purchase from multiple firms) are required for certain histochemical and immunohistochemical methods in which the final chromogen (colored reaction product) is soluble in organic solvents. Sections that are evaluated using such procedures do not need to be dehydrated after staining, so the coverslip can be applied immediately after the quality of the staining product is confirmed.

Final Thoughts

Histotechnological processing is the most important technical step in a mouse developmental pathology study. Seemingly minor deviations from established protocols or inattention to the positioning of the specimen may substantially impact the quality of the final sections, thereby reducing the amount of information that can be acquired from any given sample. Novice investigators and laboratories that lack experience with histotechnological practice as applied to developing mouse tissues should always consult a qualified histotechnologist and an expert comparative pathologist before collecting samples to ensure that tissue preparation is appropriate for the question being explored.

REFERENCES

1. Bancroft JD, Gamble G (2007). *Theory and Practice of Histological Techniques*, 6th edn. New York: Churchill Livingstone (Elsevier).
2. Callis, GM (2007). Bone. In: *Theory and Practice of Histological Techniques*, 6th edn, Bancroft JD, Gamble G (eds.). New York: Churchill Livingstone (Elsevier); pp. 333–363.
3. Carson FL, Hladik C (2009). *Histotechnology: A Self-Instructional Text*, 3rd edn. Chicago, IL: American Society for Clinical Pathology Press.
4. Junqueira LC, Carneiro J (1983). *Basic Histology*, 4th edn. Los Altos, CA: LANGE Medical Publications.
5. Kaufman MH (1992). *The Atlas of Mouse Development*, 2nd edn. San Diego, CA: Academic Press.

6. Kaufman MH, Schnebelen MT (1986). The histochemical identification of primordial germ cells in diploid parthenogenetic mouse embryos. *J Exp Zool* **238** (1): 103–111.

7. Kiernan JA (1999). Strategies for preventing detachment of sections from glass slides. *Microsc Today* **99** (6): 20, 22, 24.

8. Kiernan JA (2008). *Histological and Histochemical Methods: Theory and Practice*, 4th edn. Woodbury, NY: Cold Spring Harbor Laboratory Press.

9. Kristensen HK (1948). An improved method of decalcification. *Biotech Histochem* **23** (3): 151–154.

10. Latendresse JR, Warbrittion AR, Jonassen H, Creasy DM (2002). Fixation of testes and eyes using a modified Davidson's fluid: Comparison with Bouin's fluid and conventional Davidson's fluid. *Toxicol Pathol* **30** (4): 524–533.

11. Prophet EB, Mills R, Arrington JB, Sobin LH (1992). *Armed Forces Institute of Pathology Laboratory Methods in Histotechnology*. Washington, DC: American Registry of Pathology.

12. Relyea MJ, Miller J, Boggess D, Sundberg JP (2000). Necropsy methods for laboratory mice: Biological characterization of a new mutation. In: *Systematic Approach to Evaluation of Mouse Mutations*, Sundberg JP, Boggess D (eds.). Boca Raton, FL: CRC Press; pp. 57–89.

10

Localization of Gene Expression in Developing Mice

Jennifer M. Mele and Brad Bolon

CONTENTS

Modern biology is founded on an increasingly sophisticated understanding of genomic mechanisms that specify cellular and tissue anatomy and function during both health and disease. Abnormal gene structure and localization, especially in animals that have been deliberately engineered to have null mutations, are a major cause of embryolethality in developing mice. Therefore, developmental pathologists must be well versed in the *in situ* molecular pathology methods used to examine anomalous genetic sequences and expression patterns. The main technique utilized for the region-specific assessment of gene distribution in mouse tissues is *in situ* hybridization (ISH) to reveal mRNA from transcribed genes.[3,7,9,12] This procedure may be performed effectively on tissue sections or whole mount (intact) embryos or isolated (intact) organs.

This chapter introduces standard ISH protocols used to explore spatial and temporal differences in the distribution of mRNA in developing mice. These basic techniques are derived from the authors' experience. Numerous protocols are also available for review on the Internet. Basic ISH techniques for tissue sections (Protocol 10.1) and whole mount embryos (Protocol 10.2) are given to demonstrate the flow of work and potential options in such projects. Assistance in modifying protocols for your own setting may be obtained from standard references for ISH.[1,8,18,20,25,29] Procedures to define the extent to which mRNA of a given gene has been produced in a homogenized tissue (e.g., northern blots) fall outside the scope of a pathology volume. Detailed protocols for such procedures have been published elsewhere.[11,22,28]

Fundamentals of Evaluating Gene Expression

The ISH technique depends on the specific annealing of a labeled nucleic acid probe to complementary RNA sequences in a specimen. The primary decision when designing an ISH probe is the choice of nucleic acid backbone. An ISH probe is usually either a cloned DNA sequence (typical range, 100–500 base pairs [bp] in length); a riboprobe (50–400 bp); or a synthetic oligonucleotide (DNA or RNA, generally about 20–30 bp). Single-stranded probes are more sensitive than double-stranded DNA probes because the paired strands of the latter may re-anneal, thereby reducing the number of probe elements available to bind with the tissue. A primary advantage of the longer probes is that the greater number of bp improves their specificity, because even 1 or 2 mismatched bp will prevent effective annealing; incompletely bound probes are removed during ISH, which substantially reduces nonspecific background labeling. The main benefit of synthetic *oligomer* probes is that they can be designed and fabricated cheaply and quickly. Probes are typically generated in the investigator's laboratory, but synthetic oligomers are available for purchase from some commercial firms.

The second major decision when designing an ISH procedure is the choice of isotopic (radioactive) versus nonisotopic labeling for the probe. Isotopic probes retain a natural configuration because the label produces no (for ^{3}H- or ^{33}P-labeled) or minimal (for ^{35}S-labeled) distortion in the nucleic acid structure. Radioactive reagents are sensitive and specific, but require additional precautions when working with and disposing of reagents and experimental materials. In our experience, probes made with ^{33}P, while more expensive, provide better resolution than those incorporating ^{3}H or ^{35}S due to their higher energy and less extensive scattering of particles (yielding more discrete grains). Nonisotopic probes are made by incorporating a labeled nucleic acid (usually conjugated to biotin, digoxigenin [DIG], or a fluorescent dye)[14,16,17,19] in place of some of its unlabeled counterparts. Such nonradioactive reagents are detected in a fashion comparable to a routine immunohistochemistry (IHC) procedure and therefore involve fewer inherent hazards relative to isotopic procedures. However, the presence of the protruding label may reduce the sensitivity of nonisotopic methods as side chains can distort the fit between the probe and its target.

Technical Principles of *In Situ* Hybridization in Tissue Sections

The initial step in exploring gene expression during mouse developmental pathology studies is to obtain suitable tissue sections for ISH. Fundamental histotechnological steps to prepare acceptable tissue sections—fixation, dehydration (used for paraffin sections only), embedding, and sectioning—are reviewed in Chapter 9. In our experience, all specimens dedicated to gene expression analysis should be fixed using an aldehyde-based solution (e.g., neutral buffered 10% formalin [NBF]) because acetic acid-containing fluids tend to degrade nucleic acids (Figure 10.1). Furthermore, it is imperative that all RNA-degrading enzymes (ribonucleases, or RNases) be inactivated in reagents and on surfaces used to process ISH sections. Solutions are rendered RNase-free by adding diethylpyrocarbonate (DEPC; Sigma–Aldrich, St. Louis, MO) at 0.1% v/v, incubation at 37°C for 1–2 h, and then autoclaving for 15–30 min. Instruments and surfaces are cleansed of RNases by swabbing with RNase AWAY® (Life Technologies, Inc., Grand Island, NY).

In general, sections should be cut fresh (within 24 h of the ISH experiment) for best results. However, paraffin-embedded sections may be cut and archived at −80°C for up to 2 weeks. Frozen sections are removed from the freezer and warmed to RT. Sections should be kept in closed containers while warming to prevent water condensation and dust accumulation on the unfixed tissue. Slides are immersed in fixative for 30 s to 1 min and then washed three times for 5 min in RT buffer. Typical fixatives are acetone, 70% ethanol, 4% formaldehyde, or 4% paraformaldehyde; the choice of fixative will depend on the degree to which cellular morphology must be preserved and prior experience (as reported in the literature or experienced in your laboratory) with given combinations of tissue and fixatives. In general, aldehydes preserve the structural integrity of tissues more effectively. Alternatively, paraffin-embedded sections are deparaffinized in an organic solvent (typically toluene or xylene) and then rehydrated through a series of graded alcohols and finally water so that the aqueous ISH reagents can gain entry to all portions of the tissue.

The next step is to synthesize the probe. The exact procedure used to prepare the probe will depend on its nature. Means for fabricating suitable riboprobe and oligomer reagents are given later. However, details for each of the large number of possibilities for making a probe are beyond the scope of this pathology book, and instead must be sought in molecular biology references.[2,4,8,17,25,28,29]

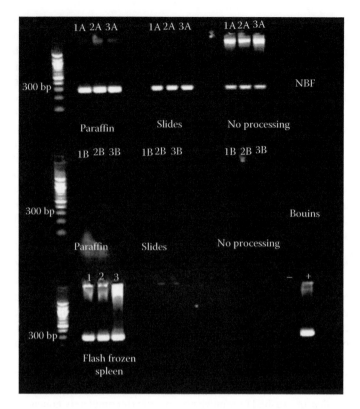

FIGURE 10.1 Nucleic acids preservation depends on tissue handling. Relative to unfixed, flash-frozen specimens (bottom row), northern blot analysis demonstrated that RNA transcripts of an unidentified housekeeping gene are retained intact in tissue that has been fixed only in aldehyde (top row, neutral-buffered 10% formalin [NBF]) but are lost in samples fixed in complex solutions containing mixtures of acids and aldehydes (middle row, Bouin's solution). The transcripts can be obtained from NBF-fixed, paraffin-infiltrated blocks (lanes labeled *paraffin*) and tissues sections (lanes designated *slides*) as well as from NBF-fixed, flash-frozen tissue (lanes termed *no processing*). Tissue preparation: spleens from three pregnant wild-type C57BL/6 dams were subdivided and either fixed for 24 h at room temperature by immersion in NBF (top row; samples 1A, 2A, 3A) or Bouin's solution (middle row; samples 1B, 2B, 3B) or frozen immediately (bottom row; samples 1, 2, 3). Specimens were either embedded in paraffin or covered with water-soluble gel (Tissue-Tek® Optimal Cutting Temperature [OCT™] compound; Sakura Finetek USA, Torrance, CA). Left lane, ladder for estimating transcript size. Control lanes (right side of bottom row), lacking (–) or containing (+) the transcripts.

The next step is to apply pre-hybridization treatments to optimize the cellular and molecular features of the sections for mRNA detection. One or more such steps may be appropriate, with the specific choice depending on the nature of the research question and the investigator's experience with ISH. The first pre-hybridization step is an enzyme treatment to increase the permeability of the section to reagents. A common choice is incubation in proteinase K (1 μg/mL) in 0.1 M Tris–HCL buffer (pH 8.0), containing 50 mM ethylenediaminetetraacetic acid (EDTA), for 30 min at 37°C. The proteinase K stock solution (10 mg/mL) is stored in aliquots at –20°C; enzyme aliquots are thawed immediately prior to use. Slides are then washed in DEPC-distilled water (two times for 1 min each) followed by 0.1 M triethanolamine–HCL (TEA) buffer (pH 8.0; two times for 2 min each). Sections are then acetylated with 0.25% (v/v) acetic anhydride in TEA buffer at room temperature (RT) by agitating the slides rapidly up and down in the solution for 30 s and then allowing them to stand for 10 min. Slides may be retained for several days by washing them in 2X saline–sodium citrate (SSC) buffer for 1 min, dehydrating through graded ethanols (10–15 s each in 30%, 50%, 70% 80%, 95%, 100%, 100%) without passing to organic solvent, then storage in a vertical orientation in a dry, dust-free container at 25°C. Alternatively, slides may pass immediately from the acetylation pretreatment to the hybridization step.

The hybridization process begins by defining the amount of probe that is needed. The location of all sections to be hybridized is determined, and if necessary marked so that hard-to-see tissue is not missed. The section dimensions (in cm) are used to calculate the amount of probe that is required; the volume (in µL) of probe solution used per section is the tissue area multiplied by 4. An additional volume (10%–20%) should be made to compensate for handling and evaporative losses. Next, the hybridization cocktail is prepared in sufficient quantity to constitute all test and control probe solutions, and the appropriate probe mix is added. The final composition of a probe solution for ^{3}H- or ^{33}P-labeled probes will be 80% hybridization cocktail and 20% probe mixture, where the probe mixture proportions are 50% yeast tRNA (10 mg/mL), 0.3 ng/µL/kb of probe, and water. For ^{35}S-labeled probes, the amount of probe mix added must be lowered to 15% to accommodate the inclusion of the redox reagent dithiothreitol, which will make up 5% of the final probe solution. Each probe mixture is heated to 80°C for 2 min to ensure uniform dispersion and then placed on ice for 5 min. Hybridization cocktail is then added to generate the probe solution, which must be agitated vigorously on a vortex mixer for 30 s and then centrifuged for 30 s to collect the fluid and remove air bubbles. The radioactivity of each probe solution is then calculated in a 1 µL aliquot using a standard scintillation counter, where

$$\text{Probe quantity (ng/µL)} = \text{cpm/µL} \times \text{ng Probe/cpm} \times \% \text{Precipitability}.$$

Subsequently, the probe solution is applied to the center of the section, using 4 µL/cm^2 of section area (i.e., enough to cover the section when spread). A silanized coverslip is added to promote uniform dispersal of the fluid over the section. The reaction chamber is made by sealing the edge of the coverslip with a thin bead of rubber cement, after which slides are placed (in horizontal orientation) in a humidified chamber at RT for 4 h to overnight to permit hybridization. The appropriate hybridization temperature will vary with the sequence of the probe but is generally in the range 42°C–50°C. For example, the optimal incubation for a probe with average (~40%) GC content is 45°C.

Next, hybridized slides are washed to remove nonspecific (incompletely annealed) probe. The stringency of these washes is dictated by the nature of the gene being investigated. For small families of closely related genes, washes must be fairly stringent to ensure that the desired target mRNA is the only one capable of specific annealing. In most cases, however, the major source of background labeling is spurious, incomplete binding (seemingly by electrostatic interactions rather than annealing of complimentary nucleic acid sequences) of probe to random connective tissue and/or cellular components. For each slide, the reaction chamber is opened by peeling away the rubber cement seal without moving the coverslip. The slides are incubated in 4X SSC at RT (three times for 5 min each); the coverslips should detach without any manipulation during the first or second wash. For probes labeled with ^{35}S, the 4X SSC should contain 10 mM dithiothreitol, and the time for each wash should be extended to 20 min. The vast majority of radioactivity will be removed in the first two washes. Slides are then treated with RNase A (20 µg/mL) in RNase buffer (0.5 M NaCl, 10 mM Tris–HCl, 1 mM EDTA, pH 8.0) at 37°C for 30 min to digest and release unpaired (nonspecifically bound) probe. It is imperative that containers exposed to RNase be manipulated and stored separately from those used during preceding steps of the ISH process to avoid RNase contamination and indiscriminate destruction of specifically annealed probe along with nonspecifically attached probe. Next, slides are washed at 37°C in 4–5 changes of RNase buffer (5–10 min and 200–250 mL per wash for each 10 slides) followed by 4–5 changes in 2X SSC (5–10 min and 200–250 mL per wash for each 10 slides) to remove all RNase. Finally, slides are passed through a RNase-free 0.1X SSC wash (500 mL per 20 slides) at 50°C for 15 min and then dehydrated through graded alcohols (30%, 50%, 70%, 95%, 100%, 100%, where the first three solutions contain 0.3 M ammonium acetate to prevent denaturation of the hybrids) and air-dried.

Autoradiography is used to demonstrate probe annealing to the hybridized slides. Detection of an abundant RNA (0.5%–1% of message) requires approximately 1 week of exposure, while detection of a rare RNA (0.0005%–0.01% of message) generally requires 3–4 weeks. The first step in autoradiography is to prepare a stock emulsion by heating a suitable reagent (e.g., Kodak NTB-2 emulsion; Eastman Kodak, Rochester, NY) to 45°C, diluting it by adding an equal volume of 600 mM ammonium acetate (prewarmed to 45°C), and then mixing it by gentle swirling (to avoid introduction of air bubbles). The stock is split into 12 mL aliquots, wrapped in aluminum foil to prevent light exposure, and then stored in a light-tight container at 4°C. For each experiment (conducted in the dark or under red light), sufficient emulsion is melted in a water bath at 45°C and gently decanted into a dipping chamber. Air bubbles are allowed to surface for 10–20 min, after which slides are dipped in emulsion (two times, 2 s per dip). Excess solution is removed from the slide edges by gentle blotting, and slides are placed in vertical orientation in racks and allowed to air-dry at RT for 30 min. If message is rare, slide racks are then transferred to a moist chamber and allowed to stand at RT for 3–6 h to remove latent grains in the emulsion. Coated slides are air-dried at RT for 30–60 min and then placed in light-tight plastic boxes with a small packet of desiccant. Boxes are placed in sealable plastic bags (e.g., Ziploc®) along with a larger desiccant packet and then maintained at either 4°C or −20°C for an appropriate period of time. To develop the emulsion, black boxes still sealed in their bags are allowed to warm to 15°C (15–20 min at RT). Slides are then moved at 15°C through Kodak D-19 developer (Eastman Kodak) for 2.5 min, a 2% acetic acid (in distilled water) bath for 10 s to stop the reaction, and Kodak fixer for 5 min. Developed slides are washed in distilled water for 15 min and then cold running tap water for 30 min.

Developed sections are stained to demonstrate structural features of the tissue section. Lighter stains (eosin, hematoxylin, or hematoxylin and eosin [H&E]) are preferable as they give good contrast but do not obscure the dark grains. Reaction times at each staining step will typically be double those used for routine H&E staining to permit solutions time to penetrate the emulsion. Extended exposure to acidic solutions should be minimized as developed grains may be lost at low pH.

Technical Principles of *In Situ* Hybridization in Whole Mount (Intact) Specimens

The fundamental concepts for assessing gene expression in whole mount preparations are equivalent to those described earlier for tissue sections. Fixation usually is performed overnight at 4°C or for 2–4 h at RT. All other steps typically are completed at RT. The main technical difference is that probes are labeled with a chemical side chain (e.g., biotin, DIG, or a fluorescent dye) rather than an isotope. In our experience, whole mount ISH may be applied successfully to intact mouse embryos up to gestational day (GD) 10 or sometimes GD 11, and to isolated organs from older developing mice.

The main benefit to whole mount ISH compared to the similar procedure on a tissue section is that whole mount preparations rapidly display in three dimensions the spatial relationship between structures expressing a given gene and those that do not. Other advantages include the ability to process more whole mount samples in parallel and the reduced amount of labor needed for processing. The major drawbacks to whole mount methods are that localizing a gene to a specific cell type (rather than to large regions) is more difficult, and the techniques cannot be used effectively in older embryos (which have thicker torsos and impermeable skin).

Multiple Staining Experiments

Assessment of gene interactions is often best evaluated during co-localization studies of two or more molecules. Such studies may utilize dual ISH-based endpoints,[13,27,30] although ISH may also be paired with immunohistochemistry (IHC; see Chapter 11) to evaluate protein distribution.[24]

Two general approaches are employed for such experiments. The first applies both probes simultaneously, where each probe is linked to a unique side chain (i.e., biotin or DIG or a fluorophore). The second approach performs the ISH technique for each RNA sequentially. The entire ISH method is completed for the first transcript. Then, instead of counterstaining, the specimens are washed in buffer, and the second probe is added. Care must be taken to ensure that all relevant blocking solutions are applied at the correct time, and that all reagents are added in the appropriate order.

Control Materials for Experiments Investigating Gene Expression

Several controls must be included in each ISH experiment to ensure that binding of labeled probes is are specific, as spurious binding of probe is a common occurrence.[15] Tissue controls, reaction controls, reagent controls, and treatment controls are all necessary (Table 10.1).

TABLE 10.1

Common Controls Used for *In Situ* Hybridization Procedures

Type of Control	Control Material
Reaction control	*Concentration control*
	Concentration of sense probe is increased to assess spurious labeling.
	Concentration of random probe of the same length (used in place of the antisense probe) is increased to examine background labeling.
	Replacement control
	Random probe substituted for antisense probe.
Tissue control	*Positive control* (known to express messenger RNA)
	Internal (located within test specimen) (preferred)
	External—tissue > cell pellet ≫ protein linked to polymer
	On same slide as test specimen (preferred)
	On different slide
	Negative control (known not to express messenger RNA)
	Internal (located within test specimen) (preferred)
	External—tissue > cell pellet
	On same slide as test specimen (preferred)
	On different slide

Tissue controls are used to ensure that RNA for the gene of interest will be detected where it is expressed, and cannot be observed where it does not exist. Control samples must be collected, fixed, and processed in the same fashion as the test specimens. The ideal procedure for staining control tissues is to mount them on the same slide, either within a single section or as adjacent positive and negative sections; this approach exposes both controls to the same reagents for the same length of time, and thus provides the most rigorous comparison between them. If actual control tissues cannot be obtained, surrogate control tissues may include pellets of cells in which the gene of interest has been expressed.

Reaction controls confirm that the design of the ISH protocol was able to detect the RNA of interest, and to show that the procedure is performed consistently over time. The most common reaction control is to replace the antisense (AS) probe (which detects the RNA of interest) with either the sense sequence (a length of nucleic acid complementary to the AS probe) or a nonsense nucleic acid of similar size and random sequence. In general, omission of probe is not a sufficient reaction control.

Reagent controls are used to ensure that the AS probe is specific for its target. These controls are often performed in detail only when establishing a novel ISH method by testing the probe over a range of dilutions against control tissues known to express or not express the target. In this fashion, the optimal dilution for using the probe is defined while simultaneously confirming the specificity of the probe sequence. Once a method has been perfected, tissue controls may be used as proxies for reagent controls.

When performing multiple labeling procedures, appropriate controls for both target molecules must be undertaken at each step to ensure that deposition of the final products on the dual-labeled sections is specific.

Storage of Reagents and Tissues

Materials used in gene expression studies must be retained in such a fashion that their integrity (in the cellular, chemical, and molecular senses) is preserved for extended periods. The key factors in preventing degradation during storage are to maintain a constant low temperature and to avoid RNase contamination.

PROTOCOL 10.1   BASIC *IN SITU* HYBRIDIZATION
METHOD FOR TISSUE SECTIONS

Solutions[2]

Solutions are made using nuclease-free water in autoclaved, sterile glassware or RNase-free plastic ware and are autoclaved or filter-sterilized before use.

1. DEPC water
   a. Make a 0.1% v/v solution of diethylpyrocarbonate (DEPC) in double-distilled water (ddH$_2$O). (*Note*: Deionized water is also acceptable.)
   b. Incubate the solution for a minimum of 1–2 h at 37°C.
   c. Autoclave (on the liquid cycle) for 15–30 min.

Preparation of Radiolabeled Riboprobe

1. Insert the sequence for a gene of interest into an appropriate vector (i.e., one that contains a promoter for either the SP6 or the T7 RNA polymerase).
2. Isolate the inserted probe sequence.
   a. Linearize the DNA plasmid downstream of the coding sequence (details of which will vary with the nature of the restriction enzyme sites on the plasmid).
   b. Extract the linearized DNA using a standard phenol/chloroform protocol.
      i. In a microcentrifuge tube, add an equal volume of buffer-saturated phenol–chloroform (1:1) to the DNA solution.
      ii. Mix well by vortexing for 10 s.
      iii. Spin in a microcentrifuge at 16,000g for 3 min.
      iv. Carefully remove the aqueous layer to a new microcentrifuge tube, being careful to avoid the interface. (*Note*: Steps i–iv can be repeated as often as needed until an interface no longer can be seen.)
      v. Mix chloroform 1:1 with the aqueous layer to remove trace phenol contamination.
      vi. Spin in a microcentrifuge at 16,000g for 3 min.
      vii. Remove aqueous layer to new tube.
   c. Concentrate DNA using a standard ethanol precipitation protocol.
      i. Measure the volume of the DNA sample.
      ii. Adjust the salt concentration in the sample by adding 1 volume of 3 M sodium acetate, pH 5.2, to 10 volumes of sample (to yield a final salt concentration of 0.3 M).
      iii. Mix well by vortexing for 10 s.
      iv. Add three volumes of cold 100% ethanol.
      v. Mix well.
      vi. Hold at 4°C (on ice) or at −20°C for at least 20 min.
      vii. Spin at 16,000g (maximum speed) in a microcentrifuge for 10–15 min.
      viii. After carefully decanting the supernatant, add 1 mL of 70% ethanol.
      ix. Mix well.
      x. Spin at 16,000g in a microcentrifuge for 5 min.
   d. Wash the pellet.
      i. After carefully decanting the supernatant, add 1 mL of 70% ethanol.
      ii. Mix well by vortexing for 10 s.
      iii. Spin at 16,000g in a microcentrifuge for 3 min.
   e. After carefully decanting the supernatant, air-dry or briefly vacuum dry the pellet.
   f. Resuspend the DNA pellet in sterile H$_2$O at a concentration of 1 μg/μL.

3. Prepare 20 μL of reaction mix. The recipe for this solution is
   a. 5X transcription/hybridization buffer, 4.0 μL, where the buffer composition (after reconstitution from a working stock of four parts formamide to one part buffer) is
       i.   Piperazine-*N,N'-bis*(2-ethanesulfonic acid) (PIPES), 200 mM, pH 6.4
       ii. NaCl, 2 M
       iii. Ethylenediaminetetraacetic acid (EDTA), 5 mM
   b. Dithiothreitol (DTT), 0.2 μL of 1 M stock
   c. RNase inhibitor, 60 U
   d. Nucleoside triphosphates (NTPs)—working from 10 mM stocks
       i.   Radiolabeled ribonucleotide: 10.0 μL of UTP, labeled with either ^{33}P or ^{35}S. (The P-labeled probes yield less background, but they are much more expensive than the S-labeled versions.)
       ii. Unlabeled deoxyribonucleotides (dNTPs): 1.0 μL each of dATP, dCTP, and dGTP.
   e. Precipitated plasmid DNA, 1 μL (of a 1 mg/mL solution)
   f. RNA polymerase (SP6 or T7), 16 U
4. Incubate the reaction mix at 37°C for 30 min.
5. Add more RNA polymerase (16 U), and incubate at 37°C for another 40 min.
6. Expand the reaction mix. Add the following reagents:
   a. Ribonuclease inhibitor, 60 U
   b. Carrier RNA, 2.0 μL of a 10 mg/mL solution
   c. DNase I, 1.0 μL
7. Incubate the reaction mix at 37°C for 10 min.
8. Calculate the radioactivity of the riboprobe.
   a. Add the following reagents to the reaction mix:
       i.   DTT, 0.8 μL of a 1 M solution
       ii. Sterile H$_2$O, 63.0 μL
       iii. Sodium acetate, 10.0 μL of 3 M solution
   b. Remove 1.0 μL and measure the counts per minute (cpm) using a standard scintillation counter to define the cpm/μL.
   c. Calculate the quantity of probe (ng/μL) using the formula:
       Probe Quantity (ng/μL) = cpm/μL × ng Probe/cpm × % Precipitability
9. Add 36.4 μL of 7.5 M ammonium acetate to the reaction mix (to achieve a final salt concentration of 2 M).
10. Prepare the radiolabeled probe for use:
    a. Add carrier tRNA (50–100 mg) and ice-cold 100% ethanol (272 μL).
    b. Precipitate the RNA pellet for 10 min on dry ice.
    c. Carefully decant the supernatant and wash the pellet with 272 μL of ice-cold 70% ethanol.
    d. After carefully decanting the supernatant, air-dry or briefly vacuum dry the pellet.
    e. Resuspend the pellet in 100 μL of 10 mM DTT.
    Note: Probe may be retained for 2 weeks at −20°C.

In Situ Hybridization Procedure for a Radiolabeled Riboprobe

1. Deparaffinize sections:
   a. Xylene, two times for 5 min each
   b. 1:1 Mixture of xylene:100% ethanol, 5 min
   c. 100% Ethanol, 5 min each

2. Rehydrate sections:
   a. 100% Ethanol, 5 min
   b. 100% Ethanol, 5 min
   c. 95% Ethanol, 5 min
   d. 80% Ethanol, 5 min
   e. 70% Ethanol, 5 min
   f. 50% Ethanol, 5 min
   g. 30% Ethanol, 5 min
   h. DEPC-distilled water, two times for 5 min each
3. Proteinase K pretreatment (to permeabilize cells)
   a. Place slides in 0.1 M Tris–HCl, containing 50 mM EDTA, pH 8.0, prewarmed to 37°C
   b. Add proteinase K (1 µg/mL)
   c. Incubate at 37°C for 30 min
4. Wash in DEPC-distilled water, two times for 1 min each.
5. Wash in 0.1 M triethanolamine–HCL (TEA) buffer (pH 8.0), two times for 2 min each.
6. Acetylation pretreatment (to reduce background binding during *in situ* hybridization)
   a. Place 0.25% (v/v) acetic anhydride in TEA buffer (0.25 mL of anhydride per 100 mL TEA) in a staining dish at room temperature (RT)
   b. Add slides and agitate them rapidly up and down in the solution for 30 s
   c. Allow slides to incubate for 10 min at RT
7. Hybridization
   a. Apply probe to the section (at 4 µL/cm^2 of tissue)
   b. Place silanized coverslip over section, allowing fluid to spread by capillary action (without pressure)
   c. Seal reaction chamber using rubber cement to tack down the edges of the coverslip
   d. Place slides in horizontal orientation within a humidified container for 4 h to overnight at the appropriate annealing temperature (generally 45°C)
8. Posthybridization treatments (to remove nonspecifically bound probe)
   a. Remove the rubber cement seal without displacing slide
   b. Immerse slides in 4X SSC (saline–sodium citrate buffer) at RT, three times for 5 min each; cover slips will detach in the first wash
   c. RNase A treatment (to remove nonspecifically bound probe): immerse in RNase buffer containing RNase A (20 µg/mL) at 37°C for 30 min
9. Washes (to remove RNase activity)
   a. Wash in RNase buffer at 37°C, four to five times for 5–10 min each (using 200–250 mL per wash for each 10 slides)
   b. Wash in RNase-free 0.1X SSC at 50°C for 15 min (500 mL per 20 slides)
10. Dehydrate through graded alcohols:
    a. 30% Ethanol (containing 0.3 M ammonium acetate to prevent denaturation of the hybrids), 5 min
    b. 50% Ethanol (containing 0.3 M ammonium acetate), 5 min
    c. 70% Ethanol (containing 0.3 M ammonium acetate), 5 min
    d. 95% Ethanol, 5 min
    e. 100% Ethanol, 5 min
    f. 100% Ethanol, 5 min
11. Air-dry slides in vertical position, 30 min at RT in a dust-free cabinet.
12. Autoradiography
    a. Melt Kodak NTB-2 emulsion in a water bath at 45°C and gently pour it into a dipping chamber

   b. Allow air bubbles to rise to the surface for 10–20 min
   c. Dip slides in emulsion, two times for 2 s each
   d. Remove excess emulsion from slide edges by gentle blotting
   e. Arrange slides in racks in vertical orientation and allow to air-dry at RT for 30–60 min
       Note: Rarely, emulsion-coated slides are dried for a further 3–6 h at RT in a moist chamber to reduce the number of latent (artifactual) grains.
   f. Place slides in light-tight plastic boxes (with a small packet of desiccant)
   g. Place boxes in sealable plastic bags along with a larger desiccant packet
   h. Expose slides at either 4°C or −20°C for an appropriate period of time (usually 2–4 weeks)
   i. Warm slide boxes (still sealed in their bags) to 15°C for 15–20 min at RT
   j. Place slides in Kodak D-19 developer for 2.5 min at RT
   k. Move slides to 2% acetic acid (in distilled water) for 10 s
   l. Transfer slides to Kodak Fixer for 5 min at RT
   m. Wash slides in distilled water for 15 min and then cold running tap water for 30 min
13. Staining
   a. Mayer's hematoxylin, 2–3 min
   b. Wash in running tap water, 3–5 min
   c. Immerse in acid alcohol (1% HCl in 70% ethanol), two dips of 1 s each (the sections will turn red)
   d. Wash in running tap water until the sections turn blue
   e. Dehydrate slides
       i.   30% Ethanol, 5 min
       ii.   50% Ethanol, 5 min
       iii. 70% Ethanol, 5 min
       iv.   95% Ethanol, 5 min
       v.   100% Ethanol, 5 min
       vi.   Xylene, 5 min
       vii. Xylene, 5 min
14. Coverslip with mounting medium miscible with organic solvent (e.g., Permount®).

Most reagents (e.g., biotinylated or DIG-conjugated probes, photographic emulsion, SSC) are relatively stable for weeks to months when stored at 4°C. This temperature is particularly suitable for materials that are used frequently as repeated freeze-thaw cycles can destroy the molecular structure of antibodies. Freezing (generally −20°C) is permissible when reagents are to be archived for longer periods, but the material should be parsed into aliquots so that reagents are only thawed once. Commercial products should be stored according to the manufacturer's instructions. Some reagents (e.g., hybridization cocktail, radiolabeled probes) must be made fresh for each experiment for best results.

Tissues may be archived in either frozen tissue banks (stored at −20°C or colder) or as paraffin-embedded blocks. Such banks should include normal control tissues encompassing the species, genetic background, developmental ages, and (if necessary) genders of the likely test samples. Banks also may include samples from various diseases or toxicant-induced conditions, which serve as test materials or additional control tissues. In our experience, the small size of developing mice results in a frequent need to replenish control specimens, so detailed sample logs and regular inventories (at least once every 6 months) should be kept to ensure that new control tissues are acquired before the control tissues are exhausted.

PROTOCOL 10.2   BASIC WHOLE MOUNT *IN SITU* HYBRIDIZATION TECHNIQUE

Smaller specimens typically require shorter incubation times. All steps should be conducted with gentle rocking or shaking to facilitate reagent penetration and removal.

Day 1: Sample Collection and Probe Preparation

1. Harvest specimens (intact embryos [removed from placenta], intact organs of older animals).
2. Fix by immersion in 4% paraformaldehyde at 4°C for 2–4 h.
3. Wash in PBT (PBS, pH 7.4, containing 0.1% Tween 20), three times for 5 min each at 4°C.
4. Probe synthesis: digoxigenin (DIG) conjugation.
   Note: Probe may be retained for 1–2 weeks at 4°C.
   a. Prepare probes according to the manufacturer's instructions using any of a number of commercially available kits (e.g., Boehringer Mannheim).
   b. Confirm signal strength for probes:
      i. Make dilutions of 1:10, 1:100, and 1:1000 for both the probe and the reference standard.
      ii. Spot 2 µL of all probe and reference dilutions on a nylon hybridization membrane and allow to air-dry completely.
      iii. Incubate with a 1:5000 dilution of anti-DIG antibody (linked to a reporter enzyme) in a standard transcription/hybridization buffer (e.g., 5X = 200 mM PIPES, pH 6.4; 2 M NaCl, 5 mM ETDA) at RT for 30 min, with gentle rocking.
      iv. Wash with buffer, three times for 5 min each, with rocking.
      v. Incubate with appropriate chromagen at RT for 5–10 min.
      vi. Rinse for 10 s in distilled water.

A low signal on the spot test indicates a low yield of labeled RNA probe, which will likely be insufficient for the ISH procedure.

Day 2: Pretreatments and Hybridization

1. Bleach embryos in a solution containing 1:5 (v/v) 30% H_2O_2:methanol for 2 h at RT.
2. Rehydrate specimens:
   a. 70% Methanol in PBT, 20 min
   b. 50% Methanol in PBT, 20 min
   c. 25% Methanol in PBT, 20 min
   d. PBT, three times for 20 min
3. Transfer embryos to 2 mL glass tubes.
   For efficient washing, the number of embryos per tube should be adjusted based on developmental age, as follows:

E7	10–15 embryos
E8	7–10 embryos
E9	4–8 embryos
E10	1–3 embryos

4. Proteinase K pretreatment
   a. Place embryos in PBT, prewarmed to 37°C
   b. Add proteinase K (10 µg/mL)
   c. Incubate at 37°C, where the incubation time (min) equals the developmental age (in days) (e.g., embryos at gestational day [GD] 8 would be permeabilized for 8 min)

5. Prehybridization washes
    a. Decant proteinase K solution
    b. Wash in PBT at RT, two times for 30 s with gentle agitation (to avoid damaging the embryos, pour the solution down the side of the tube rather than directly onto the embryos)
    c. Wash in PBT at RT, four times for 10–15 min each
    d. Wash in fresh hybridization buffer (1.5 mL) at RT, one time for 2–5 min
    e. Wash in fresh hybridization buffer (1.5 mL) at 65°C, three times for 60 min each
6. Hybridization
    a. Add hybridization buffer (1.5 mL) containing DIG-labeled probe (~1 µg/mL), prewarmed to 65°C
    b. Incubate overnight at 65°C

Day 3: Posthybridization Washes and Antibody Incubation

1. Posthybridization washes
    a. Decant the probe solution
    b. Wash in hybridization buffer, prewarmed to 65°C
        i.   Two times for 3–5 min each
        ii. One time for 30 min
    c. Transfer through freshly made washing solution (50% formamide/1X SSC/0.1% Tween 20), prewarmed to 65°C
        i.   Two times for 3–5 min each
        ii. Five times for 30 min each
2. Protein blocking (to reduce nonspecific binding of anti-DIG antibody)
    a. Place samples in 2 mL of PBT containing 20% heat-inactivated nonimmune serum (from the species in which the anti-DIG antibody was raised)
    b. Incubate at RT for 1–2 h
3. Primary anti-DIG antibody
    a. Add anti-DIG antibody (dilution, 1:2000 to 1:4000) to 2 mL of PBT containing 20% heat-inactivated nonimmune serum
    b. Incubate overnight at 4°C

Day 4: Postantibody Washes and Histochemistry

1. Postantibody washes
    a. Decant the antibody solution
    b. Wash in PBT at RT
        i.   Three times for 2–3 min each
        ii.   Four times for 10–15 min each
        iii. Four times for 20–30 min each
2. Apply chromagen at RT for 20–48 h. Rock the samples for the first 20 min only.
3. Monitor color development under high magnification (e.g., stereomicroscope).
4. Once color deposition is sufficient, wash in PBT at RT (three times for 10 min each).
5. Wash in 50% glycerol/0.1% sodium azide overnight at 4°C.
6. Store embryos in glycerin until evaluation.

Interpretation of Grain Distribution

The presence of grains must be assessed carefully to ensure that it results from specific interaction with the target RNA. Such determinations require judicious comparisons to both concurrent controls (i.e., done within the same study) and, in most cases, prior knowledge regarding expression of the target molecule. When necessary, results of *in situ* gene expression studies should be confirmed by ancillary quantitative analyses in homogenized tissues such as northern blotting (to evaluate mRNA) or western blotting (for protein levels). The amount of target RNA may be quantified in an entire organ. In general, the magnitude of signal detected by northern blotting is well correlated with the outcome of ISH procedures.[31]

Cell- and Tissue-Specific Gene Analysis

The amount of target RNA also may be quantified in specific structures by using laser capture microdissection (LCM). The advantage of LCM is that discrete tissue domains may be removed, one or a few cells at a time, to avoid contamination by extraneous gene products. For example, when using laser pressure catapult (LPC) methodology, a burst of energy is applied to quickly launch the cells of interest in a small area into an inverted cap containing the RNA lysis buffer; accordingly, no contact is made with any external sources of genetic material. Another advantage of LCM is that the laser (355 nm) that excises the cells to be sampled cuts precisely, harvesting cells without harm to RNA (or DNA or protein).

In our experience, tissue destined for LCM is processed using conventional methods to produce frozen[5,6,23] or paraffin-embedded[26] sections. The lengths of time between tissue collection, preservation (by freezing or fixation), and storage are key. The best option for stabilizing RNA for ISH may be direct freezing because penetration of fixative from the surface of tissue is as slow as 1 mm/h, which can permit substantial RNA degradation at the center of the sample.

The desired site is harvested according to the manufacturer's instructions (Figure 10.2). Once the tissue is sectioned onto the slide, it should not be cover-slipped as it needs to be exposed for LCM. Frozen blocks may be sectioned up to 2 weeks prior to LCM onto special microscope slides coated with a RNase-pretreated polyethylene naphthalate (PEN) membrane; these slides should be stored in air-tight containers at −80°C. Frozen sections should be dehydrated using a series of RNase-free, ice-cold ethanols (70%, 85%, 95%, 100%). Any stains used to visualize cells of interest also must be RNase-free. We recommend that the quality of RNA in tissue blocks should be tested prior to collecting sections for LCM. Since RNA can degrade easily at RT (i.e., the usual conditions for LCM because it provides higher yields), a better RNA yield will be obtained if RNA extraction from sampled cells is undertaken in <30 min for paraffin embedded tissue and <60 min for frozen tissue.[10] Once cells are harvested and lysed, routine methods of RNA extraction and expression analysis in small specimens (gene microarray,[23] polymerase chain reaction [PCR],[21] etc.) also are effective in LCM samples.

Troubleshooting Gene Expression Methods

In our experience, a common problem with examining gene expression in mouse developmental pathology studies is a low signal-to-noise ratio. The likely explanations for this difficulty are inadequate proteinase K digestion (thus preventing entry of the probe to sites with target RNA), excessive nonspecific binding of probe, stretching or folding of the emulsion (which allows grains to collect at points of stress), or the absence of the RNA during the developmental stage being examined. A conclusion that the gene is not expressed may require additional confirmation that the ISH procedure is not at fault, such as supporting evidence by quantitative RNA analysis and/or inclusion of additional control tissues (i.e., adult organs in which the expression pattern of the target gene is known). If a gene is demonstrated in the sample by northern blotting, a missing or weak signal may be augmented by choosing a different fixative, reducing the length and/or temperature of fixation, replacing old reagents with fresh and/or less dilute ones, by utilizing a longer and/or less stringent annealing step, or by reducing the number and length of the washing steps. Finally, specimens may disintegrate during handling if the proteinase K digestion is too long. Common indications of excessive proteolysis are release of peripheral placental fragments and the ragged appearance of surface tissues, especially along the neural tube.

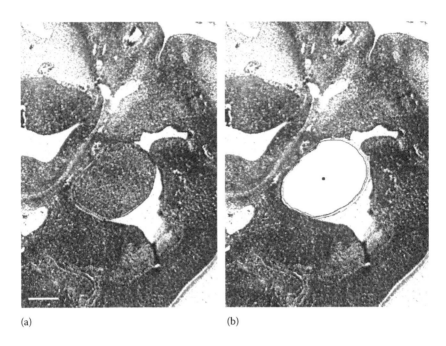

(a)   (b)

FIGURE 10.2 Laser capture microdissection (LCM) permits a single early embryo, typically gestational day (GD) 9.5 or younger, to be used for both microscopic analysis of embryonic structure and molecular assessment of an individual's genotype or site-specific molecular biochemistry. Routinely processed conceptuses (e.g., formalin-fixed, paraffin-embedded) are sectioned and stained with hematoxylin and eosin in a routine fashion except that sections are applied to slides coated on one side with polyethylene naphthalate (PEN; Carl Zeiss Microscopy, Jena, Germany). Specimens are evaluated by bright-field microscopy to define regions to be harvested (a), after which the operator outlines the desired site and the LCM instrument is used to collect it—leaving a hole in the section where the tissue was removed (b). *Tissue preparation*: Wild-type FVB mouse embryo at GD 11.5, tissue removed by laser pressure catapult (LPC) methodology using a PALM MicroBeam instrument (Carl Zeiss Microscopy). White bar = 200 μm.

Additional troubleshooting considerations and techniques have been tabulated in many ISH resources. In our experience, use of quality-controlled commercial kits according to the manufacturer's instructions is a reliable way of minimizing the need for extensive troubleshooting.

ACKNOWLEDGMENTS

The authors thank the staff of the Laser Capture Molecular Core and Nucleic Acid Shared Resource Laboratory at The Ohio State University Wexner Medical Center, Columbus, OH, for technical advice in formulating this chapter.

REFERENCES

1. Asp J, Abramsson A, Betsholtz C (2006). Nonradioactive *in situ* hybridization on frozen sections and whole mounts. *Methods Mol Biol* **326**: 89–102.
2. Ausubel FM, Brent R, Kingston RE, Moore DD, Seidman JG, Smith JA, Struhl K (1998). *Current Protocols in Molecular Biology*, Vols. 1–3. New York: John Wiley & Sons.
3. Bettenhausen B, Gossler A (1995). Efficient isolation of novel mouse genes differentially expressed in early postimplantation embryos. *Genomics* **28** (3): 436–441.
4. Brown TA (1991). *Essential Molecular Biology: A Practical Approach*, Vols. 1 and 2. Oxford, U.K.: IRL Press.

5. Brunskill EW, Aronow BJ, Georgas K, Rumballe B, Valerius MT, Aronow J, Kaimal V et al. (2008). Atlas of gene expression in the developing kidney at microanatomic resolution. *Dev Cell* **15** (5): 781–791. [Erratum in: *Dev Cell* **16** (3): 482, 2009.]

6. Chrostowski MK, McGonnigal BG, Stabila JP, Padbury JF (2009). LAT-1 expression in pre- and post-implantation embryos and placenta. *Placenta* **30** (3): 270–276.

7. Correia KM, Conlon RA (2001). Whole-mount *in situ* hybridization to mouse embryos. *Methods* **23** (4): 335–338.

8. Darby IA, Hewitson TD (2006). *In Situ Hybridization Protocols*, 3rd edn., Vol. 326: Methods in Molecular Biology. Totowa, NJ: Humana Press.

9. Duncan SA, Manova K, Chen WS, Hoodless P, Weinstein DC, Bachvarova RF, Darnell Jr JE (1994). Expression of transcription factor HNF-4 in the extraembryonic endoderm, gut, and nephrogenic tissue of the developing mouse embryo: HNF-4 is a marker for primary endoderm in the implanting blastocyst. *Proc Natl Acad Sci USA* **91** (16): 7598–7602.

10. Espina V, Wulfkuhle JD, Calvert VS, VanMeter A, Zhou W, Coukos G, Geho DH, Petricoin EF III, Liotta LA (2006). Laser-capture microdissection. *Nat Protoc* **1** (2): 586–603.

11. Golubeva Y, Salcedo R, Mueller C, Liotta LA, Espina V (2013). Laser capture microdissection for protein and nanostring RNA analysis. In: *Cell Imaging Techniques: Methods and Protocols*, 2nd edn. (Taatjes DJ, Roth J, eds.). New York: Humana Press; pp. 213–258.

12. Hargrave M, Bowles J, Koopman P (2006). *In situ* hybridization of whole-mount embryos. *Methods Mol Biol* **326**: 103–113.

13. Hecksher-Sørensen J, Hill RE, Lettice L (1998). Double labeling for whole-mount *in situ* hybridization in mouse. *Biotechniques* **24** (6): 914–916, 918.

14. Herpers B, Xanthakis D, Rabouille C (2010). ISH-IEM: A sensitive method to detect endogenous mRNAs at the ultrastructural level. *Nat Protoc* **5** (4): 678–687.

15. Johnson CW (1999). Issues in immunohistochemistry. *Toxicol Pathol* **27** (2): 246–248.

16. Komminoth P (1992). Digoxigenin as an alternative probe labeling for *in situ* hybridization. *Diagn Mol Pathol* **1** (2): 142–150.

17. Liehr T (2009). *Fluorescence In Situ Hybridization (FISH): Application Guide*. Berlin, Germany: Springer-Verlag.

18. Mahmood R, Mason I (2008). *In situ* hybridization of radioactive riboprobes to RNA in tissue sections. *Methods Mol Biol* **461**: 675–686.

19. Moorman AF, Houweling AC, de Boer PA, Christoffels VM (2001). Sensitive nonradioactive detection of mRNA in tissue sections: Novel application of the whole-mount *in situ* hybridization protocol. *J Histochem Cytochem* **49** (1): 1–8.

20. Nagy A, Gertsenstein M, Vintersten K, Behringer R (2003). *Manipulating the Mouse Embryo*, 3rd edn. Cold Spring Harbor, NY: Cold Spring Harbor Laboratory Press.

21. Nawshad A, LaGamba D, Olsen BR, Hay ED (2004). Laser capture microdissection (LCM) for analysis of gene expression in specific tissues during embryonic epithelial-mesenchymal transformation. *Dev Dyn* **230** (3): 529–534.

22. Rapley R, Manning DL (1998). *RNA Isolation and Characterization Protocols*, Vol. 86: Methods in Molecular Biology. Totowa, NJ: Humana Press.

23. Redmond LC, Dumur CI, Archer KJ, Haar JL, Lloyd JA (2008). Identification of erythroid-enriched gene expression in the mouse embryonic yolk sac using microdissected cells. *Dev Dyn* **237** (2): 436–446.

24. Rex M, Scotting PJ (1994). Simultaneous detection of RNA and protein in tissue sections by nonradioactive *in situ* hybridization followed by immunohistochemistry. *Biochemica* **11** (4): 63–65.

25. Schwarzacher T, Heslop-Harrison P (2000). *Practical In Situ Hybridization*. Oxford, U.K.: BIOS Scientific Publishers.

26. Sequeira Lopez ML, Chernavvsky DR, Nomasa T, Wall L, Yanagisawa M, Gomez RA (2003). The embryo makes red blood cell progenitors in every tissue simultaneously with blood vessel morphogenesis. *Am J Physiol Regul Integr Comp Physiol* **284** (4): R1126–R1137.

27. Son JH, Winzer-Serhan UH (2009). Signal intensities of radiolabeled cRNA probes used alone or in combination with non-isotopic *in situ* hybridization histochemistry. *J Neurosci Methods* **179** (2): 159–165.

28. Surzycki S (2000). *Basic Techniques in Molecular Biology*. Berlin, Germany: Springer-Verlag.

29. Wilkinson DG (1999). *In Situ Hybridization: A Practical Approach*, 2nd edn. Oxford, U.K.: Oxford University Press.
30. Yamagata M, Kimoto A, Michigami T, Nakayama M, Ozono K (2001). Hydroxylases involved in vitamin D metabolism are differentially expressed in murine embryonic kidney: Application of whole mount *in situ* hybridization. *Endocrinology* **142** (7): 3223–3230.
31. Yaylaoglu MB, Agbemafle BM, Oesterreicher TJ, Finegold MJ, Thaller C, Henning SJ (2006). Diverse patterns of cell-specific gene expression in response to glucocorticoid in the developing small intestine. *Am J Physiol Gastrointest Liver Physiol* **291** (6): G1041–G1050.

11

Protein Localization in Developing Mice

Brad Bolon and Efrain Pacheco

CONTENTS

Biological investigations are often performed to gain a mechanistic understanding of the chemical and molecular pathways that regulate normal and aberrant cell and tissue responses. The modern understanding that proteins are the primary elements responsible for controlling and executing most physiological functions requires that researchers be able to evaluate the distribution and quantity of specific proteins in discrete regions of biological samples. Such capabilities are absolutely critical to the rational analysis of pathological conditions that arise in developing mice. Accordingly, developmental pathologists must be well acquainted with modern molecular pathology protocols.

Proteins are major molecular targets to be examined during the course of a typical developmental pathology project. Three fundamental *in situ* approaches are used to demonstrate the distribution of proteins within specific organs, tissues, or cells. The first is to utilize immunohistochemistry (IHC) to localize proteins.[17,35,42,45] The IHC technique may be adapted to evaluate nonprotein constituents as well (e.g., 5-bromo-2′-deoxyuridine [BrdU] that has been incorporated into nuclear DNA[34]). The second is to utilize enzyme histochemistry (EHC) to define the distribution of endogenous functional molecules.[3,21] The EHC method may also be used to confirm the presence of exogenous protein markers (e.g., bacterial β-galactosidase [lacZ][6,15]) that serve as reporter proteins for the expression of genetically engineered genes. The third *in situ* approach is to pinpoint the positions of specific protein modifications, an example of which is lectin histochemistry (LHC) to reveal the type of carbohydrate side chains that were added to a protein during its post-translational processing.[13,19,23]

This chapter introduces several *in situ* molecular pathology procedures used to explore the distribution of proteins in developing mice. These basic protocols are derived from the authors' experiences. Assistance in modifying protocols for your own setting may be obtained from standard references for IHC,[9,26,36] EHC,[26,36,47] and LHC.[8,26,41] Consideration of techniques to evaluate protein quantity in the absence of identifiable morphological features (e.g., western blots of tissue homogenates or protein extracts) falls outside the scope of a pathology volume. Detailed protocols for such methods have been published elsewhere.[14,28]

Fundamentals of Evaluating Protein Distribution *In Situ*

The choice of protein localization procedure depends on both the options available for detecting a given molecule and the information that is required to answer a given research question. For example, two IHC variants are used routinely with biological specimens: immunofluorescence, which rapidly reveals protein position using an antibody linked to a fluorophore as an all-in-one reagent, and standard IHC, which usually requires more time since multiple antibodies and secondary reagents are employed sequentially to assemble an intricate marker complex at the site of the antigen. In contrast, EHC provides insight regarding the distribution of active protein, thereby offering additional confirmation that a given molecule might have functional relevance at the site to which it has been localized, while LHC evaluates post-translational modifications of protein structure.

Numerous reagents have been developed to evaluate protein distribution. Extensive compilations of antibodies for IHC are available both in print[30,33] and on the Internet (e.g., The Antibody Resource Page [http://antibodyresource.com/]). Web-based lists of lectin reagents are accessible via a number of vendors. However, to our knowledge, a single clearinghouse for lectin reagents has not been collated on the Internet.

Antibody-Based Methods

Immunofluorescence is often performed as a direct labeling procedure, where the primary antibody is linked to a fluorescent dye (fluorophore), but indirect labeling methods may also be employed where an unlabeled primary antibody is followed by a dye-conjugated secondary antibody. When excited by monochromatic light, the dye will emit light (fluoresce) at a longer wavelength. Multiple dyes with distinct excitation and emission profiles have been developed, including agents that are blue (e.g., 7-amino-4-methyl-coumarin-3-acetic acid [AMCA], 4′,6-diamidino-2-phenylindole [DAPI]); green (Alexa 488; fluorescein isothiocyanate [FITC]); orange (phycoerythrin); and red (tetrarhodamine isothiocyanate [TRITC]; Texas red). Additional theoretical details of this approach may be obtained elsewhere.[2,9]

Immunofluorescence has several advantages as a tool for defining protein position. It can be performed on frozen tissue, so the tertiary structure of the protein has not been altered by fixation and subsequent processing. Interference in the interpretation is minimal or absent because the background tissue is essentially invisible, although autofluorescence of some tissues or cells (e.g., erythrocytes [red blood cells]) can interfere with interpretation of the signal. Multiple proteins may be evaluated in a single specimen using dyes with distinctly different excitation and emission wavelengths (e.g., red paired with green) and appropriate sets of filters. However, disadvantages of this method are that fluorescent signals fade over time (*photobleaching*), and morphological features of cells and tissues are preserved less well in frozen tissues. Some commercially available mounting media help preserve fluorescence of the fluorophore (e.g., Prolong® Gold by Life Technologies, Grand Island, NY; Vectashield® by Vector Laboratories, Burlingame, CA). These drawbacks often limit the use of this procedure to those situations in which a rapid answer regarding the distribution of protein in tissues with distinct cell and tissue boundaries is well defined.

Conventional immunohistochemistry may be undertaken as a direct labeling method, where the antibody is linked to a chromagen-activating enzyme, or as an indirect labeling procedure, in which an amplification complex is assembled using sequential addition of an unlabeled primary antibody followed by one or more subsequent reagents (one of which is bound to the chromagen activator).[1] Multiple molecular systems have been adapted to IHC.[5,9] In our experience, the time-tried indirect immunoperoxidase procedure using a biotinylated secondary antibody, avidin-conjugated horseradish peroxidase (HRP), and 3,3′-diaminobenzidine (DAB, yielding a dark brown reaction product) provides suitable sensitivity for use in mouse developmental pathology studies (Figure 11.1; see also Figure 12.1). Alkaline phosphatase (ALP) with the chromagen 3-amino-9-ethylcarbazole (AEC, bright red reaction product) is used as an alternative in situations where endogenous peroxidase activity is high; it is highly sensitive, but the labeling intensity is lower than that achieved with the HRP-DAB system. Other enzymes (e.g., glucose oxidase) and chromagens (4-chloro-11-napthol [CN; blue product], 5-bromo-4-chloro-3-indolyl phosphate/nitroblue tetrazolium [BCIP/NBT; dark blue/black product]) are available,[4] but are not typically employed for mouse developmental pathology studies.

The standard IHC approach has its own spectrum of advantages and disadvantages which make this approach the preferred means of exploring many research questions in mouse developmental pathology. It may be adapted to both paraffin-embedded sections and to intact organs (or whole embryos) with equal facility. In most instances, the specimen may be fixed prior to IHC, which provides better preservation of cellular and subcellular detail while usually permitting a more exact localization of a protein to a specific cell type and/or cellular compartment. Chromagens used for IHC are stable, allowing the labeled sections to be retained as a permanent record. The choice of mounting medium is important, as some reaction products are resistant to organic solvents (DAB, BCIP/NBT) while others are removed by alcohol (AEC, CN). The main limitation to the standard IHC method is the inability of some proteins to survive even brief fixation. This shortcoming is usually surmounted by using frozen sections.

The success of IHC methods depends mainly on the availability of suitable reagents. Thousands of antibodies to hundreds of proteins have been produced during the past several decades. Published compilations of antibodies and published reports in the literature should be consulted in advance when designing mouse developmental pathology studies, because in many cases one or several reagents may be identified that have already been shown to be appropriate for a certain experimental purpose. Such preliminary searches are essential when tissues to be assessed are of mouse origin, for two reasons. First, many antibodies are derived from mouse hybridoma cell lines. Routine IHC procedures employ reagents that will bind not only to the primary antibody of interest but also to all other endogenous mouse antibodies of the same class; this situation often results in high background staining. The literature search may recommend mechanisms for overcoming nonspecific binding, such as prolonged washing (1–2 h) in buffer to extract unbound antibodies or special techniques[16,20,31] like the use of a commercial kit (e.g., ARK™ [Dako, Carpinteria, CA], HistoMouse™ [Zymed Laboratories, South San Francisco, CA], Mouse-on-Mouse™ kits [Vector Laboratories]) designed for using mouse-origin antibodies on mouse tissues. Second, compilations found in the published scientific literature or on the Internet generally give several reagent options for detecting a target molecule. In our experience, the best antibodies for use on mouse tissues are generated most reliably from certain host animal species (e.g., good secondary antibodies are of donkey, goat, or rabbit origin) or sold by certain firms. Secondary antibodies may also be absorbed against mouse tissues in advance to reduce nonspecific binding. Individual laboratories will have to confirm such preferences for their own situation. Some companies may provide data regarding antibody cross-reactivity with mouse tissue (since the primary antigen for many antibodies is usually not from a mouse protein), and a few offer exchanges or refunds if an antibody does not work. The literature review can identify researchers who may be able to provide small aliquots of a proprietary antibody at a modest cost, or who may be willing to employ their reagent on your tissues in a collaborative fashion.

One major question in developing a new IHC method is ensuring that an apparently specific signal is a true representation of the actual protein distribution in the natural state. A common approach to this problem is to compare the localization of the labeling in frozen tissue (using immunofluorescence, or less commonly standard IHC) to labeling in paraffin-embedded samples of a equivalent specimen (matched for age, gender [if necessary], and orientation). The rationale for this comparison is that frozen tissues have undergone minimal processing, and so should better reflect the protein expression pattern that exists under physiological conditions.

Enzyme-Based Methods

Enzyme histochemistry (EHC) is employed to detect functional proteins in tissues and cells. Therefore, specimens must either be stained while fresh or processed for cryosectioning to avoid inactivating endogenous proteins. The usual relevance to mouse developmental pathology projects is to detect expression of an exogenous protein (e.g., lacZ) that has been genetically engineered to be the reporter produced by transcription and translation of a targeted (*knockout*) gene.

In general, specimens used for EHC are incubated in a buffer that contains a noncolored substrate (e.g., 5-bromo-4-chloro-3-indolyl-β-ᴅ-galactoside [Xgal] for the lacZ procedure). The enzyme converts the substrate to a colored product, which is then deposited where the enzyme is found. The main advantages of this method are that it can be rendered fairly specific for a given enzyme by judicious adjustments to the components in the reaction mixture and the conditions under which the reaction is conducted[6] and that it automatically confirms that both transcription and translation of a particular gene occurred without incident. Disadvantages include the suboptimal preservation of unfixed tissues (especially when examined by light microscopy; Figure 11.2), which may preclude subcellular localization of enzyme products; the tendency for such product to disperse over time in some solutions (generally aqueous-based); and the high cost of some substrates.

Lectin-Based Methods

Lectin histochemistry [LHC] is used to detect the post-translational glycosylation pattern of proteins. Lectins (historically called agglutinins) are naturally occurring derivatives of plants or invertebrates. Their specificity for a specific carbohydrate arrangement makes them sensitive to very subtle differences in cellular glycosylation. Many plant sources yield two or more lectins, each possessing a distinct carbohydrate-binding specificity. Details of lectin specificity may be obtained elsewhere (e.g., Galab Laboratories [http://www.galab.de/technologies/technology/specificity.html], Sigma-Aldrich [http://www.sigmaaldrich.com/life-science/metabolomics/enzyme-explorer/lectin-index.html]).

Lectins to be used in histochemistry are manipulated in a fashion similar to antibodies used for IHC. The lectin may be directly bound to a fluorescent dye or biotinylated to provide a binding site for an avidin-conjugated chromagen-activating enzyme. Lectins may be applied to frozen or fixed specimens.

Technical Principles of Protein Localization in Tissue Sections

The first step in characterizing the distribution of proteins in samples of developing mice is to generate suitable tissue sections and prepare them for incubation with selected reagents. Basic histotechnological methods to prepare acceptable sections—fixation, dehydration (used for paraffin sections only), embedding, and sectioning—are reviewed in Chapter 9. For IHC, several factors must be optimized during histotechnological processing to reduce antigen degradation during storage. Specimens must be completely desiccated to prevent endogenous water residues from destroying antigens.[50] In general, sections should be cut just prior to use since antigen integrity is retained more effectively in paraffin blocks than in sections.[18,48] If tissue sections are to be cut in advance and stored until use, they should be archived at 4°C or lower if the labeling method will not be performed within a few days.[18,46]

Basic protocols for the *in situ* molecular pathology techniques to detect proteins in tissue sections are given here. All are conducted at room temperature (RT), preferably in a humidified chamber to prevent dehydration of the sections. Additional theoretical concepts and practical aspects of standard fluorescence and enzymatic IHC methods have been reported previously.[2,9,25,26] Numerous protocols are available for review on the Internet. Basic protocols for conventional enzymatic IHC of frozen (Protocol 11.1) and paraffin-embedded (Protocol 11.2) tissue sections, including antigen retrieval (Protocol 11.3), as well as whole-mount preparations (Protocol 11.4) are given later to show the flow of work and potential options in such projects. The basic procedures for EHC and LHC techniques are structured in a similar fashion (e.g., Protocol 11.5).

For immunofluorescence or conventional immunohistochemistry, frozen sections are removed from the freezer and warmed to RT. Sections should be kept within their closed containers while warming to prevent water condensation and dust accumulation on the unfixed tissue. Slides are immersed in fixative for 30 s to 10 min and then washed three times for 5 min each in RT buffer (see recipe given later). Typical fixatives are acetone, 70% ethanol, 4% formaldehyde, or 4% paraformaldehyde; the choice of fixative and the length of fixation will depend on the fragility of the protein epitope, the degree to which structural features must be preserved, and prior experience (as reported in the literature or experienced in your laboratory) with given combinations of tissue and fixatives. In general, aldehydes provide better tissue preservation but may be more liable to degrade labile antigens.

Conventional immunohistochemistry can also be performed on paraffin-embedded sections. Such material must be deparaffinized in an organic solvent (typically toluene or xylene) and then rehydrated through a series of graded alcohols and finally water. This step is required so that the aqueous antibody solutions can gain entry to all portions of the tissue.

Another step for both methods and for both types of sections is often to use a mild detergent to make cell membranes more porous to antibodies and to retrieve antigens (Protocols 11.3 and 11.4). This step is optional for extracellular and membrane-associated antigens but is usually essential for intracellular molecules. In our experience, incubation in 0.1%–1% (v/v) Triton X-100 (Sigma-Aldrich, St. Louis, MO) in buffer for 20–30 min is suitable for this purpose. This permeabilization step seems to be more critical for frozen sections, possibly because membranes retain more of their native barrier capacity since they have not been previously exposed to solvents. However, whether or not a detergent digestion is necessary for a given antibody should be tested in your own laboratory when optimizing new reagents. Alternative means for optimizing antibody entry and antigen retrieval are to include an enzyme incubation step (e.g., 0.05% protease in buffer for 10–20 min) or to apply cyclic heat (e.g., microwaving for 2–3 min at 800 W in citrate buffer [10 mM citric acid, 0.05% Tween® 20 [multiple vendors], pH 6.0; see recipe in Protocol 11.3]).[27,29] Some antigens may benefit from any retrieval procedure, while others may be revealed better by only one kind of treatment[12,24]; therefore, different methods should be tried when developing new IHC methods. Permeabilized sections are washed two or three times in buffer (1–5 min each).

Whether or not cells are permeabilized, the next step is usually to apply one or more special blocking solutions to reduce the extent of spurious background reaction that develops (Protocols 11.2 and 11.3). A routine procedure used with avidin–biotin complex (ABC) enzymatic methods is to block endogenous biotin using sequential incubations of 10–20 min each in 0.1% avidin followed by 0.01% biotin.[49] A biotin block is especially important for frozen sections; a biotin block is not required for fluorescence experiments or when using biotin-free systems as the detection molecule will not react with endogenous avidin. This blocking reagent is typically followed by one or two buffer washes (1–5 min each). Another common block required for both fluorescence and enzymatic IHC techniques is to spread a protein solution over the section to decrease the hydrophobic and ionic (electrostatic) forces that promote nonspecific binding of antibodies. Typical protein solutions used for this purpose include normal (nonimmune) serum from the species in which the primary antibody was raised, a cocktail of unrelated proteins (e.g., bovine serum albumin or casein), or a mixture of normal serum and exogenous protein. Commercial kits usually include the appropriate universal protein block (e.g., casein) at the proper concentration as one of the standard reagents. Home-made protein blocks may be produced by adjusting the concentration of each protein to 1%–3% (w/v or v/v) by diluting with buffer. The protein block is applied for 20–30 min, after which the solution is drained by gentle shaking and/or blotting around the edges of the section, but not by buffer washes.

Primary antibodies against a specific antigen (or lectins for particular carbohydrate moieties) are then applied to the blocked sections. The ideal primary antibody/lectin possesses high specificity (reacts only with a single antigen), high affinity (an exact fit between the contours of the antigen and the antibody/lectin), and high titer (amount of activity). Suitable antibodies may be either monoclonal (directed against a single epitope of the antigen) or polyclonal (directed against multiple epitopes); lectins are more or less *monoclonal* in character. The concentration of each new primary antibody/lectin needed for acceptable labeling should be tested in the laboratory as in our experience the quantity required varies from antibody to antibody and among lots produced by the same monoclonal antibody clone. Furthermore, the manufacturer's recommended titer may often be diluted by modifying the protocol (e.g., by amplifying the immunobridge), thus reducing the cost of reagents. Primary antibodies and lectins typically are applied at dilutions of between 1:100 and 1:500, or at concentrations of 0.5–2 µg/mL. Common incubation times range from 30 min to 2 h at RT, or overnight at 4°C. Unbound primary antibody/lectin is then removed by 3–5 buffer washes (2–5 min each).

At this point, another blocking procedure is employed in enzymatic IHC to reduce the endogenous peroxidase activity in certain cells (e.g., erythrocyte cytoplasm). Sections are incubated for 5–10 min in 0.3% hydrogen peroxide (H_2O_2), which is constituted by adding 1 mL of 30% H_2O_2 to 100 mL of deionized water to obtain a final concentration of 0.03% H_2O_2. This blocking step is followed by two to three buffer washes (1–5 min each). Some individuals prefer higher percentages of H_2O_2 (e.g., 1%–3%) and/or mixing the H_2O_2 with absolute methanol (e.g., one part of 3% H_2O_2 with three parts methanol). We have used these more rigorous conditions (higher H_2O_2 concentration and inclusion of methanol) with success on paraffin-embedded material. However, when working with frozen tissue sections and whole-mount preparations, such harsh treatment should be used with caution to avoid quenching the immunolabeling procedure. Furthermore, when working with frozen sections, any step that includes methanol should be undertaken only after the primary antibody has been applied to avoid loss of antibody binding.

The next step is to add the secondary antibody. By design, this reagent should react only with one or more epitopes on the primary antibody (or lectin). This ideal is readily met for antibodies made in a non-mouse host, but greater care is required in trying to detect the many current mouse-derived monoclonal primary antibodies in mouse tissue. Secondary antibodies are usually applied at dilutions of between 1:50 and 1:250 for 30 min to 1 h at RT. Unbound secondary antibody is then removed by three buffer washes (2–5 min each). The nature of this secondary reagent will dictate the next step. If the secondary antibody is linked to a fluorophore, the sections may be examined using a fluorescent microscope and appropriate filter sets immediately after the last wash. If the secondary antibody is instead linked to biotin or an enzyme, additional amplification steps are required after the buffer washes are completed.

For biotinylated secondary antibodies, which are preferred reagents in mouse developmental pathology studies, the next step is incubation with an avidin–biotin–enzyme complex (ABC) reagent. This step is performed according to the manufacturer's direction but typically involves a 30 min incubation followed by two or three buffer washes (3–5 min each). A chromogen (e.g., DAB if the enzyme is HRP, AEC if the enzyme is ALP) is then applied to the section for 2–5 min, though often the reaction is quenched once a visible (brown for DAB, red for AEC) reaction product is evident. In our experience, the incubation times in the manufacturer's instructions are suitable. Reaction times may be decreased if the ambient temperature or the solution temperature is raised.

Labeled sections are rinsed in either running tap water (2–5 min) or buffer (two washes of 3–5 min each) and then counterstained for 1–2 min. The most common counterstain for this purpose is Mayer's hematoxylin, but other stains (e.g., light green, nuclear fast red, toluidine blue) may be used as long as the reaction product is not obscured by the background color. Counterstained sections are *blued* (if necessary) in acid alcohol (1% HCl in 70% ethanol, two 1 s immersions) and washed in running tap water for 2–5 min; for hematoxylin variants that do not require bluing, the counterstained sections are moved immediately to the washing step. Sections labeled with DAB are dehydrated through graded alcohols into xylene and then coverslipped using an organic solvent-miscible medium (e.g., Permount™), while an aqueous mountant (e.g., Crystal Mount™) is utilized for AEC-labeled sections. Covered sections are now available for evaluation by brightfield light microscopy (Figure 11.1).

On occasion, a standard protocol must be altered to enhance the intensity of the reaction product, thereby facilitating analysis and photodocumentation. Factors that can be used to strengthen specific immunoreactivity while reducing background staining include the choice of fixative (certain antigens are best preserved by Bouin's solution or paraformaldehyde),[39] use of antigen retrieval procedures (enzyme digestion or microwave heating), and using additional amplification steps (e.g., tertiary antibodies, tyramide).

Technical Principles of Protein Localization in Whole-Mount (Intact) Specimens

The basic tenets for evaluating protein distribution in whole-mount preparations (i.e., unsectioned animals or organs) are comparable to those described earlier for tissue sections. Fixation is usually performed overnight at 4°C or for 1–2 h at RT. All other steps are generally completed at RT. Further details on whole-mount protein labeling procedures may be obtained elsewhere.[6,36] Many whole-mount IHC and whole-mount EHC protocols are also accessible via the Internet. Basic protocols for whole-mount IHC (Protocol 11.4) and whole-mount EHC (Protocol 11.5) are given here.

The choice of whole-mount technique for protein localization in developing mice depends chiefly on the developmental age. In our experience, whole-mount IHC may be successfully applied to intact mouse

PROTOCOL 11.1 BASIC DIRECT IMMUNOHISTOCHEMISTRY METHOD FOR FROZEN TISSUE SECTIONS

1. Air-dry slides, 5–10 min at RT.
2. Fixation (by immersion)—pick one option
   a. Acetone, 5–10 min
   b. Neutral buffered 10% formalin (~4% formaldehyde), 2–5 min
   c. 4% Paraformaldehyde, 2–5 min
3. Apply water-repellent barrier to concentrate IHC solutions over tissue by encircling sections with a PAP pen (available from multiple vendors).
4. Apply peroxidase blocking solution: 0.3% H_2O_2 (in distilled water or PBS, pH 7.4), 10–20 min.
5. Hydrate sections in phosphate buffered saline (pH 7.4), three times for 5–10 min each.
   Note: Tissues may be held overnight at RT in a humidified chamber before proceeding with the remainder of the procedure.
6. Drain excess buffer from section (by tipping slide and then blotting the edge of the PAP barrier line with an absorbent paper towel).
7. Apply biotinylated primary antibody at RT for 1 h or at 4°C overnight in a humidified chamber.
8. Wash in buffer, three to five times for 5–10 min each.
9. Assemble bridge with ABC reagent, 30 min to 1 h.
10. Wash in buffer, two to three times for 5 min each.
11. Incubate with DAB, 2–5 min (until reaction product is clearly visible in positive control tissue).
12. Wash in buffer, two times for 5 min each.
13. Counterstain
    a. Mayer's hematoxylin, 1–2 min
    b. Wash in running tap water, 3–5 min
    c. Immerse in acid alcohol (1% HCl in 70% ethanol), two dips of 1 s each (the sections will turn red)
    d. Wash in running tap water until the sections turn blue
14. Dehydrate with graded alcohols (70%, 80%, 95%, 100%, 100%) followed by organic solvent (toluene or xylene, two times), 3–5 min each.
15. Coverslip with appropriate mounting medium.

embryos up to gestational day 10 (GD 10) and to isolated organs from older developing mice. Whole-mount EHC may be utilized on intact mouse embryos up to GD 14 and on dissected organs from older animals.

The primary advantage to such whole-mount methods relative to IHC or EHC of sectioned material is that they quickly reveal the three-dimensional spatial relationship between structures expressing a given protein and those that do not (Figure 11.2). Other benefits include the ability to process more whole-mount samples in parallel and the reduced amount of labor needed for processing. The major drawbacks to whole-mount methods are that localizing a protein to specific cell types (rather than to large regions) is more difficult, and the techniques cannot be used effectively in intact older embryos (approximately GD 14.5) and fetuses, which have thicker torsos and impermeable skin.

Multiple Staining Experiments

Evaluation of protein interactions is often best served by experiments to demonstrate colocalization of two or more molecules to the same or contiguous regions. Such studies typically utilize dual IHC-based endpoints,[17] although IHC may also be paired with *in situ* hybridization (ISH; see Chapter 10) to evaluate mRNA.[40]

PROTOCOL 11.2   BASIC INDIRECT IMMUNOHISTOCHEMISTRY
METHOD FOR PARAFFIN-EMBEDDED TISSUE SECTIONS

1. Deparaffinize sections:
   a. Xylene, two times for 5 min each
   b. 1:1 Mixture of xylene:100% ethanol, 5 min
   c. 100% Ethanol, 5 min
2. Rehydrate sections:
   a. 100% Ethanol, 5 min
   b. 95% Ethanol, 5 min
   c. 80% Ethanol, 5 min
   d. 70% Ethanol, 5 min
   e. Distilled water, two times for 5 min each
   f. Wash in running tap water, 3–5 min
3. Apply water-repellent barrier to concentrate IHC solutions over tissue by encircling sections with a PAP pen (available from multiple vendors).
4. Hydrate sections in phosphate buffered saline (pH 7.4), three times for 5–10 min each.
   Note: If desired, antigen retrieval should be undertaken here before proceeding to Step 5.
   Note: Tissues may be held overnight at RT in a humidified chamber before proceeding with the remainder of the procedure.
5. Apply protein blocking solution: 1%–4% nonimmune serum (from the host species for the primary antibody), ±1% skim milk powder in phosphate buffer, pH 7.4, for 20–30 min.
6. Drain excess buffer from section (by tipping slide and then blotting the edge of the PAP barrier line with an absorbent paper towel).
7. Apply primary antibody at RT for 1 h or at 4°C overnight in a humidified chamber.
8. Apply peroxidase blocking solution: 1%–3% H_2O_2 (in distilled water or PBS, pH 7.4), 10–20 min.
9. Wash in buffer, three to five times for 5–10 min each.
10. Apply biotinylated secondary antibody at RT for 30 min in a humidified chamber.
11. Wash in buffer, two to three times for 5 min each.
12. Assemble bridge with ABC reagent, 30 min to 1 h.
13. Wash in buffer, two to three times for 5 min each.
14. Incubate with DAB, 2–5 min (until reaction product is clearly visible in positive control tissue).
15. Wash in buffer, two times for 5 min each.
16. Counterstain
    a. Mayer's hematoxylin, 1–2 min
    b. Wash in running tap water, 3–5 min
    c. Immerse in acid alcohol (1% HCl in 70% ethanol), two dips of 1 s each (the sections will turn red)
    d. Wash in running tap water until the sections turn blue
17. Dehydrate with graded alcohols (70%, 80%, 95%, 100%, 100%) followed by organic solvent (toluene or xylene, two times), 3–5 min each.
18. Coverslip with appropriate mounting medium.

Two general approaches are employed for such experiments. The first performs the IHC methods for each molecule sequentially. In this fashion, the same enzyme system may be utilized for both molecules, but with a different chromagen applied for each. The entire IHC method is completed for the first protein, including addition of the chromagen. Then, instead of counterstaining, the sections are washed in buffer (usually three times for 3–5 min at RT), and the second primary antibody is added. Care must be taken to ensure that all relevant blocking solutions are applied at the correct time, and that all antibodies

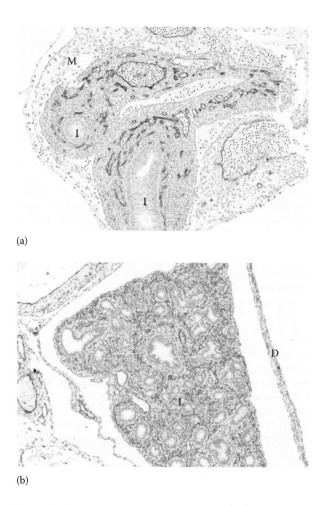

(a)

(b)

FIGURE 11.1 Immunohistochemical (IHC) detection of the endothelial cell marker CD31 in newly formed blood vessels of wild-type mouse embryos (gestational day 14). The dark brown reaction product in multiple tissues is localized to the linings of blood vessels. The section in (a) was preserved in buffered Z-fix (Anatech Ltd., Battle Creek, MI), while the section in (b) was fixed using Bouin's solution; both fixed specimens were processed routinely into paraffin (see Chapter 9 for processing details) and sectioned at 5 μm. The manual IHC procedure involved dehydration followed by sequential application of rat anti-mouse CD31 primary antibody (Catalog No. DIA 310; Dianova GmbH, Hamburg, Germany); biotinylated, mouse-adsorbed, rabbit anti-rat IgG secondary antibody (Catalog No. BA-4001; Vector); avidin–biotin-complex (ABC) linked to horseradish peroxidase (Vectastain Elite ABC Kit, Catalog No. PK-6100; Vector); 3,3′-diaminobenzidine (Catalog No. K3468; Dako); and hematoxylin. *Abbreviations*: D, diaphragm; I, intestine; L, lung; M, mesentery. (Images were kindly provided by Ms. Barbara Felder, Amgen, Inc., Thousand Oaks, CA.)

are added in the appropriate order. The second approach allows reagents to detect each molecule to be added simultaneously, such as two primary antibodies from two separate host species (i.e., requiring two separate secondary antibodies) or that have been linked to unique fluorophores (which emit light at different wavelengths). In our experience, the second approach provides better results, possibly because the simultaneous incubation reduces steric hindrance between nearby antigens and/or antibodies that may develop when one antibody has a chance to bind before the other.

Control Materials for Protein Localization

A number of concurrent controls are required during each IHC run to ensure that any visible reaction product is specific, as lack of antibody specificity is a common problem.[22] Tissue controls, reaction

PROTOCOL 11.3   BASIC PROTOCOL FOR HEAT-BASED ANTIGEN RETRIEVAL

Reagent

Sodium citrate buffer (SCB):
1. 10 mM sodium citrate, 0.05% Tween 20, pH 6.0
2. Trisodium citrate (dihydrate) 2.94 g
3. Distilled water 1000 mL

- Mix to dissolve.
- Adjust pH to 6.0 with 1 N hydrochloric acid (HCl).
- Add 0.5 mL of Tween® 20 and mix well.
- Store solution at RT for 3 months or at 4°C for 6 months.

Procedure

1. Deparaffinize sections in two changes of xylene (5 min each).
2. Rehydrate sections:
   a. 100% Ethanol, two times for 5 min each
   b. 95% Ethanol, 5 min
   c. 80% Ethanol, 5 min
   d. 70% Ethanol, 5 min
   e. Distilled water, two times for 5 min each
3. Follow all steps in one of the following lists:
   a. Microwave option
       i. Preheat SCB in a Coplin jar at the *high* setting until the solution boils
      ii. Place slides in Coplin jar, and cover with SCB
     iii. Cover jar loosely with perforated film
      iv. Microwave at *medium* setting for 3–5 min
       v. Refill Coplin jar with fresh SCB (to replace evaporative loss)
      vi. Microwave at *medium* setting for 3–5 min
     vii. Refill Coplin jar with fresh SCB (to replace evaporative loss)
    viii. Allow solution and slides to cool to RT
      Note: If desired, the microwave oven may be calibrated for greater efficiency.[43]
   b. Water bath option
       i. Preheat SCB in a Coplin jar to 100°C
      ii. Place slides in Coplin jar, and cover with SCB
     iii. Cover jar loosely with lid
      iv. Incubate in water bath at 95°C–100°C for 20–40 min
       v. Remove Coplin jar from water bath, and allow solution and slides to cool to RT
      Note: Optimal incubation time will need to be determined for each antigen for both
          heating options.
4. Rinse sections in 0.05% Tween® in phosphate buffer (pH 7.4), two times for 3–5 min.
5. Begin blocking procedures for IHC procedure.

controls, and reagent controls are all necessary (Table 11.1).[10] Comparable concurrent controls are necessary for EHC and LHC runs as well.

Tissue controls are used to ensure that antibody against the protein of interest can be detected where the protein actually is expressed, and will not be observed where it does not exist. Control tissues must be collected, fixed, and processed in the same fashion as the test specimens. The ideal procedure for staining control tissues is to mount them on the same slide, either within a single section or as adjacent positive and negative sections; this approach exposes both controls to the same reagents for the same length of time, and thus provides the most rigorous

PROTOCOL 11.4   BASIC WHOLE-MOUNT
IMMUNOHISTOCHEMISTRY TECHNIQUE

Smaller specimens typically require shorter incubation times. All steps should be conducted with gentle rocking or shaking to facilitate reagent penetration and removal.

1. Harvest specimens (intact embryos, intact organs of older animals).
2. Fixation at 4°C in either
   a. 4:1 Absolute methanol–dimethyl sulfoxide (DMSO), overnight
   b. 4% Paraformaldehyde, 2–4 h
3. Rehydrate specimens:
   a. 70% Methanol, 30 min
   b. 50% Methanol, 30 min
   c. 25% Methanol, 30 min
   d. Phosphate buffer, pH 7.4, 30 min
   e. Phosphate buffer, pH 7.4, 10–15 min
4. Apply protein blocking/detergent solution: 2%–4% skim milk powder and 0.1% (v/v) Triton X-100 in phosphate buffer, pH 7.4, for 1 h at RT.
5. Apply primary antibody at 4°C overnight.
6. Apply peroxidase blocking solution: 4:1:1 absolute methanol–DMSO–30% H_2O_2, 4 h at RT.
7. Wash in buffer:
   a. 4°C—two to three times for 30–60 min each
   b. RT—two to three times for 30–60 min each
8. Apply biotinylated secondary antibody at 4°C overnight.
9. Wash in buffer:
   a. 4°C—two to three times for 30–60 min each
   b. RT—two to three times for 30–60 min each
10. Assemble bridge with ABC reagent, 2–4 h at RT.
11. Wash in buffer two to three times for 30–60 min each at RT.
12. Incubate with 0.3 mg/mL of DAB, 0.5% nickel chloride, 0.2% bovine serum albumin (BSA), and 0.1% (v/v) of Triton X-100 in phosphate buffer (pH 7.4) for 15–20 min at RT.
13. Add H_2O_2 to 0.03% and incubate at RT until reaction product deposits darken (usually 10–15 min).
14. Wash in buffer, two times for 30 min each.
15. Dehydrate specimens:
   a. 25% Ethanol, 30 min
   b. 50% Ethanol, 30 min
   c. 70% Ethanol, 30 min
   d. 95% Ethanol, 30 min
   e. 100% Ethanol, 30 min
16. Store in 100% ethanol at 4°C.

comparison between them. If actual control tissues cannot be obtained, surrogate control tissues may include pellets of cells in which the protein of interest has been expressed, or even polymer beads to which the proteins have been linked.

Reaction controls confirm that the design of the IHC protocol was able to detect the molecule of interest and to show that the procedure is performed consistently over time. The most common reaction control is to replace the primary antibody/lectin with either another primary reagent directed against an irrelevant antigen or with nonimmune serum from the species in which the primary antibody was made. If highly purified target antigen is available, affinity absorption

PROTOCOL 11.5   BASIC WHOLE-MOUNT ENZYME
HISTOCHEMISTRY TECHNIQUE

The marker protein is bacterial β-galactosidase (lacZ). Positive control tissues may be acquired from other animals expressing lacZ in one or more tissues, or the fecal pellets in the colon of the dam may be used.

1. Harvest specimens (intact embryos, intact organs of older animals).
2. Fix for 1–4 h at 4°C in 4% paraformaldehyde in one of these buffers (pH 7.4):
   a. Phosphate buffer (PB): 100 mM sodium phosphate
   b. Phosphate buffered saline (PBS):
      100 mM sodium phosphate
      150 mM sodium chloride
   c. PIPES (piperazine-*N,N'-bis*[2-ethanesulfonic acid]):
      100 mM PIPES
      2 mM magnesium chloride
      1.25 mM EGTA (ethylene glycol tetraacetic acid)
   Note: Buffers are stable at 4°C for 6–12 months.
   Note: Use the same buffer for all subsequent solutions.
3. Wash in buffer/detergent solution two to three times for 30 min each. The buffer composition is
   a. 0.1% (w/v) sodium deoxycholate
   b. 0.02% (v/v) Tergitol™ NP-40
   c. 2 mM magnesium chloride (add only to PB or PBS)
4. Immerse specimen in staining solution (freshly reconstituted from buffer stock and substrate stock solutions) for overnight at RT or 37°C in the dark.
   a. Buffer stock (store at RT in the dark), consisting of:
      i.   5–35 mM potassium ferricyanide
      ii. 5–35 mM potassium ferrocyanide
      iii. 2 mM magnesium chloride
      iv. 0.1% (w/v) sodium deoxycholate
      v.   0.02% (v/v) Tergitol™ NP-40
   b. Substrate stock (store at 4°C or −20°C in the dark), consisting of:
      i. 40 mg X-gal (5-bromo-4-chloro-3-indolyl-β-D-galactoside)
      ii. 1 mL of *N,N*-dimethylformamide (DMF)
   c. Working solution:
      i.   Mix 100 mL of buffer stock with
      ii. 2.5 mL of substrate stock
      iii. Final X-gal concentration is 1 mg/mL
5. Postfix overnight at 4°C in neutral buffered 10% formalin (NBF)
6. Store indefinitely in 70% ethanol
   Note: Long-term storage (more than 4 weeks) in ethanol results in excessive hardening of tissues for microtomy.

of the primary antibody with the purified antigen is the ideal means for showing that the antibody is specific for the target, and that the absence of the antibody prevents specific labeling. Another reaction control that is sometimes used is to substitute nonimmune serum for the secondary antibody.

Reagent controls are used to ensure that the primary antibody/lectin, secondary antibody, and bridging elements (e.g., avidin–biotin reagents) are specific for their targets. These controls are often performed in detail only when establishing a novel IHC method by testing the reagents

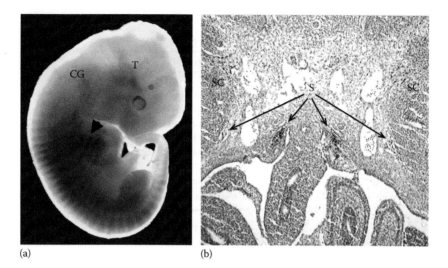

(a)   (b)

FIGURE 11.2 Enzyme histochemical (EHC) detection of the tissue distribution for a growth factor receptor in null mutant (*knockout*) mouse embryos engineered to lack *Gfrα-3* (glial cell line-derived neurotrophic factor [GDNF] family receptor, type alpha-3), the receptor for the neurotrophic factor artemin. The engineered gene contained bacterial β-galactosidase (*lacZ*) as a reporter protein; the action of this enzyme on X-gal results in deposition of blue reaction product in the peripheral nerve structures in which the Gfrα-3 protein is normally expressed. The Gfrα-3/lacZ fusion protein is localized to autonomic (cranial [superior] cervical [CG] and sympathetic chain [SC]), cranial nerve (trigeminal [T]), and sensory (dorsal root) ganglia as well as autonomic (sympathetic [S]) and somatic (arrowheads) nerves. (a) Whole-mount preparation of a knockout embryo at gestational day (GD) 10.5. (b) Slide-mounted preparation of a knockout embryo at GD 12.5, acquired through the lumbar region. Counterstain: eosin. (These images were reproduced from Bolon, B. et al., *Toxicol. Pathol.*, 32(3), 275, 2004. Courtesy of Sage Publications.)

over a range of dilutions against control tissues known to express or not express the target. In this fashion, the optimal dilution for using the reagent is defined while simultaneously confirming the specificity of the reagent. Once a method has been perfected, tissue controls are often used as proxies for reagent controls.

When performing multiple labeling procedures, appropriate controls for both target molecules must be undertaken at each step to ensure that deposition of the final products on the dual-labeled sections are specific.

Storage of Reagents and Tissues

Materials used in protein localization experiments must be retained in such a fashion that their integrity (in the cellular, chemical, and molecular senses) is preserved for extended periods. The key factor in preventing degradation during storage is to maintain a constant low temperature.

Most reagents (e.g., antibodies, lectins, commercial IHC kits) are relatively stable for weeks to months when stored at 4°C. This temperature is particularly suitable for materials that are used frequently as repeated freeze–thaw cycles can destroy the molecular structure of antibodies. Freezing (generally −20°C) is permissible when reagents are to be archived for longer periods, but the material should be parsed into aliquots so that they are only thawed once. Commercial products should be stored according to the manufacturer's instructions. If a highly diluted antibody is to be stored at 4°C for more than a day or two, we recommend that the reagent be supplemented with 1%–2% of bovine serum albumin (BSA) to inhibit denaturation of the antibody.

The diluent used to prepare antibody aliquots for storage (and use) depends on the nature of the product and should be guided by the manufacturer's instructions. Reagents prepared in phosphate-buffered saline (PBS) should be diluted with PBS, while those prepared in protein solution (e.g., 1% BSA) should

TABLE 11.1

Common Controls Used in Histochemistry Procedures

Type of Control	Control Material
Reaction control	*Concentration control*
	Concentration of spurious antibody (used in place of primary antibody) is increased to assess background labeling.
	Concentration of a different primary antibody (specific for another protein expressed in the test specimen) is raised to assess background labeling.
	Replacement control—spurious antibody of the same class (or nonimmune serum) substituted for primary antibody
Reagent control	*Adsorption control*—antibody bound by antigen prior to application
Tissue control	*Positive control* (known to express protein)
	Internal (located within test specimen) (ideal)
	External—tissue > cell pellet ≫ protein linked to polymer
	On same slide as test specimen
	On different slide
	Negative control (known not to express reaction)
	Internal (located within test specimen) (ideal)
	External—tissue > cell pellet
	On same slide as test specimen
	On different slide

be diluted with the same solution. Lyophilized (freeze-dried) materials often contain the ingredients required to produce a physiological PBS solution, and so should be reconstituted in deionized water. Pure tap water should not be used as a diluent as it will rapidly destroy protein structure.

Tissues may be archived in either frozen tissue banks (stored at $-20°C$ or colder) or as paraffin-embedded blocks. Such banks should include normal control tissues encompassing the species, genetic background, developmental ages, and (if necessary) genders of the likely test samples. Banks may also include specimens from various diseases or toxicant-induced conditions, which serve as test materials or additional control tissues. In our experience, the small size of developing mice results in a frequent need to replenish control specimens, so detailed sample logs and regular inventories (at least once every 6 months) must be kept to ensure that new control tissues are acquired before the tissue banks are exhausted.

Interpretation of Reaction Product Distribution

The presence of a reaction product must be assessed carefully to ensure that it results from specific interaction with the target protein. Such determinations require careful comparisons to both concurrent controls (i.e., done within the same study) and, in most cases, prior knowledge regarding expression of the target molecule (or *historical controls*). When necessary, results of *in situ* protein distribution studies should be confirmed by ancillary studies such as western blotting (for protein levels) or northern blotting (to evaluate mRNA). The amount of target protein may be quantified in an entire organ or, by using laser capture microdissection (LCM; see Chapter 10), within a discrete tissue domain. In general, the magnitude of signal detected by northern or western analysis is well correlated with the outcome of IHC procedures.[39] The nervous systems poses an exception to this rule because proteins expressed at distant sites in the periphery may be associated with mRNA expression in centrally located cell bodies.

Recent reports have described efforts to develop quantitative methods for measuring reaction product deposits.[32,37,44] Many factors in the IHC method can affect the amount of amplification, and thus the quantity of reaction product deposited. Therefore, in our view IHC interpretation in tissues of developing mice typically should be limited to qualitative or semi-quantitative scoring. Quantification using computer-assisted image analysis to measure the number of labeled cells or size of a labeled structure remains an acceptable application as such methods depend on the presence (rather than the amount) of positive labeling.

Troubleshooting Protein Localization Methods

In our experience, one major difficulty with evaluating protein distribution in mouse developmental pathology studies is weak or absent staining. The likely explanations for this difficulty are inadvertent omission of a reagent (usually the primary antibody) or the absence of the protein during the developmental stage being examined. An example of the latter issue is the paucity of glial proteins in mouse embryos, as myelination of the mouse nervous system does not begin in earnest until after birth. A conclusion that the protein is not present may require additional confirmation that the IHC procedure is not at fault, such as supporting evidence by quantitative protein analysis (e.g., western blotting) and/or inclusion of additional control tissues (i.e., adult organs in which the protein expression pattern is known). If a protein is shown to be present in the specimen by western blotting, a missing or weak signal may be augmented by choosing a different fixative, reducing the length and/or temperature of fixation, replacing old reagents with fresh and/or less dilute ones, by utilizing an antigen retrieval technique, or by reducing the number and length of the washing steps.

Another common difficulty in labeling tissues of developing mice is the presence of background staining. This widespread discoloration is often pale but may be darker over certain tissue types (e.g., connective tissue) or structures (epithelium with abundant cytoplasm). Multiple factors have been identified as potential causes of background staining. One is nonspecific binding of antibody reagents to connective tissues. In our experience, this artifact is infrequent in tissues of developing mice, most of which are much more cellular than their adult counterparts. Another factor is an elevated capacity by certain cell types to bind avidin (e.g., mast cells[11]) or support endogenous peroxidase activity (e.g., erythrocytes). Some cells (e.g., erythrocytes, macrophages) exhibit autofluorescence that may interfere with immunofluorescent detection methods.

Additional troubleshooting considerations and techniques have been tabulated in many IHC resources.[9,26,38] More recent innovations in IHC histotechnology have been designed to remove certain reagents that are a common source of artifactual reactions (e.g., the Dako EnVision™+ System avoids the use of avidin and biotin). In our experience, use of quality-controlled commercial kits according to the manufacturer's instructions is a reliable way of minimizing the need for extensive troubleshooting.

REFERENCES

1. Adams JC (1992). Biotin amplification of biotin and horseradish peroxidase signals in histochemical stains. *J Histochem Cytochem* **40** (10): 1457–1463.
2. Allan VJ (2000). *Protein Localization by Fluorescence Microscopy: A Practical Approach*. Oxford, U.K.: Oxford University Press.
3. Blackburn MR, Gao X, Airhart MJ, Skalko RG, Thompson LF, Knudsen TB (1992). Adenosine levels in the postimplantation mouse uterus: Quantitation by HPLC-fluorometric detection and spatiotemporal regulation by 5′-nucleotidase and adenosine deaminase. *Dev Dyn* **194** (2): 155–168.
4. Boenisch T (1989). Basic enzymology. In: *Handbook of Immunohistochemical Staining Methods*, 1st edn., Naish SJ (ed.). Carpinteria, CA: Dako Corporation; pp. 9–12.
5. Boenisch T (2006). Basic enzymology. In: *Immunohistochemical Staining Methods*, 4th edn., Key M (ed.). Carpinteria, CA: Dako Corporation; pp. 19–25.
6. Bolon B (2008). Whole mount enzyme histochemistry as a rapid screen at necropsy for expression of β-galactosidase (LacZ)-bearing transgenes: Considerations for separating specific LacZ activity from nonspecific (endogenous) galactosidase activity. *Toxicol Pathol* **36** (2): 265–276.
7. Bolon B, Jing S, Asuncion F, Scully S, Pisegna M, Van G, Hu Z et al. (2004). The candidate neuroprotective agent artemin induces autonomic neural dysplasia without preventing nerve dysfunction. *Toxicol Pathol* **32** (3): 275–294.

8. Brooks S (1997). *Lectin Histochemistry*. Oxford, U.K.: BIOS Scientific Publishers Ltd.

9. Buchwalow IB, Böcker W (2010). *Immunohistochemistry: Basics and Methods*. Berlin, Germany: Springer-Verlag.

10. Burry RW (2000). Specificity controls for immunocytochemical methods. *J Histochem Cytochem* **48** (2): 163–165.

11. Bussolati G, Gugliotta P (1983). Nonspecific staining of mast cells by avidin–biotin-peroxidase complexes (ABC). *J Histochem Cytochem* **31** (12): 1419–1421.

12. Cattoretti G, Pileri S, Parravicini C, Becker MH, Poggi S, Bifulco C, Key G et al. (1993). Antigen unmasking on formalin-fixed, paraffin-embedded tissue sections. *J Pathol* **171** (2): 83–98.

13. Cromphout K, Vleugels W, Heykants L, Schollen E, Keldermans L, Sciot R, D'Hooge R et al. (2006). The normal phenotype of Pmm1-deficient mice suggests that Pmm1 is not essential for normal mouse development. *Mol Cell Biol* **26** (15): 5621–5635.

14. Dunbar BS (1994). *Protein Blotting: A Practical Approach*. Oxford, U.K.: Oxford University Press.

15. Franco D, de Boer PA, de Gier-de Vries C, Lamers WH, Moorman AF (2001). Methods on *in situ* hybridization, immunohistochemistry and β-galactosidase reporter gene detection. *Eur J Morphol* **39** (3): 169–191.

16. Fung KM, Messing A, Lee VM-Y, Trojanowski JQ (1992). A novel modification of the avidin-biotin complex method for immunohistochemical studies of transgenic mice with murine monoclonal antibodies. *J Histochem Cytochem* **40** (9): 1319–1328.

17. Gekas C, Rhodes KE, Van Handel B, Chhabra A, Ueno M, Mikkola HK (2010). Hematopoietic stem cell development in the placenta. *Int J Dev Biol* **54** (6-7): 1089–1098.

18. Gelb AB, Freeman VA, Astrow SH (2011). Evaluation of methods for preserving PTEN antigenicity in stored paraffin sections. *Appl Immunohistochem Mol Morphol* **19** (6): 569–573.

19. Herken R, Sander B, Gabius H, Götz W (1991). Correlations between the binding of neoglycoproteins, bovine serum albumin (BSA) and lectins in 10- to 13-day-old mouse embryos. *Histochemistry* **95** (3): 297–301.

20. Hierck BP, Iperen LV, Gittenberger-de Groot AC, Poelmann RE (1994). Modified indirect immunodetection allows study of murine tissue with mouse monoclonal antibodies. *J Histochem Cytochem* **42** (11): 1499–1502.

21. Johansson S, Wide M (1994). Changes in the pattern of expression of alkaline phosphatase in the mouse uterus and placenta during gestation. *Anat Embryol (Berl)* **190** (3): 287–296.

22. Johnson CW (1999). Issues in immunohistochemistry. *Toxicol Pathol* **27** (2): 246–248.

23. Jones CJ, Kimber SJ, Illingworth I, Aplin JD (1996). Decidual sialylation shows species-specific differences in the pregnant mouse and rat. *J Reprod Fertil* **106** (2): 241–250.

24. Kanai K, Nunoya T, Shibuya K, Nakamura T, Tajima M (1998). Variations in effectiveness of antigen retrieval pretreatments for diagnostic immunohistochemistry. *Res Vet Sci* **64** (1): 57–61.

25. Key M (2006). *Immunohistochemical Staining Methods*, 4th edn. Carpinteria, CA: Dako Corporation.

26. Kiernan JA (2008). *Histological and Histochemical Methods: Theory and Practice*, 4th edn. Woodbury, NY: Cold Spring Harbor Laboratory Press.

27. Kok LP, Boone ME (1992). *Microwave Cookbook for Microscopists: Art and Science of Visualization*. Leiden, the Netherlands: Coulomb Press Leyden.

28. Kurien BT, Scofield RH (2009). *Protein Blotting and Detection: Methods and Protocols*. Totowa, NJ: Humana Press.

29. Larsson L-I (1993). Tissue preparation methods for light microscopic immunohistochemistry. *Appl Immunohistochem* **1** (1): 2–16.

30. Linscott WD (1980–2014). *Linscott's Directory of Immunological and Biological Reagents*, http://www. linscottsdirectory.com/ (last accessed November 15, 2014).

31. Lu QL, Partridge TA (1998). A new blocking method for application of murine monoclonal antibody to mouse tissue sections. *J Histochem Cytochem* **46** (8): 977–984.

32. Matkowskyj KA, Schonfeld D, Benya RV (2000). Quantitative immunohistochemistry by measuring cumulative signal strength using commercially available software Photoshop and Matlab. *J Histochem Cytochem* **48** (2): 303–312.

33. Mikaelian I, Nanney LB, Parman KS, Kusewitt DF, Ward JM, Näf D, Krupke DM et al. (2004). Antibodies that label paraffin-embedded mouse tissues: A collaborative endeavor. *Toxicol Pathol* **32** (2): 181–191.

34. Miller MW, Nowakowski RS (1988). Use of bromodeoxyuridine-immunohistochemistry to examine the proliferation, migration and time of origin of cells in the central nervous system. *Brain Res* **457** (1): 44–52.

35. Nadano D, Sugihara K, Paria BC, Saburi S, Copeland NG, Gilbert DJ, Jenkins NA, Nakayama J, Fukuda MN (2002). Significant differences between mouse and human trophinins are revealed by their expression patterns and targeted disruption of mouse trophinin gene. *Biol Reprod* **66** (2): 313–321.

36. Nagy A, Gertsenstein M, Vintersten K, Behringer R (2003). *Manipulating the Mouse Embryo*, 3rd edn. Cold Spring Harbor, NY: Cold Spring Harbor Laboratory Press.

37. Rahier J, Stevens M, de Menten Y, Henquin JC (1989). Determination of antigen concentration in tissue sections by immunodensitometry. *Lab Invest* **61** (3): 357–363.

38. Ramos-Vara JA (2005). Technical aspects of immunohistochemistry. *Vet Pathol* **42** (4): 405–426.

39. Relyea MJ, Sundberg JP, Ward JM (2000). Immunohistochemical and immunofluorescence methods. In: *Systematic Approach to Evaluation of Mouse Mutations*, Sundberg JP, Boggess D (eds.). Boca Raton, FL: CRC Press; pp. 131–144.

40. Rex M, Scotting PJ (1994). Simultaneous detection of RNA and protein in tissue sections by nonradioactive *in situ* hybridization followed by immunohistochemistry. *Biochemica* **11** (4): 63–65.

41. Rhodes JM, Milton JM (1998). *Lectin Methods and Protocols*. Totowa, NJ: Humana Press.

42. Sutherland AE, Calarco PG, Damsky CH (1993). Developmental regulation of integrin expression at the time of implantation in the mouse embryo. *Development* **119** (4): 1175–1186.

43. Tacha DE, Chen T (1994). Modified antigen retrieval procedure: Calibration technique for microwave ovens. *J Histotechnol* **17** (4): 365–366.

44. Taylor CR, Levenson RM (2006). Quantification of immunohistochemistry–Issues concerning methods, utility and semiquantitative assessment II. *Histopathology* **49** (4): 411–424.

45. Toder V, Carp H, Fein A, Torchinsky A (2002). The role of pro- and anti-apoptotic molecular interactions in embryonic maldevelopment. *Am J Reprod Immunol* **48** (4): 235–244.

46. van den Broek LJ, van de Vijver MJ (2000). Assessment of problems in diagnostic and research immunohistochemistry associated with epitope instability in stored paraffin sections. *Appl Immunohistochem Mol Morphol* **8** (4): 316–321.

47. van Noorden CJF, Frederiks WM (1993). *Enzyme Histochemistry: A Laboratory Manual of Current Methods*. Oxford, U.K.: Oxford University Press.

48. Vis AN, Kranse R, Nigg AL, van der Kwast TH (2000). Quantitative analysis of the decay of immunoreactivity in stored prostate needle biopsy sections. *Am J Clin Pathol* **113** (3): 369–373.

49. Wood GS, Warnke R (1981). Suppression of endogenous avidin-binding activity in tissues and its relevance to biotin-avidin detection systems. *J Histochem Cytochem* **29** (10): 1196–1204.

50. Xie R, Chung JY, Ylaya K, Williams RL, Guerrero N, Nakatsuka N, Badie C, Hewitt SM (2011). Factors influencing the degradation of archival formalin-fixed paraffin-embedded tissue sections. *J Histochem Cytochem* **59** (4): 356–365.

12

Quantitative Morphological Assessment in Mouse Developmental Pathology Studies

Brad Bolon and Stephen Kaufman

CONTENTS

The foundation of phenotypic analysis in developing mice is qualitative assessment for structural defects and functional deficiencies. Anatomic anomalies that can be detected in this fashion include alterations in the (1) location, (2) number, (3) shape, (4) size, and/or (5) architecture of organs, tissues, and cell populations. The first morphological information suggesting that a developmental defect is present will show that a change in one of these five parameters is present while providing an estimate of its extent. Such qualitative structural descriptions are often sufficient to address the research question of a given developmental pathology study. A common follow-on step in the evaluation is to combine a qualitative *in situ* assessment of gene expression (Chapter 10) or protein location (Chapter 11) with quantitative measurement of the same gene or protein in homogenized tissue. The technical difficulties and laborious nature of gathering quantitative data for morphological endpoints sometimes preclude such testing in mouse developmental pathology studies. Nevertheless, some research questions may only be addressed using quantitative anatomic methods, such as whether an enlarged structure is bigger due to increased cell proliferation, decreased cell death, larger cell profiles, or any combination of the three. Such studies have been simplified in recent years using computer-assisted techniques, although simple endpoints may be acquired manually.

This chapter is designed to introduce common morphological techniques for quantifying structural and functional endpoints in specimens from developing mice. Basic principles for designing studies and preparing samples for kinetic and morphometric examinations are outlined here. Individuals interested in using such methods on their own material will find multiple publications in the peer-reviewed literature to help modify such protocols for their own research programs.

Kinetics

Kinetic assays are designed to quantify relative rates of cell turnover. The most common procedures for undertaking this task are assessments of cell numbers (e.g., proliferation or elimination) or cell metabolism (e.g., DNA synthesis).

Several methods are available for evaluating cell proliferation. Pulse-labeling methods measure the number of cells that were actively proliferating (i.e., replicating DNA during the S phase of the cell cycle) at the time the pulse was given. The replication of DNA in developmental pathology specimens typically is estimated by pulse labeling due to the inherently high rate of cell proliferation in differentiating tissues. The primary disadvantage of the pulse-labeling approach is that it must be performed in a prospective manner (i.e., requires advance planning). Alternative options for assessing cell proliferation are to count the number of dividing cells (i.e., in the M phase of the cell cycle) in routinely stained sections or to assess the distribution of certain endogenous cell cycle–specific proteins by immunohistochemistry (IHC).[26,41]

Several techniques may also be employed to examine the degree of apoptosis, a critical process engaged in reorganizing the various tissues and cell populations of developing organs. Common approaches include using vital dyes to reveal dying cells,[23,51] molecular methods, or IHC to detect proteins involved in the cell death pathway.[14,46]

Cell Proliferation Measurements

Technical Principles

The two primary concerns when planning a pulse-labeling study to examine DNA synthesis during development are the length of the pulse and the choice of the DNA-labeling agent. In general, pulses in more rapidly dividing tissues must be brief so that the labeling agent is incorporated only into discrete populations of cells. The most common DNA-labeling agent in contemporary kinetic studies is 5-bromo-2′-deoxyuridine (BrdU),[41] which is incorporated during DNA synthesis in place of thymidine. This chemical is a nonradioactive alternative to tritiated (^{3}H) thymidine, the historical isotopic standard for DNA labeling.[2,29] Both BrdU and ^{3}H-thymidine yield comparable results.[13,19]

Cell proliferation may also be examined using various endogenous markers. The major advantage of these methods is that they may be used retrospectively with archived tissues. Three endogenous proteins suitable for nonisotopic measurement are proliferating cell nuclear antigen (PCNA), an acidic nuclear protein of the cyclin superfamily that is generated prior to mitosis[26,35]; the nuclear protein Ki67[48]; and phosphorylated histone H3 (pH3), a DNA structural protein that is associated with condensed chromosomes during mitosis (M phase).[17] Fewer nuclei are labeled by Ki67 than by PCNA, suggesting that Ki67 may be more specific for proliferating cells.[41] A fourth approach is to count mitotic figures in a given structure.

Nonisotopic Method: IHC Detection of BrdU Incorporation

The BrdU pulse-labeling technique is simple in both concept and execution. Pregnant dams or pups are given an intraperitoneal (IP) injection of BrdU (usually 50 μg/g body weight [BW], with a range of 20–200 μg/g) (see recipe in Protocol 12.1) in phosphate buffered saline (PBS; pH 7.4) at a specified time before tissues are harvested at necropsy. Typical pulse periods used in developing mice range from 15 min to 2 h.[30] Multiple pulses may be necessary to fully evaluate certain questions, such as patterns of

---

**PROTOCOL 12.1   PROTOCOLS FOR METHODS:
PREPARATION OF BRDU SOLUTION**

Caution! This agent is a highly effective mutagen, and thus must be prepared and handled with proper environmental controls and personal protective equipment (PPE) at all times!

Preparation of Stock Solution (10 mg/mL)

1. Mix BrdU (100 mg) into 10 mL of 10 mM Tris–Cl (pH 7.6) in a 15 mL screw-cap conical tube.
2. Heat in a water bath at 50°C, with intermittent vortexing, until the BrdU dissolves.
3. Divide the stock solution into 40 single-use aliquots of 250 µL each (yielding 2.5 mg of BrdU [a dose sufficient to administer 50 µg/g BW to a near-term, pregnant dam]).
4. Wrap vials in aluminum foil, and place in a light-tight box.
5. Store at −20°C.

---

cell migration within the various layers of the brain. In such cases, each new pulse is separated from its predecessor by a standard length of time (often 6–48 h).

Cumulative labeling protocols may be employed to evaluate the growth fraction (i.e., the proportion of cells in a given population that are proliferating) and the cell cycle length. The BrdU may be given by intermittent IP pulse dosing[33] as described earlier or by continuous infusion using an osmotic minipump implanted in the loose subcutaneous connective tissue over the dorsal cervical area.[16] The choice of pump depends on the desired infusion rate. For example, Alzet® (Durect Corporation, Cupertino, CA) Model No. 1003D delivers 1.0 µL/h for 3 days, Alzet® Model No. 1007D provides 0.5 µL/h for 7 days, and Alzet® Model No. 1002 affords 0.25 µL/h for 14 days; each of these units is 1.5 cm long × 0.6 cm in diameter. Pumps are preloaded with BrdU (25–150 mg/mL) in PBS according to the manufacturer's instructions. Pregnant dams are then anesthetized briefly with isoflurane, the haired skin is shaved and scrubbed with a topical antiseptic like povidone iodine (e.g., Betadine®, available from multiple vendors), and an incision is made through the skin between the scapulae. The pump is positioned and the incision is closed with steel staples. Typical infusion periods used to label developing mice range from 4 h to 7 days, although in some instances weeks may transpire before necropsy.[16] Morphological analysis of the labeled conceptuses must be undertaken with the knowledge that BrdU is a potent teratogen.[1,21,22,34] Therefore, BrdU administration should not be used as a means for exploring chronic patterns of cell proliferation in developing mice unless the intent of the study is to evaluate turnover in combination with induced developmental defects.

At necropsy, tissues are collected (Chapter 7), fixed (Chapters 6 and 9), and processed (Chapter 9) in a routine fashion. Developmental pathology studies commonly employ Bouin's solution, neutral buffered 10% formalin (NBF), or 4% paraformaldehyde (PFA) as the fixative, and paraffin as the embedding medium. Tissue sections (4–8 µm thick) are acquired, and BrdU within nuclear DNA is detected using conventional IHC (see generic protocol in Chapter 11) using a commercial antibody (generally with the primary anti-BrdU antibody applied at a dilution between 1:50 and 1:200) with 3,3′-diaminobenzidene (DAB) as the chromagen. Sections are usually counterstained with hematoxylin alone, or with hematoxylin and eosin (H&E), but other combinations of chromagens and counterstains are suitable. Nuclei that have incorporated BrdU are diffusely labeled with reaction product (Figure 12.1). Labeling during development may be used well into adulthood to evaluate DNA synthesis in distinct cell populations.[43]

Isotopic Method: Autoradiographic Detection of ³H-Thymidine Incorporation

The ³H-thymidine pulse-labeling method is also straightforward in concept, albeit more laborious in execution. Pregnant dams or pups receive an IP injection of the isotope (usually 1–5 µCi/g BW[41]) from 30 min to 2 h prior to necropsy.

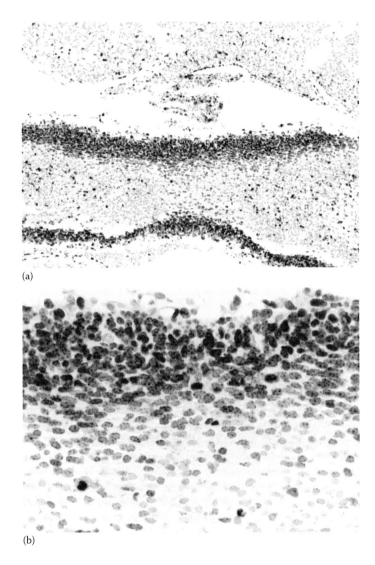

FIGURE 12.1   Cell proliferation is extensive in developing mice, but the sites exhibiting peaks of replication vary considerably with the developmental stage. In these images, the germinal cell layer on the external surface of the cerebellum anlage (primordium) of a wild-type, near-term (gestational day [GD] 17.5) mouse fetus is intensely labeled (brown reaction product) by an immunohistochemical (IHC) method to detect the endogenous marker Ki67. Sections were preserved in neutral buffered 10% formalin, processed routinely into paraffin, and sectioned at 5 μm. The manual IHC procedure involved dehydration followed by sequential application of rabbit anti-Ki67 (clone SP6, Catalog No. RM-9106-S; Lab Vision, Fremont, CA); biotinylated goat anti-rabbit IgG secondary antibody (Catalog No. BA-1000; Vector Laboratories, Burlingame, CA); avidin–biotin complex linked to horseradish peroxidase (Vectastain Elite ABC Kit, Catalog No. PK-6100; Vector); metal-enhanced 3,3'-diaminobenzidine (Catalog No. 34065; Thermo Fisher Scientific, Waltham, MA); and hematoxylin. Initial magnifications: (a) 100× and (b) 400×. (Images were kindly provided by Dr. Dorothy French, Genentech, Inc., South San Francisco, CA.)

Samples are collected, fixed, and processed into paraffin using conventional histological procedures in combination with standard precautions for handling radioactive materials. Working in the dark, sections (4–8 μm thick) are covered with a liquid photographic emulsion (e.g., NTB; Eastman Kodak, Rochester, NY) and then exposed in the dark for several weeks at either 4°C or −20°C. The labeled cells are revealed using a standard developing solution (e.g., K19; Eastman Kodak), and the sections are counterstained with hematoxylin or H&E. Nuclei associated with at least four to five grains are considered to be positive.

Other than the generation of isotopic waste, the main technical disadvantages of this approach are the long delay between sample collection and tissue analysis and the diffuse background labeling of cells and tissues. The latter problem may often be reduced or eliminated by examining the sections using a dark-field microscope. A second advantage of this method is that thymidine is the native nucleotide that is incorporated into DNA, so this agent can be used for long-term studies of cell turnover without increasing the incidence of developmental defects, as may occur with embryonic exposure to BrdU.

Nonisotopic Method: IHC Detection of Endogenous Markers

The IHC techniques for localizing PCNA, Ki67, and pH3 are equivalent to those utilized for demonstrating BrdU. Cells positive for PCNA exhibit diffusely labeled nuclei. The distribution of Ki67 depends on the stage of the cell cycle; nuclear labeling is diffuse during interphase but more punctate during mitosis, when the protein is associated with the chromosomes. The typical pH3 pattern is diffuse nuclear localization with more intense accumulation in association with chromatin.

In general, these three markers are examined in tissues preserved using an aldehyde fixative (2%–4%) in a neutral 0.1 M buffer (phosphate or cacodylate).[17,26,41] The localization of PCNA labeling is dependent on the method of fixation. All cycling cells are revealed for most fixatives except methanol, where PCNA is visible only in S-phase cells.[41] All these endogenous markers may be detected by IHC using commercially available primary antibodies and kits. The appropriate selection of chromogen and counterstain to optimize the contrast between the specific reaction product and the background facilitates automated analysis of the labeled cells.

Nonisotopic Method: Mitotic Counts

Proliferation rates may be quantified by counting the number of mitotic figures. This approach is usually undertaken in a discrete anatomic entity (e.g., an organ or tissue with well-defined margins) or for a specific population of cells using a bright-field microscope and a high magnification (usually either the 20× or 40× objective). The number of figures may be estimated (e.g., few, some, many) or enumerated; if absolute counts are obtained, a common practice is to acquire specific numbers for multiple fields (often 3–5) and then calculate an average. The primary benefits to this approach are that it requires no advance planning and thus may be performed on archived tissues and that it specifically measures a different phase of the cell cycle (M) than is assessed using the other proliferation markers. The main disadvantage is that the limited contrast between mitotic figures and adjacent formed nuclei renders automated counting quite difficult. In our experience, mitotic counts in developing tissues are best performed on sections stained lightly with hematoxylin.

Cell Elimination Measurements

Technical Principles

The most common strategy for examining cell elimination involves enumeration of apoptotic cells. Apoptosis is a frequent mechanism of physiological or *programmed cell death* by which the shape and size of tissues as well as intercellular connections are reorganized during development.[3,17,24,27,28,32] Distinct populations of developing cells have different susceptibilities to apoptotic stimuli.[25,42] Apoptotic cells may be readily recognized in H&E-stained sections using several morphological features. Early changes include chromatin condensation and nuclear contraction in association with cytoplasmic hypereosinophilia (excessive pinkness), membrane blebbing, and cell shrinkage. Shrunken cells are often surrounded by a clear pericellular halo. Late changes include karyorrhexis (nuclear fragmentation) and cytoplasmic disintegration to produce apoptotic bodies (pinched off blebs of cytoplasm). These findings affect individual cells and develop in the absence of any inflammatory response in the adjacent tissue.

Apoptosis Counts

One means of estimating the extent of apoptosis is to count the number of apoptotic cells in an H&E-stained section (Figure 12.2). Such counts are made for a defined tissue domain or for a high-power

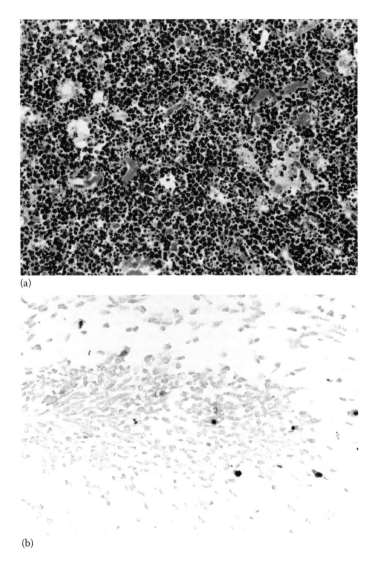

(a)

(b)

FIGURE 12.2 Programmed cell death (or apoptosis) is widespread in developing mice both during normal organogenesis and following exposure to certain toxicants. (a) When numerous, apoptotic cells are recognized easily in routine sections stained with hematoxylin and eosin (H&E) by such features as karyorrhexis and karyolysis (nuclear fragmentation and dissolution, respectively). (b) When rare, the presence of apoptotic cells is confirmed more reliably using a method to reveal cells with molecular changes associated with the death program, such as DNA fragmentation (labeled with ApopTag® [Catalog No S7100; Chemicon (Millipore), Billerica, MA] used according to the manufacturer's instructions). (a) shows the typical pattern of stress-associated apoptosis in the thymic cortex of a neonatal C57BL/6J mouse (postnatal day 1), while (b) illustrates the usual appearance of neuronal pruning in the remodeling trigeminal ganglion (cranial nerve V) of a wild-type, near-term (GD E18.0) mouse fetus. Magnification: 400×. (Image [b] was kindly provided by Dr. Dorothy French, Genentech, Inc., South San Francisco, CA.)

microscopic field (i.e., using a 20× or 40× objective). An advantage of this approach is that no special preparations are needed. However, the method is complicated by the transient lifespan of apoptotic cells in tissue sections (approximately 1 h) and the lack of contrast between the affected cells and the remainder of the tissue. Accordingly, a number of specialized approaches have been developed to enhance the visibility of apoptotic cells.[38]

Apoptosis Enhancement: Nucleic Acid Method

A means of improving contrast in apoptotic cells is via techniques that detect DNA fragmentation within disintegrating nuclei. Commercial kits for this purpose (e.g., ApopTag®, available from multiple vendors; Dead End™, Promega Corporation, Madison, WI) are based on the terminal deoxynucleotidyl transferase (TdT)-mediated 2′-deoxyuridine-5′-triphosphate (dUTP) nick-end labeling (TUNEL) method. The dUTP molecule is an intermediate in pyrimidine biosynthesis and thus is a substrate used in normal DNA metabolism. The dUTP in apoptosis-detecting kits is linked to either a fluorophore (e.g., fluorescein) or a bridging molecule (e.g., biotin, digoxigenin). Incorporation of the labeled nucleotide into fragmented DNA will color the nuclei of apoptotic cells.

Tissue collection and processing for apoptosis counts is comparable to that used for assessments of cell proliferation. In general, cells are fixed in aldehyde (2%–4%, pH 7.4) and then processed into paraffin. Sections are deparaffinized and rehydrated in the usual manner for an IHC procedure (Chapter 11). Cell and nuclear membranes are permeabilized with detergent (e.g., Triton X-100, 0.2% in 0.1 M PBS for 5 min at room temperature [RT]) and/or proteinase K (20 mg/mL for 15 min at RT). Tissues are washed twice for 2–5 min with PBS, after which equilibration buffer (supplied with the kit) is applied (75 µL for 5–10 min at RT). An aliquot of nucleotides in reaction buffer is prepared according to the manufacturer's instructions, mixed with TdT enzyme, and then placed over the section (using 10 µL/cm²). A coverslip is applied to form an incubation chamber, and the labeling reaction is allowed to proceed for 60 min at 37°C; incubations for fluorescent probes must take place in the dark. The labeling reaction is halted using 2X SSC (10–15 min at RT), after which the slides are washed three times for 5 min in PBS. The remainder of the protocol is comparable to a conventional IHC procedure (Chapter 11).

Apoptosis Enhancement: Protein Method

An alternative approach for labeling apoptotic cells is to detect a protein that participates in the apoptotic process using a standard IHC protocol (Chapter 11). Potential markers include cytochrome c, which upon release from the mitochondria will trigger the assembly and activation of the cytoplasmic apoptosome, and *c*ysteine-dependent *asp*artate-directed prote*ase* (caspase)-3, which exists as an inactive precursor until acted on by the apoptosome.[14,38,47,49] Both caspase-3 IHC and TUNEL label the same cell populations in developing mice.[47] The primary antibodies are typically used according to the manufacturer's instructions at dilutions of 1:250 to 1:1000. Both markers are present in the cytoplasm of apoptotic cells. Cytochrome c labeling may be faint or even absent since the enzyme is unstable once it is released from the mitochondria.[38]

Apoptosis Enhancement: Dye Uptake Method

A further means for demonstrating apoptotic cells is to employ dyes that will selectively label them (e.g., see Figure 14.1 in Chapter 14). Examples include cationic fluorescent dyes that collect inside the mitochondria of healthy cells but are dispersed into the cytoplasm in apoptotic ones (e.g., JC-1[50]); fluorescent analogs of caspase inhibitors that bind to activated caspases (e.g., fluorescein isothiocyanate-valyl-alanyl-aspartyl-[*O*-methyl]-fluoromethylketone [FITC-VAD-FMK, or CaspACE™, Promega][38]); and vital dyes that cross the porous membranes of disintegrating cells (e.g., Nile blue sulfate[45]). In our experience, the fluorescent dyes are utilized according to the manufacturer's instructions, performing all manipulations in the dark. Nile blue sulfate (NBS) is prepared as a 1:50,000 solution in lactated Ringer's solution or PBS (pH 7.6) into which intact embryos or isolated organs are placed for 30 min at 37°C.[6,15] All these probes will yield intense labeling of apoptotic cells and apoptotic bodies with little or no spurious staining of unaffected cell populations.

Interpretation of Kinetic Measurements

The nature of the research question will dictate the character of the analysis. Cell count data are generally acquired using either semiquantitative scoring criteria or absolute enumeration of labeled cells.

Semiquantitative assessment is a rapid and simple means of estimating the number of labeled cells in a given structure. A tiered scoring scheme is devised prior to the analysis by comparing the number of cells in control mice with those in animals that express a mutation or that were exposed to a toxicant. Ideally, readily identified inflection points will be evident for differences in the number of labeled cells between unaffected, modestly affected, and severely affected animals; the larger the number of inflection points, the more distinct scores may be developed. In our experience, either a 4-tiered (e.g., 0=unaffected, 1=slightly affected, 2=somewhat affected, 3=substantially affected) or 5-tiered (0=unaffected, 1=minimally affected, 2=mildly affected, 3=moderately affected, 4=markedly affected) scale is suitable. In some instances, a 2-tiered scale (0=unaffected, 1=affected) may be necessary. Sufficiently distinct differences in the cell populations of control and experimental (mutant or treated) animals are often suitable for statistical analysis even if the orientation of the sections exhibits modest variation among animals.

Quantitative assessment is generally a slower but more rigorous means of evaluating the effect of genotypes or treatments on cell numbers. Counts may be acquired manually or using automated software and imaging systems. The main advantage of quantitative analysis is that absolute counts may permit more subtle differences between control and treated mice to be discovered when evaluating biological endpoints that are subject to a large degree of inter-individual variability under normal conditions. The first step in obtaining quantitative data is to define the region in which measurements will be taken. Examples include an entire microscopic field that is visible through a given objective (usually 20× or 40×), the area of a structure with well-delineated borders that is evident in one or several microscopic fields of a section, or a particular length along a boundary.[4,41] The percentage of affected cells (i.e., labeled elements divided by the total number of similar cells within a certain area) is termed a *labeling index* and usually has units of mm^2 or cm^2. In principle, similar labeling indices may be calculated for three-dimensional (3D) volumes by acquiring separate counts for discrete two-dimensional (2D) sections and then combining them; the resulting values are often given units of cm^3. Numbers of affected cells (either the percentage or the absolute count) made along a well-demarcated border are termed a *unit length labeling index*, the units of which usually are expressed as the number of labeled cells per linear mm. Acquisition of useful data requires strict attention to detail in defining the sites to be assessed and ensuring that the identical or relatively homologous structures are examined in the same orientation for both the control groups and the experimental cohorts (e.g., mice with gene mutations or that have been exposed to developmental toxicants).

Morphometrics

Developmental pathology analysis often benefits by the acquisition of quantitative data. *Counts* represent quantitative measurements of objects (e.g., cells in a given structure) taken in 0 dimensions, as each object is considered to be an infinitely small point. *Morphometrics* is the preferred term for methods designed to quantify data in higher dimensions. The common morphometric measurements used for most pathology specimens are acquired in 1 (line), 2 (area), or 3 (volume) dimensions. Developmental pathology projects may add the fourth dimension (time) to any of these other endpoints by summing the evolution of a given measurement over an extended period.

Morphometric data generally are acquired for two reasons. The first is that such quantitative assessments often permit more precise estimates of biological relevance than can endpoints of merely a qualitative nature (i.e., is the conceptus normal or abnormal?) or having a semiquantitative character (i.e., is the degree of the abnormality minimal, mild, moderate, or marked?). The second reason is that morphometric measurements can demonstrate the presence of subtle defects in apparently normal individuals (Table 12.1), an outcome that has important ramifications for assessing the risk posed by agents with an unknown potential for inducing developmental toxicity.

Simple Morphometrics

The easiest approach to quantitative analysis in multiple dimensions is to measure the difference in size between two or more groups of mice with distinct genetic backgrounds (e.g., wild type vs. heterozygote vs. null mutant) or prior experiences (untreated control vs. low-dose and high-dose treatment groups).

TABLE 12.1

Morphometric Analysis Reveals Subtle Changes in Apparently Normal Individuals

		Methanol-Exposed Fetuses	
	Control Fetuses	**All Litters**	**Litters Lacking Dysraphism**
No. litters	24	16	6
No. fetuses	56	39	14
Neuroepithelium (N)	98.5±1.3	108.8±2.1**	107.1±2.5**
Intermediate cortex (I)	229.8±3.3	190.9±3.7**	193.6±3.1**
Subventricular plate (S)			
Cortical plate (C)	129.6±1.4	127.3±2.3	128.2±4.0
Cortical layer 1 (O)	30.2±0.6	22.9±0.6**	23.4±1.2**
Total thickness	488.1±3.9	449.9±5.9**	452.3±7.4**

Source: Data are reproduced from Bolon, B. et al., *Teratology*, 49(6), 497, 1994.

Notes: Data represent values obtained by assessing the thickness of various cerebrocortical layers of near-term (GD 17.5) mouse fetuses at the site shown in Figure 12.3 following
maternal inhalation of methanol (15,000 ppm) or room air (control) for 6 h/day during the
neurulation stage of development (GD 7–9). Methanol significantly altered the width of
multiple cerebrocortical layers (**, $p \leq 0.05$).

Common endpoints in developmental pathology studies in rodents are body weight, linear distance or
thickness, perimeter length, or area of a structure.[5,6] The trait to be measured may be at any level of
anatomic organization: gross, organ, tissue, cell, or subcellular.[8,12,18]

Several considerations are required to ensure the quality of simple morphometric data. First, the orientation of the structure to be measured and the location(s) where the measurements are to be taken must
be set before any data is collected. This step is required to ensure that all specimens will be processed so
that the site of interest is available for evaluation, and present in a relatively homologous orientation.[5] The
most straightforward way of assuring uniformity is to use recurring anatomic features (e.g., distance from
a universal locator point [like bregma, the junction point of the coronal and sagittal sutures on the skull
surface that is commonly used to orient stereotaxic brain sections[36,37,40]]; reliable anatomic landmarks
[external or internal]; well-defined margins between organ regions) to ensure that specimen orientation
is accurate (Figure 12.3). This check is necessary because the anatomic features of many organs change
rapidly in 2D or 3D (e.g., brain, olfactory mucosa, retina), so even minor deviations among animals may
substantially bias the results. Therefore, samples that do not reasonably meet the preset criteria regarding orientation of the site to be measured should be excluded from the quantitative portion of the study.
Consistent application of these crosschecks will assure reliable and reproducible results.

Simple morphometric measurements are most suitable for defining differences among groups based
on an apparently overt variance. Examples include large disparities in the gross size of an embryo or a
major organ or inconsistencies in the length or thickness of a tissue layer. The simplest means of quantifying such divergence typically is to define a single linear dimension of a clearly demarcated structure
(Figure 12.3). This task may be accomplished manually using a metric ruler with fine gradations (either
at necropsy or in photographs taken at a uniform size [as proven by inclusion of a scale bar]) or an optical reticule with slide-mounted calibration scale (for use when examining tissue sections on slides).
However, in most cases current morphometric studies gather such measurements more reliably and rapidly using precalibrated, semiautomated image analysis software.

More complex measurements such as perimeter, area, or length of non-linear structures (Figure 12.4)
will typically be easier if the measurements are acquired from slides or suitable photographs using automated systems. In general, such analyses are more rapid and efficient if the system software can acquire the
measurement (e.g., tracing a perimeter) without the need for human input. However, the operator must be
able to interact with the program to check the quality of the image data (i.e., does it make sense?), to refine
the quantitative algorithm by which the software selects elements for analysis, and to permit the correction
of automated errors in analysis (e.g., excluding the structures of interest, including irrelevant features).

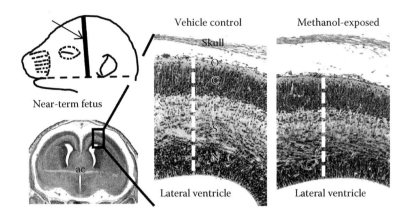

FIGURE 12.3 Morphometric data sets can be gathered easily and quickly for discrete anatomic regions in developing mice. Simple linear measurements are a common and fairly robust choice when tissue sections for all animals are relatively homologous. For example, a site at which the thicknesses of various cerebrocortical layers may be measured reliably can be accessed by making a transverse section through the rostral forebrain of a near-term mouse fetus at GD 17.5 at the plane (black line in the schematic diagram, upper left panel, indicated by an arrow) containing the rostral (anterior) commissure ("ac" in the low-magnification coronal section of brain, lower left panel). Relative to age-matched vehicle controls (middle panel), GD 17.5 fetuses from dams that had inhaled methanol (15,000 ppm for 6 h/day during neurulation [GD 7–9]) had thin cerebral cortices and altered thicknesses of most cortical layers (Table 12.1). C, cortical plate; I, intermediate zone; N, neuroepithelium; O, outer limit (cortical layer 1); S, subventricular plate. *Processing*: Immersion fixation in Bouin's solution, routine paraffin embedding, hematoxylin and eosin (H&E) stain. (Images are reproduced from Bolon, B. et al., *Teratology*, 49(6), 497, 1994.)

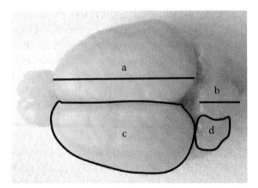

FIGURE 12.4 Quantitative morphometry can be accomplished readily and rapidly in the brains of developing mice using simple linear (a,b) or area (c,d) measurements of gross features using reproducible external landmarks.[5,39] (a) Cerebral length (from the rostral border just lateral to the olfactory bulb to the most caudal border). (b) Cerebellar length (over the midline of the vermis). (c) Cerebral area. (d) Neocerebellar area (i.e., the most evolutionarily advanced region, located lateral to the vermis). Other common measurements (not shown) include brain length (at the midline, from the rostral cerebral margin to the caudal cerebellar border) and cerebral width (widest distance across both hemispheres).

Complex Morphometrics: Stereology

Stereology is a technique for the quantitative exploration of a structure's geometric properties in 3D by using 2D specimens. The basic approach is to apply a standardized test grid to an image of sectioned material and then to enumerate only those structures that intersect with certain portions of the grid.[7,10] The entities of interest may be highlighted using special stains to facilitate the evaluation. A detailed discussion of stereological principles and protocols is beyond the scope of this book, but many resources on this topic are available.[9,20,31,44] The following discussion is designed as a brief introduction of the technique as it applies to mouse developmental pathology research.

Stereological investigations usually assess five types of quantities. These are number (or population size), length, surface (or profile) area, volume, and connectivity. The power of stereology is unleashed most effectively when evaluating irregular entities such as curved lines or bumpy surfaces. The most common estimates in contemporary studies are such quantities as the number of cells in a given tissue domain; the length of small, tortuous features like placental vessels; and the volume of irregular organs (e.g., various brain regions). The usual application in developmental pathology studies is to evaluate cell number.

Stereology is distinctly different from modern technologies for noninvasive imaging (Chapter 14). The algorithms for imaging systems gather serial 2D electronic sections of a given structure and then reconstruct a complete 3D representation. In contrast, the basis for stereology is to obtain data from a few representative 2D (planar) sections and then use statistical methods to extrapolate the structure's 3D properties. Two approaches may be used to ensure that 2D samples are genuinely *representative* of the entity being measured. The first is to assume that the specimen is homogeneous and that any randomly chosen plane of section will be typical. The second approach, which is more common in modern biomedical stereology, is to make no assumptions about the nature of the material but rather to ensure systematic uniform random sampling (SURS) or isotropic uniform random sampling (IURS) by varying the point at which specimens from different subjects are first sectioned.[20,31]

The impetus for using stereological measurements for quantitative biology (especially cell counts) is that features in standard tissue sections are distorted to a greater or lesser degree by the tissue preparation and microtomy processes. The number of cell fragments in serial 2D sections cannot be assumed to be the same as the number of cells in the original intact tissue. The Optical Fractionator is the modern method of choice for gathering unbiased estimates of cell numbers during developmental pathology and other biological applications. It is easier to implement because it uses thick (~20–50 μm) sections (thus producing fewer cell fragments) and requires no measurements of cell density. More importantly, shrinkage during histological processing (an unavoidable property of the best developmental pathology fixatives [Chapter 9]) will not impact the outcome. Nevertheless, correction factors (e.g., the diameters of maternal erythrocytes in fresh blood smears vs. within vessels of embedded specimens[11]) may be employed to approximate the measurements anticipated in fresh, unprocessed tissue. Specimens may be sectioned in any orientation as cellular shape, size, distribution, and position will not influence the outcome. The Optical Fractionator approach is typically carried out on a set of SURS sections using a high magnification (e.g., with a 40× objective or greater) with high numerical aperture (which provides a thin depth-of-field for more accurate evaluation of cells in the z-axis [*up-and-down* plane]). A single unique feature is used to count each object; for example, when counting cells the usual features are often the uppermost focal edge of the nucleus (for mononuclear elements) or the membrane that forms the outer limit of the cell itself. Design considerations that also must be addressed before the counts may be undertaken include the sizes of the counting frame and sampling grid.

Stereological specimens must be collected and processed carefully to ensure that the material is of suitable quality for unbiased quantitative analysis.[44] Samples must be removed with caution because trauma during the harvest may damage deep structures. However, the most likely time to lose tissue is when the material is being sectioned. Therefore, acquisition and preparation of tissues for stereology should be delegated to individuals with experience in serial sectioning techniques. Tips that will improve the likelihood that a given sample is suitable for stereological analysis are to gather all serial sections during the microtomy process (maintaining them in sequential order on prenumbered slides or in individual wells of a 96-well tissue culture plate[44]) and to use some landmark (internal or purposely applied) to permit unambiguous identification of the section orientation. The extra sequential sections may be substituted for damaged sections, but only if it is retained in advance.

Summary

Semiquantitative and quantitative assessments of the morphological traits that arise from kinetic and other physiological processes are a powerful means of improving our basic understanding of developmental pathology phenomena. The rise of computerized image analysis and digital imaging technology will substantially reduce the effort and expense of such studies. Furthermore, innovative software allows captured images to be immediately calibrated to real-world dimensions to facilitate statistical analysis and data interpretation. We anticipate that such improvements will increase the number of quantitative experiments performed in the mouse developmental pathology field.

ACKNOWLEDGMENT

The authors thank Mr. David Hill for assistance with ensuring the accuracy of technical details for the BrdU procedure.

REFERENCES

1. Bannigan JG, Cottell DC, Morris A (1990). Study of the mechanisms of BUdR-induced cleft palate in the mouse. *Teratology* **42** (1): 79–89.
2. Bayer SA, Wills KV, Triarhou LC, Verina T, Thomas JD, Ghetti B (1995). Selective vulnerability of late-generated dopaminergic neurons of the substantia nigra in weaver mutant mice. *Proc Natl Acad Sci USA* **92** (20): 9137–9140.
3. Blaschke AJ, Weiner JA, Chun J (1998). Programmed cell death is a universal feature of embryonic and postnatal neuroproliferative regions throughout the central nervous system. *J Comp Neurol* **396** (1): 39–50.
4. Bolon B, Dunn C, Goldsworthy TL (1996). Region-specific DNA synthesis in brains of F344 rats following a six-day bromodeoxyuridine infusion. *Cell Prolif* **29** (9): 505–511.
5. Bolon B, Garman R, Jensen K, Krinke G, Stuart B (2006). A "best practices" approach to neuropathologic assessment in developmental neurotoxicity testing—for today. *Toxicol Pathol* **34** (3): 296–313.
6. Bolon B, Welsch F, Morgan KT (1994). Methanol-induced neural tube defects in mice: Pathogenesis during neurulation. *Teratology* **49** (6): 497–517.
7. Boyce RW, Dorph-Petersen KA, Lyck L, Gundersen HJG (2010). Design-based stereology: Introduction to basic concepts and practical approaches for estimation of cell number. *Toxicol Pathol* **38** (7): 1011–1025.
8. Bush KT, Lynch FJ, DeNittis AS, Steinberg AB, Lee HY, Nagele RG (1990). Neural tube formation in the mouse: A morphometric and computerized three-dimensional reconstruction study of the relationship between apical constriction of neuroepithelial cells and the shape of the neuroepithelium. *Anat Embryol (Berl)* **181** (1): 49–58.
9. Charleston JS (2000). Estimating cell number in the central nervous system by stereological methods: The optical disector and fractionator. *Curr Protoc Toxicol* **12**: 12.16.11–12.16.19.
10. Coan PM, Conroy N, Burton GJ, Ferguson-Smith AC (2006). Origin and characteristics of glycogen cells in the developing murine placenta. *Dev Dyn* **235** (12): 3280–3294.
11. Coan PM, Ferguson-Smith AC, Burton GJ (2004). Developmental dynamics of the definitive mouse placenta assessed by stereology. *Biol Reprod* **70** (6): 1806–1813.
12. de Groot DMG, Hartgring S, van de Horst L, Moerkens M, Otto M, Bos-Kuijpers MHM, Kaufmann WSH et al. (2005). 2D and 3D assessment of neuropathology in rat brain after prenatal exposure to methylazoxymethanol, a model for developmental neurotoxicity. *Reprod Toxicol* **20** (3): 417–432.
13. del Rio JA, Soriano E (1989). Immunocytochemical detection of 5′-bromodeoxyuridine incorporation in the central nervous system of the mouse. *Brain Res Dev Brain Res* **49** (2): 311–317.
14. Duan WR, Garner DS, Williams SD, Funckes-Shippy CL, Spath IS, Blomme EA (2003). Comparison of immunohistochemistry for activated caspase-3 and cleaved cytokeratin 18 with the TUNEL method for quantification of apoptosis in histological sections of PC-3 subcutaneous xenografts. *J Pathol* **199** (2): 221–228.
15. Dunty WCJ, Chen S-Y, Zucker RM, Dehart DB, Sulik KK (2001). Selective vulnerability of embryonic cell populations to ethanol-induced apoptosis: Implications for alcohol-related birth defects and neurodevelopmental disorder. *Alcohol Clin Exp Res* **25** (10): 1523–1535.
16. Farah MH (2004). Cumulative labeling of embryonic mouse neural retina with bromodeoxyuridine supplied by an osmotic minipump. *J Neurosci Methods* **134** (2): 169–178.
17. Fernández-Terán MA, Hinchliffe JR, Ros MA (2006). Birth and death of cells in limb development: A mapping study. *Dev Dyn* **235** (9): 2521–2537.

18. Gressens P, Gofflot F, Van Maele-Fabry G, Misson JP, Gadisseux JF, Evrard P, Picard JJ (1992). Early neurogenesis and teratogenesis in whole mouse embryo cultures. Histochemical, immunocytological and ultrastructural study of the premigratory neuronal-glial units in normal mouse embryo and in mouse embryos influenced by cocaine and retinoic acid. *J Neuropathol Exp Neurol* **51** (2): 206–219.
19. Hayes NL, Nowakowski RS (2000). Exploiting the dynamics of S-phase tracers in developing brain: Interkinetic nuclear migration for cells entering versus leaving the S-phase. *Dev Neurosci* **22** (1–2): 44–55.
20. Howard CV, Reed MG (2005). *Unbiased Stereology*, 2nd edn. New York: Garland Science.
21. Kalter H (1985). Experimental teratological studies with the mouse CNS mutations cranioschisis and delayed splotch. *J Craniofac Genet Dev Biol Suppl* **1**: 339–342.
22. Kolb B, Pedersen B, Ballermann M, Gibb R, Whishaw IQ (1999). Embryonic and postnatal injections of bromodeoxyuridine produce age-dependent morphological and behavioral abnormalities. *J Neurosci* **19** (6): 2337–2346.
23. Kotch LE, Sulik KK (1992). Experimental fetal alcohol syndrome: Proposed pathogenic basis for a variety of associated facial and brain anomalies. *Am J Med Genet* **44** (2): 168–176.
24. Kuan C-Y, Roth KA, Flavell RA, Rakic P (2000). Mechanisms of programmed cell death in the developing brain. *Trends Neurosci* **23** (7): 291–297.
25. Kuida K, Zheng TS, Na S, Kuan C, Yang D, Karasuyama H, Rakic P, Flavell RA (1996). Decreased apoptosis in the brain and premature lethality in CPP32-deficient mice. *Nature* **384** (6607): 368–372.
26. LeCouter JE, Kablar B, Whyte PF, Ying C, Rudnicki MA (1998). Strain-dependent embryonic lethality in mice lacking the retinoblastoma-related p130 gene. *Development* **125** (3): 4669–4679.
27. Lo AC, Houenou LJ, Oppenheim RW (1995). Apoptosis in the nervous system: Morphological features, methods, pathology, and prevention. *Arch Histol Cytol* **58** (2): 139–149.
28. Majno G, Joris I (1995). Apoptosis, oncosis, and necrosis. An overview of cell death. *Am J Pathol* **146** (1): 3–15.
29. Martí J, Wills KV, Ghetti B, Bayer SA (2002). A combined immunohistochemical and autoradiographic method to detect midbrain dopaminergic neurons and determine their time of origin. *Brain Res Brain Res Protoc* **9** (3): 1197–1205.
30. Menezes JR, Dias F, Garson AV, Lent R (1998). Restricted distribution of S-phase cells in the anterior subventricular zone of the postnatal mouse forebrain. *Anat Embryol (Berl)* **198** (3): 205–211.
31. Mouton PR (2002). *Principles and Practices of Unbiased Stereology: An Introduction for Bioscientists.* Baltimore, MD: The Johns Hopkins University Press.
32. Narayanan V (1997). Apoptosis in development and disease of the nervous system: 1. Naturally occurring cell death in the developing nervous system. *Pediatr Neurol* **16** (1): 9–13.
33. Nowakowski RS, Lewin SB, Miller MW (1989). Bromodeoxyuridine immunohistochemical determination of the lengths of the cell cycle and the DNA-synthetic phase for an anatomically defined population. *J Neurocytol* **18** (3): 311–318.
34. Packard DSJ, Skalko RG, Menzies RA (1974). Growth retardation and cell death in mouse embryos following exposure to the teratogen bromodeoxyuridine. *Exp Mol Pathol* **21** (3): 351–362.
35. Patel Y, Kim H, Rappolee DA (2000). A role for hepatocyte growth factor during early postimplantation growth of the placental lineage in mice. *Biol Reprod* **62** (4): 904–912.
36. Paxinos G, Ashwell KWS, Törk I (1994). *Atlas of the Developing Rat Nervous System*, 2nd edn. San Diego, CA: Academic Press.
37. Paxinos G, Halliday G, Watson C, Koutcherov Y, Wang H (2007). *Atlas of the Developing Mouse Brain at E17.5, P0, and P6.* San Diego, CA: Academic Press (Elsevier).
38. Promega (2009). Apoptosis. In: *Protocols and Applications Guide.* Madison, WI: Promega Corporation; pp. 3.1–3.22.
39. Rodier P (1978). Neuropathology as screening method for detecting injuries in the developing CNS. In: *The Effects of Foods and Drugs on the Development and Function of the Nervous System: Methods for Predicting Toxicity* (Gryder RM, Frankos VH, eds.). Washington, DC: HHS (FDA) Publication No. 80-1076; pp. 91–98.
40. Schambra U (2008). *Prenatal Mouse Brain Atlas.* New York: Springer.
41. Smith RS, Martin G, Boggess D (2000). Kinetics and morphometrics. In: *Systematic Approach to Evaluation of Mouse Mutations* (Sundberg JP, Boggess D, eds.). Boca Raton, FL: CRC Press; pp. 111–119.

42. Soleman D, Cornel L, Little SA, Mirkes PE (2003). Teratogen-induced activation of the mitochondrial apoptotic pathway in the yolk sac of day 9 mouse embryos. *Birth Defects Res A Clin Mol Teratol* **67** (2): 98–107.

43. Soriano E, Del Rio JA (1991). Simultaneous immunocytochemical visualization of bromodeoxyuridine and neural tissue antigens. *J Histochem Cytochem* **39** (3): 255–263.

44. Stereology Group (1999–2010). Stereology.info: Stereology information for the biological sciences. Available at: http://www.stereology.info/ (last accessed November 1, 2014).

45. Sulik KK, Cook CS, Webster WS (1988). Teratogens and craniofacial malformations: Relationships to cell death. *Development* **103** (Suppl.): 213–231.

46. Toder V, Carp H, Fein A, Torchinsky A (2002). The role of pro- and anti-apoptotic molecular interactions in embryonic maldevelopment. *Am J Reprod Immunol* **48** (4): 235–244.

47. Umpierre CC, Little SA, Mirkes PE (2001). Co-localization of active caspase-3 and DNA fragmentation (TUNEL) in normal and hyperthermia-induced abnormal mouse development. *Teratology* **63** (3): 134–143.

48. Xue M, Balasubramaniam J, Buist RJ, Peeling J, Del Bigio MR (2003). Periventricular/intraventricular hemorrhage in neonatal mouse cerebrum. *J Neuropathol Exp Neurol* **62** (11): 1154–1165.

49. Yamauchi H, Katayama K, Ueno M, Uetsuka K, Nakayama H, Doi K (2004). Involvement of p53 in 1-β-D-arabinofuranosylcytosine-induced trophoblastic cell apoptosis and impaired proliferation in rat placenta. *Biol Reprod* **70** (6): 1762–1767.

50. Zamzami N, Métivier D, Kroemer G (2000). Quantitation of mitochondrial transmembrane potential in cells and in isolated mitochondria. *Methods Enzymol* **322**: 208–213.

51. Zucker RM, Hunter ES III, Rogers JM (1999). Apoptosis and morphology in mouse embryos by confocal laser scanning microscopy. *Methods* **18** (4): 473–480.

13

Ultrastructural Evaluation of Developing Mice

Brad Bolon, Stephen Kaufman, and Richard Montione

CONTENTS

The cause of some phenotypes may be self-evident in the pattern of structural lesions observed at necropsy or, more commonly, after a combination of additional histopathologic assessment, functional analysis, and/or biochemical or molecular testing. In some instances, however, more subtle changes in organelle anatomy warrant further morphologic evaluation at the subcellular level. Such ancillary studies involve ultrastructural (electron microscopic [EM]) evaluation. The information provided by examining organs and tissues using EM promotes a deeper understanding of biological events by revealing the cytoarchitectural basis for normal cell and tissue function as well as more exact explanations about causes and consequences of disease.

The advantage of EM analysis is that techniques for preparing samples for EM examination are designed to maintain cellular and subcellular structures in a state that is as close as possible to the conditions that pertain during life. Accordingly, techniques for collecting and processing such specimens are specialized to provide optimal fixation and atraumatic handling. The quality of tissue preservation is ultimately predicated on the attention to detail that is paid to all processing steps.

This chapter discusses basic considerations and useful protocols for EM examination of tissues and cells from developing mice.* More detailed discussions of EM techniques for use with biological specimens may be found in specialty references for this field.[1,3,11,14] Researchers and pathologists engaged in EM studies of embryo, fetal, and/or placental pathology on a routine basis should have access to appropriate references that describe and illustrate normal features of subcellular anatomy as well as common changes that occur during various disease conditions.[6,9,10]

---

* These protocols briefly recapitulate the features of the general necropsy procedure given in Chapter 7, concentrating on those characteristics required for optimizing specimen collection for EM analysis.

Fundamentals of Ultrastructural Evaluation

Electron microscopes produce a greatly magnified image using an electron beam focused with electromagnetic "lenses". The maximum magnification attained using a modern electron microscope (1,000,000 times, or 0.1 nm resolution) significantly exceeds that which can be obtained using a conventional brightfield light microscope (1000 times, or 0.2 μm). This marked enhancement in resolution results from the much shorter wavelength of electrons relative to photons of visible light.

Multiple EM techniques may be utilized to examine biological specimens. The most common variants used to evaluate developmental pathology specimens are transmission EM (TEM), which was the first EM variant to be invented, and scanning EM (SEM). The TEM technique permits assessment of minute *intracellular* structures at extremely high resolution by passing the electron beam through an ultrathin (~70 nm thick) section. The contrast evident in TEM images results from the interaction of electrons with many cell constituents having dissimilar densities (Figure 13.1). In contrast, the SEM technique reveals fine detail of a sample's *exterior* surface structure in three dimensions by passing the electron beam across the exterior in a raster scan pattern (Figure 13.2). The SEM technique can also be used to visualize internal structures by prior dissection of the specimen to remove external obstructing tissue or by injection of a casting medium followed by subsequent digestion (corrosion) of the surrounding tissue after the medium hardens (e.g., vascular casts).[13]

Processing Specimens for Ultrastructural Evaluation

Examination of cells and subcellular organelles necessitates a series of specialized steps to prevent tissue degradation and stabilize fine structures and their molecular constituents. Conventional procedures for processing EM specimens demand a precise and carefully controlled succession of stages to achieve this purpose. These steps are discussed below in the order in which they are undertaken when preparing samples. Given the number and intricacy of these steps, it is imperative for investigators to consult an experienced EM technologist *before* designing ultrastructural experiments to ensure that specimens are harvested and handled in an appropriate manner. Failure to devise a careful experimental plan will undoubtedly ruin the samples, thus precluding EM analysis and in most cases necessitating that the study be repeated.

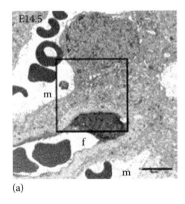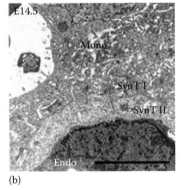

(a)　　　　　(b)

FIGURE 13.1 Transmission electron micrographs (TEM) of mouse placenta at gestational day (GD) 14.5 demonstrating trophoblast cell subtypes in the labyrinthine layer. (a) The large cells protruding into maternal sinusoids (m) are mononuclear trophoblasts (Mono). (b) Two layers of syncytiotrophoblasts (SynT I and SynT II) separate the Mono layer from the endothelial (Endo)-lined fetal capillaries (f). Bar=5 μm. (Reprinted from *Dev. Biol.*, 284(1), Simmons, D.G. and Cross, J.C., Determinants of trophoblast lineage and cell subtype specification in the mouse placenta, 12–24, Copyright 2005, with permission from Elsevier.)

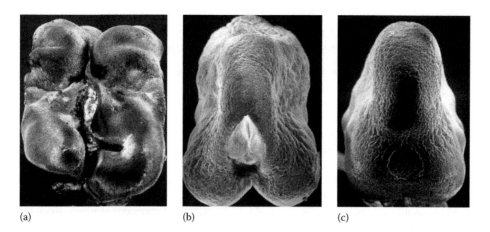

(a)   (b)   (c)

FIGURE 13.2 Scanning electron micrographs (SEM) of CD-1 mouse embryos at GD 9.25, shortly after the stage of organogenesis during which the rostral (anterior or cranial) neuropore closes during initial formation of the neural axis of the central nervous system. The image is focused at the crown of the cephalic region, which at this stage actually consists of the mesencephalon (i.e., primitive midbrain). (a) The irregular surface of the specimen is indicative of cephalic dysraphism (failed closure of the rostral neuropore; colloquial designation—neural tube defect [NTD]). The rough surface represents the exposed linings of the mesencephalon at the top and the telencephalon (primordial cerebral hemispheres) at the bottom. The likely presentation in a near-term fetus would be *exencephaly*. (b) The mid-axial opening in the rostral mesencephalon represents an incompletely closed neuropore; the lesion might close in time, in which case this would be a developmental delay, or the gap might remain in the near-term fetus as an *encephalocele* or *meningocele*. (c) This control animal exhibits normal closure of the cephalic neural tube. Embryos in (a) and (b) were from litters in which the dam had inhaled methanol at high concentrations for 6 h daily during the period of neurulation (GD 7–9).

Sample Collection

The first decision when designing an ultrastructural experiment is to determine how specimens will be preserved. Penetration by EM fixative solutions is particularly slow. Even minor delays in initiating fixation will yield extensive artifactual alterations of organelles that will be readily apparent at the ultrastructural level, and may confound efforts to analyze and interpret subtle subcellular lesions. Therefore, the fixation method must be chosen and the collection procedure choreographed in advance to ensure that fixation is started in a timely manner.

Two basic options are available for exposure to fixatives: immersion and perfusion. The primary advantage of immersion is the simplicity of the method (tissues are merely harvested and placed in fixative), while the main disadvantage is that the delay in fixation results in a greater degree of artifactual *post mortem* damage at the subcellular level. In contrast, perfusion introduces fixative into the core of tissues using preexisting channels (e.g., airways, blood vessels, and lumens of hollow organs), thereby halting *post mortem* disintegration more quickly, but at the cost of considerable time and effort to infuse the fixative into any single specimen. Technical aspects of perfusion fixation of developing mice are detailed in Chapter 17.

Regardless of the fixation method, specimens are processed for TEM in the following manner (Protocol 13.1). Using a stereomicroscope, the specimen of interest (an intact early [very small] embryo or an intact organ) is isolated and laid on a hard, flat surface (ideally a plastic or dense cardboard cutting board). A sharp blade is used to trim samples, which must be no larger than 1 mm³ to avoid induction of cellular damage resulting from the slow rate of fixative penetration. The blade should be wiped with a paper towel soaked in 70% ethanol prior to trimming to remove microscopic metal filings from the cutting edge before they can contaminate the tissue. If specimens are to be fixed by immersion, entry of the fixative may be accelerated by placing a large drop of ice-cold (4°C) fixative solution on the flat surface, placing the tissue to be trimmed within the drop, and then cutting the tissue while rotating it within the fluid. The small tissue blocks are then transferred immediately into vials of fresh, ice-cold fixative (0.5 mL per 1 mm³ cube). All cubes from a single site may be fixed in the same vial. If a specific

**PROTOCOL 13.1   TISSUE ACQUISITION AND FIXATION
FOR TRANSMISSION ELECTRON MICROSCOPY**

1. Prepare the necropsy area in advance. In particular, make sure that vials of fixative have been placed on ice (4°C) and that a hard cutting board (plastic or cardboard) is available.
2. Wipe the edge of a razor blade with 70% ethanol to remove microscopic metal filings.
3. Collect the sample (intact early embryos or whole organs from older animals) as rapidly as possible (ideally within no more than 5 min after death). Manipulate the specimen using a fine forceps or a plastic pipette (with the narrow tip removed); avoid using glass pipettes to prevent glass shards from contaminating the epoxy block. Do not squeeze the sample!
4. Place the specimen on the cutting board in a small puddle (several drops) of ice-cold fixative. Rotate the tissue to coat its surface in fixative.
5. Methodically cut the tissue into small blocks (maximum dimensions, 1 mm³). Keep the pieces within the pool of fixative. Do not dice the sample to excess.
6. Use a plastic pipette (with the tip removed) to transfer one or more small pieces to a glass vial filled with ice-cold fixative. Use 0.5 mL of fixative for each tissue block. Gently swirl the vial using a circular motion to dissociate the pieces, making sure that none stick to the walls.
7. Cap the vial (a screw cap is most effective at preventing leaks) and shake again to disperse the specimens. Place the vial on ice. Agitate the vial every 30–60 s for the first 5 min to improve penetration of the fixative.
8. Fix for 2 to (ideally) 24 h at 4°C.

The following steps should be performed with agitation either on a rocker or on a rotator.

9. Wash the fixed samples in the appropriate 0.1 M buffer (pH 7.4), at 4°C—two times for 15 min. (Buffer recipes are given in Table 13.1.)
10. Postfix specimens for 1–2 h in ice-cold 1% osmium tetroxide (pH 7.4). This fluid must be handled in a fume hood to prevent osmium-induced ocular and pulmonary toxicity.

orientation of the section is required, cut the tissue blocks such that one end of the cube is smaller than the other, thereby permitting its orientation to be determined by visual inspection during processing.

Buffers

Reagents used to prepare EM specimens (e.g., fixatives, washing solutions) may be prepared in several different buffers (Table 13.1), all of which may be used interchangeably. While a broad range of buffers have been used in TEM, in our experience the two most common choices are phosphate and cacodylate. As with fixative solutions, these reagents should be stored at 4°C until needed.

Phosphate buffers are the most widely used option in EM for several reasons. First, the chemical makeup is physiological, closely resembling the normal composition of tissue fluids. This resemblance to regular body chemistry is instrumental in reducing artifactual damage during fixation. Next, phosphate buffers function effectively at pH values near the physiological range (pH 6.8–7.6, where the actual physiologic pH of tissue fluid is ~7.4); maintaining the pH of EM solutions in the physiological range prevents tissue distortion that might develop from altered tissue pH. Finally, phosphate buffers are not toxic. The major disadvantage of phosphate for EM is that over time the solution supports microbial growth; such contaminants may precipitate onto specimens during processing, thereby compromising tissue morphology. Accordingly, phosphate buffers should be remade fresh every 4–6 weeks.

Cacodylate is an excellent buffer for EM applications and is considered by many to be the gold standard. It also functions well at physiological pH. An additional benefit is that it can be stored for months at 4°C. The drawback to this buffer is that it contains arsenic, and thus represents both a hazard to personnel and a disposal problem.

TABLE 13.1

Buffers for Use with Tissues Designated for Ultrastructural Analysis

Cacodylate Buffer, 0.2 M	
Caution! Cacodylate buffer contains arsenic and is poisonous. Personal protective equipment should be used when handling this buffer and its ingredients, and work with it should be performed in a fume hood.	
Materials	
Solution A	
Sodium cacodylate.	4.28 g
Distilled or deionized water.	100 mL
Solution B	
Concentrated hydrochloric acid (HCl; 36%–38%).	1.66 mL
Distilled or deionized water.	100 mL
Stock solution	
Solution A—mix sodium cacodylate with water until dissolved.	
Solution B—mix HCl into water.	
Working solution	0.2 M cacodylate buffer
Slowly add solution B to solution A, while mixing, until the desired pH is reached (approximately pH 7.2–7.4).	
Washing Solution	0.1 M cacodylate buffer
Mix working solution 1:1 with distilled water.	
Storage: Working solution can be stored at 4°C for up to 6 months.	
Phosphate Buffer, 0.2 M	
Note: Sodium phosphate is an irritant as a solid. Wear personal protective equipment when handling.	
Materials:	
Solution A	
Sodium phosphate monobasic (NaH$_2$PO$_4$·H$_2$O).	2.78 g
Distilled or deionized water.	100 mL
Solution B	
Sodium phosphate dibasic (Na$_2$HPO$_4$·7H$_2$O).	5.36 g
Distilled or deionized water.	100 mL
Stock solution	
Solution A—mix sodium phosphate monobasic with water until dissolved.	
Solution B—mix sodium phosphate dibasic with water until dissolved.	
Working solution	0.2 M phosphate buffer
Slowly add solution B to solution A, while mixing, until the desired pH is reached (approximately pH 7.2–7.4).	
Washing solution	0.1 M phosphate buffer
Mix working solution 1:1 with distilled water.	
Storage: Working solution can be stored at 4°C for 4–6 weeks.	

Fixatives

Reagent Options

Solutions used to fix tissues slated for EM analysis usually are cross-linking chemicals (e.g., aldehydes, osmium tetroxide) rather than denaturing agents (e.g., acids, alcohols). These fixatives build cross-links between proteins and other macromolecules, thereby preventing them from being extracted or substantially distorted during subsequent tissue processing. Details of major fixatives used in mouse developmental pathology studies are given in Chapters 6 and 9. However, the typical choices used for ultrastructural studies are briefly reviewed here as they differ in certain critical respects from solutions utilized for light microscopy.

Good primary fixatives for EM penetrate rapidly and produce extensive protein cross-linking. The best EM fixatives for tissues of developing mice incorporate two different aldehydes mixed together in a single solution (Table 13.2). One is usually glutaraldehyde, because it has two reactive groups that can form cross-links (as opposed to formaldehyde and paraformaldehyde, which have only one). Molecule for molecule, glutaraldehyde is a more effective cross-linking chemical. The second agent is usually formaldehyde (e.g., McDowell and Trump's solution) or paraformaldehyde (e.g., Karnovsky's solution), both of which infiltrate more quickly than glutaraldehyde since they have only one reactive moiety. All these fixatives must be *research grade* reagents rather than *commercial grade* as the methanol (up to 15%) included in commercial formulations as a stabilizing agent will produce more subcellular artifactual damage than will solutions containing only aldehydes. Acrolein, a glutaraldehyde precursor, may be employed in place of glutaraldehyde as it penetrates more rapidly while still providing effective nonreversible cross-linking.[2]

In general, for EM analysis primary fixation is followed by an additional secondary fixation step. The purpose of secondary fixation is to stabilize structures (e.g., lipid-rich membranes) that are not well preserved by aldehydes. The reagent of choice for postfixation in EM normally is osmium tetroxide (OsO_4). This agent not only helps to cross-link proteins, but it also enters lipid-rich membranes where its metallic nature contributes greatly to improving the contrast of membranous structures in images.

Fixation Principles for EM

Several physicochemical properties dictate the manner in which EM fixatives are employed. First, EM fixatives usually are used at 4°C because this temperature retards artifactual damage within tissues (e.g., autolysis, molecular extraction, shrinkage) more than it slows penetration of the fixative.[2] Next, fixatives are typically made fresh before each use because the aldehyde groups usually convert to acidic moieties over time when constituted in aqueous buffers. Fresh solutions may be made by dissolving aldehyde crystals or by aliquoting small volumes of concentrated stock into buffer (e.g., see recipes for paraformaldehyde solutions below). Preparations made directly from crystals must be prepared well in advance because it requires several hours of heating (65°C) to disperse the solid. Concentrated solutions may be kept for weeks at 4°C.

A final consideration in fixative selection is the electron microscopic endpoints to be assessed in a given sample. Conventions for routine ultrastructural analysis are well known, and are easily accommodated by the techniques given below. More recently, new methods have been developed to permit ultrastructural localization of protein epitopes (immuno EM) or mRNA (*in situ* EM) by labeling the probe molecule (an antibody or cDNA) with ultrasmall gold particles. Standard methods for immunohistochemistry and *in situ* hybridization have been adapted for use in TEM sections, including dual labeling (where reagents are conjugated to gold particles of different sizes) to detect mRNA and proteins in the same sections.[12] Great care is required to ensure that tissue morphology is retained during the application of the probe molecules.[16] A detailed discussion of these specialized protocols is outside the scope of this volume. Instead, interested readers are directed to current references for additional information on immuno EM[1,12,20,24] and *in situ* EM.[12,18]

Fixation Practices for EM

In our experience, primary fixatives that are well suited for EM applications contain 1%–5% glutaraldehyde along with 2.5%–4% of a second aldehyde (see fixative recipes below). We prefer McDowell and Trump's solution (1–2.5% glutaraldehyde and ~4% formaldehyde) or modified Karnovsky's solution (2–2.5% glutaraldehyde and 2% paraformaldehyde) for biological specimens. For immersion fixation, tissues are typically submersed for 2 h to overnight at 4°C. In contrast, perfusion fixation is performed using either ice-cold (~4°C) or room temperature (~22°C) solutions and followed by overnight immersion in fresh, ice-cold fixative. Perfusion is considered to be complete when the tissues blanch (more readily visible in neonates and juveniles) and the limbs become rigidly extended.

TABLE 13.2

Fixatives for Use with Tissues Designated for Ultrastructural Analysis

Caution! These agents are irritants and potential toxicants. Use in a well-ventilated area or fume hood, and wear personal protective equipment.

McDowell and Trump's Fixative

Materials and preparation—to make 100 mL, add materials in the following order:

Phosphate buffer, 0.2 M, pH 7.2–7.4	88 mL
Sodium hydroxide	0.27 g
Formaldehyde, 40% (electron microscopy [EM] grade)	10.0 mL
Glutaraldehyde, 50% (EM grade)	2.0 mL

Adjust pH to 7.2–7.4

Final composition of this fixative is 3.7% formaldehyde and 2.5% glutaraldehyde in 0.1 M phosphate buffer.

Storage: This fixative has a shelf life of 3 months when stored at 4°C.

Recipe adapted from Elizabeth McDowell and Benjamin Trump, "Histological fixatives for diagnostic light and electron microscopy," *Arch Pathol Lab Med* **100** (8) (1976): 405–414

Modified Karnovsky's Fixative

Paraformaldehyde working solution

Reagents

Paraformaldehyde (electron microscopy [EM] grade, prill)	4 g
Distilled or deionized water	50 mL
Sodium hydroxide (NaOH), 1 N	

Preparation

Mix paraformaldehyde with 50 mL of water in an Erlenmeyer flask.

Heat to 60°C while stirring.

Slowly add 1 N NaOH, usually 2–4 drops, and allow the solution to clear.

Cool and filter.

Modified Karnovsky's solution—to make 100 mL, perform the following steps:

Add materials to the paraformaldehyde working solution in the following order:

Glutaraldehyde, 25% (research grade)	10 mL

   Note: Research grade solutions of glutaraldehyde can be purchased in sealed ampoules under inert gas for a long shelf life.

Buffer, 0.2 M (cacodylate OR phosphate), pH 7.2–7.4.	50 mL
Calcium chloride ($CaCl_2$).	0.01 g

   Note: This reagent is added only if using cacodylate buffer.

Adjust pH to 7.2–7.4.

Final composition of this fixative is 2.5% glutaraldehyde and 2% paraformaldehyde in 0.1 M buffer. (*Note*: The original Karnovsky's fixative is 5% glutaraldehyde and 4% paraformaldehyde and thus is very hypertonic.)

Storage: This fixative has a shelf life of 6 months when stored at 4°C.

Osmium Tetroxide

Caution! This fixative must be prepared and used in a fume hood due to the high volatility and toxicity of this metal. Personal protective equipment must include close-fitting goggles and gloves to avoid surface deposition of metal on the corneas and skin.

Stock solution: 2% osmium tetroxide (aq).

Reagents

Osmium tetroxide (OsO_4; crystalline).	1 g
Distilled or deionized water.	50 mL

Preparation

Dissolve OsO_4 overnight in water at room temperature.

(Continued)

TABLE 13.2 (*CONTINUED*)

Fixatives for Use with Tissues Designated for Ultrastructural Analysis

Working solution: 1% osmium tetroxide (aq).

   Reagents

      Osmium tetroxide stock solution.

      Diluent—pick among these options, depending on the desired application.

         Buffer, 0.2 M (cacodylate OR phosphate), pH 7.2–7.4.

         Distilled or deionized water.

   Preparation

      Dilute the stock solution 1:1 with the chosen diluent.

Storage: This fixative has a shelf life of 6 months when stored at 4°C. Solution should be stored in a glass bottle sealed with Parafilm® and then placed in a second cushioned container. Solution should be discarded if it shows any signs of darkening.

Secondary fixation with OsO_4 usually involves immersion in a 1%–2% solution (unbuffered, or buffered to pH 7.4) for 1–2 h at 4°C. Specimens must be washed before they are placed into OsO_4 as osmium reacts extensively with any residual glutaraldehyde in the tissue. Washing typically is carried out in buffer at room temperature while being agitated; common strategies range from 2 to 3 short washes (e.g., 15 min) so that specimens move more rapidly into the membrane stabilization step with OsO_4 up to overnight to ensure essentially complete removal of glutaraldehyde. Individuals working with OsO_4 must remember at all times that it may be quite toxic as this agent volatizes easily and binds rapidly to the cornea and skin (forming opaque black deposits). Therefore, OsO_4 should only be used in a fume hood and handled while wearing appropriate personal protective equipment (especially form-fitting goggles and gloves). Ruthenium tetroxide substituted in place of OsO_4 yields comparable results for some applications,[23] though to our knowledge it has not been evaluated in the developmental pathology setting.

Dehydration

Specimens are moved through a series of dehydrating baths followed by one or more organic solvents to gradually replace all water with a substance that will be miscible with the embedding medium. For both TEM (Protocol 13.2) and SEM (Protocol 13.3) samples, the volume of solution needed to bathe the sample in each dehydration step should greatly exceed the size of the specimen (approximately 0.5 mL per 1 mm^3 block). Each dehydration bath lasts from 10 to 20 min. Steps are conducted at room temperature.

Dehydration is required for two reasons. The first is that the epoxy resins used as EM embedding media do not polymerize fully if contaminated with residual water, but rather remain sticky. Such blocks cannot be cut or salvaged. The second reason is that the sample chambers on EM instruments are held at a high vacuum; the violent extraction of residual water under such conditions can damage delicate tissues beyond repair. The most common dehydrating agents are alcohols (usually ethanol but sometimes methanol), acetone, or acetonitrile. Each agent has its own benefits and disadvantages. Ethanol is relatively innocuous and is available commercially in a variety of volumes and premade concentrations, but it readily removes lipids. Acetone removes water more readily than an alcohol but is more toxic. Acetonitrile preserves membrane phospholipids and makes an effective substitute for alcohol when lipid-rich structures are potential targets for developmental pathology.[8]

Dehydration for TEM Specimens

In general, TEM samples are transferred progressively through a graded series of solvents (usually starting at 50%), ending with several exchanges of 100% pure solvent. Wash durations of 15 or 20 min in

PROTOCOL 13.2 TISSUE PROCESSING FOR
TRANSMISSION ELECTRON MICROSCOPY

1. Wash osmium tetroxide (OsO_4) postfixed samples in ice-cold 0.1 M buffer (pH 7.4) two times for 15 min each.
   Optional: *En bloc* staining with uranyl acetate to provide extra contrast is done at this step in the following manner:
   a. Place the tissue in 2% uranyl acetate in 10% ethanol (see Protocol 13.4 for stain recipe) for 1 h at room temperature.
   b. Rinse the tissue in 50% ethanol one time for 5 min.
2. Dehydrate specimens by sequential incubations at room temperature (RT) in
   a. 50% ethanol (in distilled or deionized water)—15–30 min
   b. 70% ethanol—15–30 min
   c. 80% ethanol—15–30 min
   d. 95% ethanol—15–30 min
   e. 100% ethanol—15–30 min
   f. 100% ethanol—15–30 min
   g. 100% propylene oxide—5–10 min
   h. 100% propylene oxide—5–10 min
3. Infiltrate specimens at RT in
   a. Propylene oxide: epoxy resin (1:1 mixture)—1–2 h
   b. Propylene oxide: epoxy resin (1:2 mixture)—2 h to overnight
   c. Epoxy resin—2 h
   d. Epoxy resin—2 h
4. Embed in pure epoxy resin as follows:
   a. Place a label (precut to size) with identifying information in a commercially available capsule (e.g., BEEM®, available from many vendors) or flat silicon mold.
   b. Add approximately 0.5 mL of resin to each embedding mold.
   c. Place one specimen into each resin-filled mold.
   d. Using a wooden applicator, gently move the tissue to the tip (i.e., bottom) of the mold. Remove any air bubbles trapped in the resin, especially if they are located near the tissue.
   e. Fill the remainder of the mold with epoxy resin.
   f. Cure the resin by heating for at least 8 h at 60°C–70°C.

each dehydration stage ensure that all water has been replaced by the dehydrating agent. If the agent is ethanol, the last 100% ethanol step is usually followed by two passages of 20 min each through a transfer fluid like propylene oxide, which is more miscible with most EM embedding media. The reason that propylene oxide is not used for the entire dehydration process is because of its high extraction rate for cellular components.

The thin sections have already been enhanced by metal deposition during postfixation with OsO_4, but in many instances an additional heavy metal should be applied to further boost contrast (i.e., serve as a *stain*) and stabilize biomaterials so that they are not destroyed by the electron beam. The most common option is *en bloc* staining, in which a metal is applied via one of the ethanol washes. We usually perform this step using 1%–2% uranyl acetate in 10% ethanol followed by a single rinse in 50% ethanol before starting dehydration in 50% ethanol. Care must be taken not to overstain as too much metal will impede uniform infiltration by the epoxy resin, thereby yielding blocks of irregular hardness that are difficult or impossible to section.

Dehydration for SEM Specimens

For SEM, dehydration through graded ethanols is equivalent to the approach used for TEM, but the initial solvent step is 25% (rather than 50%) as higher concentrations may cause regional collapse of rapidly dehydrated tissues and promote precipitation of buffer salts on the surface of the specimens. The propylene oxide passages are omitted because the sample is to be subjected to critical point drying, which slowly and gradually displaces all remaining fluids (solvents and residual water) with liquid carbon dioxide (CO_2) while under high pressure. As the critical point is reached (i.e., a temperature and pressure where a clear distinction between liquid and gaseous phases ceases to exist), the liquid CO_2 is converted gently to the gaseous state and departs the specimen, leaving it completely desiccated.

Chemical alternatives to critical point drying may be used for very fragile samples. One example is hexamethyldisilazane (HMDS),[4,5] which can be evaporated to complete dryness in a fume hood without sample distortion.

Plastic Embedding for TEM

Specimens destined for TEM analysis are embedded in epoxy resin (i.e., hard plastic) for several reasons. The most important reflect the desirable physical properties of such media. Epoxy resins readily

---

**PROTOCOL 13.3   TISSUE PROCESSING FOR
SCANNING ELECTRON MICROSCOPY**

1. Wash the fixed samples three times for 15–60 min in ice-cold 0.1 M buffer (pH 7.4).
2. Dehydrate specimens by sequential 30 min incubations at room temperature (RT) in
   a. 25% ethanol (in distilled or deionized water)
   b. 50% ethanol
   c. 70% ethanol
   d. 95% ethanol
   e. 100% ethanol
   f. 100% ethanol
3. Perform critical point drying to fully desiccate the sample according to the vendor's directions. The specific steps vary with the manufacturer.
   a. Load samples, in small vessels and immersed in 100% ethanol, into the critical point drying chamber, being careful not to let them become exposed and begin drying during the transfer.
   b. Flush the chamber gently 3–8 times (depending on the number and size of samples) for 5–10 min with liquid carbon dioxide (CO_2) to remove the ethanol. When there is no more ethanol leaving the chamber, close all the portals to the chamber. Close the valve on the CO_2 tank.
   c. Turn on the chamber heater so that the temperature and pressure will start to gradually rise.
   d. Slowly increase the pressure in the chamber to 1200 psi. (The critical point of CO_2 is 31.1°C at 1072 psi.) Do not allow the pressure to exceed 1300 psi. On older dryers, the chamber may have to be vented to prevent this situation.
   e. Monitor the temperature until it reaches approximately 32°C (i.e., past the critical point).
   f. Vent the critical point dryer VERY slowly (10–12 min).
   g. Remove samples and store in a desiccator until observation.

penetrate properly fixed and dehydrated cells. Furthermore, upon polymerization these materials form extremely hard blocks that prevent damage to delicate subcellular structures during sectioning. In addition, epoxy resins resist the disintegrating effects induced by repeated exposure to the electron beam. Finally, epoxy resins are available as commercial kits, and the ingredients have very long shelf lives. Options include Araldite®, Epon®, and Spurr resin.[7] The main difference among them is the ease with which they may be sectioned: Araldite® tends to be softest, while Epon® is the hardest. All are used according to the manufacturer's instructions. If desired, resin mixtures may be used (e.g., Epon–Araldite) to alter the cutting properties of the polymerized blocks; mixtures yielding a medium hardness are suitable for almost all mouse developmental pathology specimens as they contain no or little hard tissues such as mineralized bone.[15] The choice of fixative may dictate the most appropriate choice of resin.[22]

Our usual choice for TEM during mouse developmental pathology studies is to process fixed tissues into epoxy resin (e.g., Eponate 12™; Ted Pella, Inc., Redding, CA). The specimen is infiltrated for 2–4 h at RT in 100% polymer and then moved to fresh resin for overnight incubation at RT. A single specimen accompanied by a small identifying label (paper with pencil- or printer-generated markings) is then placed in an embedding capsule, and the capsule is filled with epoxy resin (Protocol 13.2). If a specific orientation is desired, flat silicon molds are available from most vendors of EM supplies. Samples are placed in the molds and oriented, under a dissecting microscope, using an instrument such as a thin tungsten wire mounted on a wooden applicator stick. Capsules/molds are then placed in an oven at 60°C to polymerize for 48 h.

Cutting and Staining TEM Sections

Polymerized blocks are sectioned on an ultramicrotome equipped with a diamond or glass knife. Acquisition of suitable sections requires considerable experience and ample reserves of patience.

The TEM evaluation of biological specimens is conducted in two stages. The blocks are first screened by cutting one or more *semi-thin* sections (1–2 μm thick) to assess whether or not the desired structures will be available for examination. In general, these sections are stained with toluidine blue. Histopathologic resolution of these sections is increased over that which is available using standard *thick* paraffin sections (5–8 μm thick), so these semi-thin sections often are subjected to histopathological analysis at the light microscopic level as an adjunct to the TEM analysis. Once a block is certified to contain the region of interest, the block face is further trimmed and multiple *ultrathin* sections (generally 75–90 nm) are cut. These sections are collected off the surface of the water in the knife trough onto copper or nickel grids, placed on filter paper in a Petri dish, and allowed to air-dry prior to staining.

A heavy metal stain is applied to the grid-mounted sections at this time (Protocol 13.4). The usual procedure, especially if a heavy metal was not applied *en bloc* during the dehydration stage, is to use uranyl acetate and/or lead citrate. A suitable protocol for biological specimens is to stain at RT in 2% uranyl acetate (aq) for 3–10 min, followed by three rinses for 30 s each in distilled/deionized water. Grid-mounted sections then are stained again with Reynolds lead citrate for 3–4 min, followed by three more 30 s rinses in distilled/deionized water. Fully stained grids are air-dried and then placed in grid boxes or small capsules so that they remain clean and undamaged until imaging.

Contrast Enhancement for SEM Samples

Desiccated samples are mounted on metal stubs using a conducting glue or tape. Mounted specimens are covered by a thin layer of heavy metal atoms, generally gold or gold and palladium, in a process referred to as *sputter coating* (Protocol 13.5). This step is required to improve the ability of the tissue to conduct electricity and emit secondary electrons. Biological specimens that lack such a coating absorb electrons and consequently will suffer visible structural damage at the cellular and eventually the gross level.

PROTOCOL 13.4 HEAVY METAL TREATMENTS FOR USE WITH TISSUES DESIGNATED FOR TRANSMISSION ELECTRON MICROSCOPY

Caution! Heavy metals are toxic, and uranyl acetate is mildly radioactive. Personal protective equipment should be worn at all times.

Reynolds lead citrate (for staining grids)

Reagents

Lead nitrate	1.33 g
Sodium citrate	1.76 g
Sodium hydroxide (NaOH), 1 N	8.0 mL
Distilled or deionized water	~45 mL
Triton® X-100 (non-ionic surfactant)	

Preparation of Lead Citrate Staining Solution

1. Perform all steps at room temperature (RT) in a fume hood.
2. Dissolve solid reagents.
   a. Add lead nitrate to 15 mL of water.
   b. Add sodium citrate to 15 mL of water.
3. Mix lead nitrate and sodium citrate solutions in a 50 mL volumetric flask.
4. Add 8.0 mL of 1 N NaOH and agitate until all solids have dissolved.
5. Dilute to 50 mL with distilled water.
6. Add three drops of Triton® X-100, mix thoroughly, and filter.
7. The solution should be clear. (If not, centrifuge in capped tubes at a slow speed to clear.)

Storage: This fixative has a shelf life of 3 months when stored at 4°C in an air-tight glass bottle. (Exposure to air reduces the stability.)

Adapted from Edward S. Reynolds, "The use of lead citrate at high pH as an electron-opaque stain in electron microscopy," *J Cell Biol* **17** (1963): 208–212

Lead Citrate Staining Method

Perform all steps at RT in a fume hood.

1. *Prepare the staining chamber.*
   a. Place several pellets of NaOH at one side of a plastic Petri dish and replace the lid to create a low-CO_2 environment.
   b. Remove the lid of the dish. Using a pipette, place one or more drops of lead citrate staining solution on the floor of the Petri dish (one drop for each grid that is to be stained). Close the lid.
2. *Prepare the rinsing chamber.* Remove the lid of a second Petri dish. Using a pipette, place multiple drops of distilled or deionized water on the floor of the Petri dish (three drops for each grid that is to be stained). Close the lid.
3. *Stain the grid.* Raise the lid of the staining chamber. With a pair of fine forceps, float a grid—section side down—on a drop of stain. Close the lid. Incubate for 3–5 min.
4. *Rinse the grid.*
   a. Raise the lid of the staining chamber. With a pair of fine forceps, retrieve the grid. Close the lid.
   b. Raise the lid of the rinsing chamber. With a pair of fine forceps, float a grid—section side down—on a drop of water. Rinse for 30 s in three drops of water.
   c. Retrieve the grid, and close the lid.

5. Gently draw off the water from the grid with the edge of a pointed piece of filter paper. Do not touch the section with the paper.
6. Place the grid on a filter paper (taking care that the section not touch the paper) in a covered dish and allow to air-dry at RT.
7. Store grids in a grid box or small capsule at RT.

1% Uranyl acetate (for staining grids)

Reagents

Uranyl acetate	0.5 g
Distilled or deionized water	~50 mL
Triton® X-100 (non-ionic surfactant)	

Preparation of 1% Uranyl Acetate Staining Solution

1. Perform all steps at room temperature (RT) in a fume hood.
2. Dissolve uranyl acetate in water on a stir plate.
3. Add three drops of Triton® X-100, mix thoroughly, and filter.
4. The solution should be clear. (If not, centrifuge in capped tubes at a slow speed to clear.)

Storage: This fixative has a shelf life of 3 months when stored at 4°C in a light-tight bottle. The solution should be discarded if a precipitate forms.

1% Uranyl Acetate Staining Method

Perform all steps at RT in a fume hood.

1. *Prepare the staining chamber.* Remove the lid of the dish. Using a pipette, place one or more drops of uranyl acetate staining solution on the floor of the Petri dish (one drop for each grid that is to be stained). Close the lid.
2. *Prepare the rinsing chamber.* Remove the lid of a second Petri dish. Using a pipette, place multiple drops of distilled or deionized water on the floor of the Petri dish (three drops for each grid that is to be stained). Close the lid.
3. *Stain the grid.* Raise the lid of the staining chamber. With a pair of fine forceps, float a grid—section side down—on a drop of stain. Close the lid. Incubate for 3–5 min.
4. *Rinse the grid.*
   a. Raise the lid of the staining chamber. With a pair of fine forceps, retrieve the grid. Close the lid.
   b. Raise the lid of the rinsing chamber. With a pair of fine forceps, float a grid—section side down—on a drop of water. Rinse for 30 s in three drops of water.
   c. Retrieve the grid, and close the lid.
5. Gently draw off the water from the grid with the edge of a pointed piece of filter paper. Do not touch the section with the paper.
6. Place the grid on a filter paper (taking care that the section not touch the paper) in a covered dish and allow to air-dry at RT.
7. Store grids in a grid box or small capsule at RT.

2% Uranyl acetate in 10% ethanol (for *en bloc* staining)

Reagents

Uranyl acetate	1.0 g
10% ethanol (in distilled or deionized water)	50 mL

Preparation of 2% Uranyl Acetate Staining Solution

1. Perform all steps at room temperature (RT) in a fume hood.
2. Dissolve uranyl acetate in 10% ethanol on a stir plate.
3. Filter before use.
4. The solution should be clear. (If not, centrifuge in capped tubes at a slow speed to clear.)

Storage: This fixative has a shelf life of 3 months when stored at 4°C in a light-tight bottle. The solution should be discarded if it precipitates.

The 2% uranyl acetate staining method is given above Step 1 of in Protocol 13.2.

PROTOCOL 13.5 METAL ENHANCEMENT (SPUTTER COATING) FOR USE WITH SPECIMENS DEDICATED FOR SCANNING ELECTRON MICROSCOPY

1. Attach the desiccated specimen to a clean aluminum SEM stub using a conductive adhesive (available from multiple vendors).
2. Use sputter coating to cover the specimen(s) with a gold/palladium layer (20–40 nm thick). The particular procedure varies with the manufacturer of the equipment. In general, the following steps are performed:
   a. Insert the samples (up to 8 per coating run) into the coating chamber and initiate a vacuum.
   b. Flush the chamber several times with pure argon gas, leaving a specified argon atmospheric pressure (in most instances approximately 5.25^{-2} torr).
   c. Energize the gold/palladium target and adjust the current to ~17 mA to begin the coating process.
   d. Coating only takes 120 s.
3. Release the pressure and vent the coating chamber slowly (1–2 min).
4. Remove coated samples and store in covered boxes (to prevent dust contamination) in a bell jar containing desiccant under mild vacuum until use.

Final Thoughts

Electron microscopy is a powerful tool for detailed analysis of biological specimens. However, these techniques are expensive, labor intensive, and prolonged. Furthermore, considerable experience is required before useful data may be acquired. Accordingly, these methods usually function as an ancillary approach that provides an extra level of analysis in lesions first observed during the gross examination or by histopathology. In our experience, electron microscopy should not be used to assess tissues from developing animals if tissue alterations are not evident by light microscopy.

REFERENCES

1. Allen TD (2008). *Introduction to Electron Microscopy for Biologists*. Burlington, MA: Academic Press (Elsevier).
2. Bechtold LS (2000). Ultrastructural evaluation of mouse mutations. In: *Systematic Approach to Evaluation of Mouse Mutations* (Sundberg JP, Boggess D, eds.). Boca Raton, FL: CRC Press, pp. 121–129.
3. Bozzola JJ, Russell LD (1998). *Electron Microscopy*, 2nd edn. Sudbury, MA: Jones & Bartlett Publishers.
4. Braet F, De Zanger R, Wisse E (1997). Drying cells for SEM, AFM and TEM by hexamethyldisilazane: A study on hepatic endothelial cells. *J Microsc* **186** (Pt 1): 84–87.

5. Bray DF, Bagu J, Koegler P (1993). Comparison of hexamethyldisilazane (HMDS), Peldri II, and critical-point drying methods for scanning electron microscopy of biological specimens. *Microsc Res Tech* **26** (6): 489–495.

6. Cheville NF (2009). *Ultrastructural Pathology: The Comparative Cellular Basis of Disease*, 2nd edn. Hoboken, NJ: Wiley-Blackwell.

7. Coutinho AR, Mendes CM, Caetano HV, Nascimento AB, Oliveira VP, Hernadez-Blazquez FJ, Sinhorini IL, Visintin JA, Assumpção ME (2007). Morphological changes in mouse embryos cryopreserved by different techniques. *Microsc Res Tech* **70** (4): 296–301.

8. Edwards HH, Yeh YY, Tarnowski BI, Schonbaum GR (1992). Acetonitrile as a substitute for ethanol/propylene oxide in tissue processing for transmission electron microscopy: Comparison of fine structure and lipid solubility in mouse liver, kidney, and intestine. *Microsc Res Tech* **21** (1): 39–50.

9. Ghadially FN (1997). *Ultrastructural Pathology of the Cell and Matrix,* 4th edn. Burlington, MA: Butterworth-Heinemann.

10. Ghadially FN (1998). *Diagnostic Ultrastructural Pathology: A Self-Evaluation and Self-Teaching Manual*, 2nd edn. London, U.K.: Hodder Arnold.

11. Hayat MA (2000). *Principles and Techniques of Electron Microscopy: Biological Applications*, 4th edn. Cambridge, U.K.: Cambridge University Press.

12. Herpers B, Xanthakis D, Rabouille C (2010). ISH-IEM: A sensitive method to detect endogenous mRNAs at the ultrastructural level. *Nat Protoc* **5** (4): 678–687.

13. Kondo S (1998). Microinjection methods for visualization of the vascular architecture of the mouse embryo for light and scanning electron microscopy. *J Electron Microsc (Tokyo)* **47** (2): 101–113.

14. Kuo J (2007). *Electron Microscopy: Methods and Protocols*, 2nd edn. (*Methods Mol Biol*, Vol. 369). Totowa, NJ: Human Press.

15. Lynch CM, Johnson J, Vaccaro C, Thurberg BL (2005). High-resolution light microscopy (HRLM) and digital analysis of Pompe disease pathology. *J Histochem Cytochem* **53** (1): 63–73.

16. Macville MV, Van Dorp AG, Wiesmeijer KC, Dirks RW, Fransen JA, Raap AK (1995). Monitoring morphology and signal during non-radioactive *in situ* hybridization procedures by reflection-contrast microscopy and transmission electron microscopy. *J Histochem Cytochem* **43** (7): 665–674.

17. McDowell E, Trump B (1976). Histological fixatives for diagnostic light and electron microscopy. *Arch Pathol Lab Med* **100** (8): 405–414.

18. Morel G, Cavalier A, Williams L (2001). *In Situ Hybridization in Electron Microscopy*. Boca Raton, FL: CRC Press.

19. Reynolds ES (1963). The use of lead citrate at high pH as an electron-opaque stain in electron microscopy. *J Cell Biol* **17**: 208–212.

20. Schwartzbach SD, Osafune T (2010). *Immunoelectron Microscopy: Methods and Protocols*. Totowa, NJ: Humana Press.

21. Simmons DG, Cross JC (2005). Determinants of trophoblast lineage and cell subtype specification in the mouse placenta. *Dev Biol* **284** (1): 12–24.

22. Smart Y, Millard PR (1985). Different fixative-resin combinations and the immunocytochemical demonstration of immunoglobulins (kappa light chains) in semi-thin sections. *Histochem J* **17** (12): 1337–1345.

23. Swartzendruber DC, Burnett IH, Wertz PW, Madison KC, Squier CA (1995). Osmium tetroxide and ruthenium tetroxide are complementary reagents for the preparation of epidermal samples for transmission electron microscopy. *J Invest Dermatol* **104** (3): 417–420.

24. Webster P, Schwarz H, Griffiths G (2008). Preparation of cells and tissues for immuno EM. In: *Introduction to Electron Microscopy for Biologists* (*Meth Cell Biol*, Vol. 88; Allen TD, ed.). San Diego, CA: Academic Press (Elsevier), pp. 45–58.

14

Three-Dimensional Imaging in Mouse Developmental Pathology Studies

Brad Bolon, Kathleen Gabrielson, Sara Cole, Olga N. Kovbasnjuk, Kimerly A. Powell, and David Weinstein

CONTENTS

Historically, the foundation for assessing lethal phenotypes in developing mice has been assessment for structural defects and, to a lesser degree, functional deficiencies. Routine techniques for characterizing anatomic abnormalities include qualitative macroscopic and light microscopic examinations, supplemented as needed by electron microscopy (Chapter 13) and quantitative methods to measure the dimensions,[7,13] volume,[89] or cell numbers[12] in particular structures (Chapter 12). The primary difficulties with such classic approaches to structural analysis are that a thorough assessment of any given organ is (1) quite laborious due to the requirement for extensive interval (*step*) or consecutive (*serial*) sectioning, which in turn (2) necessitates destruction of the physical specimen to produce a series of two-dimensional (2D) levels showing the structure in a specific planar orientation. Therefore, a complete appreciation of anatomic lesions traditionally has entailed the tedious fabrication of a three-dimensional (3D) construct by mental or physical realignment (*stacking*) and merging of multiple 2D images.

In recent years, the application of noninvasive imaging to mouse developmental pathology has exploded with the emergence of instruments and procedures capable of providing the high resolutions needed to view the tiny organs of embryos and fetuses. Such techniques have several substantial advantages over routine macroscopic and microscopic evaluations. The main advantage is that noninvasive imaging permits structural analysis without destruction (and generally with minimal distortion) of the sample. Furthermore, some noninvasive imaging modalities may be employed to follow the progression of developmental events over time. Third, the output is inherently both 3D and digital, which greatly facilitates stacking, automated analysis, and data sharing. Certain imaging techniques also permit rapid mapping of molecular locations in 3D. Finally, the time needed to render a complete 3D representation of the entire specimen is substantially reduced because extensive sectioning and slide-by-slide comparisons are unnecessary; indeed, many mouse conceptuses may be processed in parallel.[6,93] These imaging techniques are unlikely to replace traditional optical (microscopic) methods, but instead will serve as ever more useful complementary methods, especially for high-throughput phenogenomic analysis.[11,82,93]

This chapter introduces common noninvasive or minimally invasive methods for the nondestructive 3D evaluation of structure and function in developing mice (Table 14.1). The text emphasizes theoretical principles and references for exploring the utility of these techniques rather than giving detailed experimental protocols for their use. This strategy was selected for two related reasons: (1) most investigators will utilize such tools in consultation with experts in their institution's imaging facility, who will provide current recommendations on sample processing, and (2) the rapid evolution of such technologies will likely result in substantial procedural changes in the very near future.

Classic Tomography: Physical Methods

Tomography is the study of internal body structures in 3D. Such assessments involve the evaluation of spatial relationships along three distinct axes (x, y, and z) at the same time. By convention, the x-axis and y-axis are used for the height and width dimensions that are evident in a standard 2D plane (such as a tissue section). The z-axis is associated with specimen depth and is typically ignored in conventionally processed samples due to the thinness of the section (1–8 μm, depending on the choice of embedding medium).

Classic developmental pathology methods for tomographic analysis rely on physical manipulation of specimens. For example, renal development has been assessed in 3D by reconstruction of serial histological sections,[27] electron microscopy,[20] or microdissection of individual nephron segments.[46–49] The most common method, serial sectioning, leads to the exhaustive slicing of embedded tissues to create a series of thin 2D levels in which the internal organs are exposed for evaluation; the levels are then stacked to form a 3D reconstruction of the original sample.[61] A primary problem in such evaluations is to attain correct alignment of adjacent sections during reassembly of the specimen. The typical approach to counter this difficulty is to anchor one or more external markers in the block near the sample. These markers are sectioned along with the tissue, and thus can be used as reliable reference points to ensure that anatomic features in sections are congruent.[4,76,87]

Even with the advent of digital imaging, such reconstructions are quite laborious if the images are acquired as a series of individual files. Some labor-saving innovations are episcopic microscopy[89] and surface imaging microscopy,[21] in which the block face is photographed digitally after each stroke of the microtome. As long as the camera is not repositioned, the images can be aligned automatically using software. However, the specimen still has to be destroyed by sectioning to acquire data along the z-axis.

These physical methods produce tissue sections that are sufficiently thin for light microscopic analysis. Thus, internal structures may be viewed at very high resolutions (1 μm or below) regardless of the subject's developmental age. In addition, conventional histopathology permits sections to be stained with many different dyes (Chapter 9) or labeled to demonstrate the presence of one or more genes (Chapter 10) and/or proteins (Chapter 11). The high resolution coupled to the flexible processing options will sustain the traditional importance of such physical sectioning methods in mouse developmental pathology studies.

Other physical methods that may be used to assess internal structures of developing mice are stereomicroscopy (Chapter 12)[96] and scanning electron microscopy (SEM; Chapter 13).[77] Stereological analysis may be performed using either bright-field or dark-field[96] light microscopic or electron microscopic images. Two primary problems must be faced when using stereomicroscopy for 3D analysis. The first is that internal structures may be obscured by the progressive opacity of deep tissues over time; the technique works best with translucent tissue and thus is most useful at early gestational ages (gestational day [GD] 10.5 or earlier). More importantly, the large specimens are too thick to provide clear focus on all visceral planes at once, so bigger organs cannot be evaluated except by focusing up and down. Digitized images of such structures will be blurred, and thus are often not useful for 3D reconstruction. An alternative suitable for both stereomicroscopy[96] and SEM[33] is to microdissect fixed specimens to expose internal organs for regional analysis.

TABLE 14.1

Comparison of Common Three-Dimensional Imaging Modalities Available for Use in Mouse Development Pathology Studies

	Conventional Sectioning	CLSM	MPM	MRM	μCT	OPT	UBM	US
Modality physics	None	Fluorescence	Fluorescence	Magnetic field	X-rays	Fluorescence	Sound waves	Sound waves
Invasiveness	*Ex vivo*	*Ex vivo*	*Ex vivo*	*Ex vivo*	*In situ or ex vivo*	*Ex vivo*	*In situ*	*In situ*
Destructiveness	Yes	Maybe	No	No	No	No	No	No
Simultaneous analysis of multiple specimens	Yes	No	No	Yes	Yes	No	No	No
Progressive assessment	No	No	Maybe	No	No	No	Yes	Yes
Time to 3D image	Days	1–15 min	5–100 min	30–60 min	15 min	30 min	5–10 min	5–10 min
Typical resolution (μm)	0.5	0.4–0.8	0.4–0.8	20–40	25–50	5	30	200
Structure analysis	Yes	Yes	Yes	Yes	Yes	Yes	Yes	Yes
Functional analysis	No	No	No	No	No	No	Yes	Yes
Quantitative analysis	Yes	Yes	Yes	Yes	Yes	Yes	Yes	Yes
Gene expression	Yes	Yes	Yes	No	No	Yes	No	No
Instrumentation cost	+	+	+++	+++	++	+++	+	+
Anesthesia required	No	No	No	No	No	No	Yes	Yes

Notes: CLSM, confocal laser scanning microscopy; min, minutes; MPM, multiphoton microscopy; MRM, magnetic resonance microscopy; μCT, microscopic computer-assisted tomography; OPT, optical projection tomography; UBM, ultrasound biomicroscopy; US, (conventional) ultrasound.

Optical Tomography

The invention of optical microscopy techniques is one recent solution to the problems posed by physical sectioning methods. Some common optical tomography approaches include confocal laser scanning microscopy (CLSM; Figure 14.1),[94,96] multiphoton microscopy (MPM),[56] and optical projection tomography (OPT).[70,79] Optical tomography is appropriate for fluorescing or reflecting tissues, but it is best suited to tissues with little autofluorescence that have been prelabeled with fluorophore-conjugated reagents to reveal the localized distribution of cells or molecules. The use of optical clearing techniques in OPT can be employed to increase resolution and contrast, provided that the stability of the fluorophore is preserved during processing.[23] The judicious use of fluorescent proteins and cellular dyes may either obviate the need for subsequent immunohistochemistry (IHC) or be undertaken in combination with IHC to increase the number of targets identified in a single embryo. An advantage of using fluorophore markers is that they may be introduced into cells or tissue prior to fixation. For example, a recent study treated green fluorescent protein (GFP)-positive, fixed tissue samples with an optical clearing agent followed by MPM to achieve a much greater depth of imaging in brains of mouse embryos.[26] However, care must be taken when choosing cellular dyes for application prior to fixation since some will be lost during the fixation step (e.g., LysoTracker Red®; Life Technologies, Grand Island, NY).

These methods employ tightly focused beams of light to probe the deep structures of biological specimens in a nondestructive manner. Therefore, optical tomography techniques typically require that samples be translucent, so the methods are more suited to small embryos (GD 11.5 or earlier). Infrared (IR) light (longer wavelength, so lower energy) used for excitation in MPM permits deeper penetration into tissues relative to the visible light (shorter wavelengths, so higher energy) used in CLSM. In addition, IR light is less susceptible to light absorption and backscattering, thereby producing less photobleaching and phototoxicity in MPM when compared to CLSM.[45] These optical characteristics together with the growing numbers of novel genetically encoded fluorescent protein tags[10,25] make MPM a preferred imaging modality for studying cellular detail in living embryos.[5] The OPT approach minimizes the impact of scattered light on image resolution by suspending the specimen in a fluid that has a similar

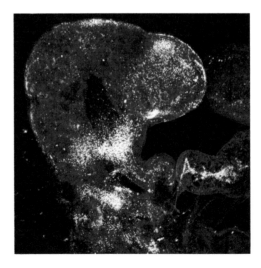

FIGURE 14.1 Confocal laser scanning microscopy (CLSM) reconstruction of an ethanol-exposed gestational day (GD) 10 mouse embryo derived from 30 CLSM sections, demonstrating abundant cell death in the prosencephalon, optic stalk, otic vesicle, and ganglia of cranial nerves V–X. Bar = 200 μm. Isolated embryos were stained with LTR (Lysotracker Red; Molecular Probes, Eugene, OR). (Dunty Jr., W.C., Chen, S.-Y., Zucker, R.M., Dehart, D.B., Sulik, K.K.: Selective vulnerability of embryonic cell populations to ethanol-induced apoptosis: Implications for alcohol-related birth defects and neurodevelopmental disorder. *Alcohol. Clin. Exp. Res.* 2001. 25(10). 1523–1535. Copyright Wiley-VCH Verlag GmbH & Co. KGaA. Reprinted with permission of the Research Society on Alcoholism [color image courtesy of Dr. K.K. Sulik, University of North Carolina, Chapel Hill, NC].)

optical index. Both MPM and OPT can detect nonfluorescent markers such as the chromagens used in conventional enzyme histochemistry (e.g., bacterial β-galactosidase [lacZ]) and immunohistochemistry (e.g., 3,3'-diaminobenzidine [DAB]).

The ability to scan the deep tissues of intact specimens without physical thin sectioning allows the use of thicker slabs (up to 100 μm for CLSM) or even thick blocks (1 mm or more for MPM, 1 cm or greater for OPT), thus preserving structural integrity with less distortion than typically occurs in conventional histological sections. Modern CSLM objectives with longer working distances allow structures to be scanned over depths up to several hundred microns.[28] However, in practice the significant inherent light scattering of biological tissues limits the depth of confocal imaging to 100 μm when a high degree of spatial resolution is required.[64]

The exquisite focus of optical sections depends on several principles. One is to vary the amount of tissue that is probed by a single scanning pass. Typically, the light beam is moved (*rastered*) across or around the sample to collect data from numerous x–y points in *one-at-a-time* fashion, but the character of the scan in the z-dimension varies with the technique. For example, OPT scans a single line along the entire z-axis (*1 dimension*), calculating distance as the time it takes for light to travel down and back up from an opaque object in the specimen. In contrast, CLSM and MPM sample the tissue at a single point along the z-axis (effectively *0 dimensions*, although the point is really a minute 3D volume). An additional means of fine-focus control is to incorporate a mechanism to reduce the detection of extraneous light. For CLSM, a pinhole aperture is used to exclude light from all tissue layers above or below the desired focal plane; contrariwise, MPM dispenses with the confocal aperture[16] and instead requires the simultaneous action of more than one photon with a fluorescent molecule at a discrete spot in the focal plane to incite a localized emission of light with essentially no excitation elsewhere along the beam.[90] The elimination of extraneous light in MPM can render this technique more sensitive than CLSM when analyzing large, heterogeneous structures, with the added benefit that photobleaching of the sample is minimal.[56]

Once light reaches the detector, automated software gathers information about the tissue-specific refraction of light. A deconvolution algorithm based on the presumptive light scatter properties is then employed to digitally remove *noise* (produced by the random backscatter of light) from the data. The amount of noise in large specimens may interfere with the clarity of the image, so bigger specimens must be analyzed with more tightly focused beams (e.g., CLSM). Many individual measurements are rapidly summed to build a 2D or 3D image of the specimen. An important advance is that software can be used to project the entire collection of stacked optical images into a single digitized image, so that structures throughout the tissue can be viewed in focus at the same time.

The exact optical resolution of optical tomographic methods depends on many factors. For example, important characteristics for CLSM include the numerical aperture of the objective lens; the refractive indices of the immersion medium, coverslip, mounting medium, and specimen; and the excitation and emission wavelengths (which vary with the fluorescent marker). Typical resolutions reported in the literature for instruments used in mouse developmental pathology studies are 0.4–0.8 μm for CLSM,[95] and MPM,[59,90] and 5 μm for OPT.[62] Note that the axial or *z* resolution of a CLSM instrument varies with the optical system and can be influenced by a number of factors including (but not limited to) proper alignment of the optical system components, the magnification and numerical aperture of the objective lens, and the size of the pinhole (focal point for scanning). For this reason, the z-resolution should be determined for each confocal imaging system using standard samples (typically fluorescent beads) prior to analyzing critical samples.[95]

Wave-Based Tomography

The adaptation of medical diagnostic imaging techniques to basic biomedical research is another recent innovation in mouse developmental pathology analysis. These methods use various energy waves to illuminate the margins among tissues with different physical properties. The most common applications for mouse phenotyping are cardiography (undertaken with a conventional ultrasound [US] instrument),[91] magnetic resonance microscopy (MRM),[58] microscopic computer-assisted tomography (μCT),[32,36,37] and ultrasound biomicroscopy (UBM). Though these methods lack the ability to readily capture concomitant molecular expression data, they possess a strong potential for high-throughput experimental designs focused on gross anatomical form.[32,51] Other methods used in human clinical practice, such as positron emission tomography (PET), single-photon emission computed tomography (SPECT), and optical imaging, do not provide sufficient resolution at this time for use with developing mice.

Different physical principles underpin these wave-based tomography methods. For example, MRM relies on quantum-level mechanical changes in atomic nuclei positioned within a strong magnetic field, while μCT depends on tissue-specific variations in the passage of x-rays. In contrast, both US and UBM employ reflected sound waves to outline deep tissues. The sharpness of embryonic and fetal images is directly related to the energetic principle, as the sound-based technologies generally provide images with lower image quality. However, major advantages of the sound-based platforms are that they are rapid and do not disrupt cellular biochemistry (as in MRM) or gene integrity (as in μCT), and thus may be used longitudinally to take multiple measurements throughout the course of development.

Conventional Ultrasound

As implied in the name, conventional US (Figure 14.2) may be utilized for *in vivo* evaluation of heart structure and function.[65] This approach complements *ex vivo* techniques that require isolation of the conceptus, such as MRM,[72,73] SEM,[85,86] and trans-illumination microscopy.[18,81] Cardiac activities that can be

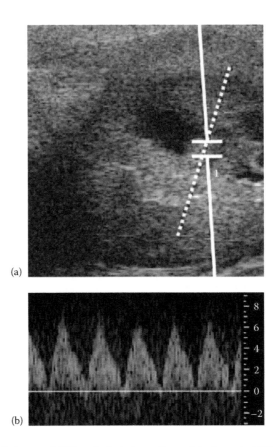

(a)

(b)

FIGURE 14.2 Conventional ultrasound (US) analysis (*echocardiography*) of a control GD 10 mouse embryo. Panel (a) is a conventional US image. The lines cross over the umbilical cord, which connects the placenta (half-circle) below and the irregular embryo above. The US beam (dotted line) was parallel to the blood flow in the umbilical vessels, while the probe (solid line) crossed the umbilical cord at a shallow angle (θ; <60°). Panel (b) depicts the regular heart contractions recorded by the US probe.

assessed by US in mouse embryos, fetuses, and neonates include changes in cardiac output,[57] heart wall compliance,[57] and intravascular blood flow [34,57,78] as well as valve regurgitation and abnormal connectivity between heart chambers.[57,71,92]

Commercially available US machines readily measure murine heart function beginning about GD 14.5,[74] while ultra-high-frequency probes on specially constructed instruments may identify heart contraction at GD 8.5.[75] The resolution of US images in embryonic mice is 200–500 μm, which for practical purposes restricts US in mouse developmental pathology to conceptuses at GD 12.5 or older. This resolution may be attained with scanning times of 15–20 min.[38]

Application of this method to developing mice presents many technical challenges. The most prominent physiological difficulties are the animals' small size (usually 15–20 mm at birth); rapid but variable heart rate (between 180 and 200 beats per minute at GD 14.5, which is slower than a typical adult mouse); and the tendency for many external factors to alter the heart rate (e.g., anesthesia, temperature). In addition, appropriate orientation of the conceptus may be difficult to achieve *in utero*, which can preclude functional (M-mode) measurements. One solution to the latter problem is to exteriorize the uterus of the anesthetized dam to ensure accurate positioning of the progeny,[29,57] but the ramifications of such manipulations on the physiology of the conceptus are not well characterized. Finally, the blood of developing mice may produce its own echoes during mid-gestation due to the large proportion of nucleated erythrocytes. This tendency is reported to peak about GD 13.5,[29,57] which makes differentiation between blood cells and the heart walls quite complex prior to GD 15.5.

Magnetic Resonance Microscopy

Conventional (*structural*) MRM delineates the margins between adjacent tissues based on divergence in their water content and the physical properties of macromolecular–water interactions. Functional MRM employs contrast associated with different physiological states to define soft tissue borders.[42,60,83] The physiological basis for inter-tissue contrast includes local variations in hemodynamic properties (e.g., relative blood flow, blood volume)[42] or molecular constituents (e.g., blood protein levels, oxygenation).[60,83] Therefore, MRM generally yields greater contrast between different soft tissues, making it especially useful for identifying the boundaries of viscera in the cardiovascular, digestive, nervous, pulmonary, and urogenital systems.

The principle behind MRM involves the detection of energy fluctuation among atomic nuclei. Application of a strong magnetic field causes unpaired protons (hydrogen nuclei, especially in water molecules) to align, after which radiofrequency (RF) fields are used to systematically alter the alignment by elevating nuclei to a high energy state. Over time, the excited nuclei will emit RF waves and lose energy. The abrupt transitions between energy states differ depending on the organ composition, and such tissue-specific variations can be used to define internal structures. Magnetic field gradients are used to map distinct spatial locations within the specimen. Tissue-specific differences in contrast can be emphasized by different image-weighting schemes (e.g., T1, T2, proton density). The weighting is achieved by altering the duration and/or temporal sequence in which the sample is excited and scanned.[35]

The primary drawback preventing more widespread application of MRM in mouse developmental pathology is the expense associated with high-field-strength research magnets. Instruments used for MRM applications have field strengths that are 5–10 times higher than those of diagnostic magnetic resonance imaging (MRI) systems used in clinical medicine.[35] This enhanced power permits resolutions over a hundred times greater than those for diagnostic MRI instruments,[31] which is absolutely necessary when evaluating specimens as small as mouse embryos.

Remarkably detailed images are possible with high-resolution MRM (Figure 14.3). However, in addition to high field strengths, small RF coils and fast gradients also are required to obtain high-resolution MRM images while minimizing the scan time.[30] Additionally, MRM sensitivity may be greatly enhanced using contrast agents[2] or stains.[54] Literature reports indicate that MRM assessment of mouse conceptuses generally will require magnetic field strengths of 9 tesla (T) or higher and scanning times of 8 h or more to achieve a resolution of approximately 20 μm.[67,68,73] Newer MRM techniques permit simultaneous imaging of multiple embryos with no loss of resolution.[69]

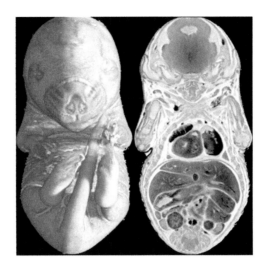

FIGURE 14.3   Magnetic resonance microscopy (MRM) volume-rendered composite (left) and *dissected* image of a GD 17.5 mouse fetus. The image was acquired at 20 μm using a magnetic field strength of 9.4 T. (Courtesy of Dr. G. Allan Johnson, Center for *In Vivo* Microscopy, Duke University, Durham, NC [production of which was supported by NIH P4105959]; Petiet, A.E. et al., *Proc. Natl. Acad. Sci. U.S.A.*, 105(34), 12331, 2008.)

Microscopic Computer-Assisted Tomography

The aim of μCT in phenotypic analysis of developing mice is to examine and, if feasible, quantify morphological variation,[8,51,66] often in a high-throughput setting.[41] Since μCT distinguishes among tissues based on their differing densities and chemical compositions, this modality offers exceptional contrast between hard and soft tissues. Therefore, μCT is an ideal choice for evaluating bone.[44,84] The modality is somewhat more effective when applied to live-born neonates and juveniles, where the presence of air in the lungs and (after suckling) gases in the digestive tract usually improves the ability to discern between viscera in the thorax and abdomen relative to those of unborn conceptuses. Nevertheless, μCT has been successfully used in fixed mouse embryos stained *en bloc* with osmium tetroxide.[1,32] The technique can also be employed for 3D quantitative analysis of macroscopic features.[66]

Application of μCT to mouse developmental pathology studies offers several benefits compared to other wave-based imaging modalities. Instruments are relatively inexpensive and simple to operate, and multiple embryos may be scanned at once. The overall image quality attainable using μCT (Figure 14.4) exceeds that obtained with other inexpensive imaging modalities, such as US (Figure 14.2) and UBM (Figure 14.5). However, as with MRM, better μCT image quality in terms of contrast and spatial resolution is dependent upon a number of factors, particularly acquisition time. For example, μCT imaging of mid-gestation (GD 9.5 to GD 12.5) mouse embryos acquired at a spatial resolution of 25 μm would require a 2 h scan time, whereas an image resolution of 8 μm would necessitate a significantly longer acquisition time of 12 h.[32]

Ultrasound Biomicroscopy

This modality permits the *in utero* assessment of organ development in living conceptuses. Published *in vivo* uses include functional evaluation of the developing heart[29,57] and arteries[39] during gestation and also after birth[14], progressive assessment of ocular development[22], and morphometric assessment of placental growth.[40] The orientation of UBM images from developing mice can be readily matched with those in routine tissue sections,[22] but UBM images (Figure 14.5) typically are more grainy relative to scans acquired using other methods of 3D reconstruction such as MRM (Figure 14.3) and μCT (Figure 14.4). The typical resolution provided by UBM is 30 μm, which requires a scanning time of approximately 5–10 min. In general, UBM works best for mouse embryos at E7.5 or older. Nonetheless, the ability to

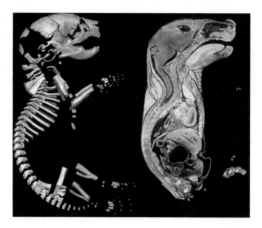

FIGURE 14.4 Microscopic computer-assisted tomography (µCT) images of a neonatal mouse (postnatal day [PND] 0) demonstrating the ability to differentially emphasize the skeleton (left panel, unstained specimen) or soft tissues (right panel, sample processed with the proprietary Virtual Histology™ staining protocol [Numira Biosciences, Inc., Salt Lake City, UT]).

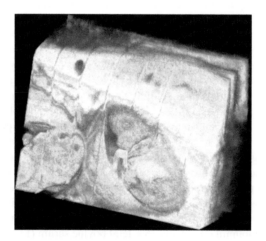

FIGURE 14.5 Ultrasound biomicroscopy (UBM) depiction of two GD 13 mouse embryos *in utero*. The image was acquired using a Vevo 770 Micro-imaging System (VisualSonics Inc., Toronto, Ontario, Canada) operated at 40 MHz. (Courtesy of Dr. Kelli Boyd, Vanderbilt University, Nashville, TN.)

evaluate organ development *in vivo* over time and the relatively low equipment cost make this modality an important addition to the tool kit for high-throughput phenotypic analysis of developing mice.

Technical Aspects of Three-Dimensional Imaging for Mouse Developmental Pathology Studies

Optical Microscopy

Tissue processing requirements for CLSM[19] and MPM[56] are comparable to those employed for conventional immunofluorescence (Chapter 11). The choice of processing method and the fluorophores should be determined and tested in advance before use on critical specimens.[9] In our experience, next-generation fluorophores (e.g., Alexa Fluor® dyes; Invitrogen Corp., Carlsbad, CA) provide superior performance and stability relative to more traditional fluorophores (e.g., fluorescein, rhodamine).

The first step for freshly harvested specimens is fixation. This step is generally performed either at room temperature (RT) for 1–4 h or at 4°C for 2–24 h. The most common fixatives (Chapter 9) are neutral buffered 10% formalin (NBF, research grade [i.e., without methanol as a preservative]) or 4% paraformaldehyde (PFA, freshly reconstituted in phosphate buffer, pH 7.4), but in some cases glutaraldehyde and/or picric acid (0.1%–0.25% of each) may be included to further improve tissue integrity. For example, a recent report indicates that fixation with either 4% PFA containing 0.2% glutaraldehyde at 25°C (i.e., RT) or with ice-cold methanol for 30 min (for a single layer of cultured cells) or longer (for tissue samples) is best for immobilizing molecules in membranes.[80] Similar approaches are utilized for MPM[56] and OPT.[79,88] Thicker specimens, such as those used for MPM, may benefit if tissue preservation is undertaken by intravascular perfusion; this is especially true where fixative solutions contain only aldehydes (e.g., PFA), which tend to penetrate slowly. Alternatively, thicker specimens may be fixed effectively by immersion when using composite solutions that contain one or more constituents capable of more rapid tissue penetration (e.g., methanol, a common stabilizer in commercial formulations of NBF).

After fixation, tissue samples for CLSM must be cut into thin sections to increase the efficiency of antibody penetration. Tissues that have been fixed using PFA often are frozen following infiltration with sucrose (to prevent the formation of large ice crystals that can damage tissue ultrastructure)[9] and then sectioned on a cryostat. Alternatively, NBF-fixed tissues typically are dehydrated in a series of organic solvents and then infiltrated and embedded in paraffin wax; in some instances, an antigen retrieval step, which usually involves heating the sections in an acidic or basic buffer,[9] will be necessary in paraffin-embedded tissues (Chapter 11).

Following sectioning, the majority of specimens have to be permeabilized to permit the entry of primary antibodies (needed to localize intracellular epitopes) and fluorophore-labeled reagents. Straightforward but harsh options are (1) two freeze–thaw cycles or (2) dehydration through an ascending series of alcohols into absolute alcohol followed by rehydration through a descending series of alcohols. Chemical treatment with detergents like digitonin, saponin, or Triton X-100 is a suitable option. The detergents are usually applied as a 0.1%–1.0% (v/v) solution in phosphate-buffered saline (PBS), pH 7.4, for 5–15 min at RT; such agents may be included in the buffer for a primary antibody (at 0.1%–0.25%), but the specimen will need to be thoroughly washed in buffer before application of secondary fluorophore-conjugated reagents. Brief incubations with enzymes (such as proteinase K) may be used to open membrane-bound structures. Finally, large sections may be penetrated directly by microdissection and/or thick sectioning (40 μm or more), usually performed on frozen tissue via vibratome.[56] Often the detergent permeabilization step may be combined with a treatment to inhibit nonspecific binding of the antibody reagents. This nonspecific localization is usually blocked by incubating specimens in PBS containing 10% bovine serum albumin (BSA) and 15% fetal bovine serum (FBS) for 1 h at RT (for samples <30 μm thick) or at 4°C overnight (for thicker tissues). An alternative means for enhancing reagent entry into thick specimens (e.g., for MPM) is to perform these and subsequent steps on floating sections so that the materials may contact all sides of the sample.

Primary antibodies, usually diluted 1:100 in PBS, are applied from 1 h to overnight using RT for short incubations and 4°C for longer periods. Mixtures of monoclonal and polyclonal primary antibodies can be used simultaneously. Even after permeabilization, dense tissues may block the entrance of larger protein reagents. A common solution to this difficulty is to employ fluorophore-conjugated F(ab) or F(ab')₂ fragments rather than intact immunoglobulins. After immunostaining is complete, the specimen is washed three times for 15 min in PBS to remove the unbound primary antibody. The secondary fluorophore-labeled antibody should be applied for 30 min to 1 h at RT to minimize nonspecific binding. After three more 15 min PBS washes, the tissue can be mounted. Most modern fluorophores (e.g., Alexa Fluor® dyes) do not require the use of an anti-fade solution to prevent quenching of the signal during analysis, but if desired, an anti-fade solution (e.g., ProLong® Gold, Vectashield®) may be used according to the manufacturer's instructions when mounting the specimen. Anti-fade reagents containing *p*-phenylenediamine (PPD) have been reported to increase autofluorescence over time,[9] so the mounting procedure should be tested before critical samples are processed.

The processing conditions for OPT differ from those utilized for other optical microscopy modalities.[63] Briefly, specimens are stained, washed to remove any remaining reagent (which could be reactivated if exposed to ultraviolet light), and then fixed in 4% PFA for 2 h at RT or overnight at 4°C. Next, fixed samples are embedded in low-melting-point agarose (1% in water). Embedded specimens may be stored at 4°C until scanning. On the day of OPT imaging, the agarose is trimmed to yield a small cube containing the sample positioned in the desired orientation. The cube is dehydrated in 100% methanol, cleared in BABB solution (a 2:1 mixture of benzyl alcohol and benzyl benzoate), mounted, and scanned.[63] Following OPT, the embedded specimen may be stored in BABB at 4°C for possible reanalysis, keeping in mind that most dyes and fluorophores are soluble in BABB and so will be lost with time.

Magnetic Resonance Microscopy

Standard fixatives may be used to preserve mouse conceptuses for MRM. One useful protocol is to immerse specimens in 20 parts Bouin's solution spiked with 1 part gadoteridol (Prohance®; Bracco Diagnostics Inc., Princeton, NJ) overnight at RT.[24,43,50,53] Following fixation, tissues are held in a 200:1 mixture of PBS with Prohance for 48–72 h. An alternative fixation protocol is 4% PFA in PBS containing 2 mM of the contrast agent gadopentetate dimeglumine (Magnevist®; Bayer HealthCare Pharmaceuticals, Inc., Wayne, NJ)[58]; specimens may be fixed in this solution for several days at 4°C. The use of contrast-enhancing agents balances the reduction in the nuclear magnetic resonance signal produced by fixation.[54,58]

Microscopic Computer-Assisted Tomography

Fixation of conceptuses destined for µCT may be completed using standard solutions, such as NBF, mixtures of 4% formaldehyde and glutaraldehyde (commonly 1%–5%), 4% PFA, and Bouin's solution (Chapter 9).[1,36,37,66] Fixation-related artifacts may be substantial,[3,52] so the desired fixation protocol should be verified prior to use in definitive experiments. In general, consistent processing of all specimens followed by imaging in a standard orientation is presumed to result in regular artifacts among samples, thereby permitting the assumption that no systematic error has been introduced during processing.[66] This step is typically performed either at RT for 1–4 h or at 4°C for 2–24 h.

The contrast among tissues during µCT may be appreciably enhanced using special stains. One common option is *en bloc* postfixation in osmium tetroxide (OsO_4),[1,32,36,37] typically as a 1%–2% solution (Chapter 9) in phosphate or cacodylate buffer (pH 7.4) for 2 h to overnight at 4°C or RT, with agitation (by rocking). Glutaraldehyde (1%) may be added to the OsO_4 solution to further enhance tissue integrity.[32] Protocols for other effective enhancement agents, such as phosphotungstic acid and iodine/potassium iodide mixtures, have been reviewed elsewhere.[15,36,37]

Following fixation and staining, conceptuses should be washed in buffer three times for 30 min at RT. Next, specimens are dehydrated in graded alcohols into absolute alcohol and then placed into 2 mL polypropylene microcentrifuge tubes for imaging. Careful consideration must be given to the endpoints to be measured, as excessive dehydration can produce such significant shrinkage (especially in younger embryos) that soft tissues may be uninterpretable; we recommend doing pilot studies of potential processing conditions when first undertaking embryo analysis by µCT to avoid this situation. After µCT imaging, it is often possible to de-stain specimens to permit traditional histological processing. Bound dye and metal ions in the specimen can be chelated and removed by immersion in 30% sucrose solution at RT for at least 48 h (for embryos) to 72 h (for neonates) while being gently agitated. De-stained samples may be embedded in paraffin or (if postfixed in OsO_4) plastic after imaging.

Sound-Based Modalities

Preparation for high-frequency sound platforms like US and UBM is straightforward. The main consideration for both modalities is to anesthetize the dam so that the instrument may be reliably positioned with respect to the uterus. Any standard inhalation anesthetic (e.g., isoflurane, 2% for 1 min for induction followed by approximately 1.5% for maintenance[38]) or injectable anesthetic (e.g., a cocktail of ketamine [80–100 mg/kg] with xylazine [5–10 mg/kg] given by the intraperitoneal [IP] route) should prove suitable for terminal procedures. In our experience, isoflurane is a suitable regimen for immobilizing the dam during a progressive study where multiple assessments will be undertaken at different points during gestation. An advantage of inhalants over injectables is that there is no possibility that the uterus and conceptuses will be traumatized as there is no IP injection. Additional stability may be attained by securing the limbs of the dam to a flat surface with rubber bands or tape for survival procedures, or push pins for terminal procedures. The dam's core body temperature may be maintained during survival studies by placing her on a heating pad that has been covered by a thin towel or by using a radiant heat lamp. For some conventional US studies, the uterus may need to be exteriorized to ensure proper orientation of the *in utero* subjects.[29,57] Such experiments are typically designed as terminal procedures.

Summary

This mini-review on 3D imaging applications in mouse developmental pathology is intended as a starting point for investigators interested in using these technologies in their own work. We estimate that these modalities will see increasing use in this field in the near future as the cost of high-resolution instruments falls. Gains will be particularly rapid for those methods that permit ready analysis of multiple specimens, as this capacity will greatly boost the productivity of high-throughput phenogenomic programs.

REFERENCES

1. Aoyagi H, Tsuchikawa K, Iwasaki S (2010). Three-dimensional observation of the mouse embryo by micro-computed tomography: Composition of the trigeminal ganglion. *Odontology* **98** (1): 26–30.
2. Artemov D (2003). Molecular magnetic resonance imaging with targeted contrast agents. *J Cell Biochem* **90** (3): 518–524.
3. Bancroft JD, Gamble G (2007). *Theory and Practice of Histological Techniques*, 6th edn. New York: Churchill Livingstone (Elsevier).
4. Beare R, Richards K, Murphy S, Petrou S, Reutens D (2008). An assessment of methods for aligning two-dimensional microscope sections to create image volumes. *J Neurosci Methods* **170** (2): 332–344.
5. Benninger RK, Hao M, Piston DW (2008). Multi-photon excitation imaging of dynamic processes in living cells and tissues. *Rev Physiol Biochem Pharmacol* **160**: 71–92.
6. Bock NA, Konyer NB, Henkelman RM (2003). Multiple-mouse MRI. *Magn Reson Med* **49** (1): 158–167.
7. Bolon B, Garman R, Jensen K, Krinke G, Stuart B (2006). A 'best practices' approach to neuropathologic assessment in developmental neurotoxicity testing—for today. *Toxicol Pathol* **34** (3): 296–313.
8. Boughner JC, Wat S, Diewert VM, Young NM, Browder LW, Hallgrímsson B (2008). Short-faced mice and developmental interactions between the brain and the face. *J Anat* **213** (6): 646–662.
9. Burry RW (2010). *Immunocytochemistry: A Practical Guide for Biomedical Research*. New York: Springer Verlag.
10. Chudakov DM, Matz MV, Lukyanov S, Lukyanov KA (2010). Fluorescent proteins and their applications in imaging living cells and tissues. *Physiol Rev* **90** (3): 1103–1163.
11. Cleary JO, Modat M, Norris FC, Price AN, Jayakody SA, Martinez-Barbera JP, Greene ND et al. (2011). Magnetic resonance virtual histology for embryos: 3D atlases for automated high-throughput phenotyping. *NeuroImage* **54** (2): 769–778.
12. Coan PM, Conroy N, Burton GJ, Ferguson-Smith AC (2006). Origin and characteristics of glycogen cells in the developing murine placenta. *Dev Dyn* **235** (12): 3280–3294.
13. Coan PM, Ferguson-Smith AC, Burton GJ (2004). Developmental dynamics of the definitive mouse placenta assessed by stereology. *Biol Reprod* **70** (6): 1806–1813.
14. Corrigan N, Brazil DP, Auliffe FM (2010). High-frequency ultrasound assessment of the murine heart from embryo through to juvenile. *Reprod Sci* **17** (2): 147–157.
15. Degenhardt K, Wright AC, Horng D, Padmanabhan A, Epstein JA (2010). Rapid 3D phenotyping of cardiovascular development in mouse embryos by micro-CT with iodine staining. *Circ Cardiovasc Imaging* **3** (3): 314–322.
16. Denk W, Piston DW, Webb WW (1995). Two-photon molecular excitation in laser-scanning microscopy. In: *Handbook of Biological Confocal Microscopy*, JB Pawley (ed.). New York: Plenum Press; pp. 445–458.

17. Dunty Jr WC, Chen S-Y, Zucker RM, Dehart DB, Sulik KK (2001). Selective vulnerability of embryonic cell populations to ethanol-induced apoptosis: Implications for alcohol-related birth defects and neuro-developmental disorder. *Alcohol Clin Exp Res* **25** (10): 1523–1535.

18. Dyson E, Sucov HM, Kubalak SW, Schmid-Schonbein GW, DeLano FA, Evans RM, Ross JJ, Chien KR (1995). Atrial-like phenotype is associated with embryonic ventricular failure in retinoid X receptor α$^{-/-}$ mice. *Proc Natl Acad Sci USA* **92** (16): 7386–7390.

19. Ekström P (2000). Confocal microscopy. In: *Current Protocols in Toxicology*. Hoboken, NJ: John Wiley & Sons; pp. 2.8.1–2.8.21.

20. Evan AP, Gattone VH, Blomgren PM (1984). Application of scanning electron microscopy to kidney development and nephron maturation. *Scan Electron Microsc* **Pt 1**: 455–473.

21. Ewald AJ, McBride H, Reddington M, Fraser SE, Kerschmann R (2002). Surface imaging microscopy, an automated method for visualizing whole embryo samples in three dimensions at high resolution. *Dev Dynam* **225** (3): 369–375.

22. Foster FS, Zhang M, Duckett AS, Cucevic V, Pavlin CJ (2003). *In vivo* imaging of embryonic development in the mouse eye by ultrasound biomicroscopy. *Invest Ophthalmol Vis Sci* **44** (6): 2361–2366.

23. Fumene Feruglio P, Vinegoni C, Gros J, Sbarbati A, Weissleder R (2010). Block matching 3D random noise filtering for absorption optical projection tomography. *Phys Med Biol* **55** (18): 5401–5415.

24. Godin EA, O'Leary-Moore SK, Khan AA, Parnell SE, Ament JJ, Dehart DB, Johnson BW, Johnson GA, Styner MA, Sulik KK (2010). Magnetic resonance microscopy defines ethanol-induced brain abnormalities in prenatal mice: Effects of acute insult on gestational day 7. *Alcohol Clin Exp Res* **34** (1): 98–111.

25. Hadjantonakis AK, Dickinson ME, Fraser SE, Papaioannou VE (2003). Technicolour transgenics: Imaging tools for functional genomics in the mouse. *Nat Rev Genet* **4** (8): 613–625.

26. Hama H, Kurokawa H, Kawano H, Ando R, Shimogori T, Noda H, Fukami K, Sakaue-Sawano A, Miyawaki A (2011). Scale: A chemical approach for fluorescence imaging and reconstruction of transparent mouse brain. *Nat Neurosci* **14** (11): 1481–1488.

27. Huber C (1905). On the development and shape of uriniferous tubules of certain higher mammals. *Am J Anat Suppl* **4**: 1–98.

28. Inoue S (1995). Foundations of confocal scanned imaging in light microscopy. In: *Handbook of Biological Confocal Microscopy*, JB Pawley (ed.). New York: Plenum Press; pp. 1–17.

29. Ji RP, Phoon CK, Aristizábal O, McGrath KE, Palis J, Turnbull DH (2003). Onset of cardiac function during early mouse embryogenesis coincides with entry of primitive erythroblasts into the embryo proper. *Circ Res* **92** (2): 133–135.

30. Johnson GA, Ali-Sharief A, Badea A, Brandenburg J, Cofer G, Fubara B, Gewalt S, Hedlund LW, Upchurch L (2007). High-throughput morphologic phenotyping of the mouse brain with magnetic resonance histology. *NeuroImage* **37** (1): 82–89.

31. Johnson GA, Cofer GP, Fubara B, Gewalt SL, Hedlund LW, Maronpot RR (2002). Magnetic resonance histology for morphologic phenotyping. *J Magn Reson Imaging* **16** (4): 423–429.

32. Johnson JT, Hansen MS, Wu I, Healy LJ, Johnson CR, Jones GM, Capecchi MR, Keller C (2006). Virtual histology of transgenic mouse embryos for high-throughput phenotyping. *PLoS Genet* **2**: e61.

33. Kaufman MH (1992). *The Atlas of Mouse Development*, 2nd edn. San Diego, CA: Academic Press.

34. Mäki JM, Räsänen J, Tikkanen H, Sormunen R, Mäkikallio K, Kivirikko KI, Soininen R (2002). Inactivation of the lysyl oxidase gene *Lox* leads to aortic aneurysms, cardiovascular dysfunction, and perinatal death in mice. *Circulation* **106**: 2503–2509.

35. Maronpot RR, Sills RC, Johnson GA (2004). Applications of magnetic resonance microscopy. *Toxicol Pathol* **32** (Suppl. 2): 42–48.

36. Metscher BD (2009). MicroCT for comparative morphology: Simple staining methods allow high-contrast 3D imaging of diverse non-mineralized animal tissues. *BMC Physiol* **9**: 11–24.

37. Metscher BD (2009). MicroCT for developmental biology: A versatile tool for high-contrast 3D imaging at histological resolutions. *Dev Dyn* **238** (3): 632–640.

38. Momoi N, Tinney JP, Liu LJ, Elshershari H, Hoffmann PJ, Ralphe JC, Keller BB, Tobita K (2008). Modest maternal caffeine exposure affects developing embryonic cardiovascular function and growth. *Am J Physiol Heart Circ Physiol* **294** (5): H2248–H2256.

39. Mu J, Adamson SL (2006). Developmental changes in hemodynamics of uterine artery, utero- and umbilicoplacental, and vitelline circulations in mouse throughout gestation. *Am J Physiol Heart Circ Physiol* **291** (3): H1421–H1428.

40. Mu J, Slevin JC, Qu D, McCormick S, Adamson SL (2008). *In vivo* quantification of embryonic and placental growth during gestation in mice using micro-ultrasound. *Reprod Biol Endocrinol* **6**: 34.

41. Nagase T, Sasazaki Y, Kikuchi T, Machida M (2008). Rapid 3-dimensional imaging of embryonic craniofacial morphology using microscopic computed tomography. *J Comput Assist Tomogr* **32** (5): 816–821.

42. Nieman BJ, Turnbull DH (2010). Ultrasound and magnetic resonance microimaging of mouse development. *Methods Enzymol* **476**: 379–400.

43. O'Leary-Moore SK, Parnell SE, Godin EA, Dehart DB, Ament JJ, Khan AA, Johnson GA, Styner MA, Sulik KK (2010). Magnetic resonance microscopy-based analyses of the brains of normal and ethanol-exposed fetal mice. *Birth Defects Res A Clin Mol Teratol* **88** (11): 953–964.

44. Oest ME, Jones JC, Hatfield C, Prater MR (2008). Micro-CT evaluation of murine fetal skeletal development yields greater morphometric precision over traditional clear-staining methods. *Birth Defects Res B Dev Reprod Toxicol* **83** (6): 582–589.

45. Oheim M, Michael DJ, Geisbauer M, Madsen D, Chow RH (2006). Principles of two-photon excitation fluorescence microscopy and other nonlinear imaging approaches. *Adv Drug Deliv Rev* **58** (7): 788–808.

46. Osathanondh V, Potter EL (1963). Development of human kidney as shown by microdissection. I. Preparation of tissue with reasons for possible misinterpretation of observations. *Arch Pathol* **76**: 271–276.

47. Osathanondh V, Potter EL (1963). Development of human kidney as shown by microdissection. III. Formation and interrelationship of collecting tubules and nephrons. *Arch Pathol* **76**: 290–302.

48. Osathanondh V, Potter EL (1966). Development of human kidney as shown by microdissection. IV. Development of tubular portions of nephrons. *Arch Pathol* **82**: 391–402.

49. Osathanondh V, Potter EL (1966). Development of human kidney as shown by microdissection. V. Development of vascular pattern of glomerulus. *Arch Pathol* **82**: 403–411.

50. Parnell SE, O'Leary-Moore SK, Godin EA, Dehart DB, Johnson BW, Johnson GA, Styner MA, Sulik KK (2009). Magnetic resonance microscopy defines ethanol-induced brain abnormalities in prenatal mice: Effects of acute insult on gestational day 8. *Alcohol Clin Exp Res* **33** (6): 1001–1011.

51. Parsons TE, Kristensen E, Hornung L, Diewert VM, Boyd SK, German RZ, Hallgrímsson B (2008). Phenotypic variability and craniofacial dysmorphology: Increased shape variance in a mouse model for cleft lip. *J Anat* **212** (2): 135–143.

52. Patten BM, Philpott R (1921). The shrinkage of embryos in the processes preparatory to sectioning. *Anat Rec (Hoboken)* **20**: 392–413.

53. Petiet A, Hedlund L, Johnson GA (2007). Staining methods for magnetic resonance microscopy of the rat fetus. *J Magn Reson Imaging* **25** (6): 1192–1198.

54. Petiet A, Johnson GA (2010). Active staining of mouse embryos for magnetic resonance microscopy. *Methods Mol Biol* **611**: 141–149.

55. Petiet AE, Kaufman MH, Goddeeris MM, Brandenburg J, Elmore SA, Johnson GA (2008). High-resolution magnetic resonance histology of the embryonic and neonatal mouse: A 4D atlas and morphologic database. *Proc Natl Acad Sci USA* **105** (34): 12331–12336.

56. Phillips CL, Arend LJ, Filson AJ, Kojetin DJ, Clendenon JL, Fang S, Dunn KW (2001). Three-dimensional imaging of embryonic mouse kidney by two-photon microscopy. *Am J Pathol* **158** (1): 49–55.

57. Phoon CK, Ji RP, Aristizabal O, Worrad DM, Zhou B, Baldwin HS, Turnbull DH (2004). Embryonic heart failure in *NFATc1⁻/⁻* mice: Novel mechanistic insights from *in utero* ultrasound biomicroscopy. *Circ Res* **95** (1): 92–99.

58. Pieles G, Geyer SH, Szumska D, Schneider J, Neubauer S, Clarke K, Dorfmeister K, Franklyn A, Brown SD, Bhattacharya S, Weninger WJ (2007). μMRI-HREM pipeline for high-throughput, high-resolution phenotyping of murine embryos. *J Anat* **211** (1): 132–137.

59. Piston DW (1999). Imaging living cells and tissues by two-photon excitation microscopy. *Trends Cell Biol* **9** (2): 66–69.

60. Plaks V, Sapoznik S, Berkovitz E, Haffner-Krausz R, Dekel N, Harmelin A, Neeman M (2010). Functional phenotyping of the maternal albumin turnover in the mouse placenta by dynamic contrast-enhanced MRI. *Mol Imaging Biol* **13** (3): 481–492.

61. Powell KA, Wilson D (2012). 3-Dimensional imaging modalities for phenotyping genetically engineered mice. *Vet Pathol* **49** (1): 106–115.

62. Quintana L, Sharpe J (2011). Optical projection tomography of vertebrate embryo development. *Cold Spring Harb Protoc* **2011** (6): 586–594.

63. Quintana L, Sharpe J (2011). Preparation of mouse embryos for optical projection tomography imaging. *Cold Spring Harb Protoc* **2011** (6): 664–669.

64. Salisbury JR (1994). Three-dimensional reconstruction in microscopical morphology. *Histol Histopathol* **9** (4): 773–780.

65. Scherrer-Crosbie M, Thibault HB (2008). Echocardiography in translational research: Of mice and men. *J Am Soc Echocardiogr* **21** (10): 1083–1092.

66. Schmidt EJ, Parsons TE, Jamniczky HA, Gitelman J, C, Boughner JC, Logan CC, Sensen CW, Hallgrímsson B (2010). Micro-computed tomography-based phenotypic approaches in embryology: Procedural artifacts on assessments of embryonic craniofacial growth and development. *BMC Dev Biol* **10**: 18 (14 pages).

67. Schneider JE, Bamforth SD, Farthing CR, Clarke K, Neubauer S, Bhattacharya S (2003). High-resolution imaging of normal anatomy, and neural and adrenal malformations in mouse embryos using magnetic resonance microscopy. *J Anat* **202** (2): 239–247.

68. Schneider JE, Bhattacharya S (2004). Making the mouse embryo transparent: Identifying developmental malformations using magnetic resonance imaging. *Birth Defects Res C Embryo Today* **72** (3): 241–249.

69. Schneider JE, Böse J, Bamforth SD, Gruber AD, Broadbent C, Clarke K, Neubauer S, Lengeling A, Bhattacharya S (2004). Identification of cardiac malformations in mice lacking *Ptdsr* using a novel high-throughput magnetic resonance imaging technique. *BMC Dev Biol* **4**: 16 (12 pages).

70. Sharpe J, Ahlgren U, Perry P, Hill B, Ross A, Hecksher-Sørensen J, Baldock R, Davidson D (2002). Optical projection tomography as a tool for 3D microscopy and gene expression studies. *Science* **296** (5567): 541–545.

71. Shen Y, Leatherbury L, Rosenthal J, Yu Q, Pappas MA, Wessels A, Lucas J et al. (2005). Cardiovascular phenotyping of fetal mice by noninvasive high-frequency ultrasound facilitates recovery of ENU-induced mutations causing congenital cardiac and extracardiac defects. *Physiol Genomics* **24**: 23–36.

72. Smith BR (2001). Magnetic resonance microscopy in cardiac development. *Microsc Res Tech* **52** (3): 323–330.

73. Smith BR, Johnson GA, Groman EV, Linney E (1994). Magnetic resonance microscopy of mouse embryos. *Proc Natl Acad Sci USA* **91** (9): 3530–3533.

74. Spurney CF, Leatherbury L, Lo CW (2004). High-frequency ultrasound database profiling growth, development, and cardiovascular function in C57BL/6J mouse fetuses. *J Am Soc Echocardiogr* **17** (8): 893–900.

75. Srinivasan S, Baldwin HS, Aristizabal O, Kwee L, Labow M, Artman M, Turnbull DH (1998). Noninvasive, *in utero* imaging of mouse embryonic heart development with 40-MHz echocardiography. *Circulation* **98** (9): 912–918.

76. Streicher J, Weninger WJ, Muller GB (1997). External marker-based automatic congruencing: A new method of 3D reconstruction from serial sections. *Anat Rec* **248** (4): 583–602.

77. Sulik KK, Schoenwolf GC (1985). Highlights of craniofacial morphogenesis in mammalian embryos, as revealed by scanning electron microscopy. *Scan Electron Microsc* **4** (Pt. 4): 1735–1752.

78. Sullivan R, Huang GY, Meyer RA, Wessels A, Linask KK, Lo CW (1998). Heart malformations in transgenic mice exhibiting dominant negative inhibition of gap junctional communication in neural crest cells. *Dev Biol* **204** (1): 224–234.

79. Summerhurst K, Stark M, Sharpe J, Davidson D, Murphy P (2008). 3D representation of Wnt and Frizzled gene expression patterns in the mouse embryo at embryonic day 11.5 (Ts19). *Brain Res Gene Expr Patterns* **8** (5): 331–348.

80. Tanaka KA, Suzuki KG, Shirai YM, Shibutani ST, Miyahara MS, Tsuboi H, Yahara M, Yoshimura A, Mayor S, Fujiwara TK, Kusumi A (2010). Membrane molecules mobile even after chemical fixation. *Nat Methods* **7** (11): 865–866.

81. Tanaka N, Mao L, DeLano FA, Sentianin EM, Chien KR, Schmid-Schonbein GW, Ross JJ (1997). Left ventricular volumes and function in the embryonic mouse heart. *Am J Physiol* **273** (3 Pt. 2): H1368–H1376.

82. Tobita K, Liu X, Lo CW (2010). Imaging modalities to assess structural birth defects in mutant mouse models. *Birth Defects Res C Embryo Today* **90** (3): 176–184.

83. Tomlinson TM, Garbow JR, Anderson JR, Engelbach JA, Nelson DM, Sadovsky Y (2010). Magnetic resonance imaging of hypoxic injury to the murine placenta. *Am J Physiol Regul Integr Comp Physiol* **298** (2): R312–R319.

84. Vasquez SX, Hansen MS, Bahadur AN, Hockin MF, Kindlmann GL, Nevell L, Wu IQ et al. (2008). Optimization of volumetric computed tomography for skeletal analysis of model genetic organisms. *Anat Rec (Hoboken)* **291** (5): 475–487.

85. Vuillemin M, Pexieder T (1989). Normal stages of cardiac organogenesis in the mouse: I. Development of the external shape of the heart. *Am J Anat* **184** (2): 101–113.

86. Vuillemin M, Pexieder T (1989). Normal stages of cardiac organogenesis in the mouse: II. Development of the internal relief of the heart. *Am J Anat* **184** (2): 114–128.

87. Vuillemin M, Pexieder T, Wong YM, Thompson RP (1992). A two-step alignment method for 3D computer-aided reconstruction based on fiducial markers and applied to mouse embryonic hearts. *Eur J Morphol* **30** (3): 181–193.

88. Walls JR, Coultas L, Rossant J, Henkelman RM (2008). Three-dimensional analysis of vascular development in the mouse embryo. *PLoS One* **3** (8): e2853 (15 pages).

89. Weninger WJ, Geyer SH (2008). Episcopic 3D imaging methods: Tools for researching gene function. *Curr Genomics* **9** (4): 282–289.

90. Williams RM, Piston DW, Webb WW (1994). Two-photon molecular excitation provides intrinsic 3-dimensional resolution for laser-based microscopy and microphotochemistry. *FASEB J* **8** (11): 804–813.

91. Yu Q, Leatherbury L, Tian X, Lo CW (2008). Cardiovascular assessment of fetal mice by *in utero* echocardiography. *Ultrasound Med Biol* **34** (5): 741–752.

92. Yu Q, Shen Y, Chatterjee B, Siegfried BH, Leatherbury L, Rosenthal J, Lucas JF et al. (2004). ENU induced mutations causing congenital cardiovascular anomalies. *Development* **131** (24): 6211–6223.

93. Zhang X, Schneider JE, Portnoy S, Bhattacharya S, Henkelman RM (2010). Comparative SNR for high-throughput mouse embryo MR microscopy. *Magn Reson Med* **63** (6): 1703–1707.

94. Zucker RM, Hunter ES III, Rogers JM (2000). Confocal laser scanning microscopy of morphology and apoptosis in organogenesis-stage mouse embryos. *Methods Mol Biol* **135**: 191–202.

95. Zucker RM, Price O (2001). Evaluation of confocal microscopy system performance. *Cytometry* **44** (4): 273–294.

96. Zucker RM, Rogers JM (2000). Embryo/fetal topographical analysis by fluorescence microscopy and confocal laser scanning microscopy. *Methods Mol Biol* **135**: 203–209.

Section IV

Analytical Practices for Mouse Developmental Pathology

15

Pathology of the Developing Mouse
from Conception to Weaning

Brad Bolon and Vinicius Carreira

CONTENTS

The analysis of lethal phenotypes in developing mice not only spans the prenatal period but also encompasses the time from birth to sexual maturity. This underappreciated fact is supported by the anatomic equivalence of the mouse brain at postnatal day (PND) 7 to that of the human infant brain at birth[67,68] as well as the myriad functional and structural changes that occur as a neonatal mouse evolves from a bald, blind, nearly sessile *pinkie* (PND 0) to a haired, mobile juvenile at weaning (PND 21 or PND 22) and finally a sexually capable animal at puberty (about PND 28 to PND 35, depending on the sex and strain/stock). Therefore, the tendency of some researchers to ignore postnatal lesions when assessing a developmental phenotype is ill-advised. The choice of parameters to examine and the methods used in the analysis will differ depending on the developmental age of the subjects, the presumed timing of the lethal event(s), and in many instances the expertise and interest of the researchers tasked with confirming the existence and defining the nature of lethal developmental phenotypes. However, the search for novel phenotypes should be undertaken by using multiple techniques—including anatomic pathology, clinical pathology, and molecular pathology tools—because more new conditions will be discovered by casting a wider net.[449]

Regardless of the timing, pathology assessments to characterize any new lethal phenotype in developing mice should be founded in a systematic analysis of structural and functional changes and their biochemical bases during one or more periods during prenatal and/or early postnatal life.[242,302,322,323,361,444] Both the embryo (or fetus)[198] and placenta[199] (see Chapter 16) should be evaluated as lethality may result from lesions in either of these entities—or in both simultaneously. Routine techniques for making macroscopic observations and taking gross measurements by direct examination (Chapter 7) or noninvasive imaging (Chapter 14) as well as methods for sampling tissues (Chapter 7) and fluids (Chapter 8) have been covered in detail earlier. Standard practices for processing specimens to assess microscopic (Chapter 9), molecular (Chapters 10 and 11), kinetic (Chapter 12), morphometric (Chapter 12), and ultrastructural features (Chapter 13) also have been reviewed previously. The current chapter completes this sequence of topics by presenting major concepts for undertaking a well-designed microscopic examination of tissues from developing mice. While not performed in some laboratories, in our experience histopathologic assessment is an essential part of a complete developmental phenotyping study since high-magnification analysis will provide both additional insights regarding the pathogenesis of grossly visible lesions and, more critically, in many cases will discern subtle phenotypes that would be missed in the absence of a microscopic review.[2]

A second purpose for the current chapter is to showcase principal categories of mutation-linked and toxicant-induced lesions that occur in developing mice that have embryonic lethal or perinatal lethal phenotypes. The text and illustrations given here should assist comparative pathologists to reliably recognize and report major developmental defects. More importantly, this material has been designed to provide an initial tutorial in mouse developmental pathology for nonpathologists. This additional focus is necessary because the only way to acquire a solid knowledge of this (or any) subject is to practice—and then practice a great deal more. While we have aimed to provide a firm foundation in this regard, we must emphasize that developmental pathology is an arena where DIY (*do it yourself*) analysis generally will be not rewarded.[191,401] Accordingly, researchers with little or no formal training in developmental anatomy and developmental pathology always should seek support for their phenotyping efforts from an expert comparative pathologist.[26]

Analysis of Cells, Tissues, and Organs from Developing Mice

Macroscopic Analysis

The developmental pathology assessment begins with the initial presentation of either a pregnant dam or one or more developing neonatal or juvenile mice for analysis. In the case of progeny still held *in utero*, maternal attributes are recorded prior to necropsy (e.g., total body weight) and for her litter after isolation of the uterus (e.g., numbers of total, live, and viable conceptuses as well as counts of implantation sites with gross defects and resorptions). Next, the embryos or fetuses are isolated from their placentas, and relevant individual animal characteristics are noted (e.g., body weight and/or crown/rump length; kind and severity of gross defects). Finally, the developing animals are relegated to different subgroups for assessment in various fashions, such as free-hand sectioning in multiple planes to permit gross examination of internal organs or double staining to showcase the progressive differentiation of the skeleton.

Major classes of developmental lesions that may be discerned by macroscopic observation are outlined in Table 15.1. For most of these lesions, the recognition that a change exists as well as preparation of a detailed description of the change are fairly straightforward. In some instances, the characteristic features of a lesion will permit assignment of a known term (see Chapter 3 for a list of common developmental pathology lesions), while the coexistence of certain lesions may suggest the presence of a particular syndrome or establish a pathogenesis. The exception to this rule is the class of *missing structures*, which may be hard to identify as the mind's eye tends to *see* absent organs (and record them as exhibiting normal features) unless specific attention is paid to their analysis.[43] For this reason, we

TABLE 15.1

Pathology Alterations in Developing Mice—Examples of Common Lesions[a]

Type of Change	Attribute	Common Significance	Citations
Cell numbers	Enhanced	Reduced apoptosis	[265,269,460]
	Reduced	Abnormal energy metabolism	[29]
		Decreased proliferation	[224,391]
Color	Green	Cell debris or long-past hemorrhage	
	Red	Fresh hemorrhage	
	White	Necrosis	
Consistency	Friable (crumbling)	Necrosis	
Shape	Swelling, generalized	Anasarca (generalized edema)	[395]
	Swelling, localized	Hydropericardium (distended pericardial sac)	[38,55,128,351]
Size (body)	Enhanced	Disrupted growth factor signaling	[117,250]
		Prolonged gestation	[116]
	Reduced	Developmental delay	[262,374,378]
		Imminent death	[158,180,301]
Size (region)	Enhanced	Macrocephaly (large head)	[17,246]
		Renal pelvic cavitation	[187]
	Reduced	Micrognathia (small jaw)	[20]
		Micromelia (small limb)	[11,157]
		Organ weights, body-to-organ weight ratios	[174,439]
Distorted structures	Malformation	Abnormal cardiac symmetry	[16,139]
		Cranioschisis/exencephaly (neural tube closure defect)	[13,38,171]
		Palatoschisis (cleft palate)	[20,364]
	Deformation	Limb defects (secondary to oligohydramnios)	[58,218]
	Disruption	Limb defects (amniotic band syndrome)	[355]
		Limb defects (maternal blood loss)	[125]
	Dysplasia	Ganglionic disorganization and fusion	[37]
		Heterotopia (ectopia)	[31,383]
	Missing structures	Cardiac septae	[129]
		Eye (anophthalmia)	[121]
		Intestine (atresia)	[386]
		Kidney (renal agenesis)	[52,240,295,333,365]
		Spleen (asplenia)	[5,270,424]
	New structures	Colonic hamartoma	[408]
		Neoplasia (kidney and liver)	[245]
	Repositioned structures	*Situs inversus*	[47,435,459]
	Retained structures	Omphalocele (persistent umbilical hernia)	[364,386]

Note: A *disruption* is a structural defect induced by an extrinsic insult that destroys normal tissue; a good illustration of this phenomenon is strangulation of the distal limbs or digits by amniotic bands, resulting in deep grooves or amputations.

[a] Citations were chosen for their good text explanations and/or illustrations of these developmental lesions.

recommend utilizing a specific checklist (Table 15.2) when undertaking mouse developmental pathology analyses to make sure that all major organs actually have received a macroscopic evaluation, and to define which ones were not examined. More details for assessing macroscopic endpoints in embryos, fetuses, and neonates are available in Chapters 7 and 17.

Techniques that are suitable for the macroscopic evaluation of the placenta are given in Chapter 16. Inclusion of this transient organ is a key element of a complete developmental pathology analysis because many lethal phenotypes in developing mice arise wholly or in part in this structure.[21,86,358,414] Major types of placental lesions are shown in Chapter 16 as well.

TABLE 15.2

Checklist of Organs and Tissues Routinely Subject to Analysis in Mouse Developmental Pathology Studies

Organ/Tissue	Examined?	Organ/Tissue	Examined?	Organ/Tissue	Examined?
Adipose—brown	Y/N	*Immune system*		*Reproductive tract*	
Adipose—white	Y/N	Lymph node	Y/N	Gonad: ovary or testis	Y/N
Cardiovascular system		Spleen	Y/N	Duct: epididymis or	Y/N
Aorta	Y/N	Thymus	Y/N	uterus	
Heart	Y/N	*Integumentary system*		Passage: cervix/vagina	Y/N
Cardiac great	Y/N	Skin	Y/N	or vas deferens	
vessels		Mammary gland	Y/N	Accessory sex glands	Y/N
Pulmonary vessels	Y/N	Vibrissae	Y/N	*Respiratory tract*	
Umbilical vessels	Y/N	*Musculoskeletal system*		Nasal cavity	Y/N
Digestive tract		Bone: femur	Y/N	Larynx	Y/N
Salivary gland	Y/N	Bone: sternum	Y/N	Trachea/bronchi	Y/N
Esophagus	Y/N	Bone: vertebrae	Y/N	Lungs	Y/N
Stomach	Y/N	Joint: femorotibial (*knee*)	Y/N	*Urinary tract*	
Duodenum	Y/N	Joint: intercarpal (*wrist*)	Y/N	Kidney	Y/N
Jejunum	Y/N	Joint: tibiotarsal (*ankle*)	Y/N	Ureter	Y/N
Ileum	Y/N	Joint: intervertebral	Y/N	Urinary bladder	Y/N
Cecum	Y/N	Muscle: diaphragm	Y/N	Urethra	Y/N
Colon	Y/N	Muscle: epaxial	Y/N	*Placenta*	
Liver	Y/N	Muscle: hypaxial	Y/N	Yolk sac	Y/N
Pancreas	Y/N	Muscle: thigh	Y/N	Amnion	Y/N
Ear	Y/N	*Nervous system*		Allantois	Y/N
Endocrine system		Brain: basal nuclei	Y/N	Chorion	Y/N
Adrenal gland	Y/N	Brain: cerebral cortex	Y/N	Labyrinth	Y/N
Pituitary gland	Y/N	Brain: cerebellum	Y/N	Junctional zone	Y/N
Thyroid gland	Y/N	Brain: corpus callosum	Y/N	Trophoblast giant cells	Y/N
Eye	Y/N	Brain: hippocampus	Y/N	Decidua (maternal zone)	Y/N
Hematopoietic system		Brain: hypothalamus	Y/N	*Maternal tissues*	
Blood	Y/N	Brain: pons	Y/N	Blood	Y/N
Bone marrow	Y/N	Brain: thalamus	Y/N	Lung	Y/N
Liver	Y/N	Spinal cord	Y/N	Mammary Gland	Y/N
Spleen	Y/N	Dorsal root ganglia	Y/N	Ovary	Y/N
Yolk sac	Y/N	Nerve (usually sciatic)	Y/N	Uterus	Y/N
				Gross lesions	Y/N

At necropsy, developmental defects may exhibit *handedness*, whereby defects occur more on one side of the body than on the other.[47] Lateral asymmetry in structural development becomes visible at about gestational day (GD) 8.5 in mice,[282,435] arising as a consequence of side-specific expression of various morphogens at an earlier stage of development. Variants of this condition include the negation[282] of handedness, where about half of all embryos would have normal placement of asymmetric structures (e.g., tail turning to the right, aorta on the left) while the other half would exhibit the opposite placement, or even its reversal (i.e., *situs inversus*).[459] While handedness is attributed chiefly to variable expression of genes,[313] the divergent position of distal appendages relative to the body wall as well as differential blood flow to the right side (indicated by the twofold greater diameter of the right umbilical vein relative to the left) also have been invoked as likely cofactors.[288] Exposure to teratogens can shift the laterality of lesions,[273] while the handedness of the animal may influence which side exhibits a teratogen-induced lesion.[47] Thus, the exact position (location AND side) of all changes should be recorded when characterizing a novel phenotype.

Microscopic Analysis

The tissue blocks produced by trimming intact animals may be processed to allow for examination of fine structural features in cells, tissues, and organs. In general, the same elements that were evaluated macroscopically as well as an additional cohort of small tissues and organs that were not clearly visible at the macroscopic level should be reviewed using routine bright-field microscopy. Again, a checklist is an important tool for ensuring that organs and tissues of interest will not be missed inadvertently during the course of the review (Table 15.2). The reasons for incorporating a comprehensive histopathologic analysis as a standard component of a developmental pathology analysis is that (1) microscopic changes often yield vital information for understanding the pathogenesis of a lesion and (2) histopathologic findings may represent the sole evidence of a new phenotype, occurring with regularity in mice that lacked clinical (either macroscopic pathology or functional) changes.[2]

Histological Processing

The means of processing specimens to be assessed by microscopy represents a compromise between convenience and precision. Multiple small embryos may be processed and examined at the same time. Preimplantation embryos (i.e., GD 4.5 and earlier) are free floating in the oviduct or uterine lumen and are viewed after they have been flushed from the reproductive tract (see Chapter 7). One option is to place the embryos from one or several litters in a vessel filled with sterile medium and then culture them for several days; the evolving morphology of embryonic cells as assessed by intermittent microscopic examination may provide clues regarding the functional integrity and viability of the embryos.[12,164,410,416,466] A second option is to concentrate isolated embryos into a loose pellet by brief centrifugation and then embed them in 1% agar followed by further embedding in either paraffin or plastic (see Chapter 9). Implanted embryos (i.e., GD 5.0 to GD 7.5) commonly are fixed *in situ*, after which the intact uterus is processed into paraffin and sectioned serially or at intervals (termed *steps*) so that all conceptuses eventually will be visible for analysis (see Figure 7.4). If funding and time permit, the best practice is to prepare serial sections so that all sections from all embryos are retained; a common practice is to start by staining every fifth or tenth serial section with hematoxylin and eosin (H&E), while the intervening sections are held as reserves for the H&E analysis or as specimens for any *in situ* molecular markers that may seem desirable after the initial H&E analysis is done. However, in a resource-limited setting a common compromise is to take sections at a limited number of steps that are likely to hold cells or organs of interest. In this scenario, the intact uterus is embedded and then sectioned until the yolk sac cavity is observed on the face of the tissue block for at least one implantation site. From this starting point, a section is taken and the microtome blade is advanced for many micrometers before another section is acquired. The width of such intervals varies with the developmental stage (i.e., size) of the embryo. In our experience, suitable intervals between steps are between 25–50 μm from GD 5.5 to GD 6.5, about 75–100 μm from GD 7.0 to GD 8.0, and approximately 150–250 μm from GD 9.5 to GD 10.5. Thereafter, a seasoned technical team (histologist and comparative pathologist) may work together to obtain sections at particular planes based on the unstained structures to be seen on the block face; a jeweler's loupe (with or without integrated light source) may be quite useful for discerning fine anatomic details. A disadvantage to step sectioning is that no single section will contain embryonic tissue from all neighboring conceptuses at the same plane (Figure 15.1). A general disadvantage to sampling implantation sites *in situ* is that the orientation of the embryo cannot be controlled precisely (Figure 15.2) relative to the well-matched features of embryos—even small ones (Figure 15.3)—that have been removed from the placenta so that they may be embedded in a specific position.

Histopathologic Assessment

A microscopic analysis of specimens from developing mice usually is undertaken in three sequential steps. The materials available for examination from most mouse developmental pathology studies are fixed, paraffin-embedded tissue sections (4–8 μm in thickness) stained with hematoxylin and eosin

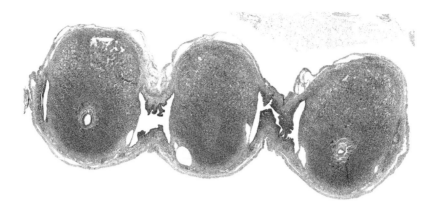

FIGURE 15.1 Photomicrograph showing a histologic preparation of the pregnant uterus (GD 7.0) that permits simultaneous assessment in a single block of many implantation sites *in situ*. Each site consists of a thick basophilic cushion of decidua (derived from the endometrial stroma) surrounding a central embryo; the absence of a visible embryo in the central site indicates that multiple serial or step sections will be needed to examine all conceptuses. The mesometrial side of the uterine horn (i.e., where the major maternal blood vessels enter the organ) is located at the upper side of the image. Histologic processing: fixation in neutral buffered 10% formalin, embedding in paraffin, H&E staining.

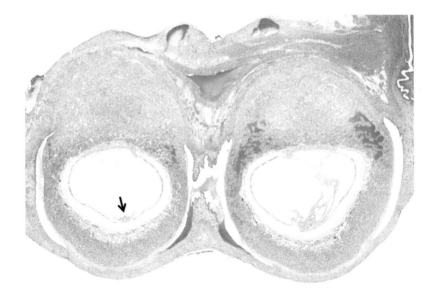

FIGURE 15.2 Photomicrograph showing a histologic preparation of the pregnant uterus (GD 8.5) demonstrating that the orientation of embryos cannot be controlled when the conceptus is embedded as an intact unit *in situ*. The left embryo is embedded in perfect cross section as shown by the bilateral symmetry of the neural tube (arrow), while symmetry is lacking in the tangentially sectioned embryo on the right. Histologic processing: neutral buffered 10% formalin, paraffin, H&E.

(H&E), although in some cases other embedding media (e.g., agar, plastic) or stains may be preferred to better preserve fine cellular structures or localize the distribution of particular molecules. A proficient developmental pathologist may undertake all three actions on a given section before stopping to record his or her findings. Less experienced investigators likely will be served better by completing each step separately to ensure that all organs and tissues of interest (Table 15.2) are subjected to an evaluation, that all elements in the section receive an appropriate level of analysis, and that all structural defects are described effectively.

Vehicle   Methanol

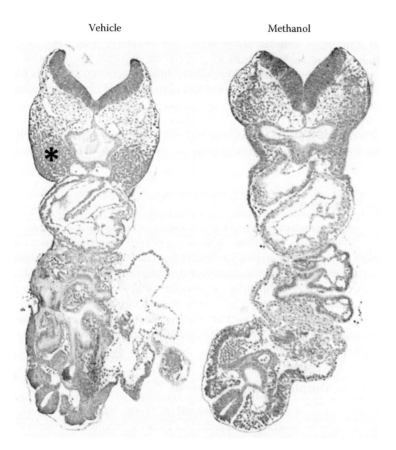

FIGURE 15.3 Photomicrograph demonstrating that equivalent positioning may be obtained in isolated embryos (shown here at GD 8.5; original length, approximately 1 mm). The vehicle-treated individual has elevating neural folds with sharp, curling tips and dense masses of mesenchyme (*) bracketing a pentagonal stomodeum (mouth primordium). The methanol-exposed embryo has flattened neural folds with blunt tips and reduced mesenchymal depots adjacent to a collapsed stomodeum; this combination of lesions presages cranioschisis (a neural tube closure defect [NTD] that affects the head). (Reproduced from Bolon, B. et al., *Teratology*, 49(6), 497, 1994, by courtesy of Wiley-Liss.)

The first step of a microscopic examination is to assess the section at a low magnification to observe the general arrangement of organs in the conceptus and regions with subdivided organs. These regions are differentiated in H&E-stained sections based on such zone-specific cytoarchitectural characteristics as the size, shape, color, and orientation of cells. This portion of the analysis will reveal overt defects affecting the blueprint or even existence of an organ. Basic features that may be observed at low magnification for each organ and the regions that comprise it include alterations in color or dimensions (i.e., thickness or area); the amount of blood and types of blood cells within blood vessels; and any well-demarcated lesions that have altered or effaced the normal architecture (e.g., foci of hemorrhage, infiltrating cells, and/or necrosis). Histopathologic evaluation using the 1×, 2×, 4×, and/or 5× objectives of a conventional research microscope is suitable for discerning such changes.

The second portion of a thorough histopathologic evaluation is to examine cellular features at an intermediate magnification to further define potential target tissues and cell populations within affected organs. Consistent cytoarchitectural differences from the appearance of similar cells in developmental stage-matched control samples provide preliminary evidence that a cell population may be abnormal. These assessments require a detailed knowledge of the unique cytoarchitectural attributes of each region and the major cell types that reside in the conceptus across many gestational and postnatal ages—knowledge that can only be gained by experience. Histopathologic evaluation for this part of the assessment typically is performed using the 4×, 5×, 10×, and/or 20× objective.

The third element of a properly designed histopathologic analysis is to fully describe the nature of the structural defects at the organ, cellular, and (if necessary) subcellular levels. This evaluation typically is performed using intermediate magnifications (i.e., the 4×, 5×, 10×, 20×, and 40× objectives), with sporadic recourse to higher magnifications (i.e., 60× or 100× objectives). The focus for this portion of the examination is to identify the location(s) of defects (e.g., affected cells and tissues) and the nature of the abnormalities using H&E-stained material. Special procedures done on serial sections (e.g., cell type–specific markers [Chapter 11] or cell kinetic assays [Chapter 12]) may be useful in defining the full list of major alterations. Successful data acquisition and interpretation generally requires comparison of potential lesions to the normal structures shown in a well-annotated rodent developmental anatomy atlas.[7,9,15,83,115,201,217,327,368,370,371,363]

Ultrastructural Evaluation

In some instances, complete understanding of a lesion and its pathogenesis may necessitate an ultrastructural examination to explore the subcellular changes responsible for creating lesions that can be seen by light microscopy. Procedures to prepare specimens from developing mice for transmission electron microscopy (TEM, typically used to assess internal structures of cells [see Figure 13.1]) or scanning electron microscopy (SEM, most often utilized to evaluate surface features [see Figure 13.2]) have been described previously (Chapter 13). Types of events examined at ultrahigh magnifications include normal developmental processes[236,277,337,398,407] as well as assessments of subtle defects in organelle structure—especially those engaged in cell metabolism[36,216,222]—and particular lesion patterns (e.g., defective closure[266]) and mechanisms (e.g., apoptosis[222,399,455]).

Special Methods for Morphological Assessment

More often, full appreciation of a novel developmental phenotype will require the use of one or more special procedures to showcase functional and molecular events occurring at the site of structural lesions. Numerous methods to explore such features have been devised, and detailed protocols for the techniques employed most frequently in conventional comparative pathology laboratories when conducting descriptive/diagnostic studies have been highlighted in previous chapters: *in situ* localization of nucleic acids (Chapter 10) and proteins (Chapter 11), and cell kinetic assays (Chapter 12). Many other applications also are utilized when undertaking hypothesis-driven studies in dedicated developmental biology and developmental toxicology laboratories, such as special microscopy instrumentation (e.g., confocal, differential interference contrast [DIC], multiphoton),[72,205,463] intravital labeling,[175,260,443] and embryo culture.[55,160] However, in our experience these options are seldom available in comparative pathology facilities, so they will not be considered further in this chapter. Interested readers may explore these topics further through a number of excellent reference books[86,92,302] and a review of the extensive literature for the developmental biology and developmental toxicology fields.

Scoring Strategies for Developmental Pathology Analyses

Macroscopic and microscopic evaluations both may be limited to qualitative interpretations (i.e., a specimen is considered as *normal* or *abnormal*) and detailed descriptions of the lesion spectrum, but in many cases a more precise assessment regarding the existence and impact of a potential phenotype may be gained by the application of a scoring scheme to generate semiquantitative data sets.[152] In our experience, scoring strategies evolve during the course of an extended mouse developmental pathology study as the criteria used to identify and score lesions tend to shift between the first and latest experiments needed to generate a statistically meaningful number of subjects. For this reason, a common practice is to use the early cohorts to find and describe major lesions that result from the experimental manipulation and then dedicate the later cohorts to defining distinct criteria needed to sort the lesions into different classes based on severity. A common scale for this purpose often includes five tiers: within normal limits (score = 0) or minimal (1+), mild (2+), moderate (3+), or marked (4+) lesions. Such a tiered scheme rapidly provides an ordinal data set suitable for testing hypotheses and discerning treatment-related differences among cohorts (see Chapter 18 for additional details regarding selection of appropriate group sizes and statistical tests for various kinds of pathology endpoints).

Scoring schemes must be well defined, reproducible, and capable of providing meaningful results.[152] Two important considerations to ensure the acceptability of the scoring scheme used in a mouse developmental pathology studies are to limit *diagnostic drift* and *measurement bias*. Diagnostic drift is the propensity for an investigator to render slightly inconsistent scores for essentially the same lesion over the length of a prolonged study or when reviewing large numbers of sections. Measurement bias is an accidental (or intentional), subjective tendency by the investigator to preferentially ascribe a certain outcome to animals in the manipulated group rather than to those in the control group. Both of these concerns may be exaggerated if a pathologist is acquainted with an animal's status (manipulated or control) when assigning a score. Fortunately, both issues may be addressed simultaneously by conducting a *coded* (*blinded* or *masked*) analysis of each affected organ or tissue at the final culmination of the study; the simplest means of completing this activity is to (1) collect all slides with sections of the structure of interest from all experiments in the study and then (2) examine them in random order, ideally in a single sitting or over a short number of days. The data set derived from this coded meta-analysis likely will require production of an additional summary report for the whole study because the scores given to animals during the coded assessment often may differ from the scores recorded in original reports made using data from the initial uncoded analysis undertaken for individual experiments within the study.

In our experience, many investigators design their mouse developmental pathology studies in a fashion that will not permit a meaningful coded analysis. This weakness usually reflects a desire to conserve research funds. The typical presentation is the placement of multiple tissue sections on a single slide, in which case unintended visualization of a lesion in a tissue not being scored may prevent assignment of an unbiased score for a lesion in a nearby organ. This situation usually is difficult to avoid in embryos, fetuses, and neonates as these routinely are processed as intact specimens rather than as isolated organs, but this measurement bias should be avoided in older animals that receive a full necropsy by limiting the number of tissues that are mounted on a single slide. An even less desirable practice is to take organ sections from all animals in the same group (i.e., manipulated or control) and place them together in rows on a single slide. This organization abolishes any opportunity for a coded assessment. Fortunately, this latter approach is rarely employed in mouse developmental pathology experiments—and never so if they have been designed with the participation of an experienced pathologist.

Resources for Structural Analysis

The greatest challenge for nonpathologists who engage in mouse pathology studies is unfamiliarity with the normal locations and appearance of tissues and organs. Atlases that illustrate the macroscopic appearance of structures in adult mice have considerable utility in defining the site, shape, size, and sometimes color of organs in nearly grown animals (e.g., near-term fetuses, neonates, and juveniles).[195,233,423] The macroscopic and microscopic depictions of embryonic and fetal characteristics in photographic[7,51,115,327,370,371,398] or hand-drawn[111,115,217,361] views within various developmental atlases often provide more relevant representations, especially for early periods of gestation (GD 12.0 and earlier) during which the structural resemblance of the evolving embryos to adults may be difficult to visualize. However, while suited to providing an initial orientation and to showing progressive changes in organ morphology at various discrete times during embryogenesis, developmental atlases do possess several limitations. The first two restrictions are that structures seen in a tissue section often do not match the features demonstrated in developmental atlases, or only are visible in a somewhat skewed orientation. Such discrepancies cast a degree of doubt onto conclusions regarding the identity of many structures. A third drawback of some developmental atlases is that they show features using stains other than H&E (e.g., cresyl violet for brain, which better reveals neuronal populations),[7,327] thereby highlighting a different set of cell populations and tissue boundaries than those that will be found on a conventional H&E-stained section. A fourth consideration is that developmental atlases have been assembled for only a few mouse strains, so in many cases structural identification represents an extrapolation made using material compiled from sections acquired for another mouse strain. This limitation is relatively minor and may be addressed efficiently and effectively by processing specimens from strain-, age-, and (where relevant) sex-matched individuals from the laboratory's mouse colony; a major added advantage of processing such concurrent control samples is that the anatomic orientation (i.e., plane of tissue trimming and sectioning) may be matched as well. A final constraint is that analysis of a complete organ set from a developing mouse typically requires evaluating multiple serial or step sections. Accordingly, proficient developmental pathologists must learn by practice to mentally assemble a 3D image of the individual's anatomy as they flip back and forth among many sections. Reliance on this laborious exercise likely will fade in the future due to the recent emergence of powerful noninvasive imaging techniques and digital developmental anatomy atlases that permit the reconstruction of computer-generated *virtual sections* into 3D whole-animal images capable of rotation around any axis.[70,115,206,332] These imaging modalities are unlikely to replace histopathologic analysis, which provides a better means of evaluating fine cytoarchitectural detail, but the ready availability of digitized and annotated 3D anatomical specimens clearly will become an important supplemental means of cross-checking phenotypic interpretations and matching the shifts in anatomic and molecular events during development.[48,93,104,115,201,217]

General Patterns of Damage Observed in Developing Mice

Theoretical consequences of damage to the developing mouse will depend on the stage at which the injury is incurred (Table 15.3).[42] Macroscopic malformations (*birth defects*) are most likely to result from damage that occurs during postimplantation development, with the nature of the defect depending on the precise stage (termed a *critical period*) at which the insult occurs (see Chapter 5 for more details on this topic). The stage dependence results from the constellation of divergent development processes taking place among distinct populations of differentiating cells in various organs[76,376,261] (see Chapter 4 for tables reporting the approximate timing of organ-specific developmental events), although other physiological processes (e.g., circadian rhythms) also play a role.[314] For example, defects due to flawed body axis specification usually follow impairment during either gastrulation (i.e., between GD 6.5 and GD 7.5) or early organogenesis (GD 8.0 to 10.0).[261] Cranioschisis (*cleft skull*, a neural tube closure defect [NTD] most often observed in mouse conceptuses as exencephaly [*exposed brain*]) due to persistent patency of the rostral neuropore (Figures 15.3 and 15.4), typically stems from cell disruption during early organogenesis (about GD 8.0 to GD 9.0),[38,237] while microcephaly (*small head*; Figures 15.3 and 15.5) with microencephaly (*small brain*) and neuronal heterotopia (ectopic neuron clusters) but no NTD is more likely if injury occurs later in organogenesis (GD 12.0 to GD 14.0).[19,31,98] Within a given organ, different regions and cell populations may have their own discrete periods of vulnerability; this principle has been effectively shown in the developing mouse brain, where such periods encompass much of gestation as well as the initial 7–10 days after birth (see Figure 5.2).[192,355,356] For a given critical period, mice with different genetic backgrounds vary in their susceptibility to injury,[183,305] possibly as a result of subtle disparities in the timing of organ development[211] or structural dissimilarities in the nature of the etiologic agent.[304] Superovulation increases the vulnerability of embryos to other insults,[280] indicating that due care must be exercised in interpreting the relevance of potential developmental lethal phenotypes that take place in transgenic experiments involving the transplantation of superovulated embryos.

Developmental injury may lead to outcome patterns other than induction of classic gross birth defects. Prior to implantation (i.e., from GD 0 to GD 4.5), embryos that experience severe damage typically will die, while milder harm may lead to developmental delays such as reduced size or slower acquisition of stage-specific features (Figure 15.6). Such delays often are reversible over time as surviving pluripotent stem cells gradually manage to compensate for the initial loss of stem cells.[338,394] After organogenesis has been completed (at GD 15.0), damage to the fetus usually will act either to slow the rate of body growth and organ enlargement or to engender microscopic changes and/or functional abnormalities rather than cause major macroscopic malformations. The exceptions to this rule are those organs (e.g., cerebellum, hippocampus) that undergo major peaks of cell proliferation during the neonatal period (see Figure 5.2),[356] disturbance of which may severely limit the structural organization of the mature organ. Postnatal exposure to some toxic agents may cause pervasive death in various cell populations, leading to microstructural changes and functional alterations but not grossly visible defects.[105,316] For this reason,

TABLE 15.3

Common Patterns of Damage during Prenatal and Postnatal Development

---

Preimplantation embryo (GD 0 to GD 4.5)

  Major developmental themes

    Conceptus consists of pluripotent stem cells

    Expansion of undifferentiated cells

    Preparation for implantation

  Consequences of damage

    Severe injury: diffuse cell death → embryonic death or malformation[163]

    Mild injury: partial cell survival → normal embryo or developmental delay

Postimplantation embryo (GD 5.0 to GD 15.0)

  Major developmental themes

    Conceptus consists of partially differentiated stem cells

    Organogenesis phase—with different critical periods for each organ (and region within the organ)

      Body axis and organ specification[324,397]

      Initial cell differentiation

  Consequences of damage

    Injury causes focal to diffuse cell death → gross malformation

    The pattern of anomalies (which organs/organ regions are affected) depends on the timing of the insult

Fetus and Neonate (GD 15.5 to ~PND 7)

  Major developmental themes

    Conceptus consists mainly of oligopotent and differentiated cells

    Growth of body and organs

    Terminal cell differentiation

  Consequences of damage

    Injury may cause cell death or delay cell adoption of mature structural and functional attributes

    Functional deficit ≫ malformation

Juvenile (~PND 7.5 to ~PND 28 [for males] or 35 [for females])

  Major developmental themes

    Conceptus consists of terminally differentiated cells, sometimes subject to replenishment by stem cells

    Cell, tissue, and organ maturation

    Addition of specialty structures (e.g., myelin, synapses)

  Consequences of damage

    Injury may inhibit cell function or cause cell death

    Functional deficit(s)

---

the morphologic assessments conducted to identify and characterize phenotypes in developing mice routinely should incorporate both macroscopic and microscopic assessments, and in many cases should be extended to include special procedures designed to provide histopathologic and mechanistic information simultaneously (e.g., immunohistochemical detection of proliferating or dying cells; see Chapters 11 and 12 for procedural details).

Common Developmental Pathology Phenotypes and Their Specific Causes and Timing

Prenatal Lethality

Disease or death may occur at any time during gestation (Table 15.4). The first clues that embryonic lethality may be occurring typically are skewing of the Mendelian ratio among the genotypes of newborn pups or weanlings and/or a modest to moderate decline in the anticipated litter size.[323] A common

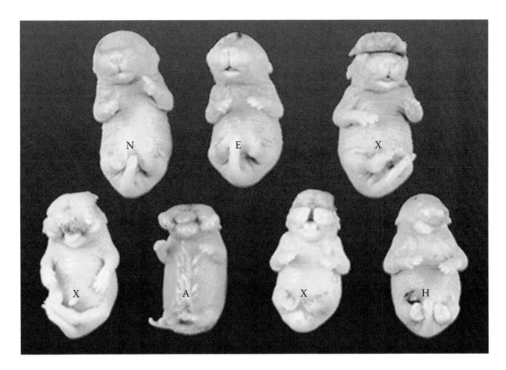

FIGURE 15.4   Neural tube closure defects (NTD) or cephalic dysraphism represents a continuum of craniofacial changes associated with abnormal sealing of the seam (*raphe*) formed by apposition of the elevated neural folds. Relative to an aged-matched control (N), near-term mouse fetuses (GD 17.5) exposed to methanol during neurulation (i.e., the period of neural tube closure, approximately GD 8.5 to GD 9.5 in the mouse) have lesions ranging from encephalocele (E; a small focal protrusion of meningeal-covered cerebral cortex on the frontal portion of the skull) to exencephaly (X; complete exposure of the cerebrum and midbrain) to anencephaly (A; total absence of the calvaria [dorsal skull], cerebrum, midbrain, and dorsal hindbrain [i.e., cerebellum]). One fetus has holoprosencephaly (H), or failure of the one original primary forebrain vesicle to differentiate into paired secondary vesicles. Compared to the narrow inverted-T conformation of the oral orifice in the control (N) animal, craniofacial defects associated with NTD usually present as either mild hypognathia of the maxillae (upper jaw), indicated by a triangular oral openings (E and X fetuses in the top row) or marked clefting of the snout (X fetuses of the lower row), aplasia of the cranium (all X and the A fetuses), and aplasia of craniofacial bones (maxillae and mandibles [lower jaws], most prominent in the H fetus) and eyes. (Reproduced from Bolon, B. et al., *Teratology*, 49(6), 497, 1994, by courtesy of Wiley-Liss.)

shift of this kind is to find a 2:1 genotypic ratio for a heterozygous (HET) mating pair (i.e., bearing one mutant [KO] and one wild-type [WT] allele), where the progeny are 67% HET and 33% WT instead of the predicted 25% WT, 50% HET, and 25% KO (a 1:2:1 ratio) typical of lines in which no *in utero* lethality is occurring. In some instances, the embryonic lethal phenotype is not fully penetrant, and some KO animals may remain viable throughout gestation (see section on *Lethality with Variable Penetrance*). Therefore, the absence of viable KO animals at birth and/or weaning indicates the existence of an embryonic lethal phenotype and will shift the focus of the analysis to the prenatal period.

The developmental stage at which embryonic lethality occurs may be estimated by the appearance of any abnormal embryos and/or the genotype of the residual extraembryonic placental tissue (collected by laser capture microdissection [see Chapter 10]) in any resorbed implantations sites. If all implantation sites contain structurally normal embryos with a WT or HET genotype, then the KO embryos likely expired before they could attach to the uterine wall and induce a decidual swelling (i.e., sometime between GD 0 and GD 4.5). If the implantation sites with embryos have a WT or HET genotype but additional implantation sites exist with either no embryo or a small, abnormal embryo, then the KO embryos likely implanted correctly only to die during the initial stages of assembling the definitive placenta (i.e., between GD 8.5 and GD 9.5). The KO embryos in the affected implantation sites often exhibit a greater proclivity to

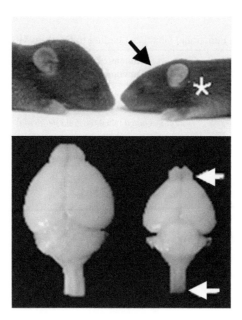

FIGURE 15.5 Microcephaly (small head) and microencephaly (small brain) in a juvenile *laggard* (*lag*) mouse, a spontaneous ataxic phenotype resulting from a splice mutation in the gene for kinesin-like protein 4 (*Kif14*). Relative to an unaffected littermate (left), the mutant (asterisk) is characterized at PND 12 by a flattened head (black arrow, upper panel), small brain (lower panel), and pallor of the olfactory bulbs and spinal cord (white arrows) due to hypomyelination. (Reproduced from Fujikura, K. et al., *PLoS One*, 8(1), e53490, 2013.)

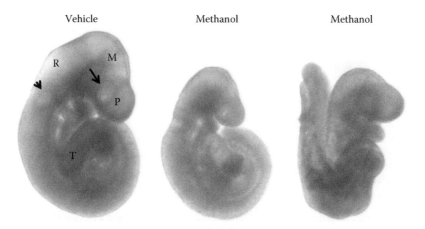

FIGURE 15.6 Variants of developmental delay observed in cultured CD-1 mouse embryos early in organogenesis (here exhibiting craniofacial features consistent with GD 9.5). The vehicle-treated embryo has a bulbous prosencephalon (P [forebrain]), mesencephalon (M [midbrain]), and rhombencephalon (R [hindbrain]) as well as prominent optic (eye [arrow]) and otic (ear [arrowhead]) anlagen. The methanol-treated embryos possess a smaller head and body (both individuals). In addition, the right embryo has not initiated *turning* as indicated by displacement of the tail (T) to a position directly behind the head rather than in its normal position at this stage (i.e., curled to the right side of the body).

TABLE 15.4

Prenatal Anatomic Lesions and Functional Changes Leading to Death in Mouse Embryos and Fetuses

Stage of Development	Gestational Days (GD)[a]	Anatomic Lesion or Functional Change	References
Pre- or peri-implantation	0–5.5	Absence of cleavage divisions	[416]
		Morula degeneration	[342,466]
		Failure to *hatch* (shed the *zona pellucida*)	[164]
		Aberrant formation/maintenance of inner cell mass (embryo anlage)	[12,164,410,466]
		Abnormal formation/maintenance of trophectoderm (placenta anlage)	[238]
		Disruption of the mitochondrial respiratory chain	[188]
		Trauma (from microinjection)	[53]
Gastrulation	6.5–9.0	Altered differentiation of one or more germ layers	[100,345]
		Defective orientation and segmentation (abnormal body plan)	[59,272,362,405]
Early organogenesis	8.0–10.5	Altered cardiac and/or vascular formation	[21,24,138,193,219,263,285, 312,406]
		Anomalous heart function	[245,456,464]
		Defective erythropoiesis (yolk sac *blood islands*)	[24,219,263,425,446,461]
		Hemorrhage	[161,312]
		Neural degeneration or dysgenesis	[28,434]
		Nutrient delivery disruption	[146]
		Visceral hypoplasia	[245]
Late organogenesis	11.0–15.0	Altered cardiac and/or vascular formation	[14,79,128,179,252,412,415]
		Anomalous heart function	[34]
		Defective erythropoiesis	[131,197,264,276,298,306,318, 330,366,367,385,409,442]
		Growth retardation (unknown cause)	[71]
		Organ-specific or multisystemic defects (key sites: brain, heart, hematopoietic)	[25,27]
Fetal growth spurt	15.5–18.5	Altered cardiac and/or vascular formation	[259,267,404]
		Anomalous heart function	[413,427]
		Defective erythropoiesis (liver, spleen, and/or bone marrow)	[57,106,241,404]
		Hemorrhage	[193]
		Intrauterine growth restriction	[88]
		Arthrogryposis	[200]
		Necrosis in critical organs	[148,439]

Source: Adapted in part from Copp, A.J., *Trends Genet.*, 11(3), 87, 1995.
[a] The plug-positive day is designated GD 0.

undergo resorption (i.e., fragmentation and eventual removal) than do the associated embryonic (tropho-blast) and maternal (decidua) contributions to the placenta.[4,320] If implantation sites are populated with WT, HET, and KO embryos at midgestation (i.e., GD 12.5), then death occurred at a later stage of gestation or immediately after birth. Dead pups commonly are cannibalized by the dam, so an assessment of viability just before birth (i.e., GD 17.5 or GD 18.5) may be needed to provide the first indication for the existence of a neonatal lethal phenotype. The range of vulnerable developmental stages often dictates that a developmental pathology study will require two or more experiments to refine the likely time of *in utero* death (see Chapter 6 for a more nuanced discussion).

Postnatal Lethality

The period from birth to puberty also is a common time frame for the manifestation of lethal developmental phenotypes (Table 15.5). The first few minutes after the placental attachment is lost represent a period of enormous physiological challenge and rapid, life-altering adaptation to a new homeostatic equilibrium. Many different disturbances may lead to a grossly normal animal's abrupt or gradual demise during the perinatal period (i.e., the first 24–96 h after birth). Death on delivery (i.e., in 1–2 h or less of birth) is a hallmark of conditions that thwart the transition from maternal support *in utero* to independent intake of oxygen, while acute but not instantaneous expiration (i.e., in 6–36 h) typically is an outcome of insufficient nutrient intake or impaired energy metabolism. Common causes of immediate death are an inability to breathe, leading to cyanosis (Figure 15.7); circulatory difficulties such as cardiovascular malformations (Figure 15.8), chronic congestion (Figure 15.9), massive internal hemorrhage (Figure 15.10), or defective hemostasis; defective glucose metabolism; an inability to nurse (Figure 15.11); and dehydration. The pathogenesis of these effects may be simple and obvious, or the mechanism may be more subtle. For example, death by respiratory insufficiency will be incontrovertible if a mutant pup possesses small[76,162] or no[223,291] lungs (Figures 15.12 and 15.13) or cannot inflate its alveoli (Figure 15.7)[474], but breathing problems may arise from subtle lesions in other systems such as neuronal reductions in brainstem autonomic centers that control respiratory rate,[33,119,190] abnormal synaptic transmission at neuromuscular junctions,[94,453] skeletal malformations (of axial bones like the ribs and vertebrae) that constrain expansion of the thorax,[215,255] and skeletal muscle paralysis.[172,321] The key to successful analysis of neonatal lethal phenotypes in engineered animals is to confirm that pups with the genotype of interest actually were present in the litter. Since dead, dying, or severely malformed pups commonly are eaten by the dam,[65,278,386,417,469] a *death watch* may be required to identify and reclaim the nonviable animals before they are lost to analysis. Unfortunately, many mice tend to deliver their litters during the dark cycle, which is scheduled at night in conventional mouse colonies. A practical means of performing a death watch without having to invert the light cycle for the entire colony is to place near-term dams in a clear cage in a room equipped with a red light (to permit viewing of the delivery and new litter without greatly disturbing the dam) and then to arrange a relay of personnel to take 3–6 h shifts in which regular checks (usually at 1 or 2 h intervals) will be made to assess the viability of newly delivered pups.

Many lethal phenotypes in the juvenile period may present as gradual but massive declines in growth as measured in stunting (Figure 15.14) and a decrease in absolute body weight or the rate of body weight gain. In such instances, death often occurs between PND 14 and PND 28. Defects in the generation and/or differentiation of many different constituents may yield this outcome, and routine anatomic pathology and clinical pathology tests of use in diagnosing and grading similar diseases in adult mice often may be adapted for the analysis of samples from juvenile animals. However, adjustments in the assays may be needed (especially for clinical pathology endpoints [Chapter 8]) that will permit meaningful data to be gained from smaller sample sizes.

Phenotypes initiated during the gestational or lactational periods need not conclude during early postnatal life. Instead, developmental lesions may cause adult mice to suffer premature senescence or early death due to degenerative[81,308,325] or neoplastic[82,419] diseases. Therefore, a phenotype characterized by a foreshortened lifespan should be investigated by examining structural, functional, and molecular changes at several time points, beginning no later than weaning, as experimental designs evaluating disease progression over time are the most straightforward means of identifying whether or not the cellular disturbances responsible for the condition arise *in utero* or soon after birth.

Strain-Based Differences in Responsiveness

Some mouse strains have inherently higher vulnerability to agents that can induce developmental defects. For example, SWV embryos are much more likely to exhibit defective neural tube closure,[134,135,305] while A/J mice tend to respond with a higher level of cleft palate.[102,103,210] The basis for such differences is thought to reside chiefly in the embryo.[138,305] The enhanced susceptibility of a given line for a particular defect may hold true for many teratogenic stimuli but likely will not be true in all instances.[183]

TABLE 15.5

Postnatal Anatomic Lesions and Functional Changes Leading to Death in Neonatal and Juvenile Mice

Stage of Development	Primary Mechanism	Postnatal Days/Hours[a]	Developmental Phenotype	References
Neonate	Cardiovascular defects	0.1–48 h	Cardiac function defects (presumed)	[90,213,231]
			Cardiac malformations	[22,23,221,312,400,427]
			Great vessel (outflow tract) defects	[73,126,130,153,189,221,268, 307,375,465]
			Hemorrhage in vital organs	[87,357,380,389,395,422]
			Lung vessel anomalies	[169,390]
			Multiorgan dysfunction	[331,457]
	Hepatic insufficiency	2–4 days	Necrosis	[142]
	Hematopoietic defect	8–72 h	Reduced definitive hematopoiesis	[22,274]
	Homeostatic disruption	8–16 h	Hypoglycemia	[75,142,162,234,244,372,441]
			Hyperglycemia	[1,208]
			Water loss (transepidermal)	[97,144,177,281,293,376]
		1–2 days	Renal insufficiency	[52,109,117,239,240,283,295, 303,334,346,365]
	Immunodeficiency	<48 h	Susceptibility to infection	[185,286]
	Inability to suckle	0.5–1.5 days	Abnormal neural regulation	[6,56,95,110,155,230,265, 284,341,426,454,473]
			Reduced mobility	[95,100,196]
			Skeletal malformations in head	[212,248,424,471]
		3–7 days	Immune-mediated oral ulcers	[257]
			Unknown	[174,185]
	Intestinal dysgenesis	8–24 h	Segmental agenesis/atresia	[386]
			Segmental aganglionosis	[52,54,117,295,334,365,381]
	Metabolic disturbance	<1 day	Hepatic dysfunction, malnutrition	[445]
	Respiratory distress	Immediate	Lung aplasia (absence)	[223,291,377]
		0.1–8 h	Lung hypoplasia (small size)	[66,77,123,162,169,343,390, 396,428]
			Skeletal malformations—face	[207]
			Skeletal malformations—thorax	[5,50,215,235,255,258, 279,317]
			Immature terminal (distal) airways	[10,74,162,169,343,396, 422,428]
			Altered respiratory muscle function	[150,172,300,321,411]
			Abnormal neurotransmission	[364]
		<12 h	Upper airway defects (major)	[107,113]
		1–24 h	Abnormal surfactant production	[69,169,289,290]
			Altered neural cell function	[33,119,122,155,156,190, 202,297,353,388,402,426, 437,467]

(Continued)

TABLE 15.5 (*CONTINUED*)

Postnatal Anatomic Lesions and Functional Changes Leading to Death in Neonatal and Juvenile Mice

Stage of Development	Primary Mechanism	Postnatal Days/Hours[a]	Developmental Phenotype	References
Neonate	Respiratory distress	1–24 h	Deficient glucocorticoid signaling (via delayed lung maturation)	[74,299]
			Neuromuscular junction anomalies (yielding respiratory failure)	[94,132,149,292,453]
		5–40 h	Inadequate fluid clearance	[186]
	Traumatic birth	Immediate	Dystocia (trauma)	[92,116,271,448]
		0.1–24 h	Defective hemostasis	[87,96,203,357]
			Hemorrhage (internal, massive)	[18,259,390,395]
	Uncertain	<1 day	Unknown	[174]
Juvenile	Cardiovascular defects	1–21 days	Hemopericardium and/or pulmonary hemorrhage	[225]
	Endocrine imbalance	7–21 days	Pituitary developmental defects	[118,271]
	Immune dysfunction	7–28 days	Exuberant inflammation	[232,243]
			Immunodeficiency	[429,472]
	Metabolic imbalance		Cachexia (severe weight loss)	[209]
	Neural dysfunction	15–30 days	Massive neuronal degeneration	[127]
	Neuromuscular deficits	15–30 days	Neuromuscular junction anomalies	[309]
	Renal failure	14–28 days	Altered glomerular filtration	[61,310]
	Respiratory dysfunction	15–30 days	Emphysema with *cor pulmonale*	[40]
	Thoracic abnormality	1–35 days	Restricted thoracic cavity volume	[374]
	Urea cycle dysfunction	15–30 days	Citrullinemia, hyperammonemia	[329]

[a] The time range gives the typical (*average*) time span until lethality.

However, heightened sensitivity with respect to one type of defect does not imply that a vulnerable strain will be more susceptible to the induction of additional defects relative to other mouse strains.[137] Therefore, the research team should understand the baseline incidence of developmental defects of interest for their mouse strain or stock before choosing animals with a given genetic background as a model for developmental pathology studies.

Sex-Based Differences in Responsiveness

In some cases, males and females exhibit differing responses to events that affect development. For example, mice with null mutations of genes for two DNA mismatch repair molecules, *Msh2* and *Trp53*, die *in utero* if they are female but live to be born before succumbing to tumors within a few months ($t_{1/2} = 73$ days) if they are male.[82] Interestingly, this female sensitivity is dependent on modifier loci near the affected genes because females born to parents from other strains also survive to be born but share the tendency to premature death from tumors ($t_{1/2} = 65$ days).[419] In addition to the genetic

WT   KO

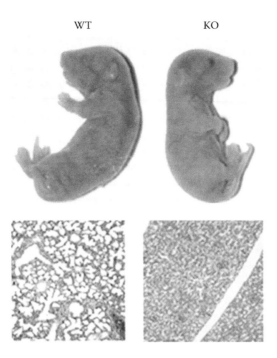

FIGURE 15.7   Cyanosis (blue discoloration) due to diffuse atelectasis (deficient alveolar inflation) and increased connective tissue in lungs of a neonatal (PND 0) mouse with a null mutation (KO) in the gene for extracellular signal–regulated kinase 3 (*Erk3*). A wild-type (WT) littermate has pink skin, indicating that the open alveolar spaces have permitted a sufficient amount of oxygen to enter the blood and reach tissues. (Reprinted from Klinger, S., Turgeon, B., Lévesque, K., Wood, G.A., Aagaard-Tillery, K.M., and Meloche, S., Loss of Erk3 function in mice leads to intrauterine growth restriction, pulmonary immaturity, and neonatal lethality, *Proc. Natl. Acad. Sci. USA*, 106(39), 16710–16715. Copyright [2009] National Academy of Sciences, U.S.A.)

background, other physiological influences such as the dosage of sex-linked genes,[194] intrauterine position relative to individuals of the opposite sex,[176] and maternal nutrient intake[228] may sway prenatal and/or postnatal phenotypes of male and female conceptuses in different ways. Toxicants may induce differential effects in male and female conceptuses; in some cases, such agents act specifically to disturb processes controlled by sex-specific endocrine pathways,[8,35] but in other instances no common motif of endocrine disruption or molecular similarity has been identified to account for the greater susceptibility of one sex.[383,431] Commonly postulated explanations for such divergence include complex sex-linked modulation of one or several genes[13,32,116,431] and sex-based differences in vulnerability of certain cell populations or expression of particular proteins.[204] Sex-related differences in the incidence of developmental defects have been recorded for spontaneous defects (e.g., cleft palate in the A/J mouse strain[214]).

Lethality with Variable Penetrance

The simplest lethal phenotypes in developing mice reliably induce a common pattern of defects at the same sites in all individuals expressing a genetic mutation or exposed to a toxicant during the organ's critical period of development. However, the real world features many cases in which occurrence of a lethal phenotype follows a more complex pattern (Figure 15.14). One partially penetrant phenotype is that many null mutant embryos will die during early or midgestation but that a small cohort of them will survive to be born,[11] and perhaps even to reach weaning or adulthood.[40,71,78,253] An alternative phenotype is that a few embryos may die at one or more stages during gestation, but attrition occurs mainly over a span of days or weeks after delivery of the pups.[225,247,271,363] In many instances, an explanation for the

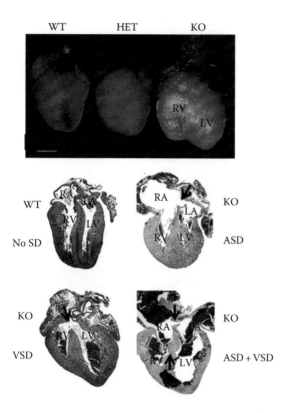

FIGURE 15.8 Defective cardiac septation in mice with an engineered null mutation (KO) of the endothelial nitric oxide synthase (*eNOS*) gene. Relative to age-matched mice with either two (wild-type [WT]) or one (heterozygous [HET]) *eNOS* alleles, KO animals with no *eNOS* alleles exhibit marked enlargement (*cardiomegaly*) of the right (RV) and left (LV) ventricles at PND 2 (upper row). Frontal heart sections revealed intact atrial and ventricular septae between cardiac chambers in WT mice, while atrial (ASD) or ventricular (VSD) septal defects (arrows), or both, were common in KO animals. LA, left atrium; RA, right atrium. (Reproduced from Feng, Q. et al., *Circulation*, 106(7), 873, 2002. With permission of the American Heart Association, Inc.)

divergent penetrance has not been defined. Some factors known to impact the penetrance include the genetic background, gene dosage, and maternal age.[109,121,127,253,256,344,411]

Dominant Phenotypes

Some genetically mediated lesions occur in heterozygous (HET [+/−]) mice as well as in homozygous null mutant (KO [−/−]) mice, indicating that the developmental phenotype is dominant (Figure 15.14). Fully dominant phenotypes have lesions that are equivalent in kind and degree in both HET and KO individuals, while semidominant phenotypes exhibit changes in HET mice that are similar in kind but less in degree relative to those observed in KO animals. Mechanisms associated with such phenotypes typically result from an abnormal protein expression pattern and/or generation of an altered protein rather than an inappropriate level of the wild-type (WT [+/+]) molecule. Specific instances include too few protein molecules to provide a complete effect (i.e., *haploinsufficiency*)[99,107,287,381,396]; a gene product that is overactive or has an extended half-life (i.e., *hypermorph*)[384,438] or hampers the effective function of its wild-type counterpart (i.e., *antimorph*, or dominant negative)[275]; or an altered protein that now plays a new role (i.e., a *neomorph*).[340,468] The key to confirming the existence of a dominant phenotype is to design developmental pathology studies so that HET mice are given the same phenotyping analysis accorded to their KO and WT littermates. Further details for investigating potential dominance may be obtained from other resources.[302,322,359]

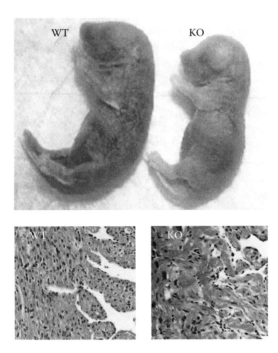

FIGURE 15.9   Null mutations in the gene for nonmuscle myosin heavy chain II-B yield knockout (KO) mice with chronic hepatic congestion (shown by a wide, dark band in the abdomen), small body size, and pallor relative to wild-type (WT) littermates. These lesions arose from congestive heart failure due to myocardial dysgenesis (indicated by the larger size and random arrangement of ventricular myofibers). The dome-shaped calvarium (skull) of the KO animal is secondary to hydrocephalus, caused by excessive accumulation of cerebrospinal fluid in the lateral ventricles; this finding was not linked to the cardiac phenotype. (Reprinted from Tullio, A., Accili, D., Ferrans, V.J., Yu, Z.-X., Takeda, K., Grinberg, A., Westphal, H., Preston, Y.A., Adelstein, R.S., Nonmuscle myosin II-B is required for normal development of the mouse heart, *Proc. Natl. Acad. Sci. U.S.A.*, 94(23), 12407–12412, Copyright 1997, National Academy of Sciences, U.S.A.)

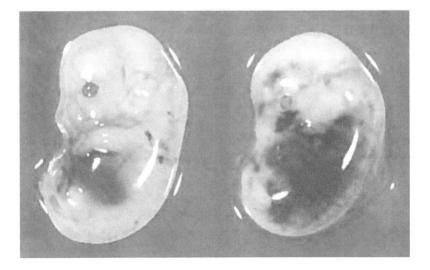

FIGURE 15.10   Freshly harvested mouse embryos (GD 12) demonstrating the divergence in appearance between a few punctate foci of hemorrhage due to capillary rupture during dissection (left) and extensive, coalescing hemorrhages resulting from vascular fragility (right) linked to an engineered disruption of a gene expressed in vascular walls.

WT KO

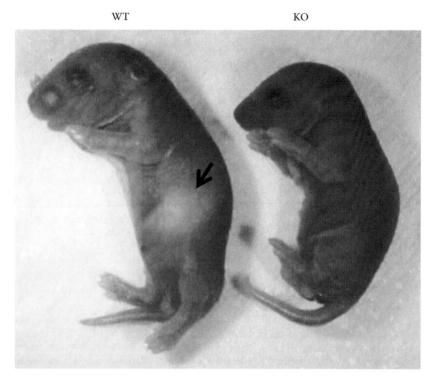

FIGURE 15.11 Mice with null mutations (KO) of cardiotrophin-like cytokine (*Clc*), one component of the heterodimeric receptor for ciliary neurotrophic factor, die by PND 1 from dehydration and starvation resulting from deficient numbers of brainstem neurons needed to mediate nursing. The KO animals have no white area on the left abdominal wall to indicate the position of the milk-engorged stomach, while wild-type (WT) littermates have visible *milk spots* (arrow). The slightly smaller size of the KO pup coupled with the distinct wrinkling of the skin over the ventral abdomen provide further evidence of hours-long starvation and dehydration, respectively. (Republished from Zou, X. et al., *Vet. Pathol.*, 46(3), 514, 2009, by courtesy of Sage Publications.)

Maternal Impact on Developmental Phenotypes

Detailed phenotypic analysis of developing mice may end inconclusively if the cause of the lethality does not reside in the impact of a gene alteration or treatment on the individual conceptus (embryo or fetus, with its placenta), neonate or juvenile. If necessary, the next phase of a developmental pathology study will involve evaluation of various maternal parameters with the capacity to influence development in the offspring (Table 15.6). Endpoints of potential interest may include anatomic features (e.g., mammary gland maturity[173,247,296]), behavioral abnormalities,[147] or molecular changes (e.g., hormonal imbalances[147]). Broad classes of etiologies by which maternal changes affect the development of the progeny include modified microenvironmental conditions,[165] inadequate oxygen delivery,[30,141,301] or altered metabolic status[134,165,254,392] at the implantation sites. Accordingly, maternal tissue specimens commonly collected for evaluation in a developmental pathology study should include blood and lung (to rule out inadequate oxygenation), mammary gland (to examine the adequacy of neonatal nutrition), and reproductive organs (to assess the maternal ability to sustain pregnancy) (see Table 15.2). Maternal toxicity,[63,220,326] stress,[49,348] and nutrition[154,229,470] also may impact the expression of developmental abnormalities in her offspring—including subtle anomalies in organ function that remain long after the progeny has reached maturity. Surprisingly, maternal immune stimulation reduces developmental damage induced by many different teratogenic stimuli.[70,178,184] A possible reason for this unexpected reduction in embryonic malformation is that modulation of the cytokine milieu within the maternal system produces a secondary readjustment of signaling cascades within the extraembryonic and embryonic tissues, thereby achieving a more harmonious cytokine balance within the conceptus as well.[178]

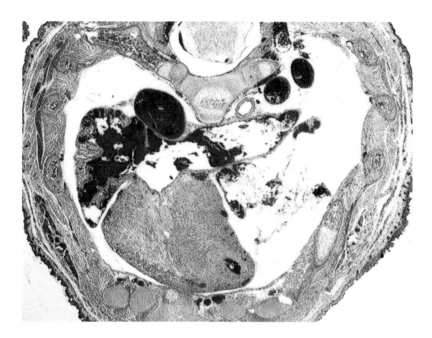

FIGURE 15.12 Transverse section through the rostral thorax of a near-term (GD 18) mouse fetus bearing a null mutation of the fibroblast growth factor 10 (*Fgf-10*) gene,[289] demonstrating the absence of lung lobes in the spaces that bracket the heart. Histologic processing: neutral buffered 10% formalin, paraffin, H&E. (Image provided courtesy of Dr. Dimitry Danilenko, Genentech, Inc., South San Francisco, CA.)

Developmental Responses to Injury

Basic Reactions to Damage during Development

A mouse conceptus essentially acts as a transient parasite toward its maternal host. The bases for its distinctive form and function, especially during the initial stages of gestation, are that embryos comprise a body of rapidly dividing totipotent and pluripotent stem cells with a broad tolerance for low-energy and low-oxygen conditions.[432] These factors allow more tolerance for cellular conditions that would be decimating to the more differentiated and oxygen-dependent tissues of the later-stage embryo and fetus. Despite these adaptations to the specialized environment within the uterus, the basic reactions of the embryo/fetus and placenta to damage are comparable to the characteristic responses that occur in cells and tissues of adult animals. Indeed, this equivalency provides grounds for confidence that a pathologist who is comfortable with evaluating genetic- and toxicant-induced lesions in adult mice also should be fairly adept at translating his or her store of knowledge to the developmental pathology arena.

At the macroscopic level, four classes of gross developmental lesions typically are defined based on the distinct mechanisms responsible for their induction. A *malformation* is a localized defect in organ structure caused by genuine (intrinsic) tissue damage leading to ablation of the cells that were to have been used to assemble the organ. Classic examples of malformations in developing mice include axial patterning abnormalities like neural tube closure defects[10,26,109] (Figures 15.3 and 15.4),[13,38,171] cleft palate[141,151,192] (Figure 15.15), skeletal patterning anomalies[114,384] (Figure 15.16), craniofacial malformations[94,362] (Figure 15.4),[108,399] and anophthalmia[121] (Figure 15.17). A *deformation* is a structural defect caused by an extrinsic mechanical force acting to distort but not destroy a nearby normal tissue. For instance, limb defects in mice with oligohydramnios (deficient amniotic fluid) have been attributed to collapse of the placental membranes against the embryo, with secondary constriction of the limbs against the body.[218] A *disruption* is a structural defect induced by an extrinsic insult that actually promotes destruction of normal tissue; a good illustration of this phenomenon is strangulation of the distal

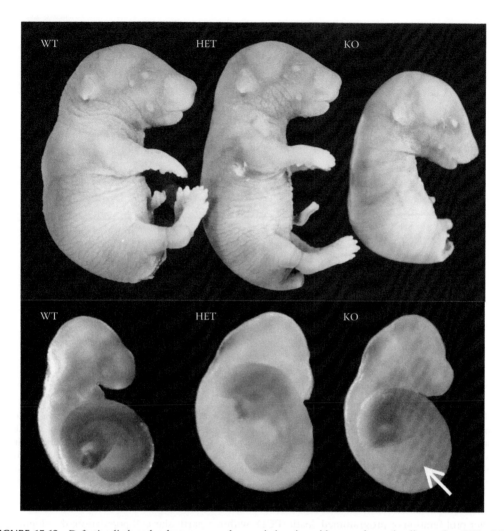

FIGURE 15.13 Defective limb and pulmonary morphogenesis in mice with an engineered null mutation (KO) of the fibroblast growth factor 10 (*Fgf-10*) gene. Relative to age-matched littermates with either two (wild-type [WT]) or one (heterozygous [HET]) *Fgf-10* allele, near-term (GD 18) KO fetuses with no *Fgf-10* alleles lack all four limbs (upper row). This absence stems from an inability to form limb buds during early gestation (here shown at GD 9.5 [lower row], with the arrow denoting the position of the forelimb bud). The sloping ventral wall of the thorax in the KO fetus provides a subtle indication that the contents of the thoracic cavity are diminished as a consequence of lung agenesis. (Reproduced from Min, H. et al., *Genes Dev.*, 12(20), 3156, 1998, by courtesy of Cold Spring Harbor Laboratory Press.)

limbs or digits by amniotic bands, thereby leading to deep grooves or amputations. A *dysplasia* occurs when the primary defect is abnormal tissue organization. Some mutant genes seem to produce defects by several of these mechanisms.[84,355] A suitable generic term for macroscopic lesions regardless of the cause is *developmental defects*.

At the microscopic level, multiple kinds of lesions may be observed in tissues of developing mice. Cell *degeneration* is a potentially reversible perturbation of cell function that often manifests as arrested cell development (Figure 15.18) with or without subtle changes to cytoplasmic and/or nuclear structure (e.g., altered staining properties, vacuolation). In contrast, *necrosis* represents an irreversible loss of a specific cell population or tissue (Figure 15.19), the absence of which will either delay or prevent further differentiation of the affected organ and any adjacent elements with which the damaged cells normally interact. *Heterotopia* (ectopic cells) are collections of misplaced cells, often as a result of abnormal cell migration within a reorganizing organ (Figure 15.20). *Kinetic defects* present as an excess or reduction

FIGURE 15.14 Phenotypic variability in three juvenile (PND 21) mice with altered expression of collagen X (Col10). The top two animals carry a null mutation (*knockout* [KO]) in the *Col10a* gene, but only the middle KO mouse exhibits a phenotype (stunted growth). The same stunting along with kyphosis (excessive dorsal curvature of the thoracic spine) is observed in the bottom mouse, which is transgenic (TG) for a *Col10a* variant that possesses dominant interference for the action of the endogenous *Col10a* alleles. Small KO and TG individuals die at three weeks of age. Bar = 1 cm. (Reproduced from Gress, C.J. and Jacenko, O., *J. Cell Biol.*, 149(4), 983, 2000, by courtesy of The Rockefeller University Press.)

in either proliferation or programmed death (PCD) within a particular cell population; developmental pathology studies generally must explore both of these processes (see Chapter 12) since phenotypes resulting from more cell proliferation resemble those due to less PCD, and lesions produced by low proliferation mimic those associated with heightened PCD. Any or all of these cell-based mechanisms as well as other causes (e.g., *hemorrhage* [Figure 15.10]) may induce macroscopic developmental defects. Interestingly, *inflammation* in response to necrosis (or other insults) is not a major response in embryonic tissues as the developing immune system does not become functionally mature until the fetal period, but a progressively more robust inflammatory reaction is mounted by neonates and juvenile animals as a means of clearing dead cells and attempting to control infectious agents.

Extensive cell death in older embryos (approximately GD 8.0 and later), fetuses, and occasionally neonates likely will induce major structural defects due to the limited ability of partly differentiated stem cells to replace differentiated cells that were decimated late in development.[60,107,112,418,433] In contrast, early embryos are comprised of totipotent and pluripotent stem cells, which in many cases can proliferate to completely reverse tissue damage induced by nonlethal stimuli. Restoration of depleted cells by compensatory growth requires time, so animals recovering from an early episode of sublethal damage often exhibit a *developmental delay*. The slowed growth often presents as alterations in the dimensions and/or weight of the whole body, one or more organs, or both,[336,338,394] but metabolic parameters also may be altered.[249] The developmental delay may be temporary (i.e., compensatory growth leads in time to complete reconstitution of the structure) or persistent (if partial regeneration is associated with a permanent reduction in size).

TABLE 15.6

Maternal Abnormalities Associated with Apparent Perinatal Lethal Phenotypes in Mouse Pups

Stage of Development	Primary Maternal Abnormality	Developmental Phenotype	References
Prenatal	Hypoxia	Anemia	[301]
		Arterial occlusion	[140]
		Low-oxygen atmosphere	[30]
	Endocrine alterations	Female hormone fluctuations	[145]
		Stress	[45,46,62,168,347,348]
	Inflammation	Bacterial molecular motifs	[339]
		Hyperthermia (fever)	[261,387,433]
	Malnutrition	Fat overload	[262]
		Protein deficiency	[392,393]
	Metabolic disease	Diabetes mellitus	[133,165,254,420]
	Uterine microenvironment	Level of growth factor–mediated signaling	[89]
Postnatal	Agalactosis (no lactation)	Abnormal mammary gland differentiation	[124,173,247,296,315,369,391,458]
		Defective milk ejection	[167,335,436,462]
		Pituitary hormone deficiency	[271]
	Behavioral problems	Cannibalism (presumably dead or badly malformed pups)	[65,278,386,417,429,469]
		Cannibalism (presumed) due to excess handling	[349]
		Nurturing deficiency (brain-based)	[44]
		Rejection (ignored or killed)	[147,428,440]
	Delivery issues	Prolonged gestation	[436]

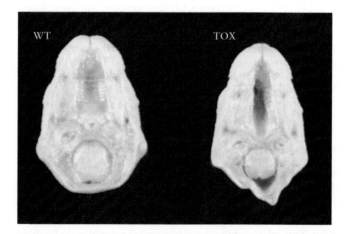

FIGURE 15.15 Palatoschisis (*cleft palate*) in a near-term (GD 17.5) mouse fetus following daily maternal inhalation of methanol throughout organogenesis (GD 6 to GD 15, a common schedule for mouse teratogenicity studies). Compared to a wild-type (WT) littermate with its intact hard palate, the absence of a palate in the exposed (TOX) animal has made the nasal cavity visible when viewed via the oral cavity. Processing: fixation in Bouin's solution.

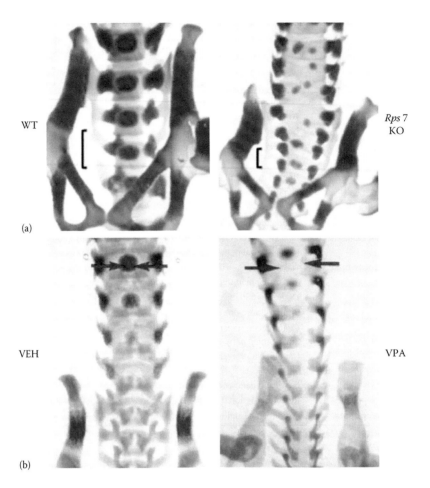

FIGURE 15.16 Axial patterning defects in the vertebral column of near-term (GD 18) mouse fetuses are highlighted using double staining with Alcian blue (for cartilage) and Alizarin red (for bone) to compare the position, shape, and differentiation state (as revealed by the distinct double staining patterns). (a) Relative to age-matched wild-type (WT) controls, fetuses with a null mutation (KO) of the ribosomal protein S7 (*Rps7*) exhibited widespread disorganization of the ossification centers (i.e., single or bipartite [divided] red foci not aligned along the midaxial plane) in the lumbar and sacral regions as well as abnormal fusion of several sacral vertebrae (brackets). (Reproduced from Watkins-Chow, D. et al., *PLoS Genet.*, 9(1), e1003094, 2013.) (b) Compared to age-matched vehicle-treated (VEH) controls, fetuses exposed to valproic acid (VPA) on GD 9 had *spina bifida occulta* as revealed by the greater distance between the margins of the vertebral arches (at arrow tips). (Ehlers, K. et al.: Valproic acid-induced spina bifida: A mouse model. *Teratology.* 1992. 45(2). 145–154. Copyright Wiley-VCH Verlag GmbH & Co. KGaA. Reproduced with permission.)

Background Incidences of Birth Defects in Developing Mice

For developmental defects, the incidence (i.e., the number of new cases that arise in the population during a given length of time) generally is low in developing animals of most mouse strains and stocks in the absence of some mutation or exposure to a developmental toxicant. The incidence of a given defect as well as the degree to which it is propagated to the next generation usually will vary with their impact on the animal's viability. This relationship allows congenital anomalies to be divided into a number of categories based on their biological significance.[319]

Minor skeletal incongruities comprise the first category of developmental defects. These changes are incidental, and thus typically are seen quite frequently when performing mouse developmental pathology studies. Common kinds of such *normal abnormalities* include renal pelvic cavitation

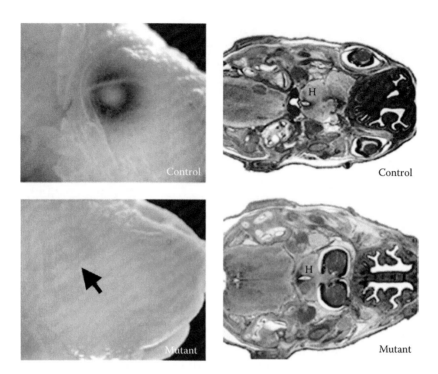

FIGURE 15.17 Anophthalmia in a mouse fetus (GD 16.5) bearing a null mutation of the gene paired-like homeodomain 2 (*Pitx2*). Relative to age-matched wild-type controls (top row), mutants (bottom row) lacking the Pitx2 homeobox protein have no eye remnants either macroscopically (left image [arrow denotes the usual location]) or microscopically (right image). (Reproduced from Evans, A. and Gage, P., *Hum. Mol. Genet.*, 14, 3347, 2005, by permission of Oxford University Press.)

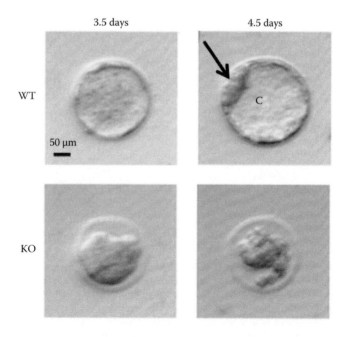

FIGURE 15.18 Preimplantation embryos exhibit lesions like morula arrest and degeneration. Relative to wild-type (WT) controls, stage-matched embryos with null mutations (KO) of RNA-binding motif protein 19 (*Rbm19*) produce a large number of cells but do not advance to become cavitated blastocysts with distinct blastocoele (cavity [C]) and inner cell mass (arrow) after one day in culture. (Reproduced from Zhang, J. et al., *BMC Dev. Biol.*, 8, 115 (14p), 2008.)

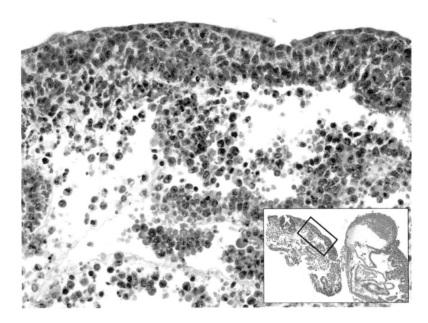

FIGURE 15.19 Photomicrograph of a GD 8.5 mutant mouse embryo (unspecified genotype and genetic background) demonstrating widespread necrosis. The yolk sac also is involved, although most of the change in this organ is acute hemorrhage (inset; the main image is from the boxed region in the inset). Embryonic disintegration with concomitant damage to extraembryonic tissues is consistent with concurrent injury to cell populations at both sites rather than a primary lesion in one that then secondarily affects the other. The recent nature of the cell death is suggested by retention of the embryonic and placental shapes as well as the relative integrity of many of the detached embryonic cells. (Specimen courtesy of Dr. Christopher Koivisto, The Ohio State University, Columbus, OH.)

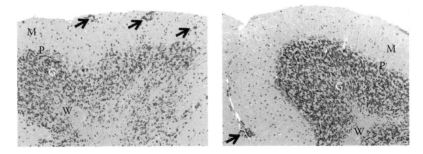

FIGURE 15.20 Neuronal heterotopia in the cerebellum of an adult, neurologically normal mouse following whole body irradiation on PND 1 to ablate bone marrow. The key findings are increased cellularity of the myelin-rich molecular layer (M) and deep white matter (W); diffuse disarray and intermingling of the Purkinje cell (P) and granule cell (G) layers; and many intraparenchymal nests of retained external granule cells (arrows), which serve as the precursors to the granule cell layer. In contrast, a nonirradiated adult mouse exhibits regular organization of the neuronal layers, low cellularity of myelin-rich regions, and one incidental focus of external granule cells beneath the meninges. (Reproduced from Bolon, B. et al., Nervous system, in: *Haschek and Rousseaux's Handbook of Toxicologic Pathology*, 3rd edn., Vol. 3, Haschek, W.M., Rousseaux, C.G., Wallig, M.A., eds., Academic Press, San Diego, CA, pp. 2005–2093, Copyright 2013, with permission from Elsevier.)

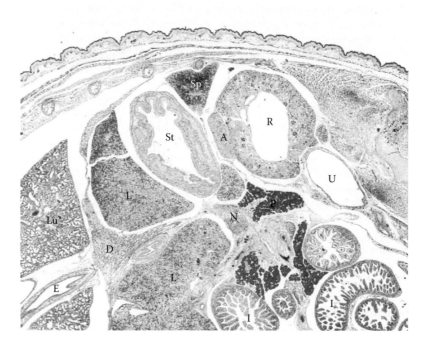

FIGURE 15.21 Hydronephrosis (renal [R] pelvic cavitation) and hydroureter (dilation of the ureter [U] lumen) of moderate degree in a wild-type, near-term (GD 17.5) mouse fetus exposed to methanol by maternal inhalation during early organogenesis (GD 7 to GD 9). Other visible organs include the adrenal gland (A); diaphragm (D); esophagus (E); liver (L), especially the darker hematopoietic cells; lung (Lu); primordial lymph node (N); pancreas (exocrine portion [P]); small intestine (I); spleen (Sp); and stomach (St).

(or physiological hydronephrosis of minimal to mild degree [Figure 15.21]), transient asymmetry or delay in cranial (occipital) and/or sternebral ossification, supernumerary ribs, and wavy ribs. In one industrial laboratory, incidences of these findings in outbred CD-1 mice exhibit a wide range within teratogenicity studies conducted over several years: approximately 7% for renal cavitation, 4% for retarded occipital ossification, 0%–20% for reduced or missing sternebral ossification, and rib irregularities ranging from 0% to 33% in the cervical region (mean = 13%) and 3%–35% in the lumbar regions (mean = 18%).[319] For a single mouse litter, the range of fetuses that carry supernumerary ribs varies from 0% to 100%.[319] The incidence of these and similar minor aberrations (e.g., tail anomalies [Figure 15.22]) may be raised by exposure to a teratogenic agent.[40,63,348] Minor variants are recognized reliably only when the colony incidence over time remains fairly low (5% or less). Otherwise, these defects likely will present as a trend in the data and not a statistically significant difference due to the small group sizes used in conventional mouse developmental pathology studies (see Chapter 18 for an explanation of this statistical outcome).

Minor visceral anomalies represent another category of developmental defect. The most frequent change of this kind is focal hemorrhage, which typically is observed in 1%–2% of animals.[319] The single most common site is the inner wall of major body cavities (abdominal or thoracic; incidence = 1.36%). Gonads may be displaced cranially in approximately 1.2% of fetuses. These changes are considered to be inconsequential except in cases where the lesion might limit the function of a nearby organ (e.g., blood loss into the pericardial sac [hemopericardium], leading to potential changes in cardiac function if pressure on the epicardial surface prevents full expansion of the ventricles).

Major malformations comprise the most important category. These anomalies occur at a low background incidence (0.84% in CD-1 mice[319]) and typically result in rapid death of the neonates either due to their inherent inability to function or due to maternal cannibalism of moribund (i.e., at the point of death)

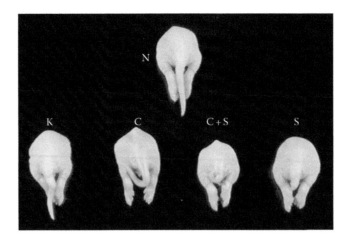

FIGURE 15.22 Minor skeletal defects—kinking (K), curling (C), and shortening (S)—affect the tails of near-term (GD 17.5) mouse fetuses following maternal inhalation of methanol throughout organogenesis (GD 6 to GD 15). In contrast, the sham-exposed control has a straight (i.e., structurally normal [N]) tail.

pups. Craniofacial defects (Figure 15.4) are the most common major defect (0.74%), followed by cleft palate (0.30%[403] to 0.46%[319]; Figure 15.15), open eyes (0.08%[403] to 0.17%[319]; Figure 15.23), exencephaly (0.06%[403] to 0.11%[319]; Figure 15.4), axial skeletal alterations (0.07%; Figure 15.16), and various cardiac anomalies (0.05%[403]; Figure 15.8). Severe lesions affecting other body parts (e.g., body wall integrity [Figure 15.23], limb orientation, assembly of the reproductive and urinary tracts, and reductions of the eye [Figure 15.17] and jaw [Figure 15.4]) are present at very low incidences (0.01%–0.02%).[319,403] The urinary bladder may be distended prominently in 0.5%[319] to 0.9%[403] of near-term animals, but while unusual this phenomenon is not considered to represent a defect unless another urinary tract abnormality is found as well. The same fetus may bear one or several major lesions. The incidence and severity of major defects may be exacerbated in animals that have been exposed to teratogenic stimuli during gestation as well as by the genotype of the progeny, the dam, or both.

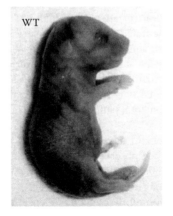

FIGURE 15.23 A near-term (GD 18.5) mouse fetus with a null mutation (KO) in the Rho-associated kinase-1 (*Rock-1*) gene has both a severe omphalocele (i.e., persistent umbilical hernia [arrow]) and open eyes (arrowhead), which are lacking in a wild-type (WT) littermate. (Reproduced from Shimizu, Y. et al., *J. Cell Biol.*, 168(6), 941, 2005. With permission of The Rockefeller University Press.)

Even in control groups, some malformations tend to cluster within litters. Examples include cleft palate and premature opening of the eye.[319] The presumed cause of such clustering in wild-type litters is a detrimental interaction between maternal and paternal genes. Therefore, investigators who regularly perform mouse developmental pathology experiments will need to define the background incidences of such lesions within their colonies (and the colonies of their animal vendor) and be ready to cull dams or sires that appear to carry genes leading to a heightened background incidence of developmental defects in order to reliably discriminate genuine developmental phenotypes produced by an experimental manipulation from expected spontaneous findings.

Linked Developmental Defects

Certain lesions in developing mice tend to occur together in the same animal. The concurrent presentation of two or more lesions may be driven in several ways. A consistent combination of developmental defects often is accorded the name *syndrome*,[226,251,294] especially if the mechanism that connects them is not well understood.

The link between developmental defects may be related to a mutual pathogenesis. For example, embryos with obvious neural tube closure defects (NTD) in the cephalic region (i.e., head) usually exhibit fairly pronounced craniofacial malformations as well (Figures 15.3 and 15.4). This outcome is to be expected because NTD arise in part from reduced production of mesenchyme in the elevating head folds,[170] which both prevents apposition of the neural fold apices (a necessary prerequisite to fusion of the folds) and also reduces the cell populations in skeletal anlagen assigned to generate the bones of the face and skull. Similarly, *spina bifida* (i.e., a caudal NTD) often involves both the vertebrae and the underlying spinal cord, although vertebrae and other axial skeletal components may develop linked defects that do not arise from failed closure[430] (Figure 15.24), and thus do not exhibit any spinal cord involvement.

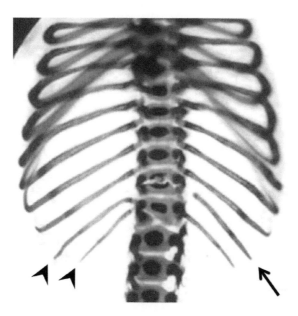

FIGURE 15.24 Linked developmental defects of the axial skeleton commonly involve the vertebrae and ribs. In this case, the thoracolumbar junction of a near-term mouse fetus from a dam treated with 2-deoxyazacytidine exhibits an unattached 12th rib on the left side (arrow) and fusion of both the 12th and 13th ribs (arrowheads) to the right side of an abnormal vertebral body. (Reproduced from Tyl, R.W. et al., *Birth Defects Res. B Dev. Reprod. Toxicol.*, 80(6), 451, 2007, by courtesy of Wiley-Liss.)

Some lesions are linked not by a common location but rather by disruption of a shared process that is essential to development. The most clear-cut example of this phenomenon is the concurrent disruption of vascular integrity in one or more placental membranes and the embryo proper due to null mutations of certain genes expressed widely in developing blood vessels.[24,263] The mechanisms leading to such widespread injury typically are alterations in the expression or a growth factor or growth factor receptor that is essential for endothelial development and maintenance. Care is required in discriminating between the existence of primary lesions in both embryonic and extraembryonic sites versus a primary defect at one location (typically the extraembryonic vessels), which secondarily impacts the function of similar structures in other places within the conceptus.[3,101]

Summary

A skewed genotypic ratio at birth or weaning and/or visible birth defects afford spectacular confirmation of developmental distress and represent the first evidence that a new developmental phenotype requires characterization. Macroscopic lesions (as shown in previous figures as well as additional illustrations in Figures 15.A.1 through 15.A.12) have been found in essentially all organs. Further details regarding these findings, the mechanisms and causes that are responsible for their induction, and their relevance as models of human birth defects may be found in various books,[181,182,311,382,450–452] book chapters,[360] journal articles on developmental phenotypes,[80,170,171,294] and websites,[421] and by perusing the growing bodies of developmental biology and developmental toxicology literature. In particular, the list of terms for specific developmental lesions (Chapter 3) provides a good starting point from which to commence or expand a search that should lead to greater understanding and growing diagnostic proficiency in mouse developmental pathology. Nonetheless, in our view—an opinion that is borne out by communal experience[191,401] with mouse phenotyping studies during the past decade—the efficient analysis and accurate interpretation of new developmental phenotypes generally requires that at least one member of the research team has some formal training in pathology.

Appendix

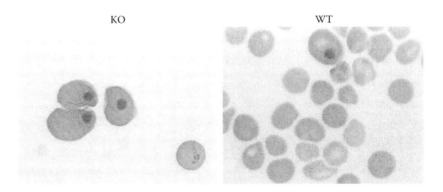

KO　　　　　　　　　　　　　　　　　　　WT

FIGURE 15.A.1　Blood from a mouse fetus (GD 15) that bears a null mutation (KO) of the cellular myeloblastosis (*c-Myb*) gene contains a few large, nucleated erythrocytes (i.e., *primitive* cells derived from yolk sac blood islands) but no small, nonnucleated erythrocytes (i.e., *definitive* cells originating in the liver), indicating that death from anemia is imminent. In contrast, an age-matched, wild-type (WT) littermate contains numerous definitive erythrocytes. (Figure reproduced from *Cell*, 65 (4), Mucenski, M., McLain, K., Kier, A.B., Swerdlow, S.H., Schreiner, C.M., Miller, T.A., Pietryga, D.W., Scott, Jr, W.J., Potter, S.S., A functional *c-myb* gene is required for normal murine fetal hepatic hematopoiesis, 677–689, Copyright 1991, with permission from Elsevier.)

WT   KO

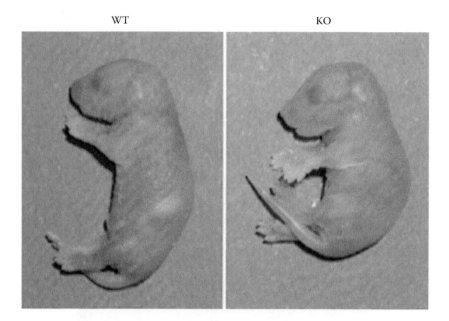

FIGURE 15.A.2   Relative to wild-type (WT) littermates, neonatal mice in which the glycine transporter 1 (*GlyT1*) gene has been removed (KO) in astrocytes develop centrally mediated neuromuscular deficits that yield severe postural defects (altered body curvature and drooping forelimbs are shown here) and respiratory distress that is lethal shortly after birth. (The images are reproduced from *Neuron*, 40(4), Gomeza, J., Hulsmann, S., Ohno, K., Eulenburg, V., Szoke, K., Richter, D., Betz, H., Inactivation of the glycine transporter 1 gene discloses vital role of glial glycine uptake in glycinergic inhibition, 785–796, Copyright 2003, with permission from Elsevier.)

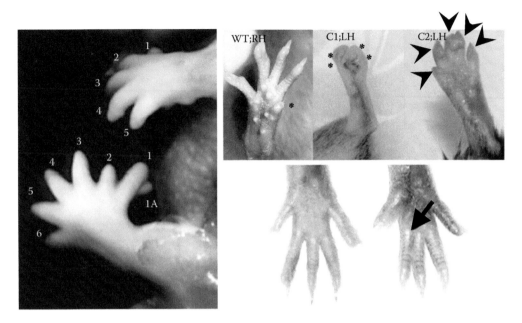

FIGURE 15.A.3 Common developmental defects of digits observed in mutant or toxicant-exposed animals include adactyly (missing digits [asterisks]), brachydactyly (short digits [arrowheads]), polydactyly (too many digits), and syndactyly (fused digits [arrow]). In some cases, extra appendages (1A [left image]) that grow from existing digits are a manifestation of polydactyly. *Abbreviations*: C, chimera; LH, left hind paw; RH, right hind paw; WT, wild-type. (Figures are reproduced from multiple papers: left panel, Sinawat, S. et al., *Hum. Reprod.*, 2003, by permission of Oxford University Press; upper right panel, Liu, W. et al., *PLoS One*, 7(3), e32331, 2012; lower right panel, Kalcheva, N., Qu, J., Sandeep, N., Garcia, L., Zhang, J., Wang, Z., Lampe, P.D., Suadicani, S.O., Spray, D.C., Fishman, G.I., Gap junction remodeling and cardiac arrhythmogenesis in a murine model of oculodentodigital dysplasia, *Proc. Natl. Acad. Sci. U.S.A.*, 104(51), 20512–20516, Copyright 2007, National Academy of Sciences, U.S.A.)

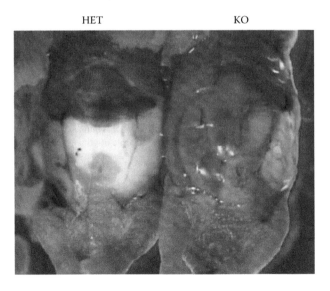

FIGURE 15.A.4 Neonatal mice (PND 0) with a null mutation (KO) of the single-minded homolog 2 (*Sim2*) gene have air-distended gastrointestinal tracts (indicated by translucent, tan/yellow walls), while unaffected heterozygous (HET) littermates have milk-filled tracts. The specimens are positioned with the abdomen oriented up and the feet located at the bottom of the image; the digestive tract is visible because the skin has been removed from the abdominal wall. (Figure reproduced from Shamblott, M.J. et al.: Craniofacial abnormalities resulting from targeted disruption of the murine *Sim2* gene. *Dev. Dyn.* 2002. 224(4). 373–380. Copyright Wiley-VCH Verlag GmbH & Co. KGaA. Reproduced with permission.)

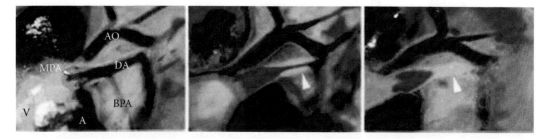

FIGURE 15.A.5 Arterial tree associated with the neonatal (PND 0) mouse heart, showing a fully patent ductus arteriosus (DA) immediately after birth (left), narrowing of the distal end within 0.5–1 h (middle), and full closure by 1–3 h (right). The white arrowhead denotes the closing DA. *Other abbreviations*: A, atrium; AO, aorta; BPA, branch pulmonary artery; MPA, main pulmonary artery; V, ventricle. (Figure reproduced from Reese, J. et al., Coordinated regulation of fetal and maternal prostaglandins directs successful birth and postnatal adaptation in the mouse, *Proc. Natl. Acad. Sci. U.S.A.*, 97(17), 9759–9764, Copyright 2000, National Academy of Sciences, U.S.A.)

WT   KO

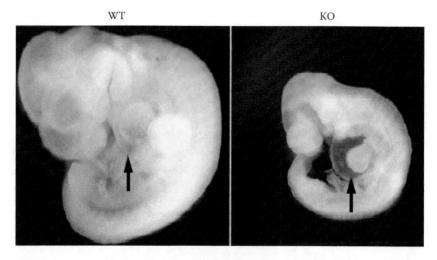

FIGURE 15.A.6 Mouse embryos (GD 9.5) bearing a null mutation (KO) of extracellular-signal-regulated kinase 5 (*Erk5*) exhibit widespread vascular defects in embryonic (shown here) and extraembryonic tissues. Relative to a stage-matched wild-type (WT) control, this dying KO embryo exhibits pronounced stunting and has an extensive pericardial effusion (i.e., hydropericardium). (Figure reproduced from Regan, C., Li, W., Boucher, D.M., Spatz, S., Su, M.S., Kuida, K., Erk5 null mice display multiple extraembryonic vascular and embryonic cardiovascular defects, *Proc. Natl. Acad. Sci. U.S.A.*, 99(14), 9248–9253, Copyright 2002, National Academy of Sciences, U.S.A.)

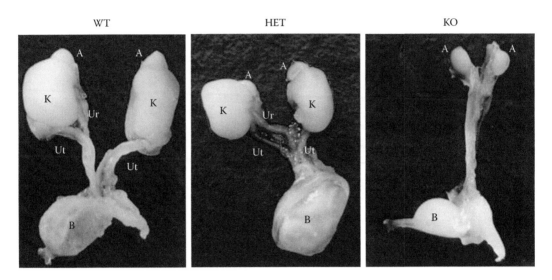

FIGURE 15.A.7   Neonatal (PND 0) mice carrying engineered null mutations (KO) of glial cell line-derived neurotrophic factor (Gdnf) family receptor alpha-1 (*Gfra1*), one component of the heterodimeric receptor for Gdnf, lack kidneys (K) and ureters (Ur), while age-matched wild-type (WT) and heterozygous (HET) controls have intact urinary tracts. *Other abbreviations*: A, adrenal gland; B, urinary bladder; Ut, uterus. (Images courtesy of Dr. Shuqian Jing, Amgen, Inc., Thousand Oaks, CA.)

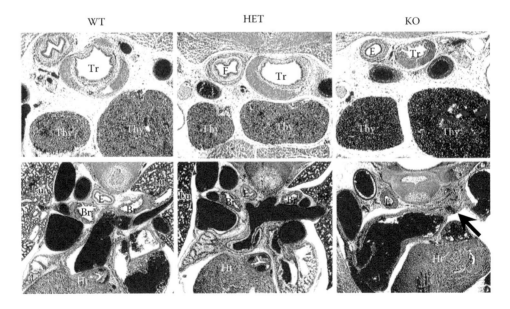

FIGURE 15.A.8   Defective lung (Lu) formation in mice with an engineered null mutation (KO) of the fibroblast growth factor-10 (*Fgf-10*) gene resulted from termination of the lower respiratory tract near the carina (i.e., that portion of the trachea [Tr] from which the main bronchi [Br] arise), which did not happen in age-matched (GD 18) fetuses with either two (wild-type [WT]) or one (heterozygous [HET]) *Fgf-10* alleles. The images represent transverse sections of the torso taken through the distal cervical region (upper row) and cranial thoracic cavity (bottom row). *Other abbreviations*: E, esophagus; H, heart; Thy, thymus. (Figure reprinted from Min, H. et al., *Genes Dev.*, 12(20), 3156, 1998. With permission of Cold Spring Harbor Laboratory Press.)

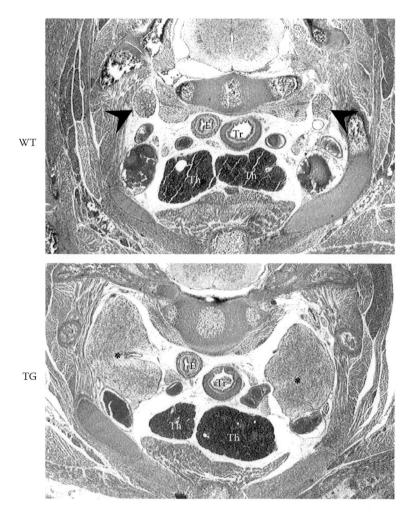

FIGURE 15.A.9 Dysgenesis characterized by abundant axonal formation in peripheral autonomic ganglia is a sequel to constitutive transgenic (TG) overexpression during organogenesis of artemin, a glial cell line–derived neurotrophic factor (Gdnf) family member. The cranial cervical ganglion (superior cervical ganglion; arrowheads) is particularly affected as indicated by the small size of this organ in age-matched (GD 14) wild-type (WT) littermates. *Abbreviations*: E, esophagus; S, spinal cord; Th, thymus; Tr, trachea. (Figure reproduced from Bolon, B. et al., *Toxicol. Pathol.*, 32(3), 275, 2004. With permission of the Society of Toxicologic Pathology and Sage Publications.)

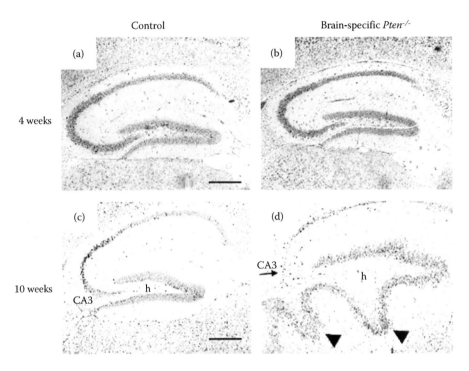

FIGURE 15.A.10 Postnatal development of mice may be associated with new dysplastic lesions that arise well after birth. For instance, mice with brain-specific deletion of phosphatase and tensin homolog (*Pten*; panels b and d) have similar anatomic features in hippocampus as do age-matched control mice at 4 weeks of age (top row), but by 10 weeks of age the dentate gyrus has an undulating and thickened granule cell layer (arrowheads) as well as sclerosis (hardening from excessive neural cell and fiber formation) of the pyramidal cell layer in *cornu ammonis* field CA3 (CA3). Histological processing conditions: fixation by immersion in neutral buffered 10% formalin, paraffin embedding, thionin staining (i.e., a dye used to highlight DNA and RNA [Nissl substance] within neurons). (Reprinted by permission from Macmillan Publishers Ltd., from Backman, S.A., et al., *Nat. Genet.*, Deletion of *Pten* in mouse brain causes seizures, ataxia and defects in soma size resembling Lhermitte-Duclos disease, 29(4), 396–403, copyright 2001.)

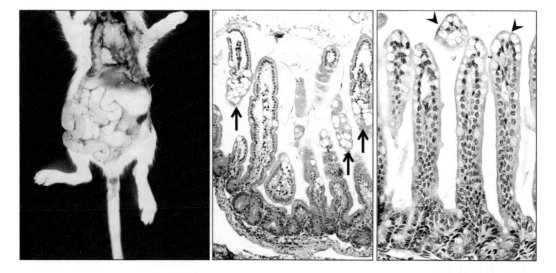

FIGURE 15.A.11 Developmental phenotypes need not arise from a genetic manipulation or a toxicant exposure, but may be due to infection with a microbial agent. The gross image (left) shows a juvenile mouse (approximately PND 14) that died as a result of rotavirus infection, a condition known as *epizootic diarrhea of mice* (EDIM). The intestinal tract contains a mixture of milk (white segments) and diarrhea (yellow to tan segments), and dried feces are evident on the proximal tail adjacent to the anus. The viral infection causes clusters of vacuoles (arrows), in various shapes and sizes, in apical enterocytes of some but not all villi; affected cells also exhibit nuclear pyknosis (a degenerative condensation of the nuclear contents). Rotaviral lesions must be discriminated from the clear, uniformly sized vacuoles (arrowheads) containing milk lipoproteins within the cytoplasm of enterocytes that line the tips of most villi in nursing mice; these vacuoles may contain an eosinophilic protein droplet, and the nuclei exhibit their normal features. (The macroscopic figure is reproduced from Percy, D.H. and Barthold, S.W., *Pathology of Laboratory Rodents and Rabbits*, 3rd edn., Iowa State University Press, Ames, IA, 2007. With permission of the Iowa State University Press. The two microscopic images are reprinted from the Diseases of Research Animals (DORA) database[77] (http://dora.missouri.edu/mouse/epizootic-diarrhea-of-infant-mice-edim/) courtesy of the University of Missouri, Comparative Medicine Program, Columbia, MO, and IDEXX BioResearch, Columbia, MO.)

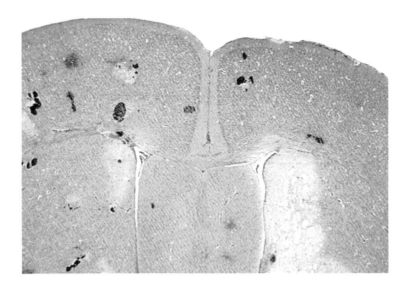

FIGURE 15.A.12 Developmental phenotypes may be induced unintentionally by certain experimental manipulations. This photomicrograph shows the brain of a 7-day-old mouse exhibiting neurological signs from a project investigating the effects of a novel brain-targeted null mutation. The animal developed seizures and was humanely euthanized, at which time meningeal vessels over the brain were seen to be engorged and in some cases associated with multiple hemorrhages. Numerous small, irregular microabscesses (blue foci) were scattered randomly within the cerebral cortex and basal nuclei (BN), while two large ischemic infarcts (white regions of necrosis) were evident in the BN. These lesions resulted from tail biopsy on PND 6 (to accelerate the speed of the phenotyping program), which allowed a virulent (β-hemolytic) strain of the bacterium *Streptococcus* sp. to enter the blood and colonize the brain. The pale nature of the infarcts shows that they likely resulted from total occlusion rather than rupture of the affected arteries, which would have led to extensive hemorrhage into the brain parenchyma.

REFERENCES

1. Accili D, Drago J, Lee EJ, Johnson MD, Cool MH, Salvatore P, Asico LD, José PA, Taylor SI, Westphal H (1996). Early neonatal death in mice homozygous for a null allele of the insulin receptor gene. *Nat Genet* **12** (1): 106–109.
2. Adissu HA, Estabel J, Sunter D, Tuck E, Hooks Y, Carragher DM, Clarke K et al. (2014). Histopathology reveals correlative and unique phenotypes in a high-throughput mouse phenotyping screen. *Dis Model Mech* **7** (5): 515–524.
3. Agah R, Prasad KS, Linnemann R, Firpo M, Quertermous T, Dichek DA (2000). Cardiovascular overexpression of transforming growth factor-β_1 causes abnormal yolk sac vasculogenesis and early embryonic death. *Circ Res* **86** (10): 1024–1030.
4. Airhart MJ, Robbins CM, Knudsen TB, Church JK, Skalko RG (1993). Occurrence of embryotoxicity in mouse embryos following *in utero* exposure to 2′-deoxycoformycin (pentostatin). *Teratology* **47** (1): 17–27.
5. Akazawa H, Komuro I, Sugitani Y, Yazaki Y, Nagai R, Noda T (2000). Targeted disruption of the homeobox transcription factor *Bapx1* results in lethal skeletal dysplasia with asplenia and gastroduodenal malformation. *Genes Cells* **5** (6): 499–513.
6. Alexander WS, Rakar S, Robb L, Farley A, Willson T, Zhang J-G, Hartley L et al. (1999). Suckling defect in mice lacking soluble haemopoietin receptor NR6. *Curr Biol* **9** (11): 605–608.
7. Allen Institute (2009). Allen Institute for Brain Science Atlas Portal. http://www.brain-map.org/ (last accessed December 1, 2014).
8. Allgeier SH, Lin TM, Moore RW, Vezina CM, Abler LL, Peterson RE (2010). Androgenic regulation of ventral epithelial bud number and pattern in mouse urogenital sinus. *Dev Dyn* **239** (2): 373–385.
9. Altman J, Bayer SA (1995). *Atlas of Prenatal Rat Brain Development*. Boca Raton, FL: CRC Press.
10. Arai T, Kasper JS, Skaar JR, Ali SH, Takahashi C, DeCaprio JA (2003). Targeted disruption of *p185/Cul7* gene results in abnormal vascular morphogenesis. *Proc Natl Acad Sci USA* **100** (17): 9855–9860.

11. Arikawa-Hirasawa E, Watanabe H, Takami H, Hassell JR, Yamada Y (1999). Perlecan is essential for cartilage and cephalic development. *Nat Genet* **23** (3): 354–358.
12. Arman E, Haffner-Krausz R, Chen Y, Heath JK, Lonai P (1998). Targeted disruption of fibroblast growth factor (FGF) receptor 2 suggests a role for FGF signaling in pregastrulation mammalian development. *Proc Natl Acad Sci USA* **95** (9): 5082–5087.
13. Armstrong JF, Kaufman MH, Harrison DJ, Clarke AR (1995). High-frequency developmental abnormalities in *p53*-deficient mice. *Curr Biol* **5** (8): 931–936.
14. Arnold TD, Zang K, Vallejo-Illarramendi A (2013). Deletion of integrin-linked kinase from neural crest cells in mice results in aortic aneurysms and embryonic lethality. *Dis Model Mech* **6** (5): 1205–1212.
15. Ashwell KWS, Paxinos G (2008). *Atlas of the Developing Rat Nervous System*, 3rd edn. San Diego, CA: Academic Press (Elsevier).
16. Aune CN, Chatterjee B, Zhao XQ, Francis R, Bracero L, Yu Q, Rosenthal J, Leatherbury L, Lo CW (2008). Mouse model of heterotaxy with single ventricle spectrum of cardiac anomalies. *Pediatr Res* **63** (1): 9–14.
17. Backman SA, Stambolic V, Suzuki A, Haight J, Elia A, Pretorius J, Tsao MS, Shannon P, Bolon B, Ivy GO, Mak TW (2001). Deletion of *Pten* in mouse brain causes seizures, ataxia and defects in soma size resembling Lhermitte-Duclos disease. *Nat Genet* **29** (4): 396–403.
18. Bader BL, Rayburn H, Crowley D, Hynes RO (1998). Extensive vasculogenesis, angiogenesis, and organogenesis precede lethality in mice lacking all αv integrins. *Cell* **95** (4): 507–519.
19. Bannigan J, Langman J (1979). The cellular effect of 5-bromodeoxyuridine on the mammalian embryo. *J Embryol Exp Morphol* **50**: 123–135.
20. Bannigan JG, Cottell DC, Morris A (1990). Study of the mechanisms of BUdR-induced cleft palate in the mouse. *Teratology* **42** (1): 79–89.
21. Barak Y, Nelson MC, Ong ES, Jones YZ, Ruiz-Lozano P, Chien KR, Koder A, Evans RM (1999). PPARγ is required for placental, cardiac, and adipose tissue development. *Mol Cell* **4** (4): 585–595.
22. Bard-Chapeau EA, Szumska D, Jacob B, Chua BQ, Chatterjee GC, Zhang Y, Ward JM et al. (2014). Mice carrying a hypomorphic Evi1 allele are embryonic viable but exhibit severe congenital heart defects. *PLoS One* **9** (2): e89397 (14 pages).
23. Bartram U, Molin DG, Wisse LJ, Mohamad A, Sanford LP, Doetschman T, Speer CP, Poelmann RE, Gittenberger-de Groot AC (2001). Double-outlet right ventricle and overriding tricuspid valve reflect disturbances of looping, myocardialization, endocardial cushion differentiation, and apoptosis in TGF-β₂-knockout mice. *Circulation* **103** (22): 2745–2752.
24. Baudino TA, McKay C, Pendeville-Samain H, Nilsson JA, Maclean KH, White EL, Davis AC, Ihle JN, Cleveland JL (2002). c-Myc is essential for vasculogenesis and angiogenesis during development and tumor progression. *Genes Dev* **16** (19): 2530–2543.
25. Beg AA, Sha WC, Bronson RT, Ghosh S, Baltimore D (1995). Embryonic lethality and liver degeneration in mice lacking the RelA component of NF-κB. *Nature* **376** (6536): 167–170.
26. Benirschke K (1987). You need a sympathetic pathologist!: The borderline of embryology and pathology revisted. *Teratology* **36** (3): 389–393.
27. Bertolino P, Radovanovic I, Casse H, Aguzzi A, Wang Z-Q, Zhang C-X (2003). Genetic ablation of the tumor suppressor menin causes lethality at mid-gestation with defects in multiple organs. *Mech Dev* **120** (5): 549–560.
28. Bérubé N, Jagla M, Smeenk C, De Repentigny Y, Kothary R, Picketts D (2002). Neurodevelopmental defects resulting from ATRX overexpression in transgenic mice. *Hum Mol Genet* **11** (3): 253–261.
29. Binder NK, Mitchell M, Gardner DK (2012). Parental diet-induced obesity leads to retarded early mouse embryo development and altered carbohydrate utilisation by the blastocyst. *Reprod Fertil Dev* **24** (6): 804–812.
30. Blackburn RO, Clegg EJ (1979). The survival and implantation of mouse blastocysts at varying degrees of reduced atmospheric pressure. *Teratology* **20** (3): 441–446.
31. Blackshear PJ, Silver J, Nairn AC, Sulik KK, Squier MV, Stumpo DJ, Tuttle JS (1997). Widespread neuronal ectopia associated with secondary defects in cerebrocortical chondroitin sulfate proteoglycans and basal lamina in MARCKS-deficient mice. *Exp Neurol* **145** (1): 46–61.
32. Blackshear PJ, Stumpo DJ, Carballo E, Lawrence Jr JC. (1997). Disruption of the gene encoding the mitogen-regulated translational modulator PHAS-I in mice. *J Biol Chem* **272** (50): 31510–31514.

33. Blanchi B, Kelly LM, Viemari JC, Lafon I, Burnet H, Bevengut M, Tillmanns S, Daniel L, Graf T, Hilaire G, Sieweke MH (2003). MafB deficiency causes defective respiratory rhythmogenesis and fatal central apnea at birth. *Nat Neurosci* **6** (10): 1091–1100.

34. Blaschke RJ, Hahurij ND, Kuijper S, Just S, Wisse LJ, Deissler K, Maxelon T et al. (2007). Targeted mutation reveals essential functions of the homeodomain transcription factor *Shox2* in sinoatrial and pacemaking development. *Circulation* **115** (14): 1830–1838.

35. Block K, Kardana A, Igarashi P, Taylor HS (2000). *In utero* diethylstilbestrol (DES) exposure alters Hox gene expression in the developing mullerian system. *FASEB J* **14** (9): 1101–1108.

36. Bnait KS, Seller MJ (1995). Ultrastructural changes in 9-day old mouse embryos following maternal tobacco smoke inhalation. *Exp Toxicol Pathol* **47** (6): 453–461.

37. Bolon B, Jing S, Asuncion F, Scully S, Pisegna M, Van G, Hu Z et al. (2004). The candidate neuroprotective agent artemin induces autonomic neural dysplasia without preventing nerve dysfunction. *Toxicol Pathol* **32** (3): 275–294.

38. Bolon B, Welsch F, Morgan KT (1994). Methanol-induced neural tube defects in mice: Pathogenesis during neurulation. *Teratology* **49** (6): 497–517.

39. Bolon B et al. (2013). Nervous system. In: *Haschek and Rousseaux's Handbook of Toxicologic Pathology*, 3rd edn., Vol. 3, Haschek, W.M., Rousseaux, C.G., Wallig, M.A., eds. San Diego, CA: Academic Press, pp. 2005–2093.

40. Boström H, Willetts K, Pekny M, Leveen P, Lindahl P, Hedstrand H, Pekna M et al. (1996). PDGF-A signaling is a critical event in lung alveolar myofibroblast development and alveogenesis. *Cell* **85** (6): 863–873.

41. Branch S, Rogers JM, Brownie CF, Chernoff N (1996). Supernumerary lumbar rib: Manifestation of basic alteration in embryonic development of ribs. *J Appl Toxicol* **16** (2): 115–119.

42. Brent RL, Fawcett L, Beckman DA (2006). Principles of teratology. In: *Prenatal Diagnosis*, Evans MI, Johnson MP, Yaron Y, Drugan A (eds.). New York: McGraw-Hill, pp. 79–112.

43. Bronson RT (2001). How to study pathologic phenotypes of knockout mice. In: *Methods in Molecular Biology: Gene Knockout Protocols*, Tymms MJ, Kola I (eds.). Totowa, NJ: Humana, pp. 155–180.

44. Brown JR, Ye H, Bronson RT, Dikkes P, Greenberg ME (1996). A defect in nurturing in mice lacking the immediate early gene *fosB*. *Cell* **86** (2): 297–309.

45. Brown KS, Hackman RM (1985). The significance of receptor physiology for corticosterone-induced cleft palate in A/J mice. *J Craniofac Genet Dev Biol Suppl* **1**: 299–304.

46. Brown KS, Johnston MC, Niswander JD (1972). Isolated cleft palate in mice after transportation during gestation. *Teratology* **5** (1): 119–124.

47. Brown NA, Hoyle CI, McCarthy A, Wolpert L (1989). The development of asymmetry: The sidedness of drug-induced limb abnormalities is reversed in *situs inversus* mice. *Development* **107** (3): 637–642.

48. Brunskill EW, Aronow BJ, Georgas K, Rumballe B, Valerius MT, Aronow J, Kaimal V et al. (2008). Atlas of gene expression in the developing kidney at microanatomic resolution. *Dev Cell* **15** (5): 781–791. [Erratum in: *Dev Cell* **16** (3): 482, 2009.]

49. Burkus J, Cikos S, Fabian D, Kubandova J, Czikkova S, Koppel J (2013). Maternal restraint stress negatively influences growth capacity of preimplantation mouse embryos. *Gen Physiol Biophys* **32** (1): 129–137.

50. Bussen M, Petry M, Schuster-Gossler K, Leitges M, Gossler A, Kispert A (2004). The T-box transcription factor Tbx18 maintains the separation of anterior and posterior somite compartments. *Genes Dev* **18** (10): 1209–1221.

51. Butler H, Juurlink BHJ (1987). *An Atlas for Staging Mammalian and Chick Embryos.* Boca Raton, FL: CRC Press.

52. Cacalano G, Farinas I, Wang LC, Hagler K, Forgie A, Moore M, Armanini M et al. (1998). GFRα1 is an essential receptor component for GDNF in the developing nervous system and kidney. *Neuron* **21** (1): 53–62.

53. Canseco RS, Sparks AE, Page RL, Russell CG, Johnson JL, Velander WH, Pearson RE, Drohan WN, Gwazdauskas FC (1994). Gene transfer efficiency during gestation and the influence of co-transfer of nonmanipulated embryos on production of transgenic mice. *Transgenic Res* **3** (1): 20–25.

54. Cantrell VA, Owens SE, Chandler RL, Airey DC, Bradley KM, Smith JR, Southard-Smith EM (2004). Interactions between *Sox10* and *EdnrB* modulate penetrance and severity of aganglionosis in the *Sox10^{Dom}* mouse model of Hirschsprung disease. *Hum Mol Genet* **13** (19): 2289–2301.

55. Chaineau E, Binet S, Pol D, Chatellier G, Meininger V (1990). Embryotoxic effects of sodium arsenite and sodium arsenate on mouse embryos in culture. *Teratology* **41** (1): 105–112.

56. Chan EYW, Nasir J, Gutekunst C-A, Coleman S, Maclean A, Maas A, Metzler M, Gertsenstein M, Ross CA, Nagy A, Hayden MR (2002). Targeted disruption of *Huntingtin-associated protein-1* (*Hap1*) results in postnatal death due to depressed feeding behavior. *Hum Mol Genet* **11** (8): 945–959.

57. Chan JY, Kwong M, Lu R, Chang J, Wang B, Yen TS, Kan YW (1998). Targeted disruption of the ubiquitous CNC-bZIP transcription factor, Nrf-1, results in anemia and embryonic lethality in mice. *EMBO J* **17** (6): 1779–1787.

58. Chang HH, Schwartz Z, Kaufman MH (1996). Limb and other postcranial skeletal defects induced by amniotic sac puncture in the mouse. *J Anat* **189** (Pt 1): 37–49.

59. Chawengsaksophak K, de Graaff W, Rossant J, Deschamps J, Beck F (2004). *Cdx2* is essential for axial elongation in mouse development. *Proc Natl Acad Sci USA* **101** (20): 7641–7645.

60. Chen BY, Chang HH, Chen ST, Tsao ZJ, Yeh SM, Wu CY, Lin DP (2009). Congenital eye malformations associated with extensive periocular neural crest apoptosis after influenza B virus infection during early embryogenesis. *Mol Vis* **15**: 2821–2828.

61. Chen J, Futami K, Petillo D, Peng J, Wang P, Knol J, Li Y et al. (2008). Deficiency of FLCN in mouse kidney led to development of polycystic kidneys and renal neoplasia. *PLoS One* **3** (10): e3581 (8 pages).

62. Chernoff N, Miller DB, Rosen MB, Mattscheck CL (1988). Developmental effects of maternal stress in the CD-1 mouse induced by restraint on single days during the period of major organogenesis. *Toxicology* **51** (1): 57–65.

63. Chernoff N, Rogers EH, Gage MI, Francis BM (2008). The relationship of maternal and fetal toxicity in developmental toxicology bioassays with notes on the biological significance of the "no observed adverse effect level". *Reprod Toxicol* **25** (2): 192–202.

64. Chernoff N, Rogers JM, Turner CI, Francis BM (1991). Significance of supernumerary ribs in rodent developmental toxicity studies: Postnatal persistence in rats and mice. *Fundam Appl Toxicol* **17** (3): 448–453.

65. Chiang C, Litingtung Y, Lee E, Young KE, Corden JL, Westphal H, Beachy PA (1996). Cyclopia and defective axial patterning in mice lacking *Sonic hedgehog* gene function. *Nature* **383** (6599): 407–413.

66. Chuang P-T, Kawcak T, McMahon AP (2003). Feedback control of mammalian Hedgehog signaling by the Hedgehog-binding protein, Hip1, modulates Fgf signaling during branching morphogenesis of the lung. *Genes Dev* **17** (3): 342–347.

67. Clancy B, Charvet CJ, Darlington RB, Finlay BL, Workman A (2013). Translating time (across developing mammalian brains). http://translatingtime.net/ (last accessed December 1, 2014).

68. Clancy B, Kersh B, Hyde J, Darlington RB, Anand KJ, Finlay BL (2007). Web-based method for translating neurodevelopment from laboratory species to humans. *Neuroinformatics* **5** (1): 79–94.

69. Clark JC, Wert SE, Bachurski CJ, Stahlman MT, Stripp BR, Weaver TE, Whitsett JA (1995). Targeted disruption of the surfactant protein B gene disrupts surfactant homeostasis, causing respiratory failure in newborn mice. *Proc Natl Acad Sci USA* **92** (17): 7794–7798.

70. Cleary JO, Modat M, Norris FC, Price AN, Jayakody SA, Martinez-Barbera JP, Greene ND et al. (2011). Magnetic resonance virtual histology for embryos: 3D atlases for automated high-throughput phenotyping. *NeuroImage* **54** (2): 769–778.

71. Cloud JE, Rogers C, Reza TL, Ziebold U, Stone JR, Picard MH, Caron AM, Bronson RT, Lees JA (2002). Mutant mouse models reveal the relative roles of E2F1 and E2F3 *in vivo*. *Mol Cell Biol* **22** (8): 2663–2672.

72. Cody SH, Xiang SD, Layton MJ, Handman E, Lam MH, Layton JE, Nice EC, Heath JK (2005). A simple method allowing DIC imaging in conjunction with confocal microscopy. *J Microsc* **217** (Pt 3): 265–274.

73. Coggins KG, Latour A, Nguyen MS, Audoly L, Coffman TM, Koller BH (2002). Metabolism of PGE$_2$ by prostaglandin dehydrogenase is essential for remodeling the ductus arteriosus. *Nat Med* **8** (2): 91–92.

74. Cole TJ, Blendy JA, Monaghan AP, Krieglstein K, Schmid W, Aguzzi A, Fantuzzi G, Hummler E, Unsicker K, Schutz G (1995). Targeted disruption of the glucocorticoid receptor gene blocks adrenergic chromaffin cell development and severely retards lung maturation. *Genes Dev* **9** (13): 1608–1621.

75. Collombat P, Mansouri A, Hecksher-Sørensen J, Serup P, Krull J, Gradwohl G, Gruss P (2003). Opposing actions of Arx and Pax4 in endocrine pancreas development. *Genes Dev* **17** (20): 2591–2603.

76. Colvin JS, White AC, Pratt SJ, Ornitz DM (2001). Lung hypoplasia and neonatal death in *Fgf9*-null mice identify this gene as an essential regulator of lung mesenchyme. *Development* **128** (11): 2095–2106.

77. Comparative Medicine Program, University of Missouri, and IDEXX BioResearch (2013). Diseases of Research Animals (DORA). http://dora.missouri.edu/mouse/epizootic-diarrhea-of-infant-mice-edim/ (last accessed December 1, 2014).

78. Conway SJ, Henderson DJ, Kirby ML, Anderson RH, Copp AJ (1997). Development of a lethal congenital heart defect in the *splotch* (*Pax3*) mutant mouse. *Cardiovasc Res* **36** (2): 163–173.

79. Conway SJ, Kruzynska-Frejtag A, Kneer PL, Machnicki M, Koushik SV (2003). What cardiovascular defect does my prenatal mouse mutant have, and why? *Genesis* **35** (1): 1–21.

80. Copp AJ (1995). Death before birth: Clues from gene knockouts and mutations. *Trends Genet* **11** (3): 87–93.

81. Coral-Vazquez R, Cohn RD, Moore SA, Hill JA, Weiss RM, Davisson RL, Straub V et al. (1999). Disruption of the sarcoglycan–sarcospan complex in vascular smooth muscle: A novel mechanism for cardiomyopathy and muscular dystrophy. *Cell* **98** (4): 465–474.

82. Cranston A, Bocker T, Reitmar A, Palazzo J, Wilson T, Mak T, Fishel R (1997). Female embryonic lethality in mice nullizygous for both *Msh2* and *p53*. *Nat Genet* **17** (1): 114–118.

83. Crawford LW, Foley JF, Elmore SA (2010). Histology atlas of the developing mouse hepatobiliary system with emphasis on embryonic days 9.5–18.5. *Toxicol Pathol* **38** (6): 872–906.

84. Crosby JL, Varnum DS, Washburn LL, Nadeau JH (1992). Disorganization is a completely dominant gain-of-function mouse mutation causing sporadic developmental defects. *Mech Dev* **37** (3): 121–126.

85. Cross JC, Werb Z, Fisher SJ (1994). Implantation and the placenta: Key pieces of the development puzzle. *Science* **266** (5190): 1508–1518.

86. Croy BA, Yamada AT, DeMayo FJ, Adamson SL (2014). *The Guide to Investigation of Mouse Pregnancy*. London, U.K.: Academic Press (Elsevier).

87. Cui J, O'Shea KS, Purkayastha A, Saunders TL, Ginsburg D (1996). Fatal haemorrhage and incomplete block to embryogenesis in mice lacking coagulation factor V. *Nature* **384** (6604): 66–68.

88. Dackor J, Caron KM, Threadgill DW (2009). Placental and embryonic growth restriction in mice with reduced function epidermal growth factor receptor alleles. *Genetics* **183** (1): 207–218.

89. Dackor J, Li M, Threadgill DW (2009). Placental overgrowth and fertility defects in mice with a hypermorphic allele of epidermal growth factor receptor. *Mamm Genome* **20** (6): 339–349.

90. Danielsson B, Sköld AC, Azarbayjani F, Öhman I, Webster W (2000). Pharmacokinetic data support pharmacologically induced embryonic dysrhythmia as explanation to fetal hydantoin syndrome in rats. *Toxicol Appl Pharmacol* **163** (2): 164–175.

91. Daston P (1997). *Molecular and Cellular Methods in Developmental Toxicology*. Boca Raton, FL: CRC Press.

92. Davis VL, Couse JF, Goulding EH, Power SG, Eddy EM, Korach KS (1994). Aberrant reproductive phenotypes evident in transgenic mice expressing the wild-type mouse estrogen receptor. *Endocrinology* **135** (1): 379–386.

93. de Boer BA, Ruijter JM, Voorbraak FP, Moorman AF (2009). More than a decade of developmental gene expression atlases: Where are we now? *Nucleic Acids Res* **37** (22): 7349–7359.

94. DeChiara TM, Bowen DC, Valenzuela DM, Simmons MV, Poueymirou WT, Thomas S, Kinetz E et al. (1996). The receptor tyrosine kinase MuSK is required for neuromuscular junction formation *in vivo*. *Cell* **85** (4): 501–512.

95. DeChiara TM, Vejsada R, Poueymirou WT, Acheson A, Suri C, Conover JC, Friedman B et al. (1995). Mice lacking the CNTF receptor, unlike mice lacking CNTF, exhibit profound motor neuron deficits at birth. *Cell* **83** (2): 313–322.

96. Denis C, Methia N, Frenette PS, Rayburn H, Ullman-Cullere M, Hynes RO, Wagner DD (1998). A mouse model of severe von Willebrand disease: Defects in hemostasis and thrombosis. *Proc Natl Acad Sci USA* **95** (16): 9524–9529.

97. Descargues P, Deraison C, Bonnart C, Kreft M, Kishibe M, Ishida-Yamamoto A, Elias P, Barrandon Y, Zambruno G, Sonnenberg A, Hovnanian A (2005). *Spink5*-deficient mice mimic Netherton syndrome through degradation of desmoglein 1 by epidermal protease hyperactivity. *Nat Genet* **37** (1): 56–65.

98. Devi PU, Baskar R (1996). Influence of gestational age at exposure on the prenatal effects of γ-radiation. *Int J Radiat Biol* **70** (1): 45–52.

99. Di Cristofano A, Pesce B, Cordon-Cardo C, Pandolfi PP (1998). *Pten* is essential for embryonic development and tumour suppression. *Nat Genet* **19** (4): 348–355.

100. Di Paolo G, Moskowitz HS, Gipson K, Wenk MR, Voronov S, Obayashi M, Flavell R, Fitzsimonds RM, Ryan TA, De Camilli P (2004). Impaired PtdIns(4,5)P$_2$ synthesis in nerve terminals produces defects in synaptic vesicle trafficking. *Nature* **431** (7007): 415–422.

101. Dickson MC, Martin JS, Cousins FM, Kulkarni AB, Karlsson S, Akhurst RJ (1995). Defective haematopoiesis and vasculogenesis in transforming growth factor-β1 knock out mice. *Development* **121** (6): 1845–1854.

102. Diewert VM (1982). A comparative study of craniofacial growth during secondary palate development in four strains of mice. *J Craniofac Genet Dev Biol* **2** (4): 247–263.

103. Diewert VM, Pratt RM (1981). Cortisone-induced cleft palate in A/J mice: Failure of palatal shelf contact. *Teratology* **24** (2): 149–162.

104. Diez-Roux G, Banfi S, Sultan M, Geffers L, Anand S, Rozado D, Magen A et al. (2011). A high-resolution anatomical atlas of the transcriptome in the mouse embryo. *PLoS Biol* **9** (1): e1000582 (13 pages).

105. Dikranian K, Qin YQ, Labruyere J, Nemmers B, Olney JW (2005). Ethanol-induced neuroapoptosis in the developing rodent cerebellum and related brain stem structures. *Brain Res Dev Brain Res* **155** (1): 1–13.

106. DiMartino JF, Selleri L, Traver D, Firpo MT, Rhee J, Warnke R, O'Gorman S, Weissman IL, Cleary ML (2001). The Hox cofactor and proto-oncogene Pbx1 is required for maintenance of definitive hematopoiesis in the fetal liver. *Blood* **98** (3): 618–626.

107. Dixon J, Brakebusch C, Fässler R, Dixon MJ (2000). Increased levels of apoptosis in the prefusion neural folds underlie the craniofacial disorder, Treacher Collins syndrome. *Hum Mol Genet* **9** (10): 1473–1480.

108. Dixon J, Dixon MJ (2004). Genetic background has a major effect on the penetrance and severity of craniofacial defects in mice heterozygous for the gene encoding the nucleolar protein Treacle. *Dev Dyn* **229** (4): 907–914.

109. Doyonnas R, Kershaw DB, Duhme C, Merkens H, Chelliah S, Graf T, McNagny KM (2001). Anuria, omphalocele, and perinatal lethality in mice lacking the CD34-related protein podocalyxin. *J Exp Med* **194** (1): 13–27.

110. Dragatsis I, Zeitlin S, Dietrich P (2004). Huntingtin-associated protein 1 (Hap1) mutant mice bypassing the early postnatal lethality are neuroanatomically normal and fertile but display growth retardation. *Hum Mol Genet* **13** (24): 3115–3125.

111. Drews U (1995). *Color Atlas of Embryology.* New York: Thieme Medical Publishers.

112. Dunty WCJ, Chen S-Y, Zucker RM, Dehart DB, Sulik KK (2001). Selective vulnerability of embryonic cell populations to ethanol-induced apoptosis: Implications for alcohol-related birth defects and neurodevelopmental disorder. *Alcohol Clin Exp Res* **25** (10): 1523–1535.

113. Dupe V, Matt N, Garnier JM, Chambon P, Mark M, Ghyselinck NB (2003). A newborn lethal defect due to inactivation of retinaldehyde dehydrogenase type 3 is prevented by maternal retinoic acid treatment. *Proc Natl Acad Sci USA* **100** (24): 14036–14041.

114. Ehlers K, Sturje H, Merker HJ, Nau H (1992). Valproic acid-induced spina bifida: A mouse model. *Teratology* **45** (2): 145–154.

115. EMAP (e-MOUSE Atlas Project) (2013). e-Mouse Atlas, v3.5. http://www.emouseatlas.org/emap/home. html (last accessed December 1, 2014).

116. Embree-Ku M, Boekelheide K (2002). Absence of p53 and FasL has sexually dimorphic effects on both development and reproduction. *Exp Biol Med* **227** (7): 545–553.

117. Enomoto H, Araki T, Jackman A, Heuckeroth RO, Snider WD, Johnson Jr EM, Milbrandt J (1998). GFRα1-deficient mice have deficits in the enteric nervous system and kidneys. *Neuron* **21** (2): 317–324.

118. Erdmann G, Schütz G, Berger S (2008). Loss of glucocorticoid receptor function in the pituitary results in early postnatal lethality. *Endocrinology* **149** (7): 3446–3451.

119. Erickson JT, Conover JC, Borday V, Champagnat J, Barbacid M, Yancopoulos G, Katz DM (1996). Mice lacking brain-derived neurotrophic factor exhibit visceral sensory neuron losses distinct from mice lacking NT4 and display a severe developmental deficit in control of breathing. *J Neurosci* **16** (17): 5361–5371.

120. Essien FB, Haviland MB, Naidoff AE (1990). Expression of a new mutation (*Axd*) causing axial defects in mice correlates with maternal phenotype and age. *Teratology* **42** (2): 183–194.

121. Evans AL, Gage PJ (2005). Expression of the homeobox gene *Pitx2* in neural crest is required for optic stalk and ocular anterior segment development. *Hum Mol Genet* **14** (22): 3347–3359.
122. Fan G, Egles C, Sun Y, Minichiello L, Renger JJ, Klein R, Liu G, Jaenisch R (2000). Knocking the NT4 gene into the BDNF locus rescues BDNF deficient mice and reveals distinct NT4 and BDNF activities. *Nat Neurosci* **3** (4): 350–357.
123. Fan G, Xiao L, Cheng L, Wang X, Sun B, Hu G (2000). Targeted disruption of NDST-1 gene leads to pulmonary hypoplasia and neonatal respiratory distress in mice. *FEBS Lett* **467** (1): 7–11.
124. Fata JE, Kong YY, Li J, Sasaki T, Irie-Sasaki J, Moorehead RA, Elliott R et al. (2000). The osteoclast differentiation factor osteoprotegerin-ligand is essential for mammary gland development. *Cell* **103** (1): 41–50.
125. Fawcett LB, Buck SJ, Brent RL (1998). Limb reduction defects in the A/J mouse strain associated with maternal blood loss. *Teratology* **58** (5): 183–189.
126. Feiner L, Webber AL, Brown CB, Lu MM, Jia L, Feinstein P, Mombaerts P, Epstein JA, Raper JA (2001). Targeted disruption of semaphorin 3C leads to persistent truncus arteriosus and aortic arch interruption. *Development* **128** (16): 3061–3070.
127. Felbor U, Kessler B, Mothes W, Goebel HH, Ploegh HL, Bronson RT, Olsen BR (2002). Neuronal loss and brain atrophy in mice lacking cathepsins B and L. *Proc Natl Acad Sci USA* **99** (12): 7883–7888.
128. Feng Q, Di R, Tao F, Chang Z, Lu S, Fan W, Shan C, Li X, Yang Z (2010). PDK1 regulates vascular remodeling and promotes epithelial-mesenchymal transition in cardiac development. *Mol Cell Biol* **30** (14): 3711–3721.
129. Feng Q, Song W, Lu X, Hamilton JA, Lei M, Peng T, Yee SP (2002). Development of heart failure and congenital septal defects in mice lacking endothelial nitric oxide synthase. *Circulation* **106** (7): 873–879.
130. Feng X, Krebs LT, Gridley T (2010). Patent ductus arteriosus in mice with smooth muscle-specific *Jag1* deletion. *Development* **137** (24): 4191–4199.
131. Feng Y, Yang Y, Ortega MM, Copeland JN, Zhang M, Jacob JB, Fields TA, Vivian JL, Fields PE (2010). Early mammalian erythropoiesis requires the Dot1L methyltransferase. *Blood* **116** (22): 4483–4491.
132. Ferguson SM, Bazalakova M, Savchenko V, Tapia JC, Wright J, Blakely RD (2004). Lethal impairment of cholinergic neurotransmission in hemicholinium-3-sensitive choline transporter knockout mice. *Proc Natl Acad Sci USA* **101** (23): 8762–8767.
133. Fine EL, Horal M, Chang TI, Fortin G, Loeken MR (1999). Evidence that elevated glucose causes altered gene expression, apoptosis, and neural tube defects in a mouse model of diabetic pregnancy. *Diabetes* **48** (12): 2454–2462.
134. Finnell RH, Bennett GD, Karras SB, Mohl VK (1988). Common hierarchies of susceptibility to the induction of neural tube defects in mouse embryos by valproic acid and its 4-propyl-4-pentenoic acid metabolite. *Teratology* **38** (4): 313–320.
135. Finnell RH, Moon SP, Abbott LC, Golden JA, Chernoff GF (1986). Strain differences in heat-induced neural tube defects in mice. *Teratology* **33** (2): 247–252.
136. Finnell RH, Shields HE, Taylor SM, Chernoff GF (1987). Strain differences in phenobarbital-induced teratogenesis in mice. *Teratology* **35** (2): 177–185.
137. Fischer A, Schumacher N, Maier M, Sendtner M, Gessler M (2004). The Notch target genes *Hey1* and *Hey2* are required for embryonic vascular development. *Genes Dev* **18** (8): 901–911.
138. Fleming A, Copp AJ (2000). A genetic risk factor for mouse neural tube defects: Defining the embryonic basis. *Hum Mol Genet* **9** (4): 575–581.
139. Francis RJB, Christopher A, Devine WA, Ostrowski L, Lo C (2012). Congenital heart disease and the specification of left–right asymmetry. *Am J Physiol Heart Circ Physiol* **302** (10): H2102–H2111.
140. Franklin JB, Brent RL (1964). The effect of uterine vascular clamping on the development of rat embryos three to fourteen days old. *J Morphol* **115**: 273–290.
141. Fraser FC, Chew D, Verrusio AC (1967). Oligohydramnios and cortisone-induced cleft palate in the mouse. *Nature* **214** (5086): 417–418.
142. Fruman DA, Mauvais-Jarvis F, Pollard DA, Yballe CM, Brazil D, Bronson RT, Kahn CR, Cantley LC (2000). Hypoglycaemia, liver necrosis and perinatal death in mice lacking all isoforms of phosphoinositide 3-kinase p85α. *Nat Genet* **26** (3): 379–382.
143. Fujikura K et al. (2013). *Kif14* mutation causes severe brain malformation and hypomyelination. *PLoS One* **8** (1): e53490 (17 pages).

144. Furuse M, Hata M, Furuse K, Yoshida Y, Haratake A, Sugitani Y, Noda T, Kubo A, Tsukita S (2002). Claudin-based tight junctions are crucial for the mammalian epidermal barrier: A lesson from claudin-1-deficient mice. *J Cell Biol* **156** (6): 1099–1111.

145. Butterstein GM, Morrison J, Mizejewski GJ (2003). Effect of α-fetoprotein and derived peptides on insulin- and estrogen-induced fetotoxicity. *Fetal Diagn Ther* **18** (5): 360–369.

146. Gabriel H-D, Jung D, Bützler C, Temme A, Traub O, Winterhager E, Willecke K (1998). Transplacental uptake of glucose is decreased in embryonic lethal connexin26-deficient mice. *J Cell Biol* **140** (6): 1453–1461.

147. Gandelman R, Davis PG (1973). Spontaneous and testosterone-induced pup killing in female Rockland-Swiss mice: The effect of lactation and the presence of young. *Dev Psychobiol* **6** (3): 251–257.

148. Gao Y, Sun Y, Frank KM, Dikkes P, Fujiwara Y, Seidl K, Sekiguchi JM et al. (1998). A critical role for DNA end-joining proteins in both lymphogenesis and neurogenesis. *Cell* **95** (7): 891–902.

149. Gautam M, Noakes PG, Moscoso L, Rupp F, Scheller RH, Merlie JP, Sanes JR (1996). Defective neuromuscular synaptogenesis in agrin-deficient mutant mice. *Cell* **85** (4): 525–535.

150. Gautam M, Noakes PG, Mudd J, Nichol M, Chu GC, Sanes JR, Merlie JP (1995). Failure of postsynaptic specialization to develop at neuromuscular junctions of rapsyn-deficient mice. *Nature* **377** (6546): 232–236.

151. Gendron-Maguire M, Mallo M, Zhang M, Gridley T (1993). *Hoxa-2* mutant mice exhibit homeotic transformation of skeletal elements derived from cranial neural crest. *Cell* **75** (7): 1317–1331.

152. Gibson-Corley KN, Olivier AK, Meyerholz DK (2013). Principles for valid histopathologic scoring in research. *Vet Pathol* **50** (6): 1007–1015.

153. Gitler AD, Lu MM, Epstein JA (2004). PlexinD1 and semaphorin signaling are required in endothelial cells for cardiovascular development. *Dev Cell* **7** (1): 107–116.

154. Goad PT, Hill DE, Slikker WJ, Kimmel CA, Gaylor DW (1984). The role of maternal diet in the developmental toxicology of ethanol. *Toxicol Appl Pharmacol* **73** (2): 256–267.

155. Gomeza J, Hulsmann S, Ohno K, Eulenburg V, Szoke K, Richter D, Betz H (2003). Inactivation of the glycine transporter 1 gene discloses vital role of glial glycine uptake in glycinergic inhibition. *Neuron* **40** (4): 785–796.

156. Gomeza J, Ohno K, Hulsmann S, Armsen W, Eulenburg V, Richter DW, Laube B, Betz H (2003). Deletion of the mouse glycine transporter 2 results in a hyperekplexia phenotype and postnatal lethality. *Neuron* **40** (4): 797–806.

157. Gregg AR, Schauer A, Shi O, Liu Z, Lee CG, O'Brien WE (1998). Limb reduction defects in endothelial nitric oxide synthase-deficient mice. *Am J Physiol* **275** (6 Pt 2): H2319–2324.

158. Gress CJ, Jacenko O (2000). Growth plate compressions and altered hematopoiesis in collagen X null mice. *J Cell Biol* **149** (4): 983–993.

159. Gressens P, Gofflot F, Van Maele-Fabry G, Misson JP, Gadisseux JF, Evrard P, Picard JJ (1992). Early neurogenesis and teratogenesis in whole mouse embryo cultures. Histochemical, immunocytological and ultrastructural study of the premigratory neuronal-glial units in normal mouse embryo and in mouse embryos influenced by cocaine and retinoic acid. *J Neuropathol Exp Neurol* **51** (2): 206–219.

160. Griffin CT, Srinivasan Y, Zheng YW, Huang W, Coughlin SR (2001). A role of thrombin receptor signaling in endothelial cells during embryonic development. *Science* **293** (5535): 1666–1670.

161. Grompe M, al-Dhalimy M, Finegold M, Ou C, Burlingame T, Kennaway N, Soriano P (1993). Loss of fumarylacetoacetate hydrolase is responsible for the neonatal hepatic dysfunction phenotype of lethal albino mice. *Genes Dev* **7** (12A): 2298–2307.

162. Grunder A, Ebel TT, Mallo M, Schwarzkopf G, Shimizu T, Sippel AE, Schrewe H (2002). Nuclear factor I-B (*Nfib*) deficient mice have severe lung hypoplasia. *Mech Dev* **112** (1–2): 69–77.

163. Gu Y, Kai M, Kusama T (1997). The embryonic and fetal effects in ICR mice irradiated in the various stages of the preimplantation period. *Rad Res* **147** (6): 735–740.

164. Guidi CJ, Sands AT, Zambrowicz BP, Turner TK, Demers DA, Webster W, Smith TW, Imbalzano AN, Jones SN (2001). Disruption of Ini1 leads to peri-implantation lethality and tumorigenesis in mice. *Mol Cell Biol* **21** (10): 3598–3603.

165. Gutierrez JC, Hrubec TC, Prater MR, Smith BJ, Freeman LE, Holladay SD (2007). Aortic and ventricular dilation and myocardial reduction in gestation day 17 ICR mouse fetuses of diabetic mothers. *Birth Defects Res A Clin Mol Teratol* **79** (6): 459–464.

166. Gutierrez JC, Prater MR, Hrubec TC, Smith BJ, Freeman LE, Holladay SD (2009). Heart changes in 17-day-old fetuses of diabetic ICR (Institute of Cancer Research) mothers: Improvement with maternal immune stimulation. *Congenit Anom (Kyoto)* **49** (1): 1–7.

167. Haaksma CJ, Schwartz RJ, Tomasek JJ (2011). Myoepithelial cell contraction and milk ejection are impaired in mammary glands of mice lacking smooth muscle alpha-actin. *Biol Reprod* **85** (1): 13–21.

168. Hackman RM, Brown KS (1972). Corticosterone-induced isolated cleft palate in A-J mice. *Teratology* **6** (3): 313–316.

169. Han RN, Babaei S, Robb M, Lee T, Ridsdale R, Ackerley C, Post M, Stewart DJ (2004). Defective lung vascular development and fatal respiratory distress in endothelial NO synthase-deficient mice: A model of alveolar capillary dysplasia? *Circ Res* **94** (8): 1115–1123.

170. Harris MJ, Juriloff DM (2007). Mouse mutants with neural tube closure defects and their role in understanding human neural tube defects. *Birth Defects Res A Clin Mol Teratol* **79** (3): 187–210.

171. Harris MJ, Juriloff DM (2010). An update to the list of mouse mutants with neural tube closure defects and advances toward a complete genetic perspective of neural tube closure. *Birth Defects Res A Clin Mol Teratol* **88** (8): 653–669.

172. Hasty P, Bradley A, Morris JH, Edmondson DG, Venuti JM, Olson EN, Klein WH (1993). Muscle deficiency and neonatal death in mice with a targeted mutation in the *myogenin* gene. *Nature* **364** (6437): 501–506.

173. Hathaway HJ, Shur BD (1996). Mammary gland morphogenesis is inhibited in transgenic mice that overexpress cell surface β1,4-galactosyltransferase. *Development* **122** (9): 2859–2872.

174. Heber S, Herms J, Gajic V, Hainfellner J, Aguzzi A, Rülicke T, von Kretzschmar H et al. (2000). Mice with combined gene knock-outs reveal essential and partially redundant functions of amyloid precursor protein family members. *J Neurosci* **20** (21): 7951–7963.

175. Hecksher-Sørensen J, Hill RE, Lettice L (1998). Double labeling for whole-mount *in situ* hybridization in mouse. *Biotechniques* **24** (6): 914–916, 918.

176. Hernandez-Tristan R, Leret ML, Almeida D (2006). Effect of intrauterine position on sex differences in the gabaergic system and behavior of rats. *Physiol Behav* **87** (3): 625–633.

177. Hewett DR, Simons AL, Mangan NE, Jolin HE, Green SM, Fallon PG, McKenzie AN (2005). Lethal, neonatal ichthyosis with increased proteolytic processing of filaggrin in a mouse model of Netherton syndrome. *Hum Mol Genet* **14** (2): 335–346.

178. Holladay SD, Sharova LV, Punareewattana K, Hrubec TC, Gogal Jr RM, Prater MR, Sharov AA (2002). Maternal immune stimulation in mice decreases fetal malformations caused by teratogens. *Int Immunopharmacol* **2** (2–3): 325–332.

179. Honda H, Oda H, Nakamoto T, Honda Z, Sakai R, Suzuki T, Saito T et al. (1998). Cardiovascular anomaly, impaired actin bundling and resistance to Src-induced transformation in mice lacking p130[Cas]. *Nat Genet* **19** (4): 361–365.

180. Hongell K, Gropp A (1982). Trisomy 13 in the mouse. *Teratology* **26** (1): 95–104.

181. Hood RD (1997). *Handbook of Developmental Toxicology*. Boca Raton, FL: CRC Press.

182. Hood RD (2012). *Developmental and Reproductive Toxicology: A Practical Approach*, 3rd edn. London, U.K.: Informa Healthcare.

183. Hovland DNJ, Machado AF, Scott WJJ, Collins MD (1999). Differential sensitivity of the SWV and C57BL/6 mouse strains to the teratogenic action of single administrations of cadmium given throughout the period of anterior neuropore closure. *Teratology* **60** (1): 13–21.

184. Hrubec TC, Prater MR, Toops KA, Holladay SD (2006). Reduction in diabetes-induced craniofacial defects by maternal immune stimulation. *Birth Defects Res B Dev Reprod Toxicol* **77** (1): 1–9.

185. Humbert PO, Rogers C, Ganiatsas S, Landsberg RL, Trimarchi JM, Dandapani S, Brugnara C, Erdman S, Schrenzel M, Bronson RT, Lees JA (2000). E2F4 is essential for normal erythrocyte maturation and neonatal viability. *Mol Cell* **6** (2): 281–291.

186. Hummler E, Barker P, Gatzy J, Beermann F, Verdumo C, Schmidt A, Boucher R, Rossier BC (1996). Early death due to defective neonatal lung liquid clearance in *aENaC*-deficient mice. *Nat Genet* **12** (3): 325–328.

187. Hunter 3rd ES, Kotch LE, Cefalo RC, Sadler TW (1995). Effects of cocaine administration during early organogenesis on prenatal development and postnatal growth in mice. *Fundam Appl Toxicol* **28** (2): 177–186.

188. Huo L, Scarpulla RC (2001). Mitochondrial DNA instability and peri-implantation lethality associated with targeted disruption of nuclear respiratory factor 1 in mice. *Mol Cell Biol* **21** (2): 644–654.

189. Iida K, Koseki H, Kakinuma H, Kato N, Mizutani-Koseki Y, Ohuchi H, Yoshioka H et al. (1997). Essential roles of the winged helix transcription factor MFH-1 in aortic arch patterning and skeletogenesis. *Development* **124** (22): 4627–4638.

190. Ikeda K, Onimaru H, Yamada J, Inoue K, Ueno S, Onaka T, Toyoda H et al. (2004). Malfunction of respiratory-related neuronal activity in Na^+, K^+-ATPase $\alpha 2$ subunit-deficient mice is attributable to abnormal Cl^- homeostasis in brainstem neurons. *J Neurosci* **24** (47): 10693–10701.

191. Ince TA, Ward JM, Valli VE, Sgroi D, Nikitin AY, Loda M, Griffey SM, Crum CP, Crawford JM, Bronson RT, Cardiff RD (2008). Do-it-yourself (DIY) pathology. *Nat Biotechnol* **26** (9): 978–979; discussion 979.

192. Inouye M, Murakami U (1978). Teratogenic effect of *N*-methyl-*N'*-nitro-*N*-nitrosoguanidine in mice. *Teratology* **18** (2): 263–267.

193. Isermann B, Hendrickson SB, Hutley K, Wing M, Weiler H (2001). Tissue-restricted expression of thrombomodulin in the placenta rescues thrombomodulin-deficient mice from early lethality and reveals a secondary developmental block. *Development* **128** (6): 827–838.

194. Ishikawa H, Rattigan A, Fundele R, Burgoyne PS (2003). Effects of sex chromosome dosage on placental size in mice. *Biol Reprod* **69** (2): 483–488.

195. Iwaki T, Yamashita H, Hayakawa T (2001). *A Color Atlas of Sectional Anatomy of the Mouse*. Braintree, MA: Braintree Scientific, Inc.

196. Iwata T, Chen L, Li C, Ovchinnikov DA, Behringer RR, Francomano CA, Deng C-X (2000). A neonatal lethal mutation in FGFR3 uncouples proliferation and differentiation of growth plate chondrocytes in embryos. *Hum Mol Genet* **9** (11): 1603–1613.

197. Jacks T, Fazeli A, Schmitt EM, Bronson RT, Goodell MA, Weinberg RA (1992). Effects of an *Rb* mutation in the mouse. *Nature* **359** (6393): 295–300.

198. Jackson Laboratory (2001–2014). Mouse phenome database—Lethal phenotypes during embryogenesis. http://www.informatics.jax.org/searches/Phat.cgi?id=MP:0001672 (last accessed December 1, 2014).

199. Jackson Laboratory (2001–2015). Mouse phenome database—Prenatal lethality due to placental pathology. http://www.informatics.jax.org/searches/Phat.cgi?id=MP:0001711 (last accessed December 1, 2014).

200. Jacobson L, Polizzi A, Vincent A (1998). An animal model of maternal antibody-mediated arthrogryposis multiplex congenita (AMC). *Ann N Y Acad Sci* **841**: 565–567.

201. Jacobwitz DM, Abbott LC (1998). *Chemoarchitectonic Atlas of the Developing Mouse Brain*. Boca Raton, FL: CRC Press.

202. Jacquin TD, Borday V, Schneider-Maunoury S, Topilko P, Ghilini G, Kato F, Charnay P, Champagnat J (1996). Reorganization of pontine rhythmogenic neuronal networks in *Krox-20* knockout mice. *Neuron* **17** (4): 747–758.

203. Jalbert LR, Rosen ED, Moons L, Chan JC, Carmeliet P, Collen D, Castellino FJ (1998). Inactivation of the gene for anticoagulant protein C causes lethal perinatal consumptive coagulopathy in mice. *J Clin Invest* **102** (8): 1481–1488.

204. Janmohamed A, Hernandez D, Phillips IR, Shephard EA (2004). Cell-, tissue-, sex- and developmental stage-specific expression of mouse flavin-containing monooxygenases (*Fmos*). *Biochem Pharmacol* **68** (1): 73–83.

205. Johnson CS, Zucker RM, Hunter 3rd ES, Sulik KK (2007). Perturbation of retinoic acid (RA)-mediated limb development suggests a role for diminished RA signaling in the teratogenesis of ethanol. *Birth Defects Res A Clin Mol Teratol* **79** (9): 631–641.

206. Johnson GA, Cofer GP, Fubara B, Gewalt SL, Hedlund LW, Maronpot RR (2002). Magnetic resonance histology for morphologic phenotyping. *J Magn Reson Imaging* **16** (4): 423–429.

207. Jones GN, Pringle DR, Yin Z, Carlton MM, Powell KA, Weinstein MB, Toribio RE, La Perle KMD, Kirschner LS (2010). Neural crest-specific loss of *Prkar1a* causes perinatal lethality resulting from defects in intramembranous ossification. *Mol Endocrinol* **24** (8): 1559–1568.

208. Joshi RL, Lamothe B, Cordonnier N, Mesbah K, Monthioux E, Jami J, Bucchini D (1996). Targeted disruption of the insulin receptor gene in the mouse results in neonatal lethality. *EMBO J* **15** (7): 1542–1547.

209. Juan TS-C, Bolon B, Lindberg RA, Sun Y, Van G, Fletcher FA (2009). Mice over-expressing murine oncostatin M (OSM) exhibit changes in hematopoietic and other organs that are distinct from those of mice over-expressing human OSM or bovine OSM. *Vet Pathol* **46** (1): 124–137.

210. Juriloff DM (1980). Genetics of clefting in the mouse. *Prog Clin Biol Res* **46**: 39–71.

211. Juriloff DM, Harris MJ, Tom C, MacDonald KB (1991). Normal mouse strains differ in the site of initiation of closure of the cranial neural tube. *Teratology* **44** (2): 225–233.

212. Kaartinen V, Voncken JW, Shuler C, Warburton D, Bu D, Heisterkamp N, Groffen J (1995). Abnormal lung development and cleft palate in mice lacking TGF-β3 indicates defects of epithelial-mesenchymal interaction. *Nat Genet* **11** (4): 415–421.

213. Kalcheva N, Qu J, Sandeep N, Garcia L, Zhang J, Wang Z, Lampe PD, Suadicani SO, Spray DC, Fishman GI (2007). Gap junction remodeling and cardiac arrhythmogenesis in a murine model of oculodentodigital dysplasia. *Proc Natl Acad Sci USA* **104** (51): 20512–20516.

214. Kalter H (1975). Prenatal epidemiology of spontaneous cleft lip and palate, open eyelid, and embryonic death in A/J mice. *Teratology* **12** (3): 245–257.

215. Karaplis AC, Luz A, Glowacki J, Bronson RT, Tybulewicz VL, Kronenberg HM, Mulligan RC (1994). Lethal skeletal dysplasia from targeted disruption of the parathyroid hormone-related peptide gene. *Genes Dev* **8** (3): 277–289.

216. Kaufman MH (1981). The role of embryology in teratological research, with particular reference to the development of the neural tube and heart. *J Reprod Fert* **62** (2): 607–623.

217. Kaufman MH (1992). *The Atlas of Mouse Development*, 2nd edn. San Diego, CA: Academic Press.

218. Kaufman MH, Chang HH (2000). Studies of the mechanism of amniotic sac puncture-induced limb abnormalities in mice. *Int J Dev Biol* **44** (1): 161–175.

219. Keith B, Adelman DM, Simon MC (2001). Targeted mutation of the murine arylhydrocarbon receptor nuclear translocator 2 (*Arnt2*) gene reveals partial redundancy with *Arnt. Proc Natl Acad Sci USA* **98** (12): 6692–6697.

220. Khera KS (1984). Maternal toxicity—A possible factor in fetal malformations in mice. *Teratology* **29** (3): 411–416.

221. Kim AJ, Francis R, Liu X, Devine WA, Ramirez R, Anderton SJ, Wong LY et al. (2013). Microcomputed tomography provides high accuracy congenital heart disease diagnosis in neonatal and fetal mice. *Circ Cardiovasc Imaging* **6** (4): 551–559.

222. Kim W-K, Mirkes PE (2003). Alterations in mitochondrial morphology are associated with hyperthermia-induced apoptosis in early postimplantation mouse embryos. *Birth Defects Res A Clin Mol Teratol* **67** (11): 929–940.

223. Kimura S, Hara Y, Pineau T, Fernandez-Salguero P, Fox CH, Ward JM, Gonzalez FJ (1996). The *T/ebp* null mouse: Thyroid-specific enhancer-binding protein is essential for the organogenesis of the thyroid, lung, ventral forebrain, and pituitary. *Genes Dev* **10** (1): 60–69.

224. Kitamura K, Yanazawa M, Sugiyama N, Miura H, Iizuka-Kogo A, Kusaka M, Omichi K et al. (2002). Mutation of ARX causes abnormal development of forebrain and testes in mice and X-linked lissencephaly with abnormal genitalia in humans. *Nat Genet* **32** (3): 359–369.

225. Kleaveland B, Zheng X, Liu JJ, Blum Y, Tung JJ, Zou Z, Sweeney SM et al. (2009). Regulation of cardiovascular development and integrity by the heart of glass-cerebral cavernous malformation protein pathway. *Nat Med* **15** (2): 169–176.

226. Kleschevnikov AM, Belichenko PV, Salehi A, Wu C (2012). Discoveries in Down syndrome: Moving basic science to clinical care. *Prog Brain Res* **197**: 199–221.

227. Klinger S, Turgeon B, Lévesque K, Wood GA, Aagaard-Tillery KM, Meloche S (2009) Loss of Erk3 function in mice leads to intrauterine growth restriction, pulmonary immaturity, and neonatal lethality. *Proc Natl Acad Sci USA* **106** (39): 16710–16715.

228. Knight BS, Pennell CE, Adamson SL, Lye SJ (2007). The impact of murine strain and sex on postnatal development after maternal dietary restriction during pregnancy. *J Physiol* **581** (Pt 2): 873–881.

229. Knight BS, Sunn N, Pennell CE, Adamson SL, Lye SJ (2009). Developmental regulation of cardiovascular function is dependent on both genotype and environment. *Am J Physiol Heart Circ Physiol* **297** (6): H2234–H2241.

230. Ko J, Humbert S, Bronson RT, Takahashi S, Kulkarni AB, Li E, Tsai LH (2001). p35 and p39 are essential for cyclin-dependent kinase 5 function during neurodevelopment. *J Neurosci* **21** (17): 6758–6771.

231. Kobayashi K, Morita S, Sawada H, Mizuguchi T, Yamada K, Nagatsu I, Hata T, Watanabe Y, Fujita K, Nagatsu T (1995). Targeted disruption of the tyrosine hydroxylase locus results in severe catecholamine depletion and perinatal lethality in mice. *J Biol Chem* **270** (45): 27235–27243.

232. Kobayashi S, Yoshida K, Ward JM, Letterio JJ, Longenecker G, Yaswen L, Mittleman B, Mozes E, Roberts AB, Karlsson S, Kulkarni AB (1999). β2-Microglobulin-deficient background ameliorates lethal phenotype of the TGF-β1 null mouse. *J Immunol* **163** (7): 4013–4019.

233. Komárek V (2007). Gross anatomy. In: *The Mouse in Biomedical Research,* 2nd edn., Vol. III. *Normative Biology, Husbandry, and Models*, Fox JG, Barthold SW, Davisson MT, Newcomer CE, Quimby FW, Smith AL (eds.). Boston, MA: Elsevier, pp. 1–22.

234. Komatsu M, Waguri S, Ueno T, Iwata J, Murata S, Tanida I, Ezaki J et al. (2005). Impairment of starvation-induced and constitutive autophagy in *Atg7*-deficient mice. *J Cell Biol* **169** (3): 425–434.

235. Komori T, Yagi H, Nomura S, Yamaguchi A, Sasaki K, Deguchi K, Shimizu Y et al. (1997). Targeted disruption of *Cbfa1* results in a complete lack of bone formation owing to maturational arrest of osteoblasts. *Cell* **89** (5): 755–764.

236. Kondo S (1998). Microinjection methods for visualization of the vascular architecture of the mouse embryo for light and scanning electron microscopy. *J Electron Microsc (Tokyo)* **47** (2): 101–113.

237. Kotch LE, Sulik KK (1992). Experimental fetal alcohol syndrome: Proposed pathogenic basis for a variety of associated facial and brain anomalies. *Am J Med Genet* **44** (2): 168–176.

238. Koutsourakis M, Langeveld A, Patient R, Beddington R, Grosveld F (1999). The transcription factor GATA6 is essential for early extraembryonic development. *Development* **126** (9): 723–732.

239. Kreidberg JA, Donovan MJ, Goldstein SL, Rennke H, Shepherd K, Jones RC, Jaenisch R (1996). Alpha 3 beta 1 integrin has a crucial role in kidney and lung organogenesis. *Development* **122** (11): 3537–3547.

240. Kreidberg JA, Sariola H, Loring JM, Maeda M, Pelletier J, Housman D, Jaenisch R (1993). WT-1 is required for early kidney development. *Cell* **74** (4): 679–691.

241. Kruger I, Vollmer M, Simmons DG, Elsasser HP, Philipsen S, Suske G (2007). *Sp1/Sp3* compound heterozygous mice are not viable: Impaired erythropoiesis and severe placental defects. *Dev Dyn* **236** (8): 2235–2244.

242. Kulandavelu S, Qu D, Sunn N, Mu J, Rennie MY, Whiteley KJ, Walls JR et al. (2006). Embryonic and neonatal phenotyping of genetically engineered mice. *ILAR J* **47** (2): 103–117.

243. Kulkarni AB, Huh CG, Becker D, Geiser A, Lyght M, Flanders KC, Roberts AB, Sporn MB, Ward JM, Karlsson S (1993). Transforming growth factor β₁ null mutation in mice causes excessive inflammatory response and early death. *Proc Natl Acad Sci USA* **90** (2): 770–774.

244. Kuma A, Hatano M, Matsui M, Yamamoto A, Nakaya H, Yoshimori T, Ohsumi Y, Tokuhisa T, Mizushima N (2004). The role of autophagy during the early neonatal starvation period. *Nature* **432** (7020): 1032–1036.

245. Kwiatkowski DJ, Zhang H, Bandura JL, Heiberger KM, Glogauer M, el-Hashemite N, Onda H (2002). A mouse model of TSC1 reveals sex-dependent lethality from liver hemangiomas, and up-regulation of p70S6 kinase activity in Tsc1 null cells. *Hum Mol Genet* **11** (5): 525–534.

246. Kwon C-H, Zhu X, Zhang J, Knoop LL, Tharp R, Smeyne RJ, Eberhart CG, Burger PC, Baker SJ (2001). Pten regulates neuronal soma size: A mouse model of Lhermitte-Duclos disease. *Nat Genet* **29** (4): 404–411.

247. Labosky PA, Winnier GE, Jetton TL, Hargett L, Ryan AK, Rosenfeld MG, Parlow AF, Hogan BL (1997). The winged helix gene, *Mf3*, is required for normal development of the diencephalon and midbrain, postnatal growth and the milk-ejection reflex. *Development* **124** (7): 1263–1274.

248. Lan Y, Ovitt CE, Cho ES, Maltby KM, Wang Q, Jiang R (2004). Odd-skipped related 2 (*Osr2*) encodes a key intrinsic regulator of secondary palate growth and morphogenesis. *Development* **131** (13): 3207–3216.

249. Langner CA, Birkenmeier EH, Ben-Zeev O, Schotz MC, Sweet HO, Davisson MT, Gordon JI (1989). The fatty liver dystrophy (*fld*) mutation. A new mutant mouse with a developmental abnormality in triglyceride metabolism and associated tissue-specific defects in lipoprotein lipase and hepatic lipase activities. *J Biol Chem* **264** (14): 7994–8003.

250. Lau MM, Stewart CE, Liu Z, Bhatt H, Rotwein P, Stewart CL (1994). Loss of the imprinted IGF2/cation-independent mannose 6-phosphate receptor results in fetal overgrowth and perinatal lethality. *Genes Dev* **8** (24): 2953–2963.

251. Layman WS, Hurd EA, Martin DM (2010). Chromodomain proteins in development: Lessons from CHARGE syndrome. *Clin Genet* **78** (1): 11–20.

252. LeCouter JE, Kablar B, Whyte PF, Ying C, Rudnicki MA (1998). Strain-dependent embryonic lethality in mice lacking the retinoblastoma-related p130 gene. *Development* **125** (3): 4669–4679.

253. Leder A, McMenamin J, Fontaine K, Bishop A, Leder P (2005). ζ⁻/⁻ Thalassemic mice are affected by two modifying loci and display unanticipated somatic recombination leading to inherited variation. *Hum Mol Genet* **14** (5): 615–625.

254. Lee AT, Plump A, DeSimone C, Cerami A, Bucala R (1995). A role for DNA mutations in diabetes-associated teratogenesis in transgenic embryos. *Diabetes* **44** (1): 20–24.

255. Leitges M, Neidhardt L, Haenig B, Herrmann BG, Kispert A (2000). The paired homeobox gene *Uncx4.1* specifies pedicles, transverse processes and proximal ribs of the vertebral column. *Development* **127** (11): 2259–2267.

256. Lemos MC, Harding B, Reed AA, Jeyabalan J, Walls GV, Bowl MR, Sharpe J, Wedden S, Moss JE, Ross A, Davidson D, Thakker RV (2009). Genetic background influences embryonic lethality and the occurrence of neural tube defects in *Men1* null mice: Relevance to genetic modifiers. *J Endocrinol* **203** (1): 133–142.

257. Lenox JM, Koch PJ, Mahoney MG, Lieberman M, Stanley JR, Radice GL (2000). Postnatal lethality of P-cadherin/desmoglein 3 double knockout mice: Demonstration of a cooperative effect of these cell adhesion molecules in tissue homeostasis of stratified squamous epithelia. *J Invest Dermatol* **114** (5): 948–952.

258. Lettice LA, Purdie LA, Carlson GJ, Kilanowski F, Dorin J, Hill RE (1999). The mouse bagpipe gene controls development of axial skeleton, skull, and spleen. *Proc Natl Acad Sci USA* **96** (17): 9695–9700.

259. Levéen P, Pekny M, Gebre-Medhin S, Swolin B, Larsson E, Betsholtz C (1994). Mice deficient for PDGF B show renal, cardiovascular, and hematological abnormalities. *Gens Dev* **8** (16): 1875–1887.

260. Lewis JD, Destito G, Zijlstra A, Gonzalez MJ, Quigley JP, Manchester M, Stuhlmann H (2006). Viral nanoparticles as tools for intravital vascular imaging. *Nat Med* **12** (3): 354–360.

261. Li ZL, Shiota K (1999). Stage-specific homeotic vertebral transformations in mouse fetuses induced by maternal hyperthermia during somitogenesis. *Dev Dyn* **216** (4–5): 336–348.

262. Liang C, Oest ME, Jones JC, Prater MR (2009). Gestational high saturated fat diet alters C57BL/6 mouse perinatal skeletal formation. *Birth Defects Res B Dev Reprod Toxicol* **86** (5): 362–369.

263. Liao HJ, Kume T, McKay C, Xu MJ, Ihle JN, Carpenter G (2002). Absence of erythrogenesis and vasculogenesis in *Plcg1*-deficient mice. *J Biol Chem* **277** (11): 9335–9341.

264. Lin C-S, Lim S-K, Dagati V, Costantini F (1996). Differential effects of an erythropoietin receptor gene disruption on primitive and definitive erythropoiesis. *Genes Dev* **10** (2): 154–164.

265. Lindsten T, Ross AJ, King A, Zong W-X, Rathmell JC, Shiels HA, Ulrich E et al. (2000). The combined functions of proapoptotic Bcl-2 family members Bak and Bax are essential for normal development of multiple tissues. *Mol Cell* **6** (6): 1389–1399.

266. Lipinski RJ, Song C, Sulik KK, Everson JL, Gipp JJ, Yan D, Bushman W, Rowland IJ (2010). Cleft lip and palate results from Hedgehog signaling antagonism in the mouse: Phenotypic characterization and clinical implications. *Birth Defects Res A Clin Mol Teratol* **88** (4): 232–240.

267. Liu Y, Jin Y, Li J, Seto E, Kuo E, Yu W, Schwartz RJ, Blazo M, Zhang SL, Peng X (2013). Inactivation of Cdc42 in neural crest cells causes craniofacial and cardiovascular morphogenesis defects. *Dev Biol* **383** (2): 239–252.

268. Loftin CD, Trivedi DB, Tiano HF, Clark JA, Lee CA, Epstein JA, Morham SG et al. (2001). Failure of ductus arteriosus closure and remodeling in neonatal mice deficient in cyclooxygenase-1 and cyclooxygenase-2. *Proc Natl Acad Sci USA* **98** (3): 1059–1064.

269. Lomaga MA, Henderson JT, Elia AJ, Robertson J, Noyce RS, Yeh W-C, Mak TW (2000). Tumor necrosis factor receptor-associated factor 6 (TRAF6) deficiency results in exencephaly and is required for apoptosis within the developing CNS. *J Neurosci* **20** (19): 7384–7393.

270. Lu J, Chang P, Richardson JA, Gan L, Weiler H, Olson EN (2000). The basic helix-loop-helix transcription factor capsulin controls spleen organogenesis. *Proc Natl Acad Sci USA* **97** (17): 9525–9530.

271. Lu Q, Hasty P, Shur BD (1997). Targeted mutation in β1,4-galactosyltransferase leads to pituitary insufficiency and neonatal lethality. *Dev Biol* **181** (2): 257–267.

272. Ludwig DL, MacInnes MA, Takiguchi Y, Purtymun PE, Henrie M, Flannery M, Meneses J, Pedersen RA, Chen DJ (1998). A murine AP-endonuclease gene-targeted deficiency with post-implantation embryonic progression and ionizing radiation sensitivity. *Mutat Res* **409** (1): 17–29.

273. Lutz J, Beck SL (2000). Caffeine decreases the occurrence of cadmium-induced forelimb ectrodactyly in C57BL/6J mice. *Teratology* **62** (5): 325–331.

274. Ma Q, Jones D, Borghesani PR, Segal RA, Nagasawa T, Kishimoto T, Bronson RT, Springer TA (1998). Impaired B-lymphopoiesis, myelopoiesis, and derailed cerebellar neuron migration in CXCR4- and SDF-1-deficient mice. *Proc Natl Acad Sci USA* **95** (16): 9448–9453.

275. MacMurray A, Shin HS (1988). The antimorphic nature of the T^c allele at the mouse *T* locus. *Genetics* **120** (2): 545–550.

276. Makita T, Hernandez-Hoyos G, Chen TH, Wu H, Rothenberg EV, Sucov HM (2001). A developmental transition in definitive erythropoiesis: Erythropoietin expression is sequentially regulated by retinoic acid receptors and HNF4. *Genes Dev* **15** (7): 889–901.

277. Manner J (2009). The anatomy of cardiac looping: A step towards the understanding of the morphogenesis of several forms of congenital cardiac malformations. *Clin Anat* **22** (1): 21–35.

278. Mansour SL, Goddard JM, Capecchi MR (1993). Mice homozygous for a targeted disruption of the proto-oncogene *int-2* have developmental defects in the tail and inner ear. *Development* **117** (1): 13–28.

279. Mansouri A, Voss AK, Thomas T, Yokota Y, Gruss P (2000). *Uncx4.1* is required for the formation of the pedicles and proximal ribs and acts upstream of *Pax9*. *Development* **127** (11): 2251–2258.

280. Martinez F, Happa J, Arias F (1985). Biochemical and morphologic effects of ethanol on fetuses from normally ovulating and superovulated mice. *Am J Obstet Gynecol* **151** (4): 428–433.

281. Matsuki M, Yamashita F, Ishida-Yamamoto A, Yamada K, Kinoshita C, Fushiki S, Ueda E et al. (1998). Defective stratum corneum and early neonatal death in mice lacking the gene for transglutaminase 1 (keratinocyte transglutaminase). *Proc Natl Acad Sci USA* **95** (3): 1044–1049.

282. McCarthy A, Brown NA (1998). Specification of left–right asymmetry in mammals: Embryo culture studies of stage of determination and relationships with morphogenesis and growth. *Reprod Toxicol* **12** (2): 177–184.

283. McDonald FJ, Yang B, Hrstka RF, Drummond HA, Tarr DE, McCray Jr PB, Stokes JB, Welsh MJ, Williamson RA (1999). Disruption of the β subunit of the epithelial Na⁺ channel in mice: Hyperkalemia and neonatal death associated with a pseudohypoaldosteronism phenotype. *Proc Natl Acad Sci USA* **96** (4): 1727–1731.

284. McEvilly RJ, Erkman L, Luo L, Sawchenko PE, Ryan AF, Rosenfeld MG (1996). Requirement for Brn-3.0 in differentiation and survival of sensory and motor neurons. *Nature* **384** (6609): 574–577.

285. McFadden DG, Barbosa AC, Richardson JA, Schneider MD, Srivastava D, Olson EN (2005). The Hand1 and Hand2 transcription factors regulate expansion of the embryonic cardiac ventricles in a gene dosage-dependent manner. *Development* **132** (1): 189–201.

286. McKercher SR, Torbett BE, Anderson KL, Henkel GW, Vestal DJ, Baribault H, Klemsz M, Feeney AJ, Wu GE, Paige CJ, Maki RA (1996). Targeted disruption of the *PU.1* gene results in multiple hematopoietic abnormalities. *EMBO J* **15** (20): 5647–5658.

287. McPherron AC, Lawler AM, Lee SJ (1999). Regulation of anterior/posterior patterning of the axial skeleton by growth/differentiation factor 11. *Nat Genet* **22** (3): 260–264.

288. Messerle K, Webster WS (1982). The classification and development of cadmium-induced limb defects in mice. *Teratology* **25** (1): 61–70.

289. Mimura N, Hamada H, Kashio M, Jin H, Toyama Y, Kimura K, Iida M et al. (2007). Aberrant quality control in the endoplasmic reticulum impairs the biosynthesis of pulmonary surfactant in mice expressing mutant BiP. *Cell Death Differ* **14** (8): 1475–1485.

290. Mimura N, Yuasa S, Soma M, Jin H, Kimura K, Goto S, Koseki H, Aoe T (2008). Altered quality control in the endoplasmic reticulum causes cortical dysplasia in knock-in mice expressing a mutant BiP. *Mol Cell Biol* **28** (1): 293–301.

291. Min H, Danilenko DM, Scully SA, Bolon B, Ring BD, Tarpley JE, DeRose M, Simonet WS (1998). *Fgf-10* is required for both limb and lung development and exhibits striking functional similarity to Drosophila branchless. *Genes Dev* **12** (20): 3156–3161.

292. Misgeld T, Burgess RW, Lewis RM, Cunningham JM, Lichtman JW, Sanes JR (2002). Roles of neurotransmitter in synapse formation: Development of neuromuscular junctions lacking choline acetyltransferase. *Neuron* **36** (4): 635–648.

293. Miyazaki M, Dobrzyn A, Elias PM, Ntambi JM (2005). Stearoyl-CoA desaturase-2 gene expression is required for lipid synthesis during early skin and liver development. *Proc Natl Acad Sci USA* **102** (35): 12501–12506.

294. Moon A (2008). Mouse models of congenital cardiovascular disease. *Curr Top Dev Biol* **84**: 171–248.

295. Moore MW, Klein RD, Farinas I, Sauer H, Armanini M, Phillips H, Reichardt LF, Ryan AM, Carver-Moore K, Rosenthal A (1996). Renal and neuronal abnormalities in mice lacking GDNF. *Nature* **382** (6586): 76–79.

296. Mori S, Nishikawa S-I, Yokota Y (2000). Lactation defect in mice lacking the helix-loop-helix inhibitor *Id2*. *EMBO J* **19** (21): 5772–5781.

297. Moseley AE, Lieske SP, Wetzel RK, James PF, He S, Shelly DA, Paul RJ et al. (2003). The Na,K-ATPase α2 isoform is expressed in neurons, and its absence disrupts neuronal activity in newborn mice. *J Biol Chem* **278** (7): 5317–5324.

298. Mucenski ML, McLain K, Kier AB, Swerdlow SH, Schreiner CM, Miller TA, Pietryga DW, Scott Jr WJ, Potter SS (1991). A functional c-*myb* gene is required for normal murine fetal hepatic hematopoiesis. *Cell* **65** (4): 677–689.

299. Muglia L, Jacobson L, Dikkes P, Majzoub JA (1995). Corticotropin-releasing hormone deficiency reveals major fetal but not adult glucocorticoid need. *Nature* **373** (6513): 427–432.

300. Nabeshima Y, Hanaoka K, Hayasaka M, Esumi E, Li S, Nonaka I (1993). Myogenin gene disruption results in perinatal lethality because of severe muscle defect. *Nature* **364** (6437): 532–535.

301. Nagao T, Saitoh Y, Yoshimura S (2000). Possible mechanism of congenital malformations induced by exposure of mouse preimplantation embryos to mitomycin C. *Teratology* **61** (4): 248–261.

302. Nagy A, Gertsenstein M, Vintersten K, Behringer R (2003). *Manipulating the Mouse Embryo*, 3rd edn. Cold Spring Harbor, NY: Cold Spring Harbor Laboratory Press.

303. Nakai S, Sugitani Y, Sato H, Ito S, Miura Y, Ogawa M, Nishi M, Jishage K, Minowa O, Noda T (2003). Crucial roles of Brn1 in distal tubule formation and function in mouse kidney. *Development* **130** (19): 4751–4759.

304. Narotsky MG, Francis EZ, Kavlock RJ (1994). Developmental toxicity and structure-activity relationships of aliphatic acids, including dose-response assessment of valproic acid in mice and rats. *Fundam Appl Toxicol* **22** (2): 251–265.

305. Naruse I, Collins MD, Scott WJJ (1988). Strain differences in the teratogenicity induced by sodium valproate in cultured mouse embryos. *Teratology* **38** (1): 87–96.

306. Neubauer H, Cumano A, Müller M, Wu H, Huffstadt U, Pfeffer K (1998). Jak2 deficiency defines an essential developmental checkpoint in definitive hematopoiesis. *Cell* **93** (3): 397–409.

307. Nguyen M, Camenisch T, Snouwaert JN, Hicks E, Coffman TM, Anderson PA, Malouf NN, Koller BH (1997). The prostaglandin receptor EP$_4$ triggers remodelling of the cardiovascular system at birth. *Nature* **390** (6655): 78–81.

308. Niedernhofer LJ (2008). Nucleotide excision repair deficient mouse models and neurological disease. *DNA Repair* **7** (7): 1180–1189.

309. Noakes PG, Gautam M, Mudd J, Sanes JR, Merlie JP (1995). Aberrant differentiation of neuromuscular junctions in mice lacking s-laminin/laminin β2. *Nature* **374** (6519): 258–262.

310. Noakes PG, Miner JH, Gautam M, Cunningham JM, Sanes JR, Merlie JP (1995). The renal glomerulus of mice lacking s-laminin/laminin β2: Nephrosis despite molecular compensation by laminin β1. *Nat Genet* **10** (4): 400–406.

311. O'Rahilly R, Müller F (2001). *Human Embryology & Teratology*, 3rd edn. New York: Wiley-Liss, Inc.

312. Offermanns S, Zhao L-P, Gohla A, Sarosi I, Simon MI, Wilkie TM (1998). Embryonic cardiomyocyte hypoplasia and craniofacial defects in Gα$_q$/Gα$_{11}$-mutant mice. *EMBO J* **17** (15): 4304–4312.

313. Oh SP, Li E (2002). Gene-dosage-sensitive genetic interactions between inversus viscerum (*iv*), *nodal*, and activin type IIB receptor (*ActRIIB*) genes in asymmetrical patterning of the visceral organs along the left–right axis. *Dev Dyn* **224** (3): 279–290.

314. Ohdo S, Watanabe H, Ogawa N, Yoshiyama Y, Sugiyama T (1996). Chronotoxicity of sodium valproate in pregnant mouse and embryo. *Jpn J Pharmacol* **70** (3): 253–258.

315. Ohtoshi A, Behringer RR (2004). Neonatal lethality, dwarfism, and abnormal brain development in *Dmbx1* mutant mice. *Mol Cell Biol* **24** (17): 7548–7558.

316. Olney JW, Tenkova T, Dikranian K, Muglia LJ, Jermakowicz WJ, D'Sa C, Roth KA (2002). Ethanol-induced caspase-3 activation in the *in vivo* developing mouse brain. *Neurobiol Dis* **9** (2): 205–219.

317. Otto F, Thornell AP, Crompton T, Denzel A, Gilmour KC, Rosewell IR, Stamp GW et al. (1997). *Cbfa1*, a candidate gene for cleidocranial dysplasia syndrome, is essential for osteoblast differentiation and bone development. *Cell* **89** (5): 765–771.

318. Ozolins TR, Fisher TS, Nadeau DM, Stock JL, Klein AS, Milici AJ, Morton D, Wilhelms MB, Brissette WH, Li B (2009). Defects in embryonic development of EGLN1/PHD2 knockdown transgenic mice are associated with induction of Igfbp in the placenta. *Biochem Biophys Res Commun* **390** (3): 372–376.

319. Palmer AK (1972). Sporadic malformations in laboratory animals and their influence on drug testing. In: *Drugs and Fetal Development*, Klingberg MA, Abramovici A, Chemke J (eds.). New York: Plenum Press, pp. 45–60.

320. Pampfer S, Donnay I (1999). Apoptosis at the time of embryo implantation in mouse and rat. *Cell Death Differ* **6** (6): 533–545.

321. Pan Y, Zvaritch E, Tupling AR, Rice WJ, de Leon S, Rudnicki M, McKerlie C, Banwell BL, MacLennan DH (2003). Targeted disruption of the *ATP2A1* gene encoding the sarco(endo)plasmic reticulum Ca²⁺ ATPase isoform 1 (SERCA1) impairs diaphragm function and is lethal in neonatal mice. *J Biol Chem* **278** (15): 13367–13375.

322. Papaioannou VE, Behringer RR (2005). *Mouse Phenotypes: A Handbook of Mutation Analysis.* Cold Spring Harbor, NY: Cold Spring Harbor Laboratory Press.

323. Papaioannou VE, Behringer RR (2012). Early embryonic lethality in genetically engineered mice: Diagnosis and phenotypic analysis. *Vet Pathol* **49** (1): 64–70.

324. Pauken CM, LaBorde JB, Bolon B (1999). Retinoic acid acts during peri-implantational development to alter axial and brain formation. *Anat Embryol* **200** (6): 645–655.

325. Paul E, Cronan R, Weston PJ, Boekelheide K, Sedivy JM, Lee SY, Wiest DL, Resnick MB, Klysik JE (2009). Disruption of *Supv3L1* damages the skin and causes sarcopenia, loss of fat, and death. *Mamm Genome* **20** (2): 92–108.

326. Paumgartten FJ (2010). Influence of maternal toxicity on the outcome of developmental toxicity studies. *J Toxicol Environ Health Pt A* **73** (13–14): 944–951.

327. Paxinos G, Halliday G, Watson C, Koutcherov Y, Wang H (2007). *Atlas of the Developing Mouse Brain at E17.5, P0, and P6.* San Diego, CA: Academic Press (Elsevier).

328. Percy DH, Barthold SW (2007). *Pathology of Laboratory Rodents and Rabbits*, 3rd edn. Ames, IA: Iowa State University Press.

329. Perez CJ, Jaubert J, Guenet JL, Barnhart KF, Ross-Inta CM, Quintanilla VC, Aubin I et al. (2010). Two hypomorphic alleles of mouse *Ass1* as a new animal model of citrullinemia type I and other hyperammonemic syndromes. *Am J Pathol* **177** (4): 1958–1968.

330. Perkins AC, Sharpe AH, Orkin SH (1995). Lethal β-thalassaemia in mice lacking the erythroid CACCC-transcription factor EKLF. *Nature* **375** (6529): 318–322.

331. Peschon JJ, Slack JL, Reddy P, Stocking KL, Sunnarborg SW, Lee D, Russell WE et al. (1998). An essential role for ectodomain shedding in mammalian development. *Science* **282** (5392): 1281–1284.

332. Petiet AE, Kaufman MH, Goddeeris MM, Brandenburg J, Elmore SA, Johnson GA (2008). High-resolution magnetic resonance histology of the embryonic and neonatal mouse: A 4D atlas and morphologic database. *Proc Natl Acad Sci USA* **105** (34): 12331–12336.

333. Pichel JG, Shen L, Sheng HZ, Granholm AC, Drago J, Grinberg A, Lee EJ et al. (1996). Defects in enteric innervation and kidney development in mice lacking GDNF. *Nature* **382** (6586): 73–76.

334. Pichel JG, Shen L, Sheng HZ, Granholm AC, Drago J, Grinberg A, Lee EJ et al. (1996). GDNF is required for kidney development and enteric innervation. *Cold Spring Harb Symp Quant Biol* **61**: 445–457.

335. Plante I, Wallis A, Shao Q, Laird DW (2010). Milk secretion and ejection are impaired in the mammary gland of mice harboring a Cx43 mutant while expression and localization of tight and adherens junction proteins remain unchanged. *Biol Reprod* **82** (5): 837–847.

336. Plum A, Winterhager E, Pesch J, Lautermann J, Hallas G, Rosentreter B, Traub O, Herberhold C, Willecke K (2001). Connexin31-deficiency in mice causes transient placental dysmorphogenesis but does not impair hearing and skin differentiation. *Dev Biol* **231** (2): 334–347.

337. Poelmann RE (1975). An ultrastructural study of implanting mouse blastocysts: Coated vesicles and epithelium formation. *J Anat* **119** (Pt 3): 421–434.

338. Power MA, Tam PPL (1993). Onset of gastrulation, morphogenesis and somitogenesis in mouse embryos displaying compensatory growth. *Anat Embryol* **187** (5): 493–504.

339. Prater MR, Johnson VJ, Germolec DR, Luster MI, Holladay SD (2006). Maternal treatment with a high dose of CpG ODN during gestation alters fetal craniofacial and distal limb development in C57BL/6 mice. *Vaccine* **24** (3): 263–271.

340. Puk O, Moller G, Geerlof A, Krowiorz K, Ahmad N, Wagner S, Adamski J, de Angelis MH, Graw J (2011). The pathologic effect of a novel neomorphic *Fgf9^{Y162C}* allele is restricted to decreased vision and retarded lens growth. *PLoS One* **6** (8): e23678.

341. Qiu Y, Pereira FA, DeMayo FJ, Lydon JP, Tsai SY, Tsai M-J (1997). Null mutation of mCOUP-TFI results in defects in morphogenesis of the glossopharyngeal ganglion, axonal projection, and arborization. *Genes Dev* **11** (15): 1925–1937.

342. Qu J, Bishop JM (2012). Nucleostemin maintains self-renewal of embryonic stem cells and promotes reprogramming of somatic cells to pluripotency. *J Cell Biol* **197** (6): 731–745.

343. Quaggin SE, Schwartz L, Cui S, Igarashi P, Deimling J, Post M, Rossant J (1999). The basic-helix-loop-helix protein Pod1 is critically important for kidney and lung organogenesis. *Development* **126** (24): 5771–5783.

344. Quinn JC, West JD, Kaufman MH (1997). Genetic background effects on dental and other craniofacial abnormalities in homozygous small eye (*Pax6^{Sey}/Pax6^{Sey}*) mice. *Anat Embryol* **196** (4): 311–321.

345. Rantakari P, Nikkila J, Jokela H, Ola R, Pylkas K, Lagerbohm H, Sainio K, Poutanen M, Winqvist R (2010). Inactivation of *Palb2* gene leads to mesoderm differentiation defect and early embryonic lethality in mice. *Hum Mol Genet* **19** (15): 3021–3029.

346. Rantanen M, Palmen T, Patari A, Ahola H, Lehtonen S, Astrom E, Floss T et al. (2002). Nephrin TRAP mice lack slit diaphragms and show fibrotic glomeruli and cystic tubular lesions. *J Am Soc Nephrol* **13** (6): 1586–1594.

347. Rasco JF, Hood RD (1994). Differential effect of restraint procedure on incidence of restraint-stress-induced rib fusion in CD-1 mice. *Toxicol Lett* **71** (2): 177–182.

348. Rasco JF, Hood RD (1995). Maternal restraint stress-enhanced teratogenicity of all-*trans*-retinoic acid in CD-1 mice. *Teratology* **51** (2): 57–62.

349. Reeb-Whitaker CK, Paigen B, Beamer WG, Bronson RT, Churchill GA, Schweitzer IB, Myers DD (2001). The impact of reduced frequency of cage changes on the health of mice housed in ventilated cages. *Lab Anim* **35** (1): 58–73.

350. Reese J et al. (2000). Coordinated regulation of fetal and maternal prostaglandins directs successful birth and postnatal adaptation in the mouse. *Proc Natl Acad Sci USA* **97** (17): 9759–9764.

351. Regan CP, Li W, Boucher DM, Spatz S, Su MS, Kuida K (2002). Erk5 null mice display multiple extraembryonic vascular and embryonic cardiovascular defects. *Proc Natl Acad Sci USA* **99** (14): 9248–9253.

352. Rengasamy P, Padmanabhan RR (2004). Experimental studies on cervical and lumbar ribs in mouse embryos. *Congenit Anom (Kyoto)* **44** (3): 156–171.

353. Rhee JW, Arata A, Selleri L, Jacobs Y, Arata S, Onimaru H, Cleary ML (2004). Pbx3 deficiency results in central hypoventilation. *Am J Pathol* **165** (4): 1343–1350.

354. Rice D, Barone SJ (2000). Critical periods of vulnerability for the developing nervous system: Evidence from humans and animal models. *Environ Health Perspect* **108** (Suppl. 3): 511–533.

355. Robin NH, Abbadi N, McCandless SE, Nadeau JH (1997). *Disorganization* in mice and humans and its relation to sporadic birth defects. *Am J Med Genet* **73** (4): 425–436.

356. Rodier PM (1980). Chronology of neuron development: Animal studies and their clinical implications. *Dev Med Child Neurol* **22** (4): 525–545.

357. Rosen ED, Chan JC, Idusogie E, Clotman F, Vlasuk G, Luther T, Jalbert LR et al. (1997). Mice lacking factor VII develop normally but suffer fatal perinatal bleeding. *Nature* **390** (6657): 290–294.

358. Rossant J, Cross JC (2001). Placental development: Lessons from mouse mutants. *Nat Rev Genet* **2** (7): 538–548.

359. Rossant J, Spence A (1998). Chimeras and mosaics in mouse mutant analysis. *Trends Genet* **14** (9): 358–363.

360. Rousseaux CG, Bolon B (2013). Embryo and fetus. In: *Handbook of Toxicologic Pathology*, 3rd edn., Vol. 3, Haschek WM, Rousseaux CG, Wallig MA (eds.). San Diego, CA: Academic Press (Elsevier), pp. 2695–2759.

361. Rugh R (1990). *The Mouse. Its Reproduction and Development.* Oxford, U.K.: Oxford University Press.

362. Ruland J, Sirard C, Elia A, MacPherson D, Wakeham A, Li L, de la Pompa JL, Cohen SN, Mak TW (2001). p53 Accumulation, defective cell proliferation, and early embryonic lethality in mice lacking *tsg101*. *Proc Natl Acad Sci USA* **98** (4): 1859–1864.

363. Sah VP, Attardi LD, Mulligan GJ, Williams BO, Bronson RT, Jacks T (1995). A subset of *p53*-deficient embryos exhibit exencephaly. *Nat Genet* **10** (2): 175–180.

364. Saito K, Kakizaki T, Hayashi R, Nishimaru H, Furukawa T, Nakazato Y, Takamori S et al. (2010). The physiological roles of vesicular GABA transporter during embryonic development: A study using knockout mice. *Mol Brain* **3**: 40 (13 pages).

365. Sanchez MP, Silos-Santiago I, Frisen J, He B, Lira SA, Barbacid M (1996). Renal agenesis and the absence of enteric neurons in mice lacking GDNF. *Nature* **382** (6586): 70–73.

366. Sasaki K, Yagi H, Bronson R, Tominaga K, Matsunashi T, Deguchi K, Tani Y, Kishimoto T, Komori T (1996). Absence of fetal liver hematopoiesis in mice deficient in transcriptional coactivator core binding factor β. *Proc Natl Acad Sci USA* **93** (22): 12359–12363.

367. Satyanarayana A, Gudmundsson KO, Chen X, Coppola V, Tessarollo L, Keller JR, Hou SX (2010). *RapGEF2* is essential for embryonic hematopoiesis but dispensable for adult hematopoiesis. *Blood* **116** (16): 2921–2931.

368. Savolainen SM, Foley JF, Elmore SA (2009). Histology atlas of the developing mouse heart with emphasis on E11.5 to E18.5. *Toxicol Pathol* **37** (4): 395–414.

369. Schaapveld RQ, Schepens JT, Robinson GW, Attema J, Oerlemans FT, Fransen JA, Streuli M, Wieringa B, Hennighausen L, Hendriks WJ (1997). Impaired mammary gland development and function in mice lacking LAR receptor-like tyrosine phosphatase activity. *Dev Biol* **188** (1): 134–146.

370. Schambra U (2008). *Prenatal Mouse Brain Atlas.* New York: Springer.

371. Schambra UB (2007). Electronic prenatal mouse brain atlas. http://www.epmba.org/ (last accessed December 1, 2014).

372. Scheuner D, Song B, McEwen E, Liu C, Laybutt R, Gillespie P, Saunders T, Bonner-Weir S, Kaufman RJ (2001). Translational control is required for the unfolded protein response and *in vivo* glucose homeostasis. *Mol Cell* **7** (6): 1165–1176.

373. Schoenwolf GC (2008). *Atlas of Descriptive Embryology*, 7th edn. San Francisco, CA: Pearson Benjamin Cummings.

374. Schoor M, Schuster-Gossler K, Roopenian D, Gossler A (1999). Skeletal dysplasias, growth retardation, reduced postnatal survival, and impaired fertility in mice lacking the SNF2/SWI2 family member ETL1. *Mechanisms Dev* **85** (1–2): 73–83.

375. Segi E, Sugimoto Y, Yamasaki A, Aze Y, Oida H, Nishimura T, Murata T et al. (1998). Patent ductus arteriosus and neonatal death in prostaglandin receptor EP4-deficient mice. *Biochem Biophys Res Commun* **246** (1): 7–12.

376. Segre JA, Bauer C, Fuchs E (1999). Klf4 is a transcription factor required for establishing the barrier function of the skin. *Nat Genet* **22** (4): 356–360.

377. Sekine K, Ohuchi H, Fujiwara M, Yamasaki M, Yoshizawa T, Sato T, Yagishita N, Matsui D, Koga Y, Itoh N, Kato S (1999). Fgf10 is essential for limb and lung formation. *Nat Genet* **21** (1): 138–141.

378. Seller MJ, Bnait KS (1995). Effects of tobacco smoke inhalation on the developing mouse embryo and fetus. *Reprod Toxicol* **9** (5): 449–459.

379. Shamblott MJ et al. (2002). Craniofacial abnormalities resulting from targeted disruption of the murine *Sim2 gene. Dev Dyn* **224** (4): 373–380.

380. Shen J, Bronson RT, Chen DF, Xia W, Selkoe DJ, Tonegawa S (1997). Skeletal and CNS defects in *Presenilin*-1-deficient mice. *Cell* **89** (4): 629–639.

381. Shen L, Pichel JG, Mayeli T, Sariola H, Lu B, Westphal H (2002). *Gdnf* haploinsufficiency causes Hirschsprung-like intestinal obstruction and early-onset lethality in mice. *Am J Hum Genet* **70** (2): 435–447.

382. Shepard TH, Lemire RJ (2010). *Catalog of Teratogenic Agents*, 13th edn. Baltimore, MD: Johns Hopkins University Press.

383. Sherman GF, Holmes LB (1999). Cerebrocortical microdysgenesis is enhanced in c57BL/6J mice exposed *in utero* to acetazolamide. *Teratology* **60** (3): 137–142.

384. Shifley ET, Cole SE (2008). Lunatic fringe protein processing by proprotein convertases may contribute to the short protein half-life in the segmentation clock. *Biochim Biophys Acta* **1783** (12): 2384–2390.

385. Shimizu R, Ohneda K, Engel JD, Trainor CD, Yamamoto M (2004). Transgenic rescue of GATA-1-deficient mice with GATA-1 lacking a FOG-1 association site phenocopies patients with X-linked thrombocytopenia. *Blood* **103** (7): 2560–2567.

386. Shimizu Y, Thumkeo D, Keel J, Ishizaki T, Oshima H, Oshima M, Noda Y, Matsumura F, Taketo MM, Narumiya S (2005). ROCK-I regulates closure of the eyelids and ventral body wall by inducing assembly of actomyosin bundles. *J Cell Biol* **168** (6): 941–953.

387. Shiota K, Shionoya Y, Ide M, Uenobe F, Kuwahara C, Fukui Y (1988). Teratogenic interaction of ethanol and hyperthermia in mice. *Proc Soc Exp Biol Med* **187** (2): 142–148.

388. Shirasawa S, Arata A, Onimaru H, Roth KA, Brown GA, Horning S, Arata S, Okumura K, Sasazuki T, Korsmeyer SJ (2000). *Rnx* deficiency results in congenital central hypoventilation. *Nat Genet* **24** (3): 287–290.

389. Shivdasani RA, Orkin SH (1995). Erythropoiesis and globin gene expression in mice lacking the transcription factor NF-E2. *Proc Natl Acad Sci USA* **92** (19): 8690–8694.
390. Shu W, Jiang YQ, Lu MM, Morrisey EE (2002). Wnt7b regulates mesenchymal proliferation and vascular development in the lung. *Development* **129** (20): 4831–4842.
391. Sicinski P, Donaher JL, Parker SB, Li T, Fazeli A, Gardner H, Haslam SZ, Bronson RT, Elledge SJ, Weinberg RA (1995). Cyclin D1 provides a link between development and oncogenesis in the retina and breast. *Cell* **82** (4): 621–630.
392. Singh J, Hood RD (1985). Maternal protein deprivation enhances the teratogenicity of ochratoxin A in mice. *Teratology* **32** (3): 381–388.
393. Singh J, Hood RD (1987). Effects of protein deficiency on the teratogenicity of cytochalasins in mice. *Teratology* **35** (1): 87–93.
394. Snow MH, Tam PPL (1979). Is compensatory growth a complicating factor in mouse teratology? *Nature* **279** (5713): 555–557.
395. Soriano P (1994). Abnormal kidney development and hematological disorders in PDGF β-receptor mutant mice. *Genes Dev* **8** (16): 1888–1896.
396. Steele-Perkins G, Plachez C, Butz KG, Yang G, Bachurski CJ, Kinsman SL, Litwack ED, Richards LJ, Gronostajski RM (2005). The transcription factor gene *Nfib* is essential for both lung maturation and brain development. *Mol Cell Biol* **25** (2): 685–698.
397. Stumpo DJ, Bock CB, Tuttle JS, Blackshear PJ (1995). MARCKS deficiency in mice leads to abnormal brain development and perinatal death. *Proc Natl Acad Sci USA* **92** (4): 944–948.
398. Sulik KK, Bream PRJ (Not given). Embryo images: Normal and abnormal mammalian development. http://www.med.unc.edu/embryo_images/ (last accessed December 1, 2014).
399. Sulik KK, Cook CS, Webster WS (1988). Teratogens and craniofacial malformations: Relationships to cell death. *Development* **103** (Suppl.): 213–231.
400. Sullivan R, Huang GY, Meyer RA, Wessels A, Linask KK, Lo CW (1998). Heart malformations in transgenic mice exhibiting dominant negative inhibition of gap junctional communication in neural crest cells. *Dev Biol* **204** (1): 224–234.
401. Sundberg JP, Ward JM, HogenEsch H, Nikitin AY, Treuting PM, Macauley JB, Schofield PN (2012). Training pathologists in mouse pathology. *Vet Pathol* **49** (2): 393–397.
402. Swiatek PJ, Gridley T (1993). Perinatal lethality and defects in hindbrain development in mice homozygous for a targeted mutation of the zinc finger gene *Krox20*. *Genes Dev* **7** (11): 2071–2084.
403. Szabo KT (1989). Appendix III: Summary data on incidence of spontaneous anomalies. In: *Congenital Malformations in Laboratory and Farm Animals*. San Diego, CA: Academic Press, pp. 294–296.
404. Tachibana K, Hirota S, Iizasa H, Yoshida H, Kawabata K, Kataoka Y, Kitamura Y et al. (1998). The chemokine receptor CXCR4 is essential for vascularization of the gastrointestinal tract. *Nature* **393** (6685): 591–594.
405. Takahashi Y, Kako K, Kashiwabara S, Takehara A, Inada Y, Arai H, Nakada K, Kodama H, Hayashi J, Baba T, Munekata E (2002). Mammalian copper chaperone Cox17p has an essential role in activation of cytochrome *c* oxidase and embryonic development. *Mol Cell Biol* **22** (21): 7614–7621.
406. Takeshima H, Komazaki S, Hirose K, Nishi M, Noda T, Iino M (1998). Embryonic lethality and abnormal cardiac myocytes in mice lacking ryanodine receptor type 2. *EMBO J* **17** (12): 3309–3316.
407. Tam PPL, Williams EA, Chan WY (1993). Gastrulation in the mouse embryo: Ultrastructural and molecular aspects of germ layer morphogenesis. *Microsc Res Tech* **26** (4): 301–328.
408. Tamai Y, Nakajima R, Ishikawa T, Takaku K, Seldin MF, Taketo MM (1999). Colonic hamartoma development by anomalous duplication in *Cdx2* knockout mice. *Cancer Res* **59** (12): 2965–2970.
409. Tamura K, Sudo T, Senftleben U, Dadak AM, Johnson R, Karin M (2000). Requirement for p38α in erythropoietin expression: A role for stress kinases in erythropoiesis. *Cell* **102** (2): 221–231.
410. Tanaka Y, Patestos NP, Maekawa T, Ishii S (1999). B-*myb* is required for inner cell mass formation at an early stage of development. *J Biol Chem* **274** (40): 28067–28070.
411. Taverna D, Disatnik MH, Rayburn H, Bronson RT, Yang J, Rando TA, Hynes RO (1998). Dystrophic muscle in mice chimeric for expression of α5 integrin. *J Cell Biol* **143** (3): 849–859.
412. Teng GQ, Zhao X, Lees-Miller JP, Quinn FR, Li P, Rancourt DE, London B, Cross JC, Duff HJ (2008). Homozygous missense N629D hERG (KCNH2) potassium channel mutation causes developmental defects in the right ventricle and its outflow tract and embryonic lethality. *Circ Res* **103** (12): 1483–1491.

413. Terada R, Warren S, Lu JT, Chien KR, Wessels A, Kasahara H (2011). Ablation of Nkx2–5 at mid-embryonic stage results in premature lethality and cardiac malformation. *Cardiovasc Res* **91** (2): 289–299.

414. Tetzlaff MT, Bai C, Finegold M, Wilson J, Harper JW, Mahon KA, Elledge SJ (2004). Cyclin F disruption compromises placental development and affects normal cell cycle execution. *Mol Cell Biol* **24** (6): 2487–2498.

415. Tevosian SG, Deconinck AE, Tanaka M, Schinke M, Litovsky SH, Izumo S, Fujiwara Y, Orkin SH (2000). *FOG-2*, a cofactor for *GATA* transcription factors, is essential for heart morphogenesis and development of coronary vessels from epicardium. *Cell* **101** (7): 729–739.

416. Thompson NA, Haefliger J-A, Senn A, Tawadros T, Magara F, Ledermann B, Nicod P, Waeber G (2001). Islet-brain1/JNK-interacting protein-1 is required for early embryogenesis in mice. *J Biol Chem* **276** (30): 27745–27748.

417. Thumkeo D, Shimizu Y, Sakamoto S, Yamada S, Narumiya S (2005). ROCK-I and ROCK-II cooperatively regulate closure of eyelid and ventral body wall in mouse embryo. *Genes Cells* **10** (8): 825–834.

418. Toder V, Carp H, Fein A, Torchinsky A (2002). The role of pro- and anti-apoptotic molecular interactions in embryonic maldevelopment. *Am J Reprod Immunol* **48** (4): 235–244.

419. Toft NJ, Arends MJ, Wyllie AH, Clarke AR (1998). No female embryonic lethality in mice nullizygous for *Msh2* and *p53*. *Nat Genet* **18** (1): 17.

420. Torchinsky A, Toder V, Carp H, Orenstein H, Fein A (1997). *In vivo* evidence for the existence of a threshold for hyperglycemia-induced major fetal malformations: Relevance to the etiology of diabetic teratogenesis. *Early Pregnancy* **3** (1): 27–33.

421. ToxNet® (2014). DART®: Developmental and reproductive toxicology database. http://toxnet.nlm.nih.gov/newtoxnet/dart.htm (last accessed December 1, 2014).

422. Tremblay LO, Nagy Kovács E, Daniels E, Wong NK, Sutton-Smith M, Morris HR, Dell A, Marcinkiewicz E, Seidah NG, McKerlie C, Herscovics A (2007). Respiratory distress and neonatal lethality in mice lacking Golgi α1,2-mannosidase IB involved in *N*-glycan maturation. *J Biol Chem* **282** (4): 2558–2566.

423. Treuting PM, Dintzis SM (2012). *Comparative Anatomy and Histology: A Mouse and Human Atlas*. San Diego, CA: Academic Press (Elsevier).

424. Tribioli C, Lufkin T (1999). The murine *Bapx1* homeobox gene plays a critical role in embryonic development of the axial skeleton and spleen. *Development* **126** (24): 5699–5711.

425. Tsai FY, Keller G, Kuo FC, Weiss M, Chen J, Rosenblatt M, Alt FW, Orkin SH (1994). An early haematopoietic defect in mice lacking the transcription factor GATA-2. *Nature* **371** (6494): 221–226.

426. Tsai G, Ralph-Williams RJ, Martina M, Bergeron R, Berger-Sweeney J, Dunham KS, Jiang Z, Caine SB, Coyle JT (2004). Gene knockout of glycine transporter 1: Characterization of the behavioral phenotype. *Proc Natl Acad Sci USA* **101** (22): 8485–8490.

427. Tullio AN, Accili D, Ferrans VJ, Yu Z-X, Takeda K, Grinberg A, Westphal H, Preston YA, Adelstein RS (1997). Nonmuscle myosin II-B is required for normal development of the mouse heart. *Proc Natl Acad Sci USA* **94** (23): 12407–12412.

428. Turgeon B, Meloche S (2009). Interpreting neonatal lethal phenotypes in mouse mutants: Insights into gene function and human diseases. *Physiol Rev* **89** (1): 1–26.

429. Tybulewicz VL, Crawford CE, Jackson PK, Bronson RT, Mulligan RC (1991). Neonatal lethality and lymphopenia in mice with a homozygous disruption of the *c-abl* proto-oncogene. *Cell* **65** (7): 1153–1163.

430. Tyl RW, Chernoff N, Rogers JM (2007). Altered axial skeletal development. *Birth Defects Res B Dev Reprod Toxicol* **80** (6): 451–472.

431. Ueta E, Kodama M, Sumino Y, Kurome M, Ohta K, Katagiri R, Naruse I (2010). Gender-dependent differences in the incidence of ochratoxin A-induced neural tube defects in the *Pdn/Pdn* mouse. *Congenit Anom (Kyoto)* **50** (1): 29–39.

432. Ufer C, Wang CC (2011). The roles of glutathione peroxidases during embryo development. *Front Mol Neurosci* **4**: 12 (14 pages).

433. Umpierre CC, Little SA, Mirkes PE (2001). Co-localization of active caspase-3 and DNA fragmentation (TUNEL) in normal and hyperthermia-induced abnormal mouse development. *Teratology* **63** (3): 134–143.

434. van der Hoeven F, Schimmang T, Volkmann A, Mattei MG, Kyewski B, Ruther U (1994). Programmed cell death is affected in the novel mouse mutant *Fused toes* (*Ft*). *Development* **120** (9): 2601–2607.

435. Van Keuren ML, Layton WM, Iacob RA, Kurnit DM (1991). Situs inversus in the developing mouse: Proteins affected by the *iv* mutation (genocopy) and the teratogen retinoic acid (phenocopy). *Mol Reprod Dev* **29** (2): 136–144.

436. Vassalli A, Matzuk MM, Gardner HA, Lee KF, Jaenisch R (1994). Activin/inhibin βB subunit gene disruption leads to defects in eyelid development and female reproduction. *Genes Dev* **8** (4): 414–427.

437. Verhage M, Maia AS, Plomp JJ, Brussaard AB, Heeroma JH, Vermeer H, Toonen RF et al. (2000). Synaptic assembly of the brain in the absence of neurotransmitter secretion. *Science* **287** (5454): 864–869.

438. Waage-Baudet H, Lauder JM, Dehart DB, Kluckman K, Hiller S, Tint GS, Sulik KK (2003). Abnormal serotonergic development in a mouse model for the Smith-Lemli-Opitz syndrome: Implications for autism. *Int J Dev Neurosci* **21** (8): 451–459.

439. Wakamiya M, Blackburn MR, Jurecic R, McArthur MJ, Geske RS, Cartwright J, Mitani K et al. (1995). Disruption of the adenosine deaminase gene causes hepatocellular impairment and perinatal lethality in mice. *Proc Natl Acad Sci USA* **92** (9): 3673–3677.

440. Wang MW, Heap RB, Taussig MJ (1992). Abnormal maternal behaviour in mice previously immunized against progesterone. *J Endocrinol* **134** (2): 257–267.

441. Wang N-D, Finegold MJ, Bradley A, Ou CN, Abdelsayed SV, Wilde MD, Taylor LR, Wilson DR, Darlington GJ (1995). Impaired energy homeostasis in C/EBPα knockout mice. *Science* **269** (5227): 1108–1112.

442. Wang VE, Schmidt T, Chen J, Sharp PA, Tantin D (2004). Embryonic lethality, decreased erythropoiesis, and defective octamer-dependent promoter activation in Oct-1-deficient mice. *Mol Cell Biol* **24** (3): 1022–1032.

443. Wang Y, Zhang J, Mori S, Nathans J (2006). Axonal growth and guidance defects in *Frizzled3* knock-out mice: A comparison of diffusion tensor magnetic resonance imaging, neurofilament staining, and genetically directed cell labeling. *J Neurosci* **26** (2): 355–364.

444. Ward JM, Elmore SA, Foley JF (2012). Pathology methods for the evaluation of embryonic and perinatal developmental defects and lethality in genetically engineered mice. *Vet Pathol* **49** (1): 71–84.

445. Ware CB, Horowitz MC, Renshaw BR, Hunt JS, Liggitt D, Koblar SA, Gliniak BC et al. (1995). Targeted disruption of the low-affinity leukemia inhibitory factor receptor gene causes placental, skeletal, neural and metabolic defects and results in perinatal death. *Development* **121** (5): 1283–1299.

446. Warren AJ, Colledge WH, Carlton MB, Evans MJ, Smith AJ, Rabbitts TH (1994). The oncogenic cysteine-rich LIM domain protein Rtbn2 is essential for erythroid development. *Cell* **78** (1): 45–57.

447. Watkins-Chow D et al. (2013). Mutation of the diamond-blackfan anemia gene *Rps7* in mouse results in morphological and neuroanatomical phenotypes. *PLoS Genet* **9** (1): e1003094 (17 pages).

448. Watt LJ (1931). Ovulation, oestrus and copulation with consequent dystocia during pregnancy, in the mouse. *Science* **73** (1881): 75–76.

449. White JK, Gerdin AK, Karp NA, Ryder E, Buljan M, Bussell JN, Salisbury J et al. (2013). Genome-wide generation and systematic phenotyping of knockout mice reveals new roles for many genes. *Cell* **154** (2): 452–464.

450. Wilson JG (1973). *Environment and Birth Defects*. New York: Academic Press.

451. Wilson JG, Fraser FC (1977). *Handbook of Teratology*, Vol. 1. *General Principles and Etiology*. New York: Plenum Publishing.

452. Wilson JG, Fraser FC (1977). *Handbook of Teratology*, Vol. 2. *Mechanisms and Pathogenesis*. New York: Plenum Publishing.

453. Wolpowitz D, Mason TB, Dietrich P, Mendelsohn M, Talmage DA, Role LW (2000). Cysteine-rich domain isoforms of the *neuregulin-1* gene are required for maintenance of peripheral synapses. *Neuron* **25** (1): 79–91.

454. Xiang M, Gan L, Zhou L, Klein WH, Nathans J (1996). Targeted deletion of the mouse POU domain gene *Brn-3a* causes selective loss of neurons in the brainstem and trigeminal ganglion, uncoordinated limb movement, and impaired suckling. *Proc Natl Acad Sci USA* **93** (21): 11950–11955.

455. Xiao R, Yu HL, Zhao HF, Liang J, Feng JF, Wang W (2007). Developmental neurotoxicity role of cyclophosphamide on post-neural tube closure of rodents *in vitro* and *in vivo*. *Int J Dev Neurosci* **25** (8): 531–537.

456. Xu W, Baribault H, Adamson E (1998). Vinculin knockout results in heart and brain defects during embryonic development. *Development* **125** (2): 327–337.

457. Xu W, Liu C, Kaartinen V, Chen H, Lu CH, Zhang W, Luo Y, Shi W (2013). TACE in perinatal mouse lung epithelial cells promotes lung saccular formation. *Am J Physiol Lung Cell Mol Physiol* **305** (12): L953–L963.

458. Yang Q, Kurotani R, Yamada A, Kimura S, Gonzalez FJ (2006). Peroxisome proliferator-activated receptor α activation during pregnancy severely impairs mammary lobuloalveolar development in mice. *Endocrinology* **147** (10): 4772–4780.

459. Yokoyama T, Copeland NG, Jenkins NA, Montgomery CA, Elder FFB, Overbeek PA (1993). Reversal of left–right asymmetry: A situs inversus mutation. *Science* **260**: 679–682.

460. Yoshida H, Kong YY, Yoshida R, Elia AJ, Hakem A, Hakem R, Penninger JM, Mak TW (1998). Apaf1 is required for mitochondrial pathways of apoptosis and brain development. *Cell* **94** (6): 739–750.

461. Yu K, Xu J, Liu Z, Sosic D, Shao J, Olson EN, Towler DA, Ornitz DM (2003). Conditional inactivation of FGF receptor 2 reveals an essential role for FGF signaling in the regulation of osteoblast function and bone growth. *Development* **130** (13): 3063–3074.

462. Yuan T, Wang Y, Pao L, Anderson SM, Gu H (2011). Lactation defect in a widely used MMTV-Cre transgenic line of mice. *PLoS One* **6** (4): e19233 (9 pages).

463. Zenclussen AC, Olivieri DN, Dustin ML, Tadokoro CE (2013). *In vivo* multiphoton microscopy technique to reveal the physiology of the mouse uterus. *Am J Reprod Immunol* **69** (3): 281–289.

464. Zhang H, Bradley A (1996). Mice deficient for BMP2 are nonviable and have defects in amnion/chorion and cardiac development. *Development* **122** (10): 2977–2986.

465. Zhang J, Lin Y, Zhang Y, Lan Y, Lin C, Moon AM, Schwartz RJ, Martin JF, Wang F (2008). *Frs2α*-deficiency in cardiac progenitors disrupts a subset of FGF signals required for outflow tract morphogenesis. *Development* **135** (21): 3611–3622.

466. Zhang J, Tomasini AJ, Mayer AN (2008). RBM19 is essential for preimplantation development in the mouse. *BMC Dev Biol* **8**: 115 (14 pp.).

467. Zhao C, Takita J, Tanaka Y, Setou M, Nakagawa T, Takeda S, Yang HW et al. (2001). Charcot-Marie-Tooth disease type 2A caused by mutation in a microtubule motor KIF1Bβ. *Cell* **105** (5): 587–597.

468. Zhao Q, Beck A, Vitale JM, Schneider JS, Terzic A, Fraidenraich D (2010). Rescue of developmental defects by blastocyst stem cell injection: Towards elucidation of neomorphic corrective pathways. *J Cardiovasc Transl Res* **3** (1): 66–72.

469. Zhao Q, Behringer RR, de Crombrugghe B (1996). Prenatal folic acid treatment suppresses acrania and meroanencephaly in mice mutant for the *Cart1* homeobox gene. *Nat Genet* **13** (3): 275–283.

470. Zhao R, Russell RG, Wang Y, Liu L, Gao F, Kneitz B, Edelmann W, Goldman ID (2001). Rescue of embryonic lethality in reduced folate carrier-deficient mice by maternal folic acid supplementation reveals early neonatal failure of hematopoietic organs. *J Biol Chem* **276** (13): 10224–10228.

471. Zhao Y, Guo Y-J, Tomac AC, Taylor NR, Grinberg A, Lee EJ, Huang S, Westphal H (1999). Isolated cleft palate in mice with a targeted mutation of the LIM homeobox gene *Lhx8*. *Proc Natl Acad Sci USA* **96** (26): 15002–15006.

472. Zhuang Y, Soriano P, Weintraub H (1994). The helix-loop-helix gene E2A is required for B cell formation. *Cell* **79** (5): 875–884.

473. Zou X, Bolon B, Pretorius JK, Kurahara C, McCabe J, Christiansen KA, Sun N et al. (2009). Neonatal death in mice lacking cardiotrophin-like cytokine (CLC) is associated with multifocal neuronal hypoplasia. *Vet Pathol* **46** (3): 514–519.

16

Pathology of the Placenta

Brad Bolon and Jerrold M. Ward

CONTENTS

Our growing understanding of gestational biology indicates that many major clinical complications of pregnancy, including implantation failure, embryonic lethality (spontaneous abortion), intrauterine restriction of fetal growth, and preeclampsia (placental ischemia), are consequences of primary placental lesions in both animals and humans. In particular, embryo lethality in mice often results from placental dysfunction rather than abnormalities in the embryo. However, placental damage is often missed by investigators. This flaw in many mouse developmental pathology studies results from three major errors. The first is a narrow analytical focus that is limited to identifying and characterizing the changes that occur in the embryo or fetus. In such cases, the extraembryonic elements are entirely ignored. The second mistake is to evaluate the placenta in a cursory fashion, noting the presence of a placental abnormality in general terms (e.g., describing the specimen as *hemorrhagic* or *pale* or *small*) without seeking to rigorously define the lesions responsible for the change at the tissue and cellular levels. The final error is to delegate the task of placental analysis to an individual with little theoretical knowledge or practical experience in mouse developmental pathology. In our research careers, many studies incorporate two or even all three of these design flaws.

Perfunctory or inexpert assessment of the placenta will preclude full characterization of embryolethal or perinatal lethal phenotypes. The extensive interdependence between an embryo/fetus and its placental support system means that failure of either element usually will end by destroying both. Primary placental abnormalities can severely disrupt the formation of many embryonic structures and delay or prevent normal physiological processes, eventually causing embryonic death.[8,118] Therefore, developmental pathology investigations that purposely ignore the placental half of the conceptus often make it impossible to determine the cause of gestational mortality. These findings are especially important for genetically engineered mouse studies, where an inactivated gene has been engineered to be expressed primarily in placental cells.

This chapter is designed as a primer for the pathologic analysis of the mouse placenta. The first portion of the chapter describes a useful strategic approach for the analysis of the mouse placenta. The remainder of the chapter considers common classes of placental lesions that have been reported in developing mice and the mechanisms thought to be responsible for their development. Additional specialized procedures (e.g., gene expression patterns [Chapter 10], protein localization [Chapter 11], quantitative morphological measurements [Chapter 12], and/or ultrastructural examination [Chapter 13]) may be needed to completely discern the causes and mechanisms that induce placental damage. However, the subjects covered in this chapter should be the foundation for all efforts to understand the origin and impact of placental diseases. More comprehensive coverage of normal placental structure and function may be found in Chapter 4 as well as in many articles,[2,20,21,23,70,72,79,84,88,91,126,128] reference books,[46,47,78,125] and websites.[11,12]

A Practical Approach to the Routine Structural Screening of Mouse Placenta

From the pathologist's perspective, anatomic evidence of placental abnormalities may be observed at the gross (macroscopic) and/or the microscopic levels. However, the sometimes divergent processing needs encountered when preparing specimens for the macroscopic vs. microscopic examination require advanced planning to prioritize tissue sampling during developmental pathology experiments. A trade-off may be necessary because a thorough gross examination (which may include weighing and/or pho-tography[5]) often will involve some delay between removal of the placenta and its fixation, while an optimal histopathological evaluation is facilitated by minimizing the time between sample collection and fixation. Two options may be used to balance these divergent requirements. The first is to provide enough litters of mice so that some are dedicated to gross evaluation, weighing, and photographic documentation while others are designated for microscopic analysis. Samples subjected to gross viewing may still be processed for microscopic evaluation, but they may be unsuitable for some purposes (e.g., morphometry, photomicrography, ultrastructural analysis) because they have a larger complement of artifacts due to delayed fixation and/or handling. This multicohort study design yields the optimal pla-cental preservation for both macroscopic and microscopic assessments. Unfortunately, in many cases multiple litters cannot be supplied due to breeding difficulties or cost. Accordingly, a second option is to perform the macroscopic and microscopic assessments on the same specimens, but to limit the length of the gross examination (typically by foregoing weighing and/or photography). Quantitative confirmation that a size difference exists then is sought by morphometric measurement of placental dimensions on relatively homologous tissue sections.

Some investigators ignore the macroscopic examination based on the erroneous assumption that overtly detectable changes in color, shape, and size will be defined and characterized even more effect-ively through histopathological evaluation. However, gross placental lesions often provide essential evi-dence that suggests possible mechanisms for embryolethal or perinatal lethal phenotypes that result from placental dysfunction. Therefore, a complete pathology analysis of placenta routinely should include both macroscopic and microscopic examinations.[72,125,126]

Macroscopic Assessment of the Mouse Placenta

At necropsy, placenta is the first portion of the conceptus that will be harvested. The explanations for this sampling order are obvious. First, the placenta must be moved in order to expose and examine the embryo or fetus. The most straightforward means of guaranteeing that the placenta does not interfere with analysis of embryonic tissues is to displace the extraembryonic derivatives completely. Second, the extensive vascularization of the placental labyrinth predisposes this site to rapid autolysis. *Postmortem* degradation of critical placental structures is reduced substantially by immediate fixation. In our experi-ence, therefore, mouse developmental pathology studies should be designed with the intent to ensure rapid removal and fixation of the placenta as the initial step in evaluating each implantation site. Options for the effective and efficient harvesting of placenta may be found in Chapter 7.

Prior to fixation, however, each placenta should be given a brief but through gross assessment. This examination may be undertaken in one of two ways. First, the placenta may be suspended in mid-air using forceps or laid on a flat surface and then evaluated quickly (usually for 10–15 s) as soon as the embryo or fetus has been extracted. With experience, such rapid analyses may afford consid-erable insight with respect to the possible timing of an embryolethal event as well as provide clues regarding potential structural changes responsible for its occurrence (see later text for basic approaches to interpreting placental lesions). Alternatively, the isolated placenta may be submerged in a suitable buffer (e.g., phosphate-buffered saline, Tyrode's buffer) and assessed using a magnifying glass or

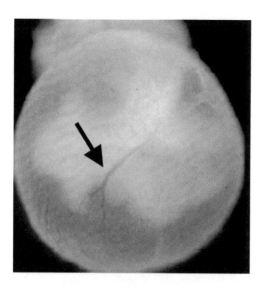

FIGURE 16.1 The yolk sac (shown here at E9.5) forms a thick, translucent membrane surrounding the embryo. The most prominent structures identifiable in the yolk sac are the thin, tortuous, branching vitelline blood vessels (arrow) that course along its surface. (Reproduced from Regan, C.P., Li, W., Boucher, D.M., Spatz, S., Su, M.S., Kuida, K., Erk5 null mice display multiple extraembryonic vascular and embryonic cardiovascular defects, *Proc. Natl. Acad. Sci. USA*, 99(14), 9248–9253, Copyright 2002 National Academy of Sciences, U.S.A.)

stereomicroscope to evaluate fine surface details (Figure 16.1). Possible benefits of this second approach are that a more systematic review may be completed over a longer period of time (1–2 min), and gross changes may be recorded photographically with less apprehension about causing handling-induced structural artifacts in delicate placental tissues. A possible drawback to this latter approach is that blood from the placental labyrinth over time will seep into the buffer, which may lead to localized deformations or collapse of the organ. Regardless of the chosen method, a macroscopic analysis of unfixed placenta must be performed carefully to minimize the development of artifactual changes that might confound the microscopic examination. Apparent qualitative differences in placental size may need to be corroborated quantitatively by acquiring organ weights.

Several options may be used in harvesting the placenta to minimize trauma. The first is to clasp the root of the umbilical cord (i.e., the part nearest the embryo or fetus) and then use the cord as a rope to hoist the whole organ. The second approach is to use a small pair of forceps to grasp one border of the placenta itself. In both of these cases, the tissue gripped by the instrument will be damaged severely, and therefore will be unsuitable for subsequent histopathological examination. A third way is to position the tips of a small pair of angled forceps under the placenta to brace it as it is lifted. The first and last methods do not harm the placenta, and thus are the recommended means for manipulating specimens.

The macroscopic appearance of the placenta attests to its highly vascular nature. Its grossly visible features fluctuate throughout development. During early postimplantation time points, the yolk sac is the primary placental element, appearing as a thick, translucent membrane that encompasses the embryo. The vitelline vessels on the yolk sac surface normally will hold visible blood (Figure 16.1); this phenomenon is appreciated more easily during later stages of development. In later gestation, the expansion of the definitive placenta results in the genesis of a dark red, discoid mass that caps one pole of the yolk sac (Figure 16.2). At necropsy, this red mass will discharge small amounts of dark red fluid (representing intermingled maternal and embryonic blood cells with tiny quantities of interstitial fluid). The umbilical cord may be tied off before placental extraction to reduce this loss.

Macroscopic placental lesions visible at necropsy commonly are related to alterations in organ color or size. A typical abnormality is pallor, characterized by a shift from the usual diffuse, dark red shade to a pale red or pink tint (Figure 16.3). This finding is indicative of less blood within the vessels.

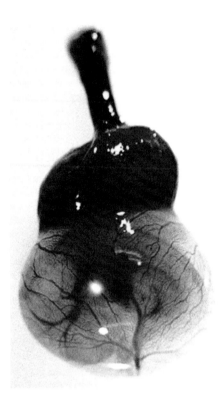

FIGURE 16.2 The definitive placenta develops as a hemisphere capping the mesometrial pole of the translucent yolk sac later during gestation (shown here at gestational day [GD] 13.5). The highly vascular nature of these structures is indicated by their deep red color (definitive placenta) and conspicuous, extensively branched vitelline vessels (yolk sac). (Reproduced from Ward, J.M. and Devor-Henneman, D.E., Gestational mortality in genetically engineered mice: Evaluating the extra-embryonal embryonic placenta and membranes, in: *Pathology of Genetically Engineered Mice*, Ward, J.M., Mahler, J.F., Maronpot, R.R., Sundberg, J.P., and Frederickson, R.M., eds., Iowa State University Press, Ames, IA, 2000, pp. 103–122, by courtesy of the Iowa State University Press.)

$p38\,\alpha^{+/+}$ $p38\,\alpha^{-/-}$

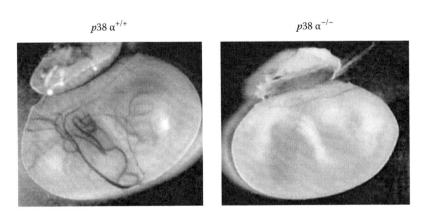

FIGURE 16.3 Placental pallor is a characteristic gross lesion associated with defective production of embryonic blood cells and/or blood vessels. Common alterations are reduced prominence or absence of the vitelline vessels of the yolk sac (as seen in the E16.5 null mutant conceptus on the right) and/or paleness of the labyrinthine portion of the definitive placenta (i.e., that part that caps the yolk sac). (Reproduced from *Cell*, 102(2), Tamura, K., Sudo, T., Senftleben, U., Dadak, A.M., Johnson, R., and Karin, M., Requirement for p38alpha in erythropoietin expression: A role for stress kinases in erythropoiesis, 221–231, Copyright 2000, with permission from Elsevier.)

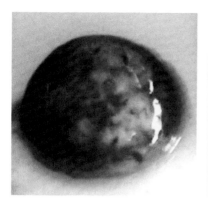
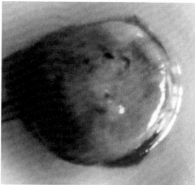

FIGURE 16.4 Discoloration of the definitive placenta is a common trait in unhealthy placentas. Relative to the normal red hue observed in control organs (left), the diffuse tan plaque covering the outer surface of the placenta (right) is consistent with marked labyrinthine inflammation and/or necrosis. This lesion resulted from infection of GD 10 mouse conceptuses with *Plasmodium berghei* (the intraerythrocytic parasite that causes malaria) 6 days previously. (Reproduced from Sharma, L. et al., *PLoS ONE*, 7(3), e32694, 2012.)

Other aberrant placental hues include tan/white/gray (Figure 16.4) or dark purple/black. Any of these non-red colors may signal that the tissue is necrotic and/or inflamed; the darker tints suggest that the degenerating tissue also contains an excessive amount of blood (often as foci of hemorrhage rather than mere congestion of some or all vessels). Another reason for placental pallor is maternal or embryonic anemia. Common size variations visible in placenta include enlargement (placentomegaly) and shrinkage (hypoplasia; Figure 16.5).[5,38,44,58,59,108,116] In our experience, these findings almost always reflect defects originating in the conceptus rather than abnormalities arising in the dam.

In most cases, a change in placental appearance is accompanied by a concomitant change in the appearance of the embryo or fetus. This coordination results from the close interdependence of the entire conceptus and the requirement for an intact placental conduit for gas and nutrient exchange to maintain viability of the embryo. When examined at necropsy, the embryo or fetus or neonate associated with an abnormal placenta may remain alive, suggesting that the placental dysfunction occurred relatively late during gestation. However, such conceptuses typically are reduced in size and may be grossly abnormal. If placental function fails at any time substantially in advance of birth, the embryo or fetus generally will die within a day or two. If the embryo or fetus is lost, the placenta will undergo necrosis almost immediately, but several days will be required for the placenta to retract (i.e., atrophy), and possibly be resorbed.

Microscopic Assessment of the Mouse Placenta

Examination of the placental tissues and cells by microscopy is similar in principle to the analysis of any other organ. For routine studies, these organs typically are fixed in an aldehyde solution and processed into paraffin. Subsequently, one or several sections are acquired and stained with hematoxylin and eosin (H&E) to highlight various tissue- and cell type–specific features. Serial sections may be cut and processed to assess the expression of various genes (Chapter 10) and proteins (Chapter 11) *in situ*, although care must be taken in designing the experiment to ensure that the tissue fixation and processing conditions will support such specialized techniques. Details regarding suitable conditions for specimen preparation are given in Chapter 9.

Assuming adequate preparation of the sample, the foremost problem in placental histopathology confronted by most investigators is unfamiliarity with the anatomic changes—many of which are sequential and transient—that transpire during development. For example, the early placenta evolves over time from largely a mass of maternal decidua encompassing a thin layer of trophectoderm (from approximately gestational day [GD] 4.5 to GD 7.0) to become a fluid-filled yolk sac and adjacent maternal-derived decidua,

Side view   Basal view

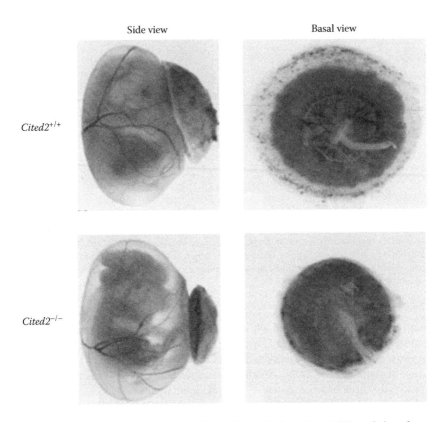

$Cited2^{+/+}$

$Cited2^{-/-}$

FIGURE 16.5 Hypoplasia of the definitive placenta frequently results from altered differentiation of one or more trophoblast lineages. Relative to wild-type ($Cited^{+/+}$) controls (top row), null mutant ($Cited^{-/-}$) conceptuses at GD 14.5 have smaller definitive placentas (bottom row) due to reduced populations of spongiotrophoblasts and trophoblast giant cells. (Reproduced from *Dev. Biol.*, 294(1), Withington, S.L., Scott, A.N., Saunders, D.N., Lopes Floro, K., Preis, J.I., Michalicek, J., Maclean, K., Sparrow, D.B., Barbera, J.P., and Dunwoodie, S.L., Loss of *Cited2* affects trophoblast formation and vascularization of the mouse placenta, 67–82, Copyright 2006, with permission from Elsevier.)

segregated from each other by a thin layer of trophoblast giant cells (from GD 7.5 to GD 10.5). Later in gestation (at about GD 12.5), the mature definitive placenta consists of an inner *labyrinth* (comprised of labyrinthine trophoblast intermingled with embryonic endothelium), a middle *junctional zone* (with spongiotrophoblast nearest the labyrinth and trophoblast giant cells peripherally), and the outer *decidua*. An educated awareness of fundamental principles concerning the function and structure of the placenta at various developmental stages will allow a pathologist to more confidently, easily, and rapidly evaluate the morphology of placental tissues and cells. In turn, such assuredness will permit the microscopist to create better explanations regarding the nature and pathogenesis of placental defects and diseases. In our experience, an additional aid for mouse placental pathologists is to cultivate a strong understanding of species differences in placental structure.[11,25,33,64,67,73,86]

Patterns of placental lesions associated with gestational or perinatal mortality have been linked to defects in many placental elements (Table 16.1; reviewed in Refs. [91,125,126,128]). The most common mechanisms by which placental development are inhibited early in gestation are circulatory abnormalities in the yolk sac or failure of chorioallantoic fusion. Later in gestation, most abnormalities are a consequence of disrupted formation of the placental labyrinth. Defects in other components of the placenta also may be responsible for the induction of lethal functional or structural anomalies in this organ (e.g., peri-implantational death is associated with defects in decidualization or decidua/trophoblast interactions[18]), but they have been documented with much less frequency. In many cases, the exact cause and pathogenesis are unknown, and the recognized diagnosis is the existence of a *small labyrinth* as

TABLE 16.1

Primary Placental Defects Associated with Embryonic Lethal Phenotypes

Affected Structure	Potential Lesions and Timing	Proposed Defect	References
Decidua	G: reduced volume H: fewer cells and vessels TD: variable (usually before GD 10.5)	Formation	[71]
Ectoplacental cone	G + H: decreased cone size TD: GD 6.5–GD 8.0	Formation	[99,137]
Yolk sac	G + H: blood vessels and blood islands fewer and smaller TD: GD 8.5–GD 10.0	Hematopoiesis Vascularization	[9,26,63,75,85,89,94,100,103,111] [9,24,26–28,30,31,43,51,52,57,63, 75,85,94,95,97,98,100,103,109, 111,124,129,138]
Chorioallantois	G + H: separation of the allantois and chorion TD: GD 9.5–GD 11.0	Chorioallantoic fusion	[27,30,37,39–41,53–55,60,62, 66,76,81,95,96,107,135,139]
Labyrinth			

Trophoblast lineages	G: altered placental size (increased or decreased)
	H: hyperplasia and/or hypoplasia of one or multiple lineages, usually leading to altered labyrinthine vascularity
	TD:
		Differentiation
		  Giant cell
		  Labyrinthine
			
			[19,50,61,87,110]
			[3,49,56,93,108,119,121]

	GD 9–GD 11 (altered expansion of trophoblast) GD 13–GD 16 (abnormal genesis of terminal labyrinth)	Spongiotrophoblasts Stem cells	[6,36,61,90,104,108,115,119,134] [3,8,60,92,135]
Developmental processes	Similar to those for trophoblast lineages	Endothelial/trophoblast interactions	[58,65]
	Similar to those for trophoblast lineages	Vascularization (vasculogenesis, angiogenesis)	[1,4,16,19,34,35,43,44,51,52,56,6 6,68,69,74,77,83,97,98,101,106,1 12,113,123,130]
	G + H: similar to trophoblast lineages, but may be none TD: any stage through birth (GD 19)	Nutrient transport	[32,133]

Notes: Structures are listed in the order in which they first gain a prominent role in placental structure and function.
G, gross (macroscopic) findings; H, histopathological (microscopic) findings; TD, time of death (approximate).

a nonspecific end-stage finding.[128] Multiple developmental placental defects also may occur, depending on the function and expression of the affected gene. Determination of the exact etiology and mechanism(s) involved in abnormal placental development in mice, especially genetically engineered mice, depend on the stages of gestation studied, gene/protein expression patterns, and the histogenesis of the elements that undergo abnormal development.

> *Circulatory abnormalities in the yolk sac* may be observed beginning at approximately GD 7.5 and become progressively more severe as gestation progresses. Such lesions are thought to reflect dysgenesis of hemangioblasts, which are the putative embryonic precursors for both endothelial cells and hematopoietic progenitors.[100] Macroscopic anomalies in the yolk sac vasculature usually are accompanied by histopathological evidence of endothelial and/or hematopoietic

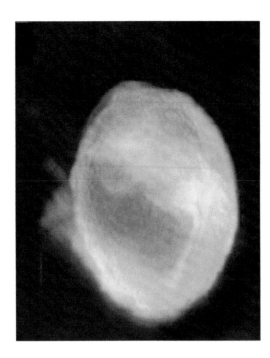

FIGURE 16.6   Defective yolk sac vasculogenesis and/or hematopoiesis is a common mechanism whereby placental dysfunction leads to early embryolethality. Relative to wild-type controls (*Erk5$^{+/+}$*) at the same developmental age (see Figure 16.1), this E9.5 conceptus engineered to lack the Erk5 mitogen-activated protein (MAP) kinase (*Erk5$^{-/-}$*) gene has a small, avascular yolk sac. (Reproduced from Regan, C.P. et al., Erk5 null mice display multiple extraembryonic vascular and embryonic cardiovascular defects, *Proc. Natl. Acad. Sci. USA*, 99(14), 9248–9253, Copyright 2002 National Academy of Sciences, U.S.A.)

disruption.[9,26,100,103,111] Mesodermal defects leading to endothelial dysfunction may reduce or prevent the formation of the vitelline vessels and their branches (Figure 16.6). Flawed hematopoiesis in the blood islands and/or a decrease in blood island numbers may culminate in production of fewer primitive erythrocytes, which will be visible in histological sections as empty or underfilled embryonic vessels and/or blood islands of decreased size (Figure 16.7). Additional changes may include dilated blood vessels[103,109] as well as increased endothelial cell numbers and proliferation.[109] Disrupted circulation in the yolk sac usually will lead to the death of the embryo during midgestation (generally between GD 8.5 and GD 10.0).

Failed chorioallantoic fusion (or *attachment*) typically arises from defects in formation of the allantois, or less frequently the chorion, which prevent the normal linking of these structures at GD 8.5. This phenomenon has been attributed to altered specification of trophoblast progenitors before[13] or shortly after implantation as well as to deficient intracellular adhesion,[62,140] with the outcome that the allantois does not approach the chorion (Figure 16.8). The consequences of failed fusion include the absence of umbilical blood vessels and reduced labyrinth vasculogenesis secondary to shortages of growth factors that usually are supplied by the fusing tissues. Defects in chorioallantoic fusion are among the most common causes responsible for embryonic lethal phenotypes that manifest in midgestation (between GD 9.5 and GD 11.0).

Several structural findings are considered to be microscopic hallmarks of altered chorioallantoic fusion. The first group of changes is indicative of primary allantoic failure. Primary allantoic defects include dysplasia[13,62,95]—characterized by compaction,[76] delayed extension,[140] dilation,[95] or thickening[62]—and an absence of blood vessels.[62,95] Related findings have been documented in other extraembryonic derivatives: hypoplasia or thickening of the amnion[62]; persistence of the proamniotic canal[140]; disordered[118] or abridged blood vessel branching in the

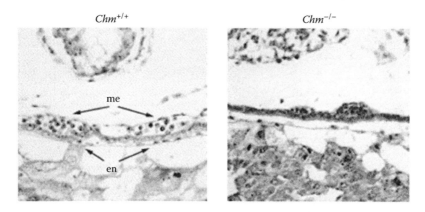

FIGURE 16.7 Circulatory abnormalities of the yolk sac often combine defects in vasculogenesis (especially reductions in vessel caliber and branching) with hematopoietic deficits (particularly smaller blood islands due to fewer erythropoietic precursors. This GD 8.5 conceptus engineered to lack the gene for the choroideremia (*Chm*[−/−]) escort protein shows both changes, while developmental age-matched wild-type controls (*Chm*[+/+]) do not. *Abbreviations:* en, endoderm; me, mesoderm. (Reproduced from *Dev. Biol.* 272(1), Shi, W., van den Hurk, J.A., Alamo-Bethencourt, V., Mayer, W., Winkens, H.J., Ropers, H.H., Cremers, F.P., and Fundele, R., Choroideremia gene product affects trophoblast development and vascularization in mouse extra-embryonic tissues, 53–65, Copyright 2004, with permission from Elsevier.)

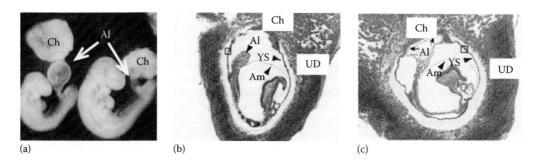

FIGURE 16.8 Failed chorioallantoic fusion is a frequent cause of placental-associated embryonic loss during early gestation. Macroscopic evidence of unsuccessful fusion (a) includes separation of the ballooned allantois (Al) from the chorion (Ch), as is evident here at GD 10.5 in a conceptus (left embryo) engineered to lack vascular cell adhesion molecule-1 (*Vcam1*[−/−]) relative to a wild-type (*Vcam1*[+/+]) littermate (right embryo). The related microscopic lesion is nonattachment of the allantois and chorion, seen here in an GD 8.5 conceptus lacking the transcription factor forkhead box O1 (*FoxO1*[−/−]; b) relative to an age-matched wild-type (*FoxO1*[+/+]) control (c). The red box denotes a blood island in the yolk sac (YS). *Abbreviations:* Am, amnion; UD, uterine decidua. (a: Adapted from Gurtner, G.C. et al., *Genes Dev.*, 9(1), 1, 1995, by permission of Cold Spring Harbor Laboratory Press; b and c: Reproduced from Ferdous, A. et al., Forkhead factor FoxO1 is essential for placental morphogenesis in the developing embryo, *Proc. Natl. Acad. Sci. USA*, 108(39), 16307–16312, Copyright 2011 National Academy of Sciences, U.S.A.)

yolk sac[39,51,62,80,95]; and abnormal thinning[118] or laminar separation[62,95] in the yolk sac. A second class of histopathological lesions represents direct indicators of abnormal chorioallantoic fusion. These changes include small fusion sites in younger embryos[7,95]; fusion sites of diminished size (suggestive of fragile, transient attachment) in older (GD 10.5) embryos[7]; and the complete absence of attachment between the allantois and chorion.[13,54,62,76,95,118,131] Interestingly, failed chorioallantoic fusion is a feature of many lethal placental phenotypes even though both the allantois and chorion seem to have evolved appropriately.[39,55,66,76,107,118]

Disrupted formation of the placental labyrinth is by far the most frequent mechanism responsible for placental failure in mice (Figure 16.9).[128] Dense trophoblast columns start assembling within the nascent labyrinth soon after chorioallantoic fusion is completed.[22]

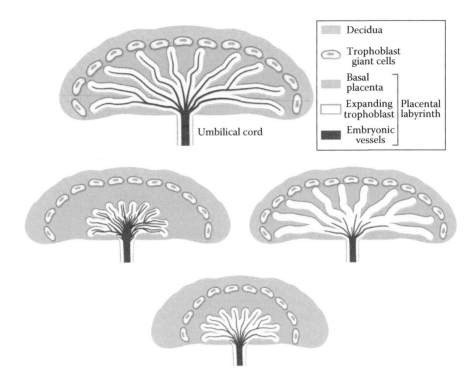

FIGURE 16.9 Schematic diagram illustrating how defective trophoblast differentiation and/or vasculogenesis (*de novo* blood vessel formation) may lead to significant lesions in the definitive placenta. In the normal labyrinth (blue region, upper row), expanding trophoblast (pale yellow) proliferates to provide the corridors into which embryonic capillaries (red) can extend. If trophoblast differentiation is abnormal (middle row, left), branches are shortened, and ingrowing embryonic blood vessels form a short, dense, highly branched network as a compensatory response. If endothelial function is aberrant (middle row, right), the trophoblast regions are of normal length, but capillary ingrowth is diminished. Concurrent defects in both trophoblast and endothelium (lower row) result in short, poorly vascularized branches, and commonly the formation of a grossly *small labyrinth*. (Adapted from Watson, E.D. and Cross, J.C., *Physiology*, 20, 180, 2005, by permission of the American Physiological Society.)

Capillaries extending from the umbilical vessels subsequently enter the cores of the expanding trophoblast,[91] after which the trophoblast and vessels begin to extend themselves in coordinated fashion; this process of branching morphogenesis persists until the end of gestation. Early generations of branches are dedicated to increasing the extent of the capillary beds within the placenta, while later generations create termini adapted to the efficient exchange of gas and nutrients.[48] Abnormalities in the early stages of labyrinth formation typically induce embryonic death near the middle of gestation (GD 9–GD 11), while disrupted production of terminal branches generally incites lethality later in gestation (by GD 13–GD 16). The lesions leading to this late failure arise after the basic labyrinthine pattern has been set, and therefore suggest that affected placentas have insufficient adaptability to near-term physiological challenges.[120]

Microscopic features that are characteristic of disrupted labyrinth formation result from defective vascular and/or villar development. Such defects usually are indicative of faulty differentiation in one or both lineages of labyrinthine trophoblasts: the multinucleated syncytiotrophoblast cells, which surround the endothelium of the embryonic capillaries, or the mononuclear trophoblast cells, which line maternal vascular sinuses. Examples include hypoplasia of endothelial cells[34] or spongiotrophoblast,[49,61,82,107] an incapacity to generate syncytiotrophoblast,[4] or overproduction of trophoblast giant cells[59,61] or the labyrinthine and spongiotrophoblast lineages.[45] Lack of normal vasculogenesis from the chorionic plate mesenchyme may prevent the development of embryonic blood vessels in the labyrinth.[35,36] Inadequate fabrication of vessels

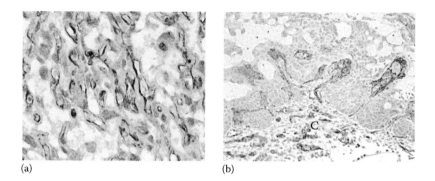

(a)   (b)

FIGURE 16.10 Abnormal vasculogenesis in the placenta is readily highlighted using anti–endothelial cell immunohistochemistry (IHC). Capillaries in the labyrinth of wild-type placentas are lined by a single layer of flat endothelial cells associated with one or two layers of trophoblast (a). In contrast, capillaries in a dysplastic placenta (here from a GD 10.5 mouse embryo lacking the *Vhl* [von Hippel Lindau] gene) exhibit clumping of both endothelial cells (brown label) and labyrinthine trophoblast (unlabeled) (b). Processing conditions: formalin fixation, paraffin embedding, goat anti-mouse CD31 IHC followed by hematoxylin as a counterstain.

AND trophoblast often leads to generation of a *small labyrinth* (i.e., reduced in area and/or thickness; Figures 16.9 and 16.10).[128] Changes linked with this end-stage lesion include differentiation defects in one or more trophoblast lineages; deficiency or absence of trophoblast zones[4] and/or their branches[39]; reduced production,[8,31,35,68] extension,[10,31,51,66,69,80,106,123] or branching[59,65,90,106] and/or disorder[90] of embryonic capillaries; and labyrinthine edema.[68,127] In most cases, such decreases arise mainly from a scarcity of labyrinthine trophoblasts.[128] Changes associated with placental dysfunction can include lesions that may be reversible, such as cell type–specific degeneration (Figure 16.11) and accumulation of substrates (Figure 16.11) as well as such irreversible processes as apoptosis, necrosis, dysplasia (Figures 16.11 through 16.13),

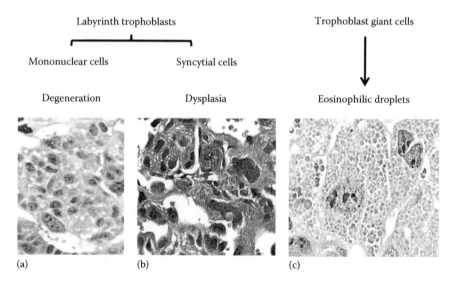

(a)   (b)   (c)

FIGURE 16.11 Cell type–specific lesions may affect any placental cell population but commonly impact one or more trophoblast lineages. Typical findings include *degeneration*, a potentially reversible condition commonly associated with clear cytoplasmic vacuolation (a); *dysplasia*, an irreversible defect caused by abnormal differentiation (b); and *accumulation* of various cell products—usually of unknown character—which appear as variably sized, colored cytoplasmic droplets (c). (a and c, Reproduced from Ward, J.M. and Devor-Henneman, D.E., Gestational mortality in genetically engineered mice: Evaluating the extraembryonal embryonic placenta and membranes, in: *Pathology of Genetically Engineered Mice*, Ward, J.M., Mahler, J.F., Maronpot, R.R., Sundberg, J.P., and Frederickson, R.M., eds., Iowa State University Press, Ames, IA, 2000, pp. 103–122, by courtesy of the Iowa State University Press.)

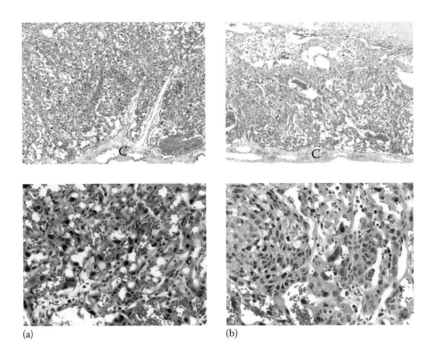

(a)   (b)

FIGURE 16.12 Trophoblast dysplasia often leads to a *small labyrinth* phenotype in affected definitive placentas. Relative to age-matched, wild-type organs (column a), placentas from mouse conceptuses at GD 13.5 that lack the genes for the transcription factors E2f7 and E2f8 (*E2f7⁻/⁻/E2f8⁻/⁻*), which regulate trophoblast differentiation, are characterized by a narrowed labyrinth comprised principally of large, coalescing aggregates of densely packed, plump trophoblast cells (column b). The distance between embryonic capillaries and maternal sinusoids is increased as a secondary consequence. C, chorionic plate. (Images of previously documented lesions[77] were acquired from formalin-fixed, paraffin-embedded, H&E-stained placenta sections kindly provided by Dr. Gustavo Leone, The Ohio State University, Columbus, OH.)

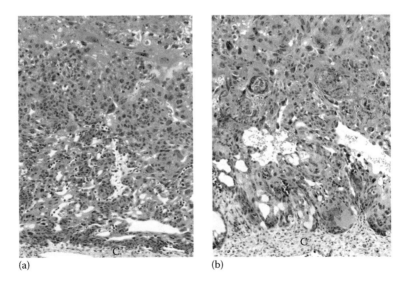

(a)   (b)

FIGURE 16.13 Excessive and ectopic proliferation of trophoblast cells may severely disrupt placental structure. Relative to an age-matched, wild-type organ (a), definitive placentas from GD 12.5 mouse conceptuses lacking the Rb tumor suppressor gene feature a markedly thin labyrinth characterized by greatly reduced numbers of poorly anastomosing capillaries separated by coalescing, densely packed clusters of dysplastic labyrinth trophoblast cells (b). C, chorionic plate. (Images of previously documented lesions[133] were obtained from formalin-fixed, paraffin-embedded, H&E-stained placenta sections kindly provided by Dr. Gustavo Leone, The Ohio State University, Columbus, OH.)

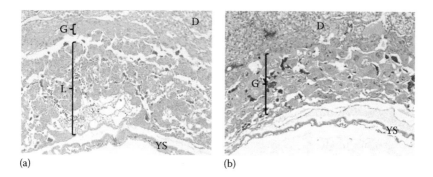

(a)   (b)

FIGURE 16.14 Placental hypoplasia often arises from abnormal patterns of trophoblast differentiation. In the normal GD 9 conceptus (a), embryonic contributions to the evolving placenta include the yolk sac (YS), the labyrinth (L), and the giant cell trophoblast (G) layer, which abuts the maternal-derived decidua (D). In an age-matched littermate with a small placenta (b), the yolk sac is attenuated, the labyrinth (trophoblast and vessels) is missing completely, and giant cell trophoblasts are more numerous than expected. (Images were obtained from formalin-fixed, paraffin-embedded, H&E-stained placenta sections generously provided by Dr. Suzanne Coberly, Amgen, Inc., South San Francisco, CA.)

and aplasia (Figure 16.14). The subsequent plunge in trophoblast area induces a secondary deficiency in labyrinthine vascular volume, which leads eventually to placental insufficiency as the metabolic needs of the growing embryo increase over time. As a separate mechanism, abnormal expansion of the labyrinthine trophoblast population (Figure 16.13) has been shown to reduce labyrinth function,[14,133] ostensibly because the enhanced trophoblast thickness hinders the movement of nutrients between embryonic capillaries and maternal sinusoids.

Primary lesions of the labyrinthine trophoblast may lead to widespread secondary disruption of other placental structures. Common findings affecting the labyrinth, decidua, or both include dilation of the embryonic[65] or maternal[66,68,90,101,127] blood vessels, tardy differentiation of hematopoietic precursors,[59] lower numbers of circulating blood cells,[69] or the occurrence of hemorrhagic foci (Figure 16.15),[7,35,68,113] thrombi (Figure 16.15),[120,127] or necrosis (Figure 16.16). Pooling of blood, fluid, or fibrin can lead to disorganization or disruption of placental layering.[49,127] Formation of thrombi within larger blood vessels may result in vascular insufficiency affecting part of the organ. These changes impede blood flow and/or diminish the available embryonic blood supply, thereby augmenting the extent of placental insufficiency.

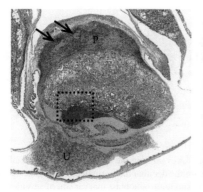

FIGURE 16.15 Circulatory disruption by thrombi in either the maternal (long arrows) or embryonic (inside box) vessels is a common finding in early resorptions. In this implantation site at GD 12.5, the chronicity of the blockage has resulted in widespread intraplacental (P) and intrauterine (U) hemorrhage and has led to the complete loss of the embryo. The image on the right is a higher-magnification view of the region within the box. Processing conditions: formalin fixation, paraffin embedding, H&E staining.

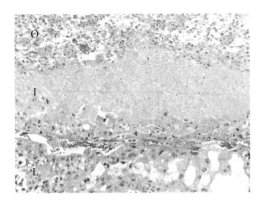

FIGURE 16.16   Placental insufficiency may result from widespread necrosis, evident here as a zone of diffuse tissue coagulation affecting the inner decidua (I) but not the outer decidua (O) or labyrinth (L) of a GD 13.5 *E2f7$^{-/-}$/E2f8$^{-/-}$* mouse embryo. (The image was obtained from a formalin-fixed, paraffin-embedded, H&E-stained placenta section kindly provided by Dr. Gustavo Leone, The Ohio State University, Columbus, OH.)

   Special techniques for morphological analysis may be necessary when investigating certain hypotheses regarding the nature and mechanisms of placental defects. Comparison of regional distributions for various cell type–specific biomarkers[42] (see Chapter 4) is an essential tool for exploring the effects of genetic alterations or xenobiotic exposures to major placental cell populations (Figure 16.17). For example, all *small labyrinth* phenotypes with a reduced density of trophoblast also will have fewer embryonic capillaries. Anatomic differences in vascularity may be assessed in a semiquantitative fashion using vascular casting to outline the extent and conformation of the embryonic and maternal vascular beds (Figure 16.18). The more accurate means of defining the primary target in such cases may be to use morphometry or stereology to compare the numbers of embryonic vessels and trophoblast.[122,133] A proportional decrease in vascular and trophoblastic elements would imply that the principal abnormality involves extension of the labyrinth, while a preferential decline in capillary density would suggest that the main defect is in the penetration and/or production of blood vessels. Similar stereological assessments are a possible way of examining the cell layers that segregate the maternal sinusoids from the embryonic capillaries.[15,105,133] Such quantitative measurements may be more informative than straightforward qualitative analyses in such cases because nutrient transfer depends not only on the labyrinth surface area but also on the distance between the maternal and embryonic vessels across which the nutrients must be transferred. In exceptional cases, evaluation of

(a)

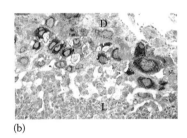

(b)

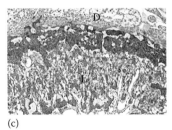

(c)

FIGURE 16.17   Cell type–specific biomarkers are a critical tool for evaluating the role of various placental cell populations in producing embryonic lethal phenotypes. Examples here are the decidua marker galectin-3 (Lgals3, also known as Mac2 [a]), the trophoblast giant cell marker prolactin (b), and the spongiotrophoblast marker son of sevenless homolog 2 (Sos2 [c]). Processing conditions: formalin fixation, paraffin embedding, indirect immunohistochemistry with commercial monoclonal antibodies followed by a hematoxylin counterstain. D, decidua; L, labyrinth. (Reproduced from Ward, J.M. and Devor-Henneman, D.E., Gestational mortality in genetically engineered mice: Evaluating the extraembryonal embryonic placenta and membranes, in: *Pathology of Genetically Engineered Mice*, Ward, J.M., Mahler, J.F., Maronpot, R.R., Sundberg, J.P., and Frederickson, R.M., eds., Iowa State University Press, Ames, IA, 2000, pp. 103–122, by courtesy of the Iowa State University Press.)

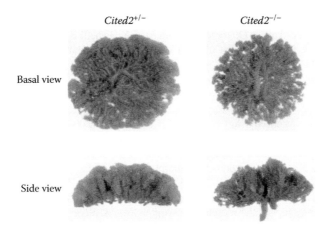

FIGURE 16.18 Altered vasculogenesis in the definitive placenta may be elegantly confirmed using vascular casting to define the extent and structure of the embryonic and maternal vascular trees. Relative to heterozygous (*Cited*[+/-]) controls (top row), null mutant (*Cited*[-/-]) conceptuses at E14.5 have smaller arterial (blue) and venous (red) beds. (Reproduced from *Dev. Biol.* 294(1), Withington, S.L., Scott, A.N., Saunders, D.N., Lopes Floro, K., Preis, J.I., Michalicek, J., Maclean, K., Sparrow, D.B., Barbera, J.P., and Dunwoodie, S.L., Loss of *Cited2* affects trophoblast formation and vascularization of the mouse placenta, 67–82, Copyright 2006, with permission from Elsevier.)

very discrete structures using electron microscopy[136] (see Chapter 13) or laser capture microdissection[117] (see Chapter 10) may be necessary to pinpoint specific subcellular lesions or affected cell populations. However, in our experience these resource-intensive procedures as well as stereology should be undertaken only after a thorough analysis at the light microscopic level.

Considerations in the Interpretation of Placental Pathology Data

Care must be taken during placental pathology investigations to consider potential confounding events and their impact on embryolethal placental phenotypes. The single best way to accomplish this task is to examine all placental components at multiple time points during gestation through a series of sequential placenta collections. This strategy is essential for several reasons. First, some mutations may yield failed chorioallantoic fusion as an early outcome followed by later trophoblastic disruption in surviving embryos,[66,118] thereby producing multiple age-dependent placental phenotypes. Second, some labyrinth phenotypes do not become evident until the latter half of gestation (after GD 12.5), when the basic organization of the labyrinth already has been set,[44,93,120,132] or even during the perinatal period.[59,68,90] Where placental dysfunction contributes to lethality late in gestation, the usual cause is thought to be insufficient metabolic exchange[17,105,127] as animals that reach term may survive to adulthood once the defective placenta is no longer the sole source of nutrition.[68,120] Third, maternal components of the placenta may play a role in some instances because a few placental phenotypes are centered in the decidua rather than in the embryonic labyrinth.[5] This outcome should be considered when the genome of the dam is heterozygous for a homozygous null mutation.

The pathologist must not jump to conclusions in interpreting structural defects until mechanistic data (i.e., gene or protein localization to specific cell lineages) have been obtained. For example, some labyrinth abnormalities originally were described as vascular defects, but subsequent gene expression studies showed that the disrupted molecular pathway actually resided in the perivascular trophoblast.[128] Experience has shown that inactivation of a gene specifically in placental cells more often than not is accompanied by gene inactivation in cells of the embryo. Similarly, inactivation of a gene in the embryonic stem cells can result in developmental defects in both the placenta (extraembryonic tissues) and the embryo. Functional defects may occur in embryonic cells that lead to embryonic death, which will be followed soon thereafter by secondary degeneration and regression or resorption of the placenta. This expected phenomenon must be differentiated from primary placental developmental defects.

REFERENCES

1. Adams RH, Porras A, Alonso G, Jones M, Vintersten K, Panelli S, Valladares A, Perez L, Klein R, Nebreda AR (2000). Essential role of p38α MAP kinase in placental but not embryonic cardiovascular development. *Mol Cell* **6** (1): 109–116.

2. Adamson SL, Lu Y, Whiteley KJ, Holmyard D, Hemberger M, Pfarrer C, Cross JC (2002). Interactions between trophoblast cells and the maternal and fetal circulation in the mouse placenta. *Dev Biol* **250** (2): 358–373.

3. Adelman DM, Gertsenstein M, Nagy A, Simon MC, Maltepe E (2000). Placental cell fates are regulated *in vivo* by HIF-mediated hypoxia responses. *Genes Dev* **14** (24): 3191–3203.

4. Anson-Cartwright L, Dawson K, Holmyard D, Fisher S, Lazzarini RA, Cross J (2000). The glial cells missing-1 protein is essential for branching morphogenesis in the chorioallantoic placenta. *Nat Genet* **25** (3): 311–314.

5. Arai T, Kasper JS, Skaar JR, Ali SH, Takahashi C, DeCaprio JA (2003). Targeted disruption of *p185/Cul7* gene results in abnormal vascular morphogenesis. *Proc Natl Acad Sci USA* **100** (17): 9855–9860.

6. Arima T, Hata K, Tanaka S, Kusumi M, Li E, Kato K, Shiota K, Sasaki H, Wake N (2006). Loss of the maternal imprint in *Dnmt3L^mat-/-* mice leads to a differentiation defect in the extraembryonic tissue. *Dev Biol* **297** (2): 361–373.

7. Barak Y, Liao D, He W, Ong ES, Nelson MC, Olefsky JM, Boland R, Evans RM (2002). Effects of peroxisome proliferator-activated receptor δ on placentation, adiposity, and colorectal cancer. *Proc Natl Acad Sci USA* **99** (1): 303–308.

8. Barak Y, Nelson MC, Ong ES, Jones YZ, Ruiz-Lozano P, Chien KR, Koder A, Evans RM (1999). PPARγ is required for placental, cardiac, and adipose tissue development. *Mol Cell* **4** (4): 585–595.

9. Baudino TA, McKay C, Pendeville-Samain H, Nilsson JA, Maclean KH, White EL, Davis AC, Ihle JN, Cleveland JL (2002). c-Myc is essential for vasculogenesis and angiogenesis during development and tumor progression. *Genes Dev* **16** (19): 2530–2543.

10. Begay V, Smink J, Leutz A (2004). Essential requirement of CCAAT/enhancer binding proteins in embryogenesis. *Mol Cell Biol* **24** (22): 9744–9751.

11. Benirschke K (2007). Comparative placentation. http://placentation.ucsd.edu/ (last accessed December 1, 2014).

12. Benirschke K, Burton GJ, Baergen RN (2012). *Pathology of the Human Placenta*, 6th edn. Berlin, Germany: Springer-Verlag.

13. Chawengsaksophak K, de Graaff W, Rossant J, Deschamps J, Beck F (2004). *Cdx2* is essential for axial elongation in mouse development. *Proc Natl Acad Sci USA* **101** (20): 7641–7645.

14. Chong J-L, Tsai S-Y, Sharma N, Opavsky R, Price R, Wu L, Fernandez SA, Leone G (2009). *E2f3a* and *E2f3b* contribute to the control of cell proliferation and mouse development. *Mol Cell Biol* **29** (2): 414–424.

15. Coan PM, Ferguson-Smith AC, Burton GJ (2004). Developmental dynamics of the definitive mouse placenta assessed by stereology. *Biol Reprod* **70** (6): 1806–1813.

16. Coan PM, Fowden AL, Constância M, Ferguson-Smith AC, Burton GJ, Sibley CP (2008). Disproportional effects of *Igf2* knockout on placental morphology and diffusional exchange characteristics in the mouse. *J Physiol* **586** (Pt 20): 5023–5032.

17. Constância M, Hemberger M, Hughes J, Dean W, Ferguson-Smith A, Fundele R, Stewart F, Kelsey G, Fowden A, Sibley C, Reik W (2002). Placental-specific IGF-II is a major modulator of placental and fetal growth. *Nature* **417** (6892): 945–948.

18. Copp AJ (1995). Death before birth: Clues from gene knockouts and mutations. *Trends Genet* **11** (3): 87–93.

19. Cowden Dahl KD, Fryer BH, Mack FA, Compernolle V, Maltepe E, Adelman DM, Carmeliet P, Simon MC (2005). Hypoxia-inducible factors 1α and 2α regulate trophoblast differentiation. *Mol Cell Biol* **25** (23): 10479–10491.

20. Cross JC (1998). Formation of the placenta and extraembryonic membranes. *Ann NY Acad Sci* **857**: 23–32.

21. Cross JC, Baczyk D, Dobric N, Hemberger M, Hughes M, Simmons DG, Yamamoto H, Kingdom JC (2003). Genes, development and evolution of the placenta. *Placenta* **24** (2–3): 123–130.

22. Cross JC, Simmons DG, Watson ED (2003). Chorioallantoic morphogenesis and formation of the placental villous tree. *Ann NY Acad Sci* **995**: 84–93.

23. Cross JC, Werb Z, Fisher SJ (1994). Implantation and the placenta: Key pieces of the development puzzle. *Science* **266** (5190): 1508–1518.

24. Cui J, O'Shea KS, Purkayastha A, Saunders TL, Ginsburg D (1996). Fatal haemorrhage and incomplete block to embryogenesis in mice lacking coagulation factor V. *Nature* **384** (6604): 66–68.

25. DeSesso JM (2006). Comparative features of vertebrate embryology. In: *Developmental and Reproductive Toxicology: A Practical Approach*, 2nd edn. (Hood RD, ed.). Boca Raton, FL: CRC Press, pp. 147–197.

26. Dickson MC, Martin JS, Cousins FM, Kulkarni AB, Karlsson S, Akhurst RJ (1995). Defective haematopoiesis and vasculogenesis in transforming growth factor-β1 knock out mice. *Development* **121** (6): 1845–1854.

27. Escalante-Alcalde D, Hernandez L, Le Stunff H, Maeda R, Lee HS, Jr Gang C, Sciorra VA, Daar I, Spiegel S, Morris AJ, Stewart CL (2003). The lipid phosphatase LPP3 regulates extra-embryonic vasculogenesis and axis patterning. *Development* **130** (19): 4623–4637.

28. Feng Y, Yang Y, Ortega MM, Copeland JN, Zhang M, Jacob JB, Fields TA, Vivian JL, Fields PE (2010). Early mammalian erythropoiesis requires the Dot1L methyltransferase. *Blood* **116** (22): 4483–4491.

29. Ferdous A, Morris J, Abedin MJ, Collins S, Richardson JA, Hill JA (2011). Forkhead factor FoxO1 is essential for placental morphogenesis in the developing embryo. *Proc Natl Acad Sci USA* **108** (39): 16307–16312.

30. Firulli AB, McFadden DG, Lin Q, Srivastava D, Olson EN (1998). Heart and extra-embryonic mesodermal defects in mouse embryos lacking the bHLH transcription factor Hand1. *Nat Genet* **18** (3): 266–270.

31. Fischer A, Schumacher N, Maier M, Sendtner M, Gessler M (2004). The Notch target genes *Hey1* and *Hey2* are required for embryonic vascular development. *Genes Dev* **18** (8): 901–911.

32. Gabriel H-D, Jung D, Bützler C, Temme A, Traub O, Winterhager E, Willecke K (1998). Transplacental uptake of glucose is decreased in embryonic lethal connexin26-deficient mice. *J Cell Biol* **140** (6): 1453–1461.

33. Georgiades P, Ferguson-Smith AC, Burton GJ (2002). Comparative developmental anatomy of the murine and human definitive placentae. *Placenta* **23** (1): 3–19.

34. Giroux S, Tremblay M, Bernard D, Cardin-Girard J-F, Aubry S, Larouche L, Rousseau S, Huot J, Landry J, Jeannotte L, Charron J (1999). Embryonic death of *Mek1*-deficient mice reveals a role for this kinase in angiogenesis in the labyrinthine region of the placenta. *Curr Biol* **9** (7): 369–372.

35. Gnarra JR, Ward JM, Porter FD, Wagner JR, Devor DE, Grinberg A, Emmertbuck MR, Westphal H, Klausner RD, Linehan WM (1997). Defective placental vasculogenesis causes embryonic lethality in VHL-deficient mice. *Proc Natl Acad Sci USA* **94** (17): 9102–9107.

36. Guillemot F, Nagy A, Auerbach A, Rossant J, Joyner AL (1994). Essential role of *Mash-2* in extraembryonic development. *Nature* **371** (6495): 333–336.

37. Gurtner GC, Davis V, Li H, McCoy MJ, Sharpe A, Cybulsky MI (1995). Targeted disruption of the murine *VCAM1* gene: Essential role of VCAM-1 in chorioallantoic fusion and placentation. *Genes Dev* **9** (1): 1–14.

38. Hemberger MC, Pearsall RS, Zechner U, Orth A, Otto S, Ruschendorf F, Fundele R, Elliott R (1999). Genetic dissection of X-linked interspecific hybrid placental dysplasia in congenic mouse strains. *Genetics* **153** (1): 383–390.

39. Hildebrand JD, Soriano P (2002). Overlapping and unique roles for C-terminal binding protein 1 (CtBP1) and CtBP2 during mouse development. *Mol Cell Biol* **22** (15): 5296–5307.

40. Hunter PJ, Swanson BJ, Haendel MA, Lyons GE, Cross JC (1999). *Mrj* encodes a DnaJ-related co-chaperone that is essential for murine placental development. *Development* **126** (6): 1247–1258.

41. Inman KE, Downs KM (2007). The murine allantois: Emerging paradigms in development of the mammalian umbilical cord and its relation to the fetus. *Genesis* **45** (5): 237–258.

42. Isaac SM, Langford MB, Simmons DG, Adamson SL (2014). Anatomy of the mouse placenta throughout gestation, In: *The Guide to Investigation of Mouse Pregnancy* (Croy BA, Yamada AT, DeMayo FJ, Adamson SL, eds.). London, U.K.: Academic Press (Elsevier), pp. 69–73.

43. Ishikawa T, Tamai Y, Zorn AM, Yoshida H, Seldin MF, Nishikawa S, Taketo MM (2001). Mouse Wnt receptor gene *Fzd5* is essential for yolk sac and placental angiogenesis. *Development* **128** (1): 25–33.

44. Itoh M, Yoshida Y, Nishida K, Narimatsu M, Hibi M, Hirano T (2000). Role of Gab1 in heart, placenta, and skin development and growth factor- and cytokine-induced extracellular signal-regulated kinase mitogen-activated protein kinase activation. *Mol Cell Biol* **20** (10): 3695–3704.

45. Kanayama N, Takahashi K, Matsuura T, Sugimura M, Kobayashi T, Moniwa N, Tomita M, Nakayama K (2002). Deficiency in p57*Kip2* expression induces preeclampsia-like symptoms in mice. *Mol Hum Reprod* **8** (12): 1129–1135.

46. Kaufman MH (1992). *The Atlas of Mouse Development*, 2nd edn. San Diego, CA: Academic Press.

47. Kaufman MH, Bard JBL (1999). *The Anatomical Basis of Mouse Development.* San Diego, CA: Academic Press.

48. Kingdom J, Huppertz B, Seaward G, Kaufmann P (2000). Development of the placental villous tree and its consequences for fetal growth. *Eur J Obstet Gynecol Reprod Biol* **92** (1): 35–43.

49. Kozak KR, Abbott B, Hankinson O (1997). ARNT-deficient mice and placental differentiation. *Dev Biol* **191** (2): 297–305.

50. Kraut N, Snider L, Chen CM, Tapscott SJ, Groudine M (1998). Requirement of the mouse *I-mfa* gene for placental development and skeletal patterning. *EMBO J* **17** (21): 6276–6288.

51. Krebs LT, Xue Y, Norton CR, Shutter JR, Maguire M, Sundberg JP, Gallahan D et al. (2000). Notch signaling is essential for vascular morphogenesis in mice. *Genes Dev* **14** (11): 1343–1352.

52. Krüger O, Plum A, Kim J-S, Winterhager E, Maxeiner S, Hallas G, Kirchhoff S, Traub O, Lamers WH, Willecke K (2000). Defective vascular development in connexin 45-deficient mice. *Development* **127** (19): 4179–4193.

53. Kwee L, Baldwin HS, Shen HM, Stewart CL, Buck C, Buck CA, Labow MA (1995). Defective development of the embryonic and extraembryonic circulatory systems in vascular cell adhesion molecule (VCAM-1) deficient mice. *Development* **121** (2): 489–503.

54. Lechleider RJ, Ryan JL, Garrett L, Eng C, Deng C, Wynshaw-Boris A, Roberts AB (2001). Targeted mutagenesis of *Smad1* reveals an essential role in chorioallantoic fusion. *Dev Biol* **240** (1): 157–167.

55. Li E, Bestor TH, Jaenisch R (1992). Targeted mutation of the DNA methyltransferase gene results in embryonic lethality. *Cell* **69** (6): 915–926.

56. Li Y, Behringer RR (1998). *Esx1* is an X-chromosome-imprinted regulator of placental development and fetal growth. *Nat Genet* **20** (3): 309–311.

57. Lin Q, Lu J, Yanagisawa H, Webb R, Lyons GE, Richardson JA, Olson EN (1998). Requirement of the MADS-box transcription factor MEF2C for vascular development. *Development* **125** (22): 4565–4574.

58. Lin SP, Coan P, da Rocha ST, Seitz H, Cavaille J, Teng PW, Takada S, Ferguson-Smith AC (2007). Differential regulation of imprinting in the murine embryo and placenta by the *Dlk1-Dio3* imprinting control region. *Development* **134** (2): 417–426.

59. Lotz K, Pyrowolakis G, Jentsch S (2004). BRUCE, a giant E2/E3 ubiquitin ligase and inhibitor of apoptosis protein of the *trans*-Golgi network, is required for normal placenta development and mouse survival. *Mol Cell Biol* **24** (21): 9339–9350.

60. Luo J, Sladek R, Bader JA, Matthyssen A, Rossant J, Giguere V (1997). Placental abnormalities in mouse embryos lacking the orphan nuclear receptor ERR-β. *Nature* **388** (6644): 778–782.

61. Ma GT, Soloveva V, Tzeng SJ, Lowe LA, Pfendler KC, Iannaccone PM, Kuehn MR, Linzer DI (2001). Nodal regulates trophoblast differentiation and placental development. *Dev Biol* **236** (1): 124–135.

62. Mahlapuu M, Ormestad M, Enerbäck S, Carlsson P (2001). The forkhead transcription factor Foxf1 is required for differentiation of extra-embryonic and lateral plate mesoderm. *Development* **128** (2): 155–166.

63. Matsumoto N, Kubo A, Liu H, Akita K, Laub F, Ramirez F, Keller G, Friedman SL (2006). Developmental regulation of yolk sac hematopoiesis by Kruppel-like factor 6. *Blood* **107** (4): 1357–1365.

64. McGeady TA, Quinn PJ, FitzPatrick ES, Ryan MT (2006). *Veterinary Embryology.* Oxford, U.K.: Blackwell Publishing.

65. Miner JH, Cunningham J, Sanes JR (1998). Roles for laminin in embryogenesis: Exencephaly, syndactyly, and placentopathy in mice lacking the laminin α5 chain. *J Cell Biol* **143** (6): 1713–1723.

66. Mo F-E, Muntean AG, Chen C-C, Stolz DB, Watkins SC, Lau LF (2002). CYR61 (CCN1) is essential for placental development and vascular integrity. *Mol Cell Biol* **22** (24): 8709–8720.

67. Moll W (1985). Physiological aspects of placental ontogeny and phylogeny. *Placenta* **6** (2): 141–154.

68. Monkley SJ, Delaney SJ, Pennisi DJ, Christiansen JH, Wainwright BJ (1996). Targeted disruption of the *Wnt2* gene results in placentation defects. *Development* **122** (11): 3343–3353.
69. Morasso MI, Grinberg A, Robinson G, Sargent TD, Mahon KA (1999). Placental failure in mice lacking the homeobox gene *Dlx3*. *Proc Natl Acad Sci USA* **96** (1): 162–167.
70. Mosaliganti K, Pan T, Ridgway R, Sharp R, Cooper L, Gulacy A, Sharma A et al. (2008). An imaging workflow for characterizing phenotypical change in large histological mouse model datasets. *J Biomed Inform* **41** (6): 863–873.
71. Nagao T, Saitoh Y, Yoshimura S (2000). Possible mechanism of congenital malformations induced by exposure of mouse preimplantation embryos to mitomycin C. *Teratology* **61** (4): 248–261.
72. Natale DRC, Starovic M, Cross JC (2006). Phenotypic analysis of the mouse placenta. In: *Placenta and Trophoblast: Methods and Protocols*, vol. 1 (Soares MJ, Hunt JS, eds.). Totowa, NJ: Humana Press, pp. 275–293.
73. Noden DM, de LaHunta A (1985). *The Embryology of Domestic Animals: Developmental Mechanisms and Malformations*. Baltimore, MD: Williams & Wilkins.
74. Ohlsson R, Falck P, Hellström M, Lindahl P, Boström H, Franklin G, Ahrlund-Richter L, Pollard J, Soriano P, Betsholtz C (1999). PDGFB regulates the development of the labyrinthine layer of the mouse fetal placenta. *Dev Biol* **212** (1): 124–136.
75. Oike Y, Takakura N, Hata A, Kaname T, Akizuki M, Yamaguchi Y, Yasue H, Araki K, Yamamura K, Suda T (1999). Mice homozygous for a truncated form of CREB-binding protein exhibit defects in hematopoiesis and vasculo-angiogenesis. *Blood* **93** (9): 2771–2779.
76. Oka C, Nakano T, Wakeham A, de la Pompa JL, Mori C, Sakai T, Okazaki S, Kawaichi M, Shiota K, Mak TW, Honjo T (1995). Disruption of the mouse *RBP-Jk* gene results in early embryonic death. *Development* **121** (10): 3291–3301.
77. Ouseph MM, Li J, Chen HZ, Pecot T, Wenzel P, Thompson JC, Comstock G et al. (2012). Atypical E2F repressors and activators coordinate placental development. *Dev Cell* **22** (4): 849–862.
78. Papaioannou VE, Behringer RR (2005). *Mouse Phenotypes: A Handbook of Mutation Analysis*. Cold Spring Harbor, NY: Cold Spring Harbor Laboratory Press.
79. Papaioannou VE, Behringer RR (2012). Early embryonic lethality in genetically engineered mice: Diagnosis and phenotypic analysis. *Vet Pathol* **49** (1): 64–70.
80. Parekh V, McEwen A, Barbour V, Takahashi Y, Rehg JE, Jane SM, Cunningham JM (2004). Defective extraembryonic angiogenesis in mice lacking LBP-1a, a member of the grainyhead family of transcription factors. *Mol Cell Biol* **24** (16): 7113–7129.
81. Parr BA, Cornish VA, Cybulsky MI, McMahon AP (2001). Wnt7b regulates placental development in mice. *Dev Biol* **237** (2): 324–332.
82. Plum A, Winterhager E, Pesch J, Lautermann J, Hallas G, Rosentreter B, Traub O, Herberhold C, Willecke K (2001). Connexin31-deficiency in mice causes transient placental dysmorphogenesis but does not impair hearing and skin differentiation. *Dev Biol* **231** (2): 334–347.
83. Qian X, Esteban L, Vass WC, Upadhyaya C, Papageorge AG, Yienger K, Ward JM, Lowy DR, Santos E (2000). The Sos1 and Sos2 Ras-specific exchange factors: Differences in placental expression and signaling properties. *EMBO J* **19** (4): 642–654.
84. Rawn SM, Cross JC (2008). The evolution, regulation, and function of placenta-specific genes. *Annu Rev Cell Dev Biol* **24**: 159–181.
85. Regan CP, Li W, Boucher DM, Spatz S, Su MS, Kuida K (2002). Erk5 null mice display multiple extra-embryonic vascular and embryonic cardiovascular defects. *Proc Natl Acad Sci USA* **99** (14): 9248–9253.
86. Rendi MH, Muehlenbachs A, Garcia RL, Boyd KL (2012). Female reproductive system. In: *Comparative Anatomy and Histology: A Mouse and Human Atlas* (Treuting PM, Dintzis SM, eds.). San Diego, CA: Academic Press (Elsevier), pp. 253–284.
87. Riley P, Anson-Cartwright L, Cross JC (1998). The Hand1 bHLH transcription factor is essential for placentation and cardiac morphogenesis. *Nat Genet* **18** (3): 271–275.
88. Rinkenberger JL, Cross JC, Werb Z (1997). Molecular genetics of implantation in the mouse. *Dev Genet* **21** (1): 6–20.
89. Robb L, Lyons I, Li R, Hartley L, Köntgen F, Harvey RP, Metcalf D, Begley CG (1995). Absence of yolk sac hematopoiesis from mice with a targeted disruption of the *scl* gene. *Proc Natl Acad Sci USA* **92** (15): 7075–7079.

90. Rodriguez TA, Sparrow DB, Scott AN, Withington SL, Preis JI, Michalicek J, Clements M, Tsang TE, Shioda T, Beddington RS, Dunwoodie SL (2004). *Cited1* is required in trophoblasts for placental development and for embryo growth and survival. *Mol Cell Biol* **24** (1): 228–244.

91. Rossant J, Cross JC (2001). Placental development: Lessons from mouse mutants. *Nat Rev Genet* **2** (7): 538–548.

92. Russ AP, Wattler S, Colledge WH, Aparicio SA, Carlton MB, Pearce JJ, Barton SC et al. (2000). *Eomesodermin* is required for mouse trophoblast development and mesoderm formation. *Nature* **404** (6773): 95–99.

93. Sachs M, Brohmann H, Zechner D, Müller T, Hülsken J, Walther I, Schaeper U, Birchmeier C, Birchmeier W (2000). Essential role of Gab1 for signaling by the c-Met receptor *in vivo*. *J Cell Biol* **150** (6): 1375–1384.

94. Satyanarayana A, Gudmundsson KO, Chen X, Coppola V, Tessarollo L, Keller JR, Hou SX (2010). *RapGEF2* is essential for embryonic hematopoiesis but dispensable for adult hematopoiesis. *Blood* **116** (16): 2921–2931.

95. Saunders DN, Hird SL, Withington SL, Dunwoodie SL, Henderson MJ, Biben C, Sutherland RL, Ormandy CJ, Watts CK (2004). *Edd*, the murine hyperplastic disc gene, is essential for yolk sac vascularization and chorioallantoic fusion. *Mol Cell Biol* **24** (16): 7225–7234.

96. Saxton TM, Cheng AM, Ong SH, Lu Y, Sakai R, Cross JC, Pawson T (2001). Gene dosage-dependent functions for phosphotyrosine-Grb2 signaling during mammalian tissue morphogenesis. *Curr Biol* **11** (9): 662–670.

97. Schorpp-Kistner M, Wang Z-Q, Angel P, Wagner EF (1999). JunB is essential for mammalian placentation. *EMBO J* **18** (4): 934–948.

98. Schreiber M, Wang Z-Q, Jochum W, Fetka I, Elliott C, Wagner EF (2000). Placental vascularisation requires the AP-1 component Fra1. *Development* **127** (22): 4937–4948.

99. Sebastiano V, Dalvai M, Gentile L, Schubart K, Sutter J, Wu GM, Tapia N et al. (2010). Oct1 regulates trophoblast development during early mouse embryogenesis. *Development* **137** (21): 3551–3560.

100. Shalaby F, Rossant J, Yamaguchi TP, Gertsenstein M, Wu X-F, Breitman ML, Schuh AC (1995). Failure of blood-island formation and vasculogenesis in Flk-1-deficient mice. *Nature* **376** (6535): 62–66.

101. Shalom-Barak T, Nicholas JM, Wang Y, Zhang X, Ong ES, Young TH, Gendler SJ, Evans RM, Barak Y (2004). Peroxisome proliferator-activated receptor γ controls *Muc1* transcription in trophoblasts. *Mol Cell Biol* **24** (24): 10661–10669.

102. Sharma L, Kaur J, Shukla G (2012). Role of oxidative stress and apoptosis in the placental pathology of *Plasmodium berghei* infected mice. *PLoS ONE* **7** (3): e32694 (8 pages).

103. Shi W, van den Hurk JA, Alamo-Bethencourt V, Mayer W, Winkens HJ, Ropers HH, Cremers FP, Fundele R (2004). Choroideremia gene product affects trophoblast development and vascularization in mouse extra-embryonic tissues. *Dev Biol* **272** (1): 53–65.

104. Sibilia M, Wagner EF (1995). Strain-dependent epithelial defects in mice lacking the EGF receptor. *Science* **269** (5221): 234–238.

105. Sibley CP, Coan PM, Ferguson-Smith AC, Dean W, Hughes J, Smith P, Reik W, Burton GJ, Fowden AL, Constância M (2004). Placental-specific insulin-like growth factor 2 (*Igf2*) regulates the diffusional exchange characteristics of the mouse placenta. *Proc Natl Acad Sci USA* **101** (21): 8204–8208.

106. Steingrímsson E, Tessarollo L, Reid SW, Jenkins NA, Copeland NG (1998). The bHLH-Zip transcription factor *Tfeb* is essential for placental vascularization. *Development* **125** (23): 4607–4616.

107. Stumpo DJ, Byrd NA, Phillips RS, Ghosh S, Maronpot RR, Castranio T, Meyers EN, Mishina Y, Blackshear PJ (2004). Chorioallantoic fusion defects and embryonic lethality resulting from disruption of *Zfp36L1*, a gene encoding a CCCH tandem zinc finger protein of the Tristetraprolin family. *Mol Cell Biol* **24** (14): 6445–6455.

108. Takahashi K, Kobayashi T, Kanayama N (2000). p57^{Kip2} regulates the proper development of labyrinthine and spongiotrophoblasts. *Mol Hum Repro* **6** (11): 1019–1025.

109. Takahashi T, Takahashi K, St John PL, Fleming PA, Tomemori T, Watanabe T, Abrahamson DR, Drake CJ, Shirasawa T, Daniel TO (2003). A mutant receptor tyrosine phosphatase, CD148, causes defects in vascular development. *Mol Cell Biol* **23** (5): 1817–1831.

110. Takahashi Y, Carpino N, Cross JC, Torres M, Parganas E, Ihle JN (2003). SOCS3: An essential regulator of LIF receptor signaling in trophoblast giant cell differentiation. *EMBO J* **22** (3): 372–384.

111. Takashima S, Kitakaze M, Asakura M, Asanuma H, Sanada S, Tashiro F, Niwa H et al. (2002). Targeting of both mouse neuropilin-1 and neuropilin-2 genes severely impairs developmental yolk sac and embryonic angiogenesis. *Proc Natl Acad Sci USA* **99** (6): 3657–3662.

112. Takeda K, Ho VC, Takeda H, Duan LJ, Nagy A, Fong GH (2006). Placental but not heart defects are associated with elevated hypoxia-inducible factor α levels in mice lacking prolyl hydroxylase domain protein 2. *Mol Cell Biol* **26** (22): 8336–8346.

113. Tamai Y, Ishikawa T, Bosl MR, Mori M, Nozaki M, Baribault H, Oshima RG, Taketo MM (2000). Cytokeratins 8 and 19 in the mouse placental development. *J Cell Biol* **151** (3): 563–572.

114. Tamura K, Sudo T, Senftleben U, Dadak AM, Johnson R, Karin M (2000). Requirement for p38α in erythropoietin expression: A role for stress kinases in erythropoiesis. *Cell* **102** (2): 221–231.

115. Tanaka M, Gertsenstein M, Rossant J, Nagy A (1997). Mash2 acts cell autonomously in mouse spongiotrophoblast development. *Dev Biol* **190** (1): 55–65.

116. Tanaka S, Oda M, Toyoshima Y, Wakayama T, Tanaka M, Yoshida N, Hattori N, Ohgane J, Yanagimachi R, Shiota K (2001). Placentomegaly in cloned mouse concepti caused by expansion of the spongiotrophoblast layer. *Biol Reprod* **65** (6): 1813–1821.

117. Tayade C, Edwards AK, Bidarimath M (2014). Laser capture microdissection. In: *The Guide to Investigation of Mouse Pregnancy* (Croy BA, Yamada AT, DeMayo FJ, Adamson SL, eds.). London, U.K.: Academic Press (Elsevier), pp. 567–575.

118. Tetzlaff MT, Bai C, Finegold M, Wilson J, Harper JW, Mahon KA, Elledge SJ (2004). Cyclin F disruption compromises placental development and affects normal cell cycle execution. *Mol Cell Biol* **24** (6): 2487–2498.

119. Threadgill DW, Dlugosz AA, Hansen LA, Tennenbaum T, Lichti U, Yee D, LaMantia C et al. (1995). Targeted disruption of mouse EGF receptor: Effect of genetic background on mutant phenotype. *Science* **269** (5221): 230–234.

120. Thumkeo D, Keel J, Ishizaki T, Hirose M, Nonomura K, Oshima H, Oshima M, Taketo MM, Narumiya S (2003). Targeted disruption of the mouse rho-associated kinase 2 gene results in intrauterine growth retardation and fetal death. *Mol Cell Biol* **23** (14): 5043–5055.

121. Uehara Y, Minowa O, Mori C, Shiota K, Kuno J, Noda T, Kitamura N (1995). Placental defect and embryonic lethality in mice lacking hepatocyte growth factor/scatter factor. *Nature* **373** (6516): 702–705.

122. Veras MM, Costa NSX, Mayhew T (2014). Best practice for quantifying the microscopic structure of mouse placenta: The stereological approach. In: *The Guide to Investigation of Mouse Pregnancy* (Croy BA, Yamada AT, DeMayo FJ, Adamson SL, eds.). London, U.K.: Academic Press (Elsevier), pp. 545–556.

123. Voss AK, Thomas T, Gruss P (2000). Mice lacking HSP90β fail to develop a placental labyrinth. *Development* **127** (1): 1–11.

124. Wang LC, Kuo F, Fujiwara Y, Gilliland DG, Golub TR, Orkin SH (1997). Yolk sac angiogenic defect and intra-embryonic apoptosis in mice lacking the Ets-related factor TEL. *EMBO J* **16** (14): 4374–4383.

125. Ward JM, Devor-Henneman DE (2000). Gestational mortality in genetically engineered mice: Evaluating the extraembryonal embryonic placenta and membranes. In: *Pathology of Genetically Engineered Mice* (Ward JM, Mahler JF, Maronpot RR, Sundberg JP, Frederickson RM, eds.). Ames, IA: Iowa State University Press, pp. 103–122.

126. Ward JM, Elmore SA, Foley JF (2012). Pathology methods for the evaluation of embryonic and perinatal developmental defects and lethality in genetically engineered mice. *Vet Pathol* **49** (1): 71–84.

127. Ware CB, Horowitz MC, Renshaw BR, Hunt JS, Liggitt D, Koblar SA, Gliniak BC et al. (1995). Targeted disruption of the low-affinity leukemia inhibitory factor receptor gene causes placental, skeletal, neural and metabolic defects and results in perinatal death. *Development* **121** (5): 1283–1299.

128. Watson ED, Cross JC (2005). Development of structures and transport functions in the mouse placenta. *Physiology* **20**: 180–193.

129. Weinstein M, Yang X, Deng C (2000). Functions of mammalian *Smad* genes as revealed by targeted gene disruption in mice. *Cytokine Growth Factor Rev* **11** (1–2): 49–58.

130. Wendling O, Chambon P, Mark M (1999). Retinoid X receptors are essential for early mouse development and placentogenesis. *Proc Natl Acad Sci USA* **96** (2): 547–551.

131. Wilson V, Rashbass P, Beddington RS (1993). Chimeric analysis of *T* (*Brachyury*) gene function. *Development* **117** (4): 1321–1331.

132. Withington SL, Scott AN, Saunders DN, Lopes Floro K, Preis JI, Michalicek J, Maclean K, Sparrow DB, Barbera JP, Dunwoodie SL (2006). Loss of *Cited2* affects trophoblast formation and vascularization of the mouse placenta. *Dev Biol* **294** (1): 67–82.

133. Wu L, de Bruin A, Saavedra HI, Starovic M, Trimboli A, Yang Y, Opavska J et al. (2003). Extra-embryonic function of Rb is essential for embryonic development and viability. *Nature* **421** (6926): 942–947.

134. Xiao X, Zuo X, Davis AA, McMillan DR, Curry BB, Richardson JA, Benjamin IJ (1999). HSF1 is required for extra-embryonic development, postnatal growth and protection during inflammatory responses in mice. *EMBO J* **18** (21): 5943–5952.

135. Xu X, Weinstein M, Li C, Naski M, Cohen RI, Ornitz DM, Leder P, Deng C (1998). Fibroblast growth factor receptor 2 (FGFR2)-mediated reciprocal regulation loop between FGF8 and FGF10 is essential for limb induction. *Development* **125** (4): 753–765.

136. Yamada AT, Lima PDA, Rätsep MT, Paffaro VA Jr, Nagata T, Zorn MT, Abrahamsohn PA (2014). Electron microscopy and immunoelectromicroscopy protocols. In: *The Guide to Investigation of Mouse Pregnancy* (Croy BA, Yamada AT, DeMayo FJ, Adamson SL, eds.). London, U.K.: Academic Press (Elsevier), pp. 557–565.

137. Yamamoto H, Flannery ML, Kupriyanov S, Pearce J, McKercher SR, Henkel GW, Maki RA, Werb Z, Oshima RG (1998). Defective trophoblast function in mice with a targeted mutation of Ets2. *Genes Dev* **12** (9): 1315–1329.

138. Yang JT, Rayburn H, Hynes RO (1993). Embryonic mesodermal defects in α5 integrin-deficient mice. *Development* **119** (4): 1093–1105.

139. Yang JT, Rayburn H, Hynes RO (1995). Cell adhesion events mediated by α4 integrins are essential in placental and cardiac development. *Development* **121** (2): 549–560.

140. Zhang H, Bradley A (1996). Mice deficient for BMP2 are nonviable and have defects in amnion/chorion and cardiac development. *Development* **122** (10): 2977–2986.

17

Mouse Developmental Pathology Assessments in High-Throughput Phenogenomic Facilities

Colin McKerlie, Susan Newbigging, and Geoffrey A. Wood

CONTENTS

High-Throughput Phenotyping of Genetically Engineered Mice (GEM)

Undoubtedly, one of the most exciting scientific challenges for the twenty-first century is to understand the function of every gene in the context of the whole organism. Such understanding is essential to a fundamental appreciation of mammalian biology as well as to the concept of translational medicine, where genomic data obtained through animal models are adapted to comprehend the molecular pathways leading to human disease and then devise new targeted therapies to correct or replace the abnormal molecules involved in these pathways. Understanding the function of individual genes, both in normal physiology and in human disease, poses a significant challenge. An additional complication is the prevailing view that many diseases are of multigenic origin and are strongly influenced by environmental factors, and thus are the result of the combined dysfunction of several gene products in a specific context. In the clinical setting, it is seldom that the underlying genetic cause of a disease is treated directly, but interventions to target and restore the affected pathway(s) or system(s) related to the disease are utilized instead to relieve symptoms. Animal experiments play a major role in defining the genes, pathways, and systems that are the most therapeutically relevant targets.

Genetically engineered mice (GEM) are by far the most accessible mammalian system for modeling human disease because of their anatomical, physiological, and genetic similarity to humans. Indeed, GEM are an ideal platform for current multinational, high-throughput phenogenomic efforts aimed at understanding the molecular basis of human disease by either individually deleting each mammalian gene and then cataloging the functional and structural phenotypes that arise from such a genetic defect (i.e., *reverse genetics*)[15] or by using mutagenic chemicals to randomly induce phenotypes that can then be linked to a specific genetic defect (i.e., *forward genetics*).[16] For example, an extensive genetic toolkit in the mouse is underpinning the work of the International Knockout Mouse Consortium (IKMC) to generate single-gene null mutations in mouse embryonic stem (mES) cells for every protein-coding gene in the mouse genome.[12] However, the large complement of mammalian genes[7,24] coupled with the extensive number of protein variants resulting from alternate splicing and posttranslational modification of proteins has complicated efforts to analyze the phenotypes associated with defects in all mouse genes, let alone appreciate the significance of such defects in their human orthologs. An efficiently coordinated and standardized high-throughput approach that permits system-wide correlation among all genes, the functional properties of their protein products, and the biological relevance of these products as drug targets is only now being established.

However, the greater challenge for modern biology is to undertake comprehensive molecular, cellular, developmental, and pathophysiological analyses of GEM models to furnish us with a *systems* view of mammalian biological networks and how perturbations lead to disease. This will require a paradigm shift from traditional single-gene/single laboratory, hypothesis-driven analysis to an increasing focus on multisystem/multifacility, biology-driven analyses of large numbers of mutant phenotypes provided by large-scale mutagenesis programs to deliver standardized phenotype datasets as the starting point for hypothesis-driven studies. Such a comprehensive functional annotation of the mouse genome presents an unparalleled opportunity to develop an integrated framework of mammalian biological systems that can drive the translational engine in biomedical science and medicine.

The ultimate goal of these coordinated high-throughput mutagenesis and phenotyping programs is to develop a comprehensive dataset of the outcomes of molecular interventions *in vivo*—from genetic lesions to small molecules. A key element of this high-throughput approach will be fostering cross talk between investigators in a wide variety of disciplines, from clinician-scientists to experimental and clinical genetics to imaging scientists to physiologists to pathologists to computational biologists. The International Mouse Phenotyping Consortium (IMPC) has been established to address this challenge. It brings together the best large-scale, centralized GEM production and phenotyping platforms worldwide with a vision to coordinate the systematic phenotyping of 20,000 mutant mouse lines over the next 10 years.

The growth of high-throughput phenogenomic laboratories has produced numerous examples of genes that cause lethality in the prenatal, perinatal, or juvenile periods of development. Accordingly, GEM have yielded remarkable insights into the gene products that control morphological development in the vertebrate embryo, so inclusion of mouse models with developmental-lethal phenotypes is an additional consideration when seeking to understand the importance of any given gene in the context of translational medicine. The methods used in phenotyping these lethal phenotypes in a high-throughput setting are essentially no different from those used in laboratories engaged in hypothesis-driven research, but for developmental investigations they tend to emphasize methods that yield quality standardized data with the highest efficiency at a more rapid rate. The remainder of this chapter reviews basic principles for the pathology analysis of abnormal mouse conceptuses, neonates, and juveniles in the high-throughput setting.

High-Throughput GEM Phenotyping Projects: Primary Considerations

If we are to fully harness the mouse to the translational medicine engine, we need to ensure that we effectively integrate the relationship between gene and phenotype. Pathology has a critical role to play in this endeavor. This discovery-based discipline, if appropriately applied to the high-throughput process, has the unique opportunity to validate and integrate all of the various genomic and phenotype datasets by linking the effects of altered genes and mutant molecules to abnormalities in function and structure. Indeed, the identification of altered structure(s) has long been a standard for demonstrating that a given genetic mutation is responsible for a certain pattern of anatomic changes. Accordingly, well-conceived experiments that correctly include the expertise of a comparative pathologist who is trained to understand and interpret deviations in structure will be critical to the success of phenogenomic efforts.

Why High-Throughput Assessments of GEM Are Necessary

The mouse is unparalleled in the insight it can offer us into the genes and pathways involved in disease. In particular, GEM offer a distinctive combination of attributes that are unmatched by any other model organism: a highly sophisticated genetic toolbox to introduce mutational changes, a wide range of standard inbred strains that underpin genetic experiments (see Chapter 2), a legacy of investigations into mammalian development and physiology (summarized in Chapter 4), low cost due to small size and ability to thrive in group housing, and robust reproductive capacity that makes large-scale genetic and phenotypic screening programs viable and affordable.

Over 70% of the 20,000 protein-coding genes in the mouse genome have been examined experimentally but do not have a corresponding GEM model, let alone an assigned phenotype. Furthermore, phenotypes in GEM can rarely be accurately predicted, even when a phenotype–genotype relationship has been previously described in humans or other animal species. It is, therefore, clear that the massive scale of the phenotyping effort required to complete a comprehensive functional annotation of the mouse genome is beyond the capacity of any individual laboratory, institution, or constituency and that a multi-site, high-throughput approach is required to achieve this goal in the next 10 years. The need for robust, efficient, and systematic phenotyping (including pathology analysis) of mouse mutants requires a degree of centralization in order to realize economies of scale, but most importantly to amass the critical breadth and depth of expertise in mouse biology available within a high-throughput platform.

Approaches to High-Throughput GEM Phenotyping

Methods of Mutagenesis for High-Throughput GEM Phenotyping Projects

The genetic tractability of the mouse genome has been exploited in large-scale, forward genetic (phenotype-driven) screening approaches with a variety of mutagens, including the commonly used chemical point mutagen *N*-ethyl-*N*-nitrosurea (ENU).[10,26] Such high-throughput, genome-wide ENU screens have been particularly effective when conducted using relatively focused phenotypic assays to identify mutants, but they have the disadvantage of inducing random and relatively uncontrolled genetic lesions that require significant effort downstream to identify and confirm the mutation responsible for the phenotype. Although many heritable mutants have been generated via these high-throughput ENU programs, the overall yield to date of identified mutant genes is modest.

In contrast, large-scale reverse genetic (genotype-driven) screens that use targeted and trapped genetic manipulation of mouse embryonic stem (mES) cells as their starting point provide efficiency and precision of mutation at the single nucleotide level.[29] Over the last 15 years, targeted and gene-trap mutations in more than 5,700 mouse genes have been described. However, the knockout phenotypes of the majority of mouse genes have yet to be assessed *in vivo*.

Phenotyping Postnatal Mouse Mutants in High-Throughput GEM Phenotyping Projects: The EMPReSS Pipeline

Conceptually, the breadth of assays and screens for phenotyping of GEM is endless. Phenotypic analysis of postnatal mice can be undertaken at many levels: from the molecular level (transcriptomic, proteomic, and metabolomic) to the ultrastructural and cellular levels, and ultimately at the physiological and tissue/pathology level. However, to date intuition and preliminary evidence confirms that in a high-throughput approach new phenotypes will be observed more frequently if the number of parameters included in the screen is increased. In high-throughput, discovery-based phenotyping programs where the goal is to identify phenotypes across many areas of biological interest, a balance is necessary between screening small cohorts of animals in a more comprehensive and systematic approach versus screening more animals with fewer tests to ultimately deliver the most informative dataset available for the least possible cost and time.

Enormous progress has been made in developing tools at all levels for mouse phenotyping. The value derived from high-throughput, hypothesis-generating mouse phenotyping studies depends on the number of analytical methods brought to bear on lines of GEM and will rely on in-depth follow-up studies by scientists with domain-specific expertise (e.g., for particular molecular pathways, organs or systems, and/or disease entities) engaged in hypothesis-driven research. The focus within the mouse genetics community toward high-throughput primary phenotyping of postnatal (juvenile or adult) mice has been on the development of phenotyping methodologies for functional, pathophysiological, and biochemical screens studying diverse features including (among others) dysmorphology, neurological/behavioral, metabolic, and cardiovascular phenotypes. In parallel, considerable focus and investment has been dedicated to the standardization of phenotyping platforms, such as the EMPReSS (*E*uropean *M*ouse *P*henotyping *Re*source of *S*tandardized *S*creens) program.[5] Harmonization of phenotyping procedures will be critical if the many organizations contributing to the process of high-throughput, comprehensive phenogenomic evaluation is to populate databases with comparable datasets. The utilization of increasingly robust, validated phenotyping platforms generating data that are reproducible across time and place is integral to the high-throughput phenotyping effort to generate standardized, scientifically relevant, and accessible datasets.[8]

The scope of pathology screening in high-throughput phenotyping is still evolving.[23,32] It is clear that pathology phenotyping has a pivotal and twofold role to play. In the first instance, morphologic analysis contributes actively to the *hit rate* of phenotypes by discovering gross and microscopic changes not identified during *in vivo* (i.e., functional) screening. Second, anatomic pathology evaluation characterizes the structural changes associated with the functional phenotypes identified during *in vivo* screening and facilitates the attribution of biological relevance to these changes. Such structural endpoints will require detailed and systematic pathology evaluation of a representative cohort of mice from each genotype. Indeed, perhaps the brightest future for mouse pathology in an era of functional annotation of the genome is the strong link between diagnostic work and the unraveling of pathogeneses and pathology phenotypes in GEM.

The EMPReSSslim phenotyping platform was recently developed and validated by the *E*uropean *U*nion *M*ouse *R*esearch for *P*ublic *H*ealth and *I*ndustrial *A*pplications (EUMORPHIA) project at four centralized mouse clinics with exceptional track records in mouse functional genomics (Wellcome Trust Sanger Centre and MRC Harwell in the United Kingdom, ICS Strasbourg in France, and GMC HelmholtzZentrum Munich in Germany).[2] EMPReSSslim (Figure 17.1) utilizes 20 phenotyping platforms to capture 406 phenotypic parameters and 155 metadata parameters to provide a comprehensive assessment across multiple organ systems and disease areas by combining a battery of assays performed on age-matched cohorts of male and female mice starting at 9 weeks of age. The platform is designed to maximize the likelihood of detecting anomalies using the most cost-effective, efficient, and robust screens assembled in a pipeline approach. EMPReSSslim has two pipelines: one tests morphology, metabolism, cardiovascular, and bone phenotypes (Pipeline 1), while the other evaluates blood, clinical chemistry, and allergy/immune phenotypes (Pipeline 2).

Phenotyping Embryonic Lethal Mouse Mutants in High-Throughput GEM Phenotyping Projects: The CMHD EMPReSSslim Pipeline

While much of the planning and implementation of large-scale phenotyping programs has focused on juvenile to adult animals, a parallel application of an equally broad and complimentary analysis at earlier developmental stages is also required to reveal diseases of both development and age-related conditions. The challenge for developmental biologists in a high-throughput phenotyping program is to systematically analyze embryonic phenotypes with altered morphology and, where possible, determine the quantitative contribution of each gene toward patterning of the embryo. Tools for this discovery and analysis must include rapid and accessible high-throughput methods like imaging (invasive or noninvasive) as well as detailed, stage-specific, *ex vivo* pathology analysis.

Initial estimates from high-throughput phenotyping programs are that approximately 30% of mutant mouse lines harbor embryonic lethal mutations. The extent of early developmental pathology provides a rich source of phenotypes and, despite certain technical challenges, provides a robust means for teasing out gene function. Nearly all areas of biology are approachable by careful comparison of normal and abnormal developing mouse embryos. While the analysis of developmental defects and their impact upon adult phenotypes has enormous potential for phenotype (and pharmaceutical) discovery, this area has not been seen as amenable to high-throughput investigation until now. One platform for such investigations is a modification of the EMPReSSslim platform developed by the *C*entre for *M*odeling *H*uman *D*isease (CMHD) in Toronto (Figure 17.2). This expansion has added a third pipeline that uses an ultrasound system (Vevo660; VisualSonics, Toronto, Ontario, Canada) to rapidly screen pregnant mice and identify the time of death and general morphology of conceptuses bearing embryonic lethal mutations.[21] Alternative image-based assays for early morphological phenotyping include iodine-stained microcomputed tomography (μCT) and optical projection tomography (OPT) with autofluorescence rendering.[9] More detailed structural analysis involves removal and dissection of embryos at appropriate stages, analysis of gene reporter expression by staining whole-mount preparations and/or slide-mounted tissue sections to localize markers, and processing for further histopathological analysis.

Pathology Phenotyping in High-Throughput GEM Phenotyping Projects

Screening Methods to Detect Postnatal (Juvenile and Adult) Phenotypes

The classic approach to diagnostic pathology for spontaneous mouse disease is dependent on well-established basic principles for pattern recognition and subsequent rote assignment of morphological and ideally etiologic diagnoses. This approach requires significant adaptation when it is applied to the discovery, description, and interpretation of an induced phenotype using GEM. Novel lesions in GEM

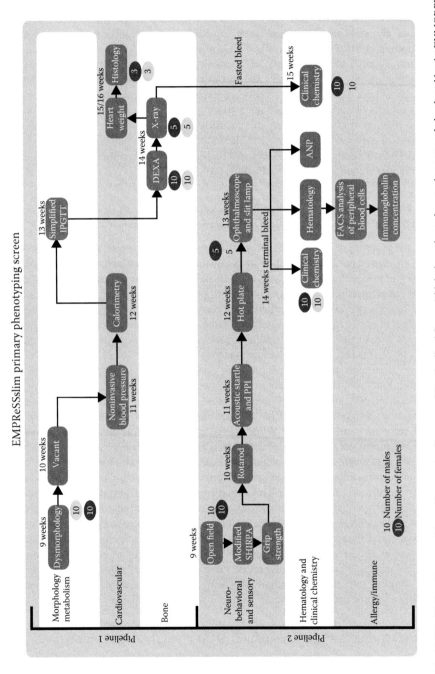

FIGURE 17.1 The EMPReSSslim (*European Mouse Phenotyping Resource of Standardized Screens*) high-throughput phenotyping protocol, developed by the EUMORPHIA (*European Union Mouse Research for Public Health and Industrial Applications*) consortium for routine morphological and functional evaluation of postnatal (juvenile to adult) mice. *Abbreviations:* ANP, atrial natriuretic peptide (levels); DEXA, dual-energy X-ray absorptiometry; FACS, fluorescence-activated cell sorting; IPGTT, intraperitoneal glucose tolerance test; PPI; prepulse inhibition; SHIRPA; SmithKline Beecham, *Harwell, Imperial College, Royal London Hospital phenotype assessment.*

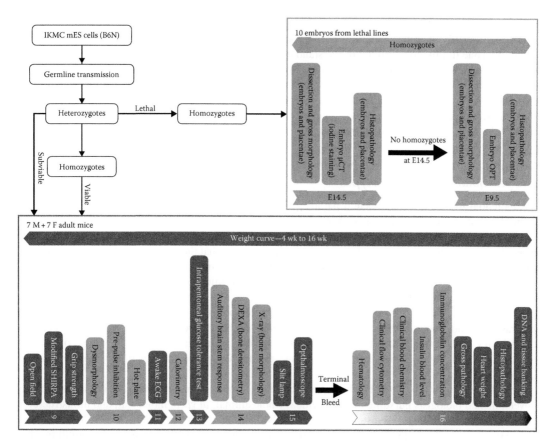

FIGURE 17.2   Schematic representation of a single-pipeline phenotyping platform. Mutant mice are subjected to a sequential series of tests that assay several clinically relevant physiological and behavioral systems. Embryos from sub-viable lines are harvested and imaged to evaluate phenotype. Adult heterozygotes from subviable lines are phenotyped to assess the effects of gene dosage. *Abbreviations:* DEXA, dual-energy X-ray absorptiometry; ECG, electrocardiogram; IMKC, International Knockout Mouse Consortium; μCT, micro-computed tomography; mES, mouse embryonic stem (cells); OPT, optical projection tomography.

typically are distributed in unusual patterns or exhibit distinctive features that may or may not mirror common changes in mice with spontaneous disease, let alone mimic the changes that occur in human pathologies. Accordingly, high-throughput pathology phenotyping of GEM necessitates an even greater appreciation for the fundamental molecular and cellular mechanisms responsible for normal development and disease. In the context of multidisciplinary phenotyping efforts to generate, describe, and utilize GEM efficiently, it is critical that the complete dataset for each mutant mouse line includes a detailed pathology analysis. Our Pathology Core laboratory functions within a multi-institutional, multidisciplinary academic biomedical research environment as the conduit through which histology, gross pathology, histopathology, and molecular pathology services are provided to a large collaborative research community supporting the EMPReSSslim high-throughput phenotyping pipelines. Our Core runs as a service-oriented, cost-recovery laboratory with opportunity for collaborations between individual scientists and the three on-site comparative (veterinary) pathologists.

A recognized deficiency of the EMPReSSslim platform (Figure 17.1) is that pathology evaluation requiring specific expertise is not widely available to the mouse functional genomics communities within typical large-scale phenotyping centers. There are international efforts underway to address this design deficiency. For example, a modification of the EMPReSSslim platform developed at the CMHD in Toronto has expanded EMPReSSslim to include a pipeline designed to better characterize the structural changes underlying the identified physiological and/or behavioral phenotypes identified in Pipelines 1 and 2 of the standard EMPReSSslim approach (Figure 17.2); an advantage of the CMHD modification is that it more readily identifies abnormal phenotypes that have not been detected by the in-life screening effort available in Pipelines 1 and 2 of the standard scheme. The modified workflow requires that mice undergo complete gross pathological examination involving macroimaging, extensive tissue collection (i.e., a routine list of 40 organs) followed by histopathological evaluation, digital photomicrographs of structural lesions, recommendations from a comparative pathologist for additional diagnostic tests that may further define the biology and pathogenesis of observed structural lesions, and correlations to comparable disease conditions in other species.

Screening Methods to Detect Prenatal (Embryonic and Neonatal) Lethal Phenotypes

The potential to detect developmental pathology in mouse mutants will continue to be dependent on keen observation supported by a systematic but disciplined analytical approach and expert knowledge of strain-dependent anatomical differences and incidental lesions. Miniaturization and high-throughput adaptation of conventional pathology methods must continue so that they can routinely provide anatomic pathology-based (i.e., structural) data concurrent with the data streams arising from other phenotyping disciplines (e.g., clinical pathology, genomic and proteomic microarray). However, the work flow in this high-throughput research environment is not identical to that in a standard diagnostic pathology laboratory; an estimated 77% of accessions that are presented for processing by the Histology Core laboratory do not subsequently arrive on the pathologist's bench, but rather are returned directly to the primary investigator. The usual requests in this histology-only category are for simple tissue collection and embedding, tissue sectioning, and special staining procedures such as histochemistry or immunohistochemistry.

Nevertheless, the availability of technical consultations from a comparative pathologist is invaluable to developmental biologists pursuing their individual whole animal, *in vitro*, molecular, and ultrastructural studies. Of the remaining 23% of cases that enter the Pathology Core laboratory for both histological processing and histopathology interpretation, an estimated 20% (33 of 191 over a 4-year period) involve embryonic or neonatal lethal phenotypes. When exploring the function of genes that cause congenital diseases and/or embryonic or neonatal lethality, it is necessary to evaluate the phenotype during the critical developmental period in which the lesion is induced (Table 17.1). Clearly, the major challenge when phenotyping mouse embryos is their small size. Mouse embryos weigh ≤1 g throughout most of gestation.

TABLE 17.1

Timing and Major Causes of Death in Mouse Embryos, Fetuses, and Neonates

Timing	Defect
At or prior to implantation (before E4.5)	Cell loss in the Inner cell mass (ICM) Defective trophoblast
Gastrulation (E7.5)	Fundamental body plan defects
Early organogenesis (E8.5–E9.5)	Abnormal establishment of the circulatory system (heart and blood vessels) and/or blood production
Placental transition (E10.5)	Formation of the chorioallantoic placenta
Late organogenesis (E11.5–E15.0) Fetal growth spurt (E15.0 to birth)	Defects in the circulatory (heart, blood vessels, placenta) and/or hematopoietic systems
Neonate	Disrupted transition to full organ function, especially in control centers (e.g., brain) or major viscera (e.g., digestive tract, kidney, liver, lung)

Source: Adapted in part from Copp, A.J., *Trends Genet.*, 11(3), 87, 1995.

Note: Developmental times are defined in terms of embryonic (E) days, where E0.5 is designated as noon of the day on which a copulation plug is observed in the vagina.

Our standardized approach for high-throughput phenotyping of developing mice ideally uses 10–20 homozygous embryos with a comparable number of appropriate controls (typically stage-matched wild-type [WT] littermates) at embryonic (gestational) day 12 (E12, where noon on the day a copulation plug is found in the vagina is designated as E0.5). This time point was chosen because the cardiovascular system is formed and already functioning by E9, and most major organ systems have been initiated and are available for analysis by E12.[3] All embryos are first given a gross evaluation using a stereomicroscope by an experienced technician to detect external gross defects. Subsequently, embryos are processed for whole-mount (gross) and section analysis of reporter gene distribution by a technician or diverted to conventional production of slide-mounted (microscopic) tissue sections stained with hematoxylin & eosin (H&E) for evaluation by a comparative pathologist. Mouse lines with no E12 homozygous mutants are rebred to produce E9 embryos, which then undergo the same evaluation. Placentas are processed in parallel with their embryos, typically by placing them within the same block. Embryos prior to E9 are less suited to high-throughput evaluation using histopathology of H&E-stained sections due to the lack of delineated tissue architecture. Instead, these early embryos are better used for whole-mount evaluation and/or imaging.

Considerable progress has been made in adapting existing and devising new technologies to provide increasingly detailed phenotype information from embryonic and neonatal mice. The major recent advances in this regard have come from improvements in noninvasive imaging (see Chapter 14). Methods for imaging mouse embryos and neonates include magnetic resonance imaging (MRI) and ultrasound for *in vivo* imaging, and MRI, vascular corrosion casts, μCT, and OPT for *postmortem* imaging.[9,21] Certain applications, such as Doppler echocardiography, can provide functional information in the context of anatomic structures.[9,18]

Each of these imaging modalities offers particular advantages when used in mouse developmental pathology studies. For example, OPT provides vibrant three-dimensional (3D) visualization of the entire vascular system during early organogenesis.[37] Similarly, ultrasound biomicroscopy (UBM) using high-frequency probes (22–50 MHz) yields two-dimensional (2D) images at resolutions up to 50 μm that can be processed into 3D images, thereby providing tissue volume data to reveal internal organ features.[36,40] For soft tissue and bone features, μCT has been used to generate 3D images from intact organs or body structures that effectively complement the impressions gained from gross pathology comparison of mutant and WT mice.[33,38] These imaging modalities may be used in advance, in tandem, or after gross pathology evaluation has identified the anatomic abnormality.

At present, imaging is often used more as a secondary screen for project-specific characterization of embryos rather than as a primary endpoint integrated into a high-throughput pipeline approach. This approach acknowledges the disadvantages of imaging techniques such as lengthy scanning times and equipment cost as well as the better microscopic resolution available by histopathologic evaluation (generally about 0.2–0.5 μm) (see Table 14.1). However, more powerful current-generation MRI and μCT scanners provide an ideal method of high-throughput embryo analysis at a subgross (i.e., organ-sized) level; this capability is especially useful in mutants with gross anatomical abnormalities (e.g., hydrocephalus, *situs inversus*) and some subgross malformations (e.g., cerebellar atrophy). Accordingly, it has been suggested that in the context of the IMPC project's plan to systematically phenotype 20,000 mouse lines in the next 10 years that the initial phenotyping pipeline should include recorded 3D images of unstained E9.5 embryos using either MRI, OPT, or contrast-enhanced μCT as a starting point for high-throughput embryonic lethal screening. This time point marks the beginning of cardiac function and its connection with the expanding vascular network, so imaging may help to detect malformations that might be difficult to discern with 2D histological sections. In this way, the screening of mutant mouse lines would be completed earlier in the pipeline, and subsequent histopathologic assessment could be deployed more selectively to characterize specific abnormalities in substructures or cell populations from those lines with *bona fide* phenotypes. This approach would counter the main disadvantage of standard mouse developmental pathology analyses: the need for step (interval) sectioning or, ideally, serial sectioning of the entire fixed and embedded embryo, which is an inherently low-throughput undertaking. A difficulty of only slightly lesser magnitude is the problem of 3D reconstruction of embryonic structures using 2D sections. Current image analysis software can readily provide 3D volume data using algorithms to stack the captured images of a given feature. Maintaining registration of the images acquired by photographing tissue sections is often problematic as ribbons of serial histological sections are delicate and need to be mounted on the glass slide relatively quickly, even though they cannot be individually manipulated into position without taking some time to separate the ribbon into segments. This dilemma, while applicable to any tissue, is increased with mouse embryo sections due to their irregular contours, small size, and fragility.

Routine Practices for High-Throughput Collection, Processing, and Evaluation of Mouse Conceptuses

Developmental biologists have their areas of organ expertise, and their individual laboratories are usually well equipped for collection, dissection, staining, and macroimaging of embryos, fetuses, and neonates. Because embryonic/neonatal samples represent 20% of the accessions to our pathology laboratory, standardized protocols are in place to accommodate this high volume. A particular issue for developmental pathology projects is that many animals can be presented in one batch. For typical projects, several breeding pairs are set up simultaneously, so multiple mutants and their controls may arrive within hours or a few days of each other. This high throughput can quickly overwhelm processing facilities and the pathologists' scheduled workloads in the absence of routine procedures capable of handling the onslaught.

Project Planning

The starting point for determining the cause(s) of embryonic and perinatal death often is influenced by the case history and amount of clinical data (acquired from previous litters) that has been gathered by the primary investigator. The pathologist often enters the picture after several litters have been delivered and long after a handful of animals having the desired genotype have been noted to have one of the following three defects: (1) found dead; (2) delivered alive but died quickly (i.e., survived no more than 1–2 days); or (3) found or suspected to have been eaten by the dam. Confounding this presentation is the usual timing of rodent deliveries. With dams often giving birth in the quieter hours of the night, animal husbandry and research staff are generally not present and thus often will miss critical perinatal events, leaving the pathologist with little information from which to refine an experimental design.

Numerous developmental anomalies are best analyzed in time-staged embryos. Such studies of embryonic lethal phenotypes require harvesting of embryos at discreet windows of development—often to a given developmental day or even a specific portion of that day. For these studies, it is important for experienced personnel to work from a standardized protocol for embryonic harvesting, fixation, and processing; the need for demonstrated expertise is that the procedures needed to properly isolate and examine near-term fetuses may be substantially different from those that best reveal the features of preimplantation and early postimplantation embryos. Experience in anesthesia of pregnant dams; dissection and handling of uterine, embryonic, and extraembryonic tissues; fixation; and macroimaging are all key aspects to making the most of available samples.

Determining the Developmental Age of Embryos, Fetuses, and Neonates

In our laboratory, females are considered to be at E0.5 at noon on the day a copulation plug is found in the vagina. However, the presence of a plug means that mating has occurred, not that the mouse is pregnant. The first indication of pregnancy is often an increase in maternal body weight, although the degree of increase varies with mouse strain. For example, by monitoring maternal weight, the presence of successful pregnancy is detectable in ICR mice by E4.5, whereas in C57BL/6 mice reliable confirmation of pregnancy takes until E8. Uterine implantation sites can usually be detected by E10.5 simply by palpating the maternal abdomen. In our laboratory, we assign newborn mice a postnatal age of P0 on the day of birth.

Euthanasia of Conceptuses

Techniques for terminating embryos and fetuses will vary depending on the developmental stage. Embryos up to E14 likely do not possess well-developed nociception pathways, so euthanasia by detachment from the placenta is sufficient and accepted.[27] Past E15, fetuses should be euthanized using any of several accepted methods.[19] In our facility's protocol, deep surgical anesthesia of the dam via intraperitoneal (IP) injection of 2,2,2,-tribromoethanol (Avertin®) is used.

Genotyping

In the high-throughput setting, conceptuses are typically genotyped using a portion of the embryo-derived placenta. The yolk sac can easily be removed starting at E9.5. We consider it to be the tissue of choice for genotyping mice at E9.5 and older.[25]

Specimen Collection and Processing

The protocol for specimen collection will depend on the nature of the sample to be evaluated. Many of our projects focus on delicate or small organs in the embryo or neonate (i.e., heart, lymph nodes). Sometimes dissecting microscopes do not provide sufficient magnification for observing and isolating the structures of interest, and it is best to process and embed the tissue and view the structures in sections. Whole body perfusion enhances preservation of deep tissues (e.g., nervous system) and often provides optimal morphological preservation. The timing of collection is also critical in order to avoid autolysis and maternal maceration of neonates and to harvest timed-pregnant embryos at the appropriate stage of development.

Adequate preservation of the specimens is the most critical step in mouse developmental pathology studies. Trauma during the dissection is a major reason that embryonic tissues are inadequate for tissue analysis. For example, tearing of umbilical vessels causes abdominal organ destruction, pressure applied to the calvaria (skull cap) often damages cortical neurons in the forebrain, and large amounts of *postmortem* hemorrhage occur at sites where tissues are cut or torn. These changes are all common artifactual findings in histological sections collected from improperly handled embryos. Expert care in collection of the gravid uterus with embryos in various stages of gestation is key.

Application of standard necropsy techniques in embryos, fetuses, and neonates is simplified from that performed on adult mice. Developmental pathology studies typically consist of an external examination and description, external macroimaging, acquisition of selected measurements (e.g., crown-rump length, embryonic and placental weights), and collection of a tissue sample for genotyping. These tasks must all be performed with careful note-taking; in the high-throughput setting, this is often conducted using semiautomated data entry. Late-stage embryos, fetuses, and neonates are typically incised longitudinally along the ventral surface of the thorax and abdomen, thereby allowing fixative to reach the internal tissues. Caution is needed here as analysis of deformities in the thoracic or abdominal organs may require less manipulation of the ventral skin and subsurface organs or structures. For example, the diaphragm may be torn when making this incision, which can artifactually pull or displace lobes of the lung or liver. In these cases, it is recommended to remove the head and allow fixative to enter via the exposed neck or to use a needle for infusion of fixative into the abdominal cavity—taking care to avoid puncturing organs (Figure 17.3). Only on very late-stage fetuses (E17–E20) and neonates can a cursory internal gross examination be performed at necropsy.

Collecting mouse conceptuses for optimal preservation is similar to collection of any other tissue. The single most common finding observed by the pathologist is inadequate preservation of internal organs. This deficiency is usually due to delayed discovery of pregnant dams or neonates found dead in the cage or alternatively because intact large fetuses were preserved by simple immersion in fixative without first opening the major body cavities, thus leading to inadequate penetration of fixative near the viscera. For embryonic time points up to E13, simple immersion of the uterus for fixation is often adequate. For specimens at ages E15 and older, the best approach is to detach the fetuses and implantation site from the uterus and sever their heads, allowing fixative to infiltrate each animal's skull and the torso for optimal fixation. Decalcifying agents are not generally needed to section the murine head until after P0.

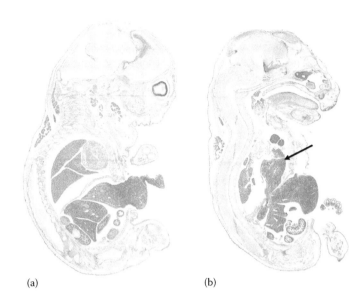

(a)   (b)

FIGURE 17.3   Representative sagittal sections of (a) a control (wild-type) mouse fetus and (b) a mutant littermate with herniation of the liver (arrow) through the diaphragm at gestational day 16.5 (E16.5).

Specimen Collection during High-Throughput Mouse Developmental Pathology Studies

This section gives the step-by-step protocol used for harvesting and evaluating embryos, fetuses, and neonates in our high-throughput Pathology Core facility. Each step includes a manipulation to perform, while many include details of our experience with the procedure.

1. Anesthetize the pregnant dam via IP injection of Avertin® at a dose of 250 mg/kg (0.5 mL) for a 25–30 g mouse. Anesthetic and analgesic effects are evident in 1–2 min. A second Avertin® dose of 250 mg/kg is then administered to ensure that complete (stage IV) anesthesia of the embryos is induced; it will lead to cardiopulmonary arrest in the dam within 2–4 min.

2. Before respiratory collapse is noted in the dam, use cervical dislocation to sever the spinal cord and induce death.

3. Arrange the dam in dorsal recumbency. Access to the uterus is eased by pinning the feet and rostrum to avoid movement of the dam's body at later steps of the necropsy.

4. Wet the ventral fur with 70% ethanol.

5. Open the abdominal wall by making a longitudinal incision along the ventral abdominal mid-line from sternum to pelvis.

6. Reflect the ventral abdominal wall to enter the abdominal cavity.

7. Fully expose both uterine horns and detach them from surrounding tissue, but leave them *in situ* at this stage.

8. Using iris scissors, carefully incise the uterine body and extend cuts bilaterally over the ventral aspect of each uterine horn, taking great care not to damage the underlying embryos (or fetuses) and their associated extraembryonic structures. This process can be aided by *tenting* the uterine wall with iris forceps.

9. Record the position of each conceptus in both horns. Start numbering the conceptuses in the right horn, moving from distal to proximal (1, 2, 3, 4, …). Then move on to number those in the left horn, again moving from proximal to distal and picking up the count with the next available number (5, 6, 7, 8, …).

10. Cut through the bones of the ventral pelvis and the overlying skin to expose the vagina.

11. Remove the uterine horns, uterine body, and vagina *in toto.*

12. For embryos up to E14, carefully remove a portion of the yolk sac from each animal using iris scissors and place it in a prelabeled microcentrifuge tube for potential genotyping.

13. Place the uterus with the attached intact embryos in a large (100 or 135 mm) Petri dish containing cold (4°C) 1x phosphate-buffered saline (PBS), pH 7.6, on ice to prevent drying and delay autolysis. Macroimage the complete uterus with the conceptuses *in situ*, and then image individual conceptuses with visible abnormalities.

14. Measure and record the crown-rump length and assess the gross appearance (e.g., structural or color abnormalities) of all conceptuses. Note the appearance and size of the placentas and their implantation sites (detachment, hemorrhage, etc.). Record any abnormalities.

15. For fetuses (E15 or older), remove the entire tail and place it in a prelabeled microcentrifuge tube for potential genotyping.

Fixation Methods for High-Throughput Mouse Developmental Pathology Studies

Immersion Fixation

Embryos and fetuses (E8–E17) are fixed prior to dissection of any organs to avoid artifactual traumatic damage to the delicate organs and organ substructures. Our preferred method for high-throughput mouse embryo phenotyping is routine immersion fixation with neutral buffered 10% formalin (NBF) using a 1:10 ratio of tissue to fixative. This method is efficient, safe, and provides adequate preservation of tissue architecture. Other fixatives that work well for high-throughput developmental pathology studies are 4% paraformaldehyde (PFA) in PBS (which is often better than NBF for slide-based assays like immunohistochemistry or *in situ* hybridization) and the combination of 2.5% glutaraldehyde and 2.0% PFA in 0.2 M PBS (for electron microscopy). We recommend removing the decidua and fixing this tissue similarly.

The advantages of immersion fixation in NBF are the cost, availability, and relative safety of the chemical versus other fixatives that are more costly (Bouin's) and more hazardous (zinc formalin). Furthermore, many antibodies are suitable for use on formalin-fixed, paraffin-embedded (FFPE) specimens. Embryos at E9 to E12 usually do not require subsequent treatment with decalcificying agents prior to embedding for histology as bone is not sufficiently mineralized until approximately E14.5–E15.0. The main disadvantage of NBF is the potential loss of fine structural detail in neuronal and reproductive tissues. If these organs are of high interest, perfusion of the embryo or tissue may be recommended on a second cohort of embryos.

Our immersion fixation procedures (Protocols 17.1 and 17.2) progress using the following routine steps. After the uterus has been removed from the dam and the embryos have been measured, examined, and imaged, two options are available.

PROTOCOL 17.1   IMMERSION *IN SITU*

1. Fix the entire uterus, with the conceptuses maintaining their positions *in situ*, by immersion in a fixative that is appropriate for the desired analytical endpoint (e.g., routine histology, immunohistochemistry, or ultrastructural examination).

2. Following 24 h of fixation, remove the uterus from the fixative and wash several times in 1x PBS (at room temperature). Carefully make a circumscribing cut through the entire uterine wall around the implantation site of each conceptus, and then remove it.

3. Weigh the embryo or fetus with its placenta and the associated implantation site in an intact and undisturbed state.

PROTOCOL 17.2   IMMERSION OF ISOLATED CONCEPTUSES

1. Dissect one conceptus and its placenta from the uterus and lay the specimen on a warmed tray to maintain body heat, and then move the tray to the stage of a dissecting microscope.
2. Remove the myometrium (muscle layer of the uterine wall).
3. Open the yolk sac.
4. Separate the embryo or fetus from its placenta.
5. Move the conceptus into fresh fixative. In order to get optimal fixation, substantially increase the ratio of tissue to fixative (1:40).

Perfusion Fixation

If time permits, older embryos (E12–E14) and fetuses (E15 to term) may be preserved even more effectively by introducing the fixative through the vascular system (Protocols 17.3 and 17.4). Perfusion fixation in developmental pathology studies is carried out using the blood vessels of the conceptus to rapidly and evenly deliver fixative to all tissues, especially for dense (e.g., brain) or highly vascularized (e.g., heart, kidneys, lungs) organs. In our laboratory, we employ perfusion techniques described in this protocol for follow-up or detailed studies requiring rapid tissue fixation. This method is the optimal means of providing good morphological detail; preserving cell organelles and sensitive epitopes for electron microscopy and immunohistochemistry, respectively; and minimizing artifacts.

Our perfusion fixation methods employ the following routine steps (Protocols 17.3 and 17.4). After the uterus has been removed from the dam and the embryos have been measured, examined, and imaged, two options are available for direct perfusion fixation of the conceptus.

High-Throughput Histology Processing for Mouse Developmental Pathology Studies

Tissue Processing

The processing of E9–E12 embryos is essentially similar to that of processing other mouse tissues which uses standard overnight protocols. Our experience has been to separate the head from the torso and place them in the same cassette. If the embryo is embedded whole or is longer than 4 mm, a 2-day processing time is necessary.

Embedding and Microtomy

Due to the unique orientation of the embryo's torso and limbs, many organs are cut in a different plane of section compared to the same organ in the adult. This will influence the orientation chosen for the embryo as it is positioned in the embedding mold. In addition, in many instances dozens of serial sections may be needed to cut into the desired area due to the minute size of organs; the histotechnologist determines that the proper plane for sampling a given organ has been reached by viewing the unstained structures on the block face or in the cut sections floating on the water bath. This process may be very time-consuming, and it will exhaust valuable specimens very quickly in laboratories that are not experienced in the microscopic anatomy and histological sectioning of mouse conceptuses.

Sections may be taken in three planes. The preferred orientation will depend on the nature of the experimental question.

1. *Sagittal sections.* Sectioning the animal from either the left or right side in the longitudinal plane *(laying the embryo on its side* in the embedding mold) allows for good visualization of many neural structures, including the diencephalon, mesencephalon, myelencephalon, ventricular system, inner ear, and spinal cord; regions associated with the central nervous system (CNS), such as the skull and intervertebral discs; as well as a number of viscera, including the gastrointestinal tract and liver. This view is very useful as a first screen of the CNS to

PROTOCOL 17.3   INTRACARDIAC PERFUSION

This technique is typically used for late embryos and fetuses (E12 or older) and also neonates. The main advantage of this protocol is that it is relatively fast. The disadvantage is that the surgical procedure will cause some structural damage to the heart.

1. Dissect one conceptus and its placenta from the uterus and lay the specimen on a warmed tray to maintain body heat, and then move the tray to the stage of a dissecting microscope.
2. Remove myometrium.
3. Open the yolk sac.
4. Separate the conceptus, leaving the placenta and umbilical cord attached but placed on the side.
5. Drop warmed PBS (45°C) on the conceptus to help it recover from cold shock and to cause the heart to start beating more vigorously.
6. Use a ventral longitudinal incision to remove the sternum and open the thoracic cavity to expose the heart.
7. Use a ventral longitudinal incision to open the abdomen. Remove the gastrointestinal tract to expose the abdominal artery and vein (Figure 17.4).
8. Insert a catheter (24 gauge) into the left ventricle and remove the inner needle.
9. Cut the abdominal artery and vein to provide an exit for flowing blood.
10. Immediately start flushing the circulatory system with PBS (100 µL/min, 2 mL total volume).
11. Switch to perfusion with fixative solution as soon as the blood flowing out starts to clear (100 µL/min, 10 mL total volume).
12. Monitor the conceptus until it is completely perfused (indicated by blanching of the body as a whole, although the area near the liver usually retains a light pink color).
13. Immerse the perfused specimen in fresh fixative for 24 h in preparation for H&E staining. However, if immunohistochemistry is to be performed, overnight fixation is adequate.

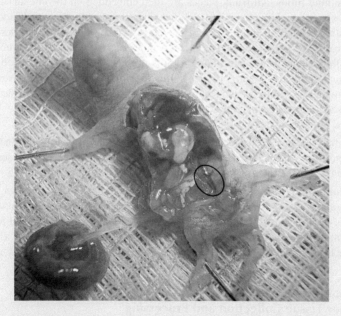

FIGURE 17.4   Isolated abdominal artery and vein (inside oval) in a mouse embryo prepared for intracardiac perfusion.

PROTOCOL 17.4   UMBILICAL ARTERY PERFUSION

This technique is typically used for younger embryos (E12 or earlier). The advantage of this approach includes no damage to organs of the embryo. The disadvantage is that this protocol requires a high degree of microsurgical skill.

1. Open the yolk sac to expose the umbilical cord.
2. Drop warmed PBS (45°C) on the embryo to let it recover from cold.
3. Insert a glass pipette needle (30–50 μm in diameter, pointing toward the embryo's body) into the umbilical artery as soon as the heart starts beating.
4. Cut the umbilical veins to provide an exit for the flowing blood.
5. Immediately start flushing the circulatory system with PBS (50–100 μL/min, 2 mL total volume).
6. Switch to perfusion with fixative solution as soon as the blood flowing out starts to clear (100 μL/min, 2–5 mL total volume).
7. Monitor the embryo until it is completely perfused (indicated by blanching of the body as a whole, although the area near the liver usually retains a light pink color).
8. Immerse the perfused specimen in fresh fixative for 24 h in preparation for H&E staining. However, if immunohistochemistry is to be performed, overnight fixation is adequate.

determine the general areas that may be abnormal, but it is not suited to evaluating symmetry in the brain. If the specimen is less than 3–4 mm in length, it can be embedded whole to produce sagittal sections. However, if longer than 4 mm, the animal should be bisected down the middle (after it has been fixed), after which both halves should be embedded.

2. *Transverse sections.* The animal is placed on its side. The head is removed, and the trunk is then cut into three pieces. One cut to the torso is made caudal to the fore limbs and another just caudal to the hind limbs. All trunk pieces are then embedded on the cut surfaces (*laying the embryo on its head*), yielding three sections that visualize most thoracic and abdominal organs. This orientation is most ideal for screening several organ systems without prior knowledge of the anatomical system that is defective. It is advantageous for revealing all components of the heart, including valves. However, this position is not ideal for analyzing much of the central nervous system in the torso; neural tissues in this region are often very soft, even with optimal fixation techniques, so cutting them results in many folded, shattered, and inadequate sections. Transverse sectioning is best achieved by cutting tissue blocks approximately 3 mm in thickness and then processing them using a standard overnight processing protocol.

   The head may be embedded with the cut surface of the neck oriented *down* in the embedding mold. This position will yield horizontal sections of the brain, and permits assessment of both brain symmetry and the integrity of white matter tracts within the rostral CNS.

3. *Coronal (frontal) sections.* Placing the specimen with its back oriented *down* in the cassette (*laying the embryo on its back*) produces sections which provide visualization of the left and right sides of the body and all organ systems *in situ*. This positioning is ideal for comparison of organ symmetry (e.g., brain substructures). However, in our experience, this orientation provides the least useful visualization of most viscera in the thoracic and abdominal cavities.

Project-Specific Mouse Developmental Pathology Requirements for Tissue Collection and Processing

Certain mouse embryology studies will require special design considerations. The nature of these distinctions will depend on the particular endpoints of primary interest to the investigator. This section relates selected features that may be needed for some common mouse developmental pathology questions.

Bone Studies: Spina Bifida and Growth Plate Closure Disorders

Genes that influence skull development require anatomical studies in mesenchymal cell migration, cartilage maturation, and bone mineralization. A mutant mouse with defects in closure of the parietal and interparietal bones in the calvaria required preliminary histological sections stained with H&E to determine which orientations gave the best view of the defects and which histological stains might provide the best information on cartilage and calcification development (G. Downey, unpublished data). Subsequently, special stains were used to explore the aspects of bone development. The safranin O histochemical method specifically stains cartilage and mucins a bright orange to red color that provides more robust visual contrast of cartilaginous features compared to a traditional H&E stain. Similarly, the von Kossa's silver stain for calcium permits the visualization of mineralized bone by providing excellent contrast between positively stained (granular, black to brown) material against a light counterstain. The suitability of such stains varies with the stage of development. For example, at E13.5 and E14.5, safranin O does not stain skeletal primordia, while at E18.5 the multiple bones are clearly evident. The von Kossa stain also works well at E18.5. Many skull components, including the nasal, frontal, parietal, interparietal, occipital and basioccipital bones as well as the first cervical vertebra, are readily visualized in sagittal sections of the head; cross sections are also suitable for frontal and parietal bones. Of most importance to this particular study was the interparietal bone, as this is small and develops later in the course of skull formation. Staining with methods other than H&E was required due to the thin nature of this bone and its pale staining with eosin. This case reveals that the orientation of small structures that develop in narrow time windows requires special training and ample experience to orient the specimen properly during the embedding step and that some pilot studies will typically have to be undertaken to devise the ideal method for visualizing the desired structure.

Brain Development

Transverse sections may be better suited for visualizing cerebrocortical layers and evaluating brain symmetry. We have found them to be more informative compared to coronal sections for high-throughput phenotyping of mouse embryos and fetuses. The stain used for preliminary morphological screening of developing mice in our high-throughput pipeline is H&E. In contrast, developmental neurobiologists prefer to go straight to cresyl violet (to demonstrate neuronal subpopulations), Luxol fast blue (for myelin), and glial fibrillary acidic protein (GFAP, for astrocytes). The choice for staining brain depends on the expertise available to the laboratory in terms of performing the method and interpreting the sections as well as the availability of suitable control materials. For optimal cell morphology, especially in the cortical plate, whole body perfusion of the dam or individual embryos is recommended. Keeping the brain in the skull provides good preservation of architecture, retains the connectivity of brain centers to cranial nerves, and minimizes artifacts such as neuropil tearing and *dark* (i.e., shrunken, basophilic) neurons that can be misinterpreted as degenerated neurons.

Heart Development

Standard sagittal and transverse sections are well suited for studies of the myocardium and pericardium. However, studies focused on endocardial cushion formation and valve development often require oblique orientations of the animal because the embryonic heart sits at an angle that is different from the adult mouse; inattention to positioning can add confusion to the terminology used for plane of section. Therefore, it is recommended that the orientation of the heart structures be related to the animal's orientation (i.e., dorsal, ventral, cranial, and caudal). Moreover, serial sections are also likely to be required to locate specific heart structures, particularly valves. We have found that keeping all the unstained slides in an archive is useful. For example, if 200 slides are produced with 12–16 sections on each slide, every fifth slide can be stained with H&E for screening and analysis. Once the slide(s) with the structures and defects of interest are located, the archived slides in that region of the set can be used for special stains (e.g., immunohistochemistry or *in situ* hybridization), or special analytical techniques such as laser capture microdissection (LCM [see Chapter 10 for technical details]).

Communication of Mouse Developmental Pathology Data from High-Throughput Studies

Histopathology

The atlases for normal mouse embryo anatomy are useful for initial orientation and demonstration of organ morphology at different time points.[17] However, these atlases have several limitations. First, in many instances, the sections available on your slides may not match what is shown in the atlas. Such discrepancies leave some level of uncertainty in making conclusions. Second, the atlas may not have all the levels in the orientation you want (i.e., sagittal, transverse, and coronal). Third, the standard mouse brain atlases (e.g., the online Allen Institute for Brain Science offering at http://mouse.brain-map.org/) include only cresyl violet-stained sections, which reveal different features than those demonstrated by the routine H&E-stained sections generally produced in a high-throughput phenotyping program. Fourth, atlases of every mouse strain are not complete, so one is not always able to compare the features of your animals to annotated materials from the same strain. Our approach to mitigate this issue is to use age-matched and, where required, sex-matched littermates to try to orient and section mutants and controls at the same tissue plane. Last, serial sections are often needed for complete review of organ architecture and analysis for histopathological change. Our pathologists also assist in the determination and annotation of the anatomical structures that are positive for various reporter genes (typically lac operon Z [*lacZ*] in our program) as some of these patterns may not look the same in mutant embryos (up to E10.5) if there are severe or subtle anatomical abnormalities. These latter interpretations are done on both whole mounts and increasingly on slide-mounted sections as finer structures are analyzed in embryos from E12 to term.

Controls

Choosing appropriate controls for any GEM being evaluated can be problematic because many mutants are a hybrid of two or more strains, and large strain-dependent differences often exist in phenotypes. Ideally, the only genetic difference between the mutant and control animals is the targeted locus or transgene, but some compromises are usually necessary. This issue is also important when studying embryos, fetuses, and neonates because of the large differences in maternal body weight, embryo weight, and placental weight among animals of different strains.

For high-throughput histopathology, pathologists ideally request side-by-side comparison of mutant and WT embryos. Initially, screening can be performed at low magnification (1× to 4× objectives) for detection of overt anatomical differences. Evaluation at higher magnification (20× and 40× objectives) is required to investigate more subtle differences (e.g., epithelial layers, specific cell types, nuclear and cytoplasmic characteristics) and may sometimes require oil immersion (100× objective) for interpretation of finer structures such as cilia. Often the pathologist is going back and forth between slides for WT or heterozygous animals and homozygous mutants, which makes comparisons cumbersome and inefficient. Taking high-resolution photomicrographs (static image or whole slide scan) of WT controls can help save time and increase efficiency. If a large monitor is used in a side-by-side format to the microscope being used for glass slide sections, tissue images from controls (WT and heterozygotes) and mutants can be visualized and compared simultaneously. Manipulating and viewing the glass slides at multiple focal points through the section, in quick comparison to controls slides, is still our recommended method of choice for high-throughput mouse embryo histopathologic analysis compared to rapid viewing of low magnification, low-resolution, whole-tissue slide scans.

Ontologies

Currently, there is an effort to formalize and standardize mouse anatomy and mouse pathology terms through concerted international resources.[34] Such tools are often used by pathologists for the entry of pathology phenotype descriptions of adult GEM, and they are adaptable to mouse developmental pathology studies. The mouse anatomy ontology can be found at Mouse Genome Informatics (http://www.informatics.jax.org/)[14] and can easily be incorporated into the user's local pathology data capture system for the entry of lesion locations. The MPATH (Mouse Pathology) ontology[28] gives a terminology for disease processes and can be used to standardize definitive diagnoses. We have developed an additional ontology for descriptors and morphological diagnoses (Figure 17.5). A purpose-developed, complete, and useful stand-alone system for ascribing complete pathology phenotypes in a high-throughput process has not been developed, particularly for lesions in developing mice. However, standardized entry of ontological data will be required to facilitate the search capabilities of databases that store results from the high-throughput international phenotyping consortium.[8,35]

Image Capture

Although static photomicrographs are the mainstay of pathology textbooks and reports, the need for whole-slide scans as captured data from large GEM phenotyping studies is a major area of potential growth. Not only will these images provide crucial data for sharing, online consultation, distribution, and archiving, they will provide higher throughput potential for image capture and image analysis. These images may require file sizes of 500 MB or larger and often require a significant support infrastructure from information technology personnel. Furthermore, these datasets portend an expanded role for the comparative pathologist as an *informatician* (for anatomic annotation) in the discipline of phenogenomics. These images need to be made accessible over the Internet via portal URLs that will serve as the repositories for open-access data generated from international high-throughput phenotyping efforts for GEM.[6,13,31]

Report Format and Controlled Ontologies

Elements required for thorough and systematic documentation of mutant embryo phenotypes include historical in-life events to the dam, embryonic genotype, and all evident lesions and artifacts. All histopathology findings are data that will add to the growing body of mutant mouse embryo findings in functional genomics—including those not infrequent instances in which overt structural phenotypes are lacking. Thus, documentation of *no significant findings* or *histologically unremarkable* are important diagnostic terms often overlooked by pathologists trained in classical anatomic pathology programs. Continued normal development in the face of an engineered mutation is a significant finding in light of the genetic manipulation imposed on the mouse embryo and, in some instances, points to the functional adaptation or compensatory ability of the mouse genome.

An example of our reporting output for embryonic phenotyping is provided in Figure 17.6. Controlled vocabularies for anatomical location and diagnoses (etiologic, morphologic, and definitive) are vital for this information to complement other mouse phenotyping databases like EuroPhenome[22] and the Mouse Phenome Database.[1]

General Approaches to Investigating Prenatal and Perinatal Lethality

In most instances, embryonic or perinatal lethality is discovered or confirmed early in the course of deriving new GEM. Depending on the breeding scheme and strain, the genotype ratios and/or raw numbers of pups born with specific genotypes will be skewed from those expected based on Mendelian inheritance. This phenomenon marks a good point for the research team to seek the input of a comparative pathologist with experience in mouse phenotyping. As outlined earlier, *in utero* phenotyping of live fetuses is possible but often requires specialized techniques, expertise, and instrumentation. New assays continue to be applied to this field, including advanced imaging[30] as well as embryo-adapted assays such as preterm glucose tolerance testing.[20] However, the expense and/or general lack of in-life phenotyping options means that *postmortem* gross and microscopic pathology evaluations are probably the most informative and cost-effective route to pursue initially as they can help direct the selection of further *in utero* assays. The pathologist can provide critical guidance at this point regarding the most appropriate specimens, collection procedures, and embryonic ages to assess. Of course, once the specimens are collected, it is too late to go back and change the collection protocols, use a different fixative, or decide to include placenta. Thus, careful planning of sample collection avoids time wasted by the pathologist

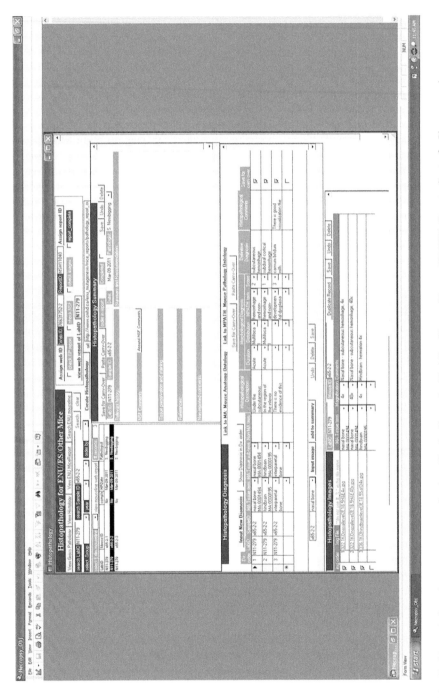

FIGURE 17.5   The histopathology data capture system developed by the Centre for Modeling Human Disease (CMHD), Toronto Centre for Phenogenomics.

in examining slides with various combinations of poor collection or fixation technique, inappropriate or inconsistent section orientation, and questionable embryonic age or genotype. The value of having a pathologist look at such slides is minimal as only the most glaringly obvious lesions can be reliably discerned under such circumstances.

Depending on the genetics and breeding scheme, investigation of the parents for potential reproductive defects may be warranted *before* extensive phenotyping of embryos begins. For example, in the case of a heterozygous knockout mouse, there may be no overt phenotype until breeding reveals abnormal fertility. One must also keep in mind that mice with normal reproductive organs and gametes can exhibit infertility due to abnormalities in nonreproductive tissues, such as musculoskeletal, sensory, neurological, or endocrine organs. The investigator also must consider maternal influences on perinatal mortality

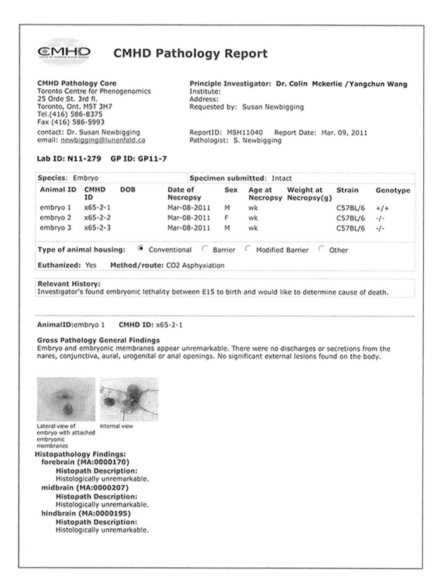

FIGURE 17.6 Example of an image-enabled report format with controlled ontologies and images for reporting mouse developmental pathology data from a high-throughput mouse phenotyping facility. Note the entry of standardized Mouse Anatomy ID (MAID) identification terms and Mouse Pathology (MPATH) anatomic pathology terms. Modifiers for morphological diagnoses are incorporated and images are included that identify and characterize numerous phenotypes (both obvious and subtle). Nonsignificant findings are also documented.   *(Continued)*

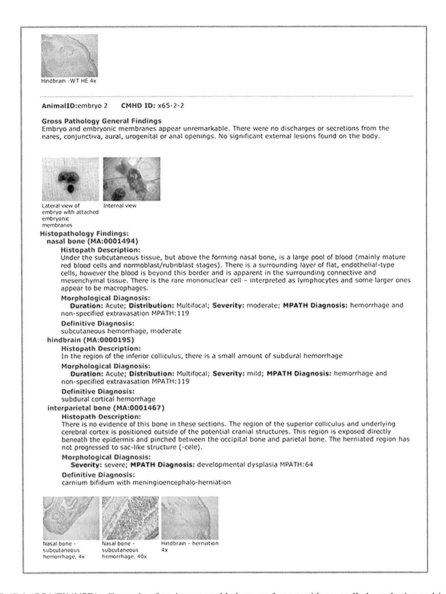

FIGURE 17.6 (CONTINUED) Example of an image-enabled report format with controlled ontologies and images for reporting mouse developmental pathology data from a high-throughput mouse phenotyping facility. Note the entry of standardized Mouse Anatomy ID (MAID) identification terms and Mouse Pathology (MPATH) anatomic pathology terms. Modifiers for morphological diagnoses are incorporated and images are included that identify and characterize numerous phenotypes (both obvious and subtle). Nonsignificant findings are also documented. (*Continued*)

since the dam is often a mutant and may have an undiscovered phenotype that affects maternal care, such as poor lactation, behavioral defects, or metabolic abnormalities. Likewise, the sire is often present when litters are born and may exhibit previously unrecognized behavioral abnormalities that contribute to perinatal death as a function of his genotype.

If the study proceeds to a histopathological investigation of prenatal or perinatal death, a major first step is to determine the time of death. Since cannibalism of dead or dying neonates is common, having someone present to observe the birth is required to determine actual numbers of pups born, whether pups are born alive or dead, and to gain information regarding the condition of the pups that are born alive but die soon after. It may be useful to video-record newborn pups rather than relying on the description

AnimalID:embryo 3   **CMHD ID:** x65-2-3

Gross Pathology General Findings
Embryo and embryonic membranes appear unremarkable. There were no discharges or secretions from the nares, conjunctiva, aural, urogenital or anal openings. No significant external lesions found on the body.

Lateral view of embryo with attached embryonic membranes   Internal view

Histopathology Findings:
nasal bone (MA:0001494)
    Histopath Description:
    Under the subcutaneous tissue, but above the forming nasal bone, is a large pool of blood (mainly mature red blood cells and normoblast/rubriblast stages). There is a surrounding layer of flat, endothelial-type cells, however the blood is beyond this border and is apparent in the surrounding connective and mesenchymal tissue. There is the rare mononuclear cell – interpreted as lymphocytes and some larger ones appear to be macrophages.
    Morphological Diagnosis:
        Duration: Acute; **Distribution:** Multifocal; **Severity:** moderate; **MPATH Diagnosis:** hemorrhage and non-specified extravasation MPATH:119
    Definitive Diagnosis:
        subcutaneous hemorrhage, moderate
hindbrain (MA:0000195)
    Histopath Description:
    In the region of the inferior colliculus, there is a small amount of subdural hemorrhage
    Morphological Diagnosis:
        Duration: Acute; **Distribution:** Multifocal; **Severity:** mild; **MPATH Diagnosis:** hemorrhage and non-specified extravasation MPATH:119
    Definitive Diagnosis:
        subdural cortical hemorrhage
interparietal bone (MA:0001467)
    Histopath Description:
    There is no evidence of this bone in these sections. The region of the superior colliculus and underlying cerebral cortex is positioned outside of the potential cranial structures. This region is exposed directly beneath the epidermis and pinched between the occipital bone and parietal bone. The herniated region has not progressed to sac-like structure (-cele).
    Morphological Diagnosis:
        Severity: severe; **MPATH Diagnosis:** developmental dysplasia MPATH:64
    Definitive Diagnosis:
        carnium bifidum with meningioencephalo-herniation

Hindbrain - meningocele - 4x

Histopathology Summary and Recommendation:
There is good visualization the occipital bone and often the basiooccipital bone and atlas in these sections. These sections reveal that there is absent interparietal bone resulting in herniation of the superior and inferior colliculi along with the local meninges between the occipital bone and parietal bone . It appears to be similar to cranium bifidum (defect in the closure of the cranial vault causing exposure of the brain) which can progress to a meningoencephalocele if both brain and meninges continues to herniate and protrude, or also encephalocele if it is only brain protruding, or meningocele if the protrusion is covered by skin is lined by meninges. In other domestic species, this can be a hereditary malformation and due to griseofulvin treatment in early gestation. The Safrinin O staining reveals the more caudal demarcation of the growing, cartilaginous occipital bone, while the von Kossa stain for mineral reveals the more rostral mineralized parietal bone.

FIGURE 17.6 (CONTINUED)   Example of an image-enabled report format with controlled ontologies and images for reporting mouse developmental pathology data from a high-throughput mouse phenotyping facility. Note the entry of standardized Mouse Anatomy ID (MAID) identification terms and Mouse Pathology (MPATH) anatomic pathology terms. Modifiers for morphological diagnoses are incorporated and images are included that identify and characterize numerous phenotypes (both obvious and subtle). Nonsignificant findings are also documented.

of the person present at the birth, which typically occurs at night in a poorly lit room. This automated data collection system avoids observer bias, especially if a phenotype is expected/hoped for based on the genetic manipulation(s). In addition, a recording provides a means for expert opinion to be sought at a later time as well as provide data that can be included as a multimedia file in a publication. Questions to consider in the perinatal period are listed in Table 17.2.

If pups with the genotype of interest are never born or are born dead, then one must work backward to determine the actual time of death *in utero* by collecting preterm litters. Resorbed embryos may be difficult to detect depending on how early in gestation death occurred, so careful dissection of the uterus is required. When collecting embryos and placentas, it is of vital importance that genotyping is done

TABLE 17.2

Questions to Consider in the Perinatal Period

How many pups were really born, and are they alive or dead at birth?

Are the pups physically normal in terms of size, anatomic structures, and movement?

Are the pups breathing normally, and is their color normal (uniformly pink)?

Do the pups suckle normally and get a full stomach of milk (as evidenced by observing
   the distended white stomach [or *milk spot*] through their thin pink skin)?

Do they display impaired locomotion, or move away from their mother?

---

accurately and that control littermates are also collected for comparison. In some breeding scenarios (e.g., homozygous knockout to homozygous knockout), offspring may carry only the experimental genotype, so careful timed breeding of control mice with the appropriate genotype and the same strain background must be undertaken to provide concurrent controls. The number of litters needed to establish the time of death *in utero* will depend partly on the number of embryos with the genotype of interest that are expected per litter. As with phenotyping of adult mice, having at least two affected mice of each sex is a minimum to begin an embryonic death investigation.

Special Cases

Mice with Multiple Genetic Alterations

More and more lines of GEM are being derived that harbor several genetic alterations, and embryonic or perinatal death investigations of these mice require special consideration. Breeding schemes can become very complex, especially if certain parental genotypes must be maintained in a heterozygous state due to health or reproductive issues. Such breeding schemes can result in only a small fraction of mice per litter carrying the genotype of interest, or worse, only one mouse in many litters. In the example shown in Figure 17.7, two genes of interest have been knocked out, but both must be maintained in the heterozygous state for the parents to be reproductively competent and healthy. If the genotype of interest is homozygous deficient for both genes, this leads to a 1 in 16 chance of such a mouse being conceived. If the same scenario is stretched to three genes, the odds of obtaining the genotype of interest are 1 in 64, making such investigations very laborious.

Lethal Phenotype Disappears (or Appears) during Backcrossing

When deriving a typical knockout mouse (unless the blastocysts and mES cells are from the same strain), the initial phenotype assessment is conducted in a mixed strain background. It is not that uncommon for embryonic or perinatal lethality to either resolve or suddenly appear after several generations of backcrossing. Significant phenotype differences in GEM occur when mice are crossed into different background strains, so it is not surprising that critical events in embryogenesis could be similarly affected by strain differences. A simple option to obtain viable mice is to backcross onto the nonlethal background

Gametes	A B	A b	a B	a b
A B	AA BB	AA Bb	Aa BB	Aa Bb
A b	AA bB	AA bb	Aa bB	Aa bb
a B	aA BB	aA Bb	aa BB	aa Bb
a b	aA bB	aA bb	aa bB	aa bb

FIGURE 17.7 Punnet square showing genetic outcome in progeny of complex crosses between parents bearing multiple mutant genes. If the parents are heterozygous for two mutations (Aa Bb genotype, where the lower case alleles are recessive), the odds of producing offspring with the desired double homozygous genotype (aa bb [shaded box]) is 1 in 16 embryos.

(or lethal background, if that is the phenotype of interest) or onto another strain altogether. A potential benefit of strain-specific lethality is that it presents an opportunity to discover important interactions between the gene of interest and strain-specific genetic variations such as single nucleotide polymorphisms and gene copy number variations.

Rare Survival of Typically Lethal Genotypes

In large breeding scenarios investigating embryonic lethality, in rare instances a mouse with a supposedly lethal genotype is born and survives. This occurrence generally sparks several rounds of regenotyping and combing through breeding records, but if the genotype is confirmed, this can be a very valuable mouse. Such mutant mice are often small and poor-doing, but they can provide in-life phenotyping data not possible *in utero*. In cases where the mouse in question is fertile, this may allow for higher ratios of the genotype of interest to be produced. For example, if the mouse is a homozygous knockout, it could be bred to a heterozygous mouse, resulting in half the embryos being homozygous knockout (rather than only a quarter as would occur in heterozygous to heterozygous matings). It is important to keep in mind that conducting a necropsy on such a rare mouse can provide very important data. Since such mice are often poor-doers, they should be monitored closely. Informative lesions due directly to the genotype can be obscured by secondary pathologies if the mouse becomes clinically ill, or by autolysis if unexpected death delays the *postmortem* examination.

Authorship by Pathologists Who Engage in High-Throughput Mouse Developmental Pathology Studies

The International Committee of Medical Journal Editors has established a set of criteria to help judge whether or not an individual qualifies for authorship.[11] In this system, authorship is based on three criteria: (1) substantial contributions to experimental conception and design, acquisition of data, or analysis and interpretation of data; (2) drafting the article or revising it critically for important intellectual content; and (3) final approval of the version to be published. Authors should meet all three criteria. The preceding sentence is of particular importance. Although a pathologist in a study can probably meet criterion #1, he or she can only meet the other two criteria if given the opportunity to participate in the writing or revising and in approval of the final draft.

The subject of authorship should be broached early in the discussions of mouse developmental pathology studies between a pathologist and the investigator so that everyone's expectations are understood. Since students, postdoctoral fellows, and collaborating scientists cannot really discuss authorship in a meaningful way, a dialogue between the pathologist and the actual principal investigator is needed. Regardless of the aforementioned criteria, authorship remains, at least to some extent, the currency of modern academic biomedical research. More and more journals are now requiring a *statement of contribution* by listed authors to discourage the inclusion of people who did not contribute substantially to the work. So how are investigators to deal with pathologists who charge for their time? Depending on their career goals, interests, criteria for promotion, and other factors, some pathologists may not be particularly interested in authorship, while others consider it so important they will do some studies at a reduced rate or without billing at all for their time. Regardless, the benefits of offering a pathologist authorship outweigh any dilution of authorship, as the final publication of studies in which gross or microscopic lesions are prominent components of the phenotype will likely be stronger, and responses to questions from peer reviewers will be smoother, if the pathologist has a vested interest in the paper.

Developmental Pathology Assessment: To Blind or Not to Blind

Looking at tissue sections without prior knowledge of the genotype (i.e., in coded [*blinded*] fashion) is the preferred method for assessing specimens from GEM but may not be practical in all situations.

Some embryos, fetuses, and newborns are grossly small or abnormal as a function of their genotype, and even a cursory examination will reveal their probable genotype immediately. When faced with evaluating only one or two mice per group, it may be more practical to decode (*unblind*) the samples in the initial assessment process to discover potentially subtle differences between genotypes. Future samples then can be looked at with the pathologist blind to the genotype, but cognizant of the subtle phenotypic lesion(s) seen previously. This approach may also be useful with a larger number of mice when the initial blinded assessment does not result in a consistent pattern of lesion(s), as would occur when a phenotype has low penetrance. Then, once a lesion(s) is discovered, the slides can be reexamined in a blinded manner.

Estimating Time and Cost for Grant Budgeting

Appropriate histopathological evaluation of embryos, fetuses, and perinatal mice generally involves looking at more slides, and taking more time per slide, than with sections from adult mice due to the rapid evolution of anatomic features over short periods of time. When the suspected lesion is an abnormality in physical structure, this requires *mentally* tracing multiple organs through many sections at higher magnifications and can take a considerable amount of time. Because of possible variations in developmental age between different litters, and due to variations in the sectioning angle among specimens, the pathologist may have to refer back to several different control mice when interpreting mutant tissues (an example of a circumstance where unblinded analysis can be quite advantageous). For the aforementioned reasons, agreeing to *just have a quick look* at one section of a mutant embryo for only the lesion of interest is often unrewarding. Worse, it could lead to the publication of a phenotype that is incorrect or incomplete compared to what a proper evaluation would have yielded.

As stated at the beginning of this chapter, it is estimated that approximately 30% of mutant mouse lines will demonstrate embryonic lethality. On an individual gene basis, lethality is relatively unpredictable. However, from a high-throughput perspective, investigators should logically be budgeting time and expenses based on a significant proportion of newly generated mouse lines having an embryonic or perinatal lethal phenotype. Yet, when pathologists first venture into a histopathological investigation of embryonic or perinatal lethality, they usually have only budgeted their time and expense based on their experience with adult mice. Obviously, this oversight can produce modest to substantial underbudgeting of histopathology costs if a new mutant mouse line is discovered to be embryonic lethal. As a very general guideline, an investigator needs to budget 60 min per whole embryo and placenta being evaluated by histopathology, and additional time if there is a specific structure(s) that calls for very close scrutiny. Often image analysis with histomorphometry is required to assign objective semiquantitative scores or acquire measurements of structures or 3D volumes, and more time must be budgeted for these and other specialized investigations.

Summary

More than a decade has passed since the initial sequencing of the human genome, which was rapidly followed by sequencing of the mouse genome. Perhaps not surprisingly, drawing biological relevance out of these large datasets continues to present a major challenge to biomedical science, with most known genes still lacking a clearly defined contribution to phenotype. A recent commentary in *Nature* laments that the majority of biomedical publications still focus on genes that were well studied before the human genome was sequenced, while newer genes are being neglected—even those associated with disease conditions.[4] In the effort to study new genes and progress from genome to phenome information in a high-throughput manner, the laboratory mouse has proven to be an efficient, cost-effective, and informative research tool for investigating and characterizing gene function. The task ahead is to achieve this level of understanding for all 20,000+ genes in our genome. Given the incredible complexity of mammalian development, it is reasonable that a large proportion of the genome is required for normal development and that disruptions in such genes will regularly lead to embryonic and neonatal lethality. Therefore, the growth of high-throughput phenogenomic laboratories to meet this challenge must take into consideration the requirements, expertise, and standardized approaches required for high-throughput phenotyping of genes that cause early lethality. Such understanding is equally important to the concept of translational medicine. Analysis of pipeline drugs between phase II clinical trials and regulatory approval from the top 10 pharmaceutical companies in the United States confirms that 85% demonstrated a sound biological rationale for the selected disease indication on the basis of a knockout mouse phenotype and that 20% of pipeline targets impacted signaling pathways where mutations result in embryonic lethality.[39]

In conclusion, all early lethalities should not be lumped together as a single phenotype. Instead, we encourage investigators to apply the approaches outlined above and seek the advice of experienced comparative pathologists in order to obtain optimal interpretation of phenotypes that impact our understanding of mammalian development.

REFERENCES

1. Bogue MA, Grubb SC, Maddatu TP, Bult CJ (2007). Mouse phenome database (MPD). *Nucleic Acids Res* **35** (Database Issue): D643–D649.
2. Brown SD, Chambon P, de Angelis MH (2005). EMPReSS: Standardized phenotype screens for functional annotation of the mouse genome. *Nat Genet* **37** (11): 1155.
3. Copp AJ (1995). Death before birth: Clues from gene knockouts and mutations. *Trends Genet* **11** (3): 87–93.
4. Edwards AM, Isserlin R, Bader GD, Frye SV, Willson TM, Yu FH (2011). Too many roads not taken. *Nature* **470** (7333): 163–165.
5. EMPReSS. EMPReSS (European Mouse Phenotyping Resource of Standardised Screens). Available at: http://empress.har.mrc.ac.uk/ (last accessed November 20, 2014).
6. Europhenome. Europhenome mouse phenotyping resource. Available at: http://www.europhenome.org/databrowser/viewer.jsp (last accessed November 20, 2014).
7. Gregory SG, Sekhon M, Schein J, Zhao S, Osoegawa K, Scott CE, Evans RS et al. (2002). A physical map of the mouse genome. *Nature* **418** (6899): 743–750.
8. Hancock JM, Mallon AM, Beck T, Gkoutos GV, Mungall C, Schofield PN (2009). Mouse, man, and meaning: Bridging the semantics of mouse phenotype and human disease. *Mamm Genome* **20** (8): 457–461.
9. Henkelman RM (2010). Systems biology through mouse imaging centers: Experience and new directions. *Annu Rev Biomed Eng* **12**: 143–166.
10. Hrabé de Angelis MH, Flaswinkel H, Fuchs H, Rathkolb B, Soewarto D, Marschall S, Heffner S et al. (2000). Genome-wide, large-scale production of mutant mice by ENU mutagenesis. *Nat Genet* **25** (4): 444–447.
11. International Committee of Medical Journal Editors (2014). Inside the uniform requirements for manuscripts. Available at: http://www.icmje.org (last accessed November 20, 2014).
12. International Mouse Knockout Consortium; Collins FS, Rossant J, Wurst W (2007). A mouse for all reasons. *Cell* **128** (1): 9–13.
13. Jackson Laboratory (2001–2014). Mouse phenome database. Available at: http://phenome.jax.org/ (last accessed November 20, 2014).
14. Jackson Laboratory (2001–2014). MGI glossary (links for the *Mouse Developmental Anatomy Browser* and *Adult Mouse Anatomy Browser*). Available at: http://www.informatics.jax.org/glossary (last accessed November 20, 2014).
15. Justice MJ (2008). Removing the cloak of invisibility: Phenotyping the mouse. *Dis Model Mech* **1** (2–3): 109–112.
16. Justice MJ, Noveroske JK, Weber JS, Zheng B, Bradley A (1999). Mouse ENU mutagenesis. *Hum Mol Genet* **8** (10): 1955–1963.
17. Kaufman MH (1992). *The Atlas of Mouse Development*, 2nd edn. San Diego, CA: Academic Press.
18. Keller BB, MacLennan MJ, Tinney JP, Yoshigi M (1996). *In vivo* assessment of embryonic cardiovascular dimensions and function in day-10.5 to -14.5 mouse embryos. *Circ Res* **79** (2): 247–255.
19. Klaunberg BA, O'Malley J, Clark T, Davis JA (2004). Euthanasia of mouse fetuses and neonates. *Contemp Top Lab Anim Sci* **43** (5): 29–34.

20. Klinger S, Turgeon B, Levesque K, Wood GA, Aagaard-Tillery KM, Meloche S (2009). Loss of Erk3 function in mice leads to intrauterine growth restriction, pulmonary immaturity, and neonatal lethality. *Proc Natl Acad Sci USA* **106** (39): 16710–16715.
21. Kulandavelu S, Qu D, Sunn N, Mu J, Rennie MY, Whiteley KJ, Walls JR et al. (2006). Embryonic and neonatal phenotyping of genetically engineered mice. *ILAR J* **47** (2): 103–117.
22. Mallon AM, Blake A, Hancock JM (2008). EuroPhenome and EMPReSS: Online mouse phenotyping resource. *Nucleic Acids Res* **36** (Database issue): D715–D718.
23. McKerlie C (2006). Cause and effect considerations in diagnostic pathology and pathology phenotyping of genetically engineered mice (GEM). *ILAR J* **47** (2): 156–162.
24. Mouse Genome Sequencing Consortium; Waterston RH, Lindblad-Toh K, Birney E, Rogers J, Abril JF, Agarwal P et al. (2002). Initial sequencing and comparative analysis of the mouse genome. *Nature* **420** (6915): 520–562.
25. Nagy A, Gertsenstein M, Vintersten K, Behringer R (2003). Production of chimeras. In: *Manipulating the Mouse Embryo: A Laboratory Manual*, 3rd edn. Cold Spring Harbor, NY: Cold Spring Harbor Laboratory Press; pp. 453–506.
26. Nolan PM, Peters J, Strivens M, Rogers D, Hagan J, Spurr N, Gray IC et al. (2000). A systematic, genome-wide, phenotype-driven mutagenesis programme for gene function studies in the mouse. *Nat Genet* **25** (4): 440–443.
27. Office of Animal Care and Use, Animal Research Advisory Committee, U.S. National Institutes of Health (2013). Guidelines for the euthanasia of rodent fetuses and neonates. Available at: http://oacu.od.nih.gov/ARAC/documents/Rodent_Euthanasia_Pup.pdf (last accessed November 20, 2014).
28. Pathbase. Mouse patholological process definitions. Available at: http://eulep.pdn.cam.ac.uk/pathbase2/pathbase2_mpath_pathology.popup.with_definitions.php (last accessed November 20, 2014).
29. Peters LL, Robledo RF, Bult CJ, Churchill GA, Paigen BJ, Svenson KL (2007). The mouse as a model for human biology: A resource guide for complex trait analysis. *Nat Rev Genet* **8** (1): 58–69.
30. Plaks V, Kalchenko V, Dekel N, Neeman M (2006). MRI analysis of angiogenesis during mouse embryo implantation. *Magn Reson Med* **55** (5): 1013–1022.
31. Schofield PN, Bard JB, Booth C, Boniver J, Covelli V, Delvenne P, Ellender M et al. (2004). Pathbase: A database of mutant mouse pathology. *Nucleic Acids Res* **32** (Database issue): D512–D515.
32. Schofield PN, Dubus P, Klein L, Moore M, McKerlie C, Ward JM, Sundberg JP (2011). Pathology of the laboratory mouse: An international workshop on challenges for high throughput phenotyping. *Toxicol Pathol* **39** (3): 559–562.
33. Sled JG, Marxen M, Henkelman RM (2004). Analysis of micro-vasculature in whole kidney specimens using micro-CT. *Proc SPIE* **5535**: 53–64.
34. Sundberg BA, Schofield PN, Gruenberger M, Sundberg JP (2009). A data-capture tool for mouse pathology phenotyping. *Vet Pathol* **46** (6): 1230–1240.
35. Sundberg JP, Sundberg BA, Schofield P (2008). Integrating mouse anatomy and pathology ontologies into a phenotyping database: Tools for data capture and training. *Mamm Genome* **19** (6): 413–419.
36. Turnbull DH (2000). Ultrasound backscatter microscopy of mouse embryos. *Methods Mol Biol* **135**: 235–243.
37. Walls JR, Coultas L, Rossant J, Henkelman RM (2008). Three-dimensional analysis of vascular development in the mouse embryo. *PLoS ONE* **3** (8): e2853 (15 pages).
38. Ward NL, Haninec AL, Van Slyke P, Sled JG, Sturk C, Henkelman RM, Wanless IR, Dumont DJ (2004). Angiopoietin-1 causes reversible degradation of the portal microcirculation in mice: Implications for treatment of liver disease. *Am J Pathol* **165** (3): 889–899.
39. Zambrowicz BP, Turner CA, Sands AT (2003). Predicting drug efficacy: Knockouts model pipeline drugs of the pharmaceutical industry. *Curr Opin Pharmacol* **3** (5): 563–570.
40. Zhou YQ, Foster FS, Qu DW, Zhang M, Harasiewicz KA, Adamson SL (2002). Applications for multi-frequency ultrasound biomicroscopy in mice from implantation to adulthood. *Physiol Genomics* **10** (2): 113–126.

18

A Statistical Analysis Primer for Mouse Developmental Pathology Studies

Brad Bolon and Colin G. Rousseaux

CONTENTS

Mouse developmental pathology assessments are commonly conducted to discern whether a particular experimental manipulation has affected anatomic, biochemical, and/or functional attributes in one of two settings. The first is to define whether or not animals having different genotypes exhibit divergent phenotypes. In the simplest case, involving a single gene with only two possible alleles (normal [+] or altered [−]), the three possible genotypic groups would be *wild-type* (+/+), *heterozygous* (+/−), and *homozygous mutant* (−/−). A variant of this scenario is the case where one cohort of animals has been genetically engineered to overexpress a foreign molecule, yielding a *transgenic* genotype (+), which will be compared to nontransgenic (−) mice that do not carry the transgene. The second instance involves distinguishing between the structure and/or function of mice that were exposed to some agent or manipulation (i.e., *treated* animals) versus others who were not subject to the same exposure (i.e., *control* animals). A standard design for this latter scenario is to treat pregnant dams with a xenobiotic (or *test article*) in a conventional developmental toxicity study, which typically uses four doses: none (vehicle only), low, middle, or high. In both settings, conclusions drawn with respect to the biological outcomes need to be interpreted with respect to their biological relevance. The general means for reaching such conclusions is first to perform an appropriate statistical analysis to evaluate the effects related to distinct genetic backgrounds or different doses of the xenobiotic treatment and then to determine whether or not any statistically significant effects also are biologically plausible.

    Analysis using statistical tests is not an absolute requirement for assessing biological data sets. If all the events of one kind are confined to a single experimental group, it is clear where the association between the biological findings and the manipulation resides (assuming that the experiment was designed and undertaken properly). Commonly, however, biological outcomes are observed to occur variably across study groups. In such instances, two potential outcomes must be distinguished: is the difference in lesion frequency and/or lesion severity among treatment groups a result of a manipulation, or is it due to inherent and uncontrollable variability among individuals (i.e., *chance*) or inadvertent differences in study conditions? The role of statistical analysis is to discriminate, objectively and with a fairly high degree of certainty, between these two possibilities.

This chapter presents basic statistical considerations that are relevant to designing, conducting, analyzing, and interpreting mouse developmental pathology studies. The authors presume that researchers will have ready access to reliable statistical software, which is available from many commercial vendors. Often, however, the software or the investigator will be unable to perform the statistical analysis for complex experiments, in which cases collaboration with a biostatistician is essential. Accordingly, the information in this chapter focuses on major principles of experimental design and analysis that will support the selection of appropriate statistical methods or enhance the interaction between the research team and the biostatistician. A more detailed consideration of basic and applied biostatistics may be obtained from other reference texts.[3–5,7,9,11,15,17,20]

Basic Statistical Concepts and Terminology

In general, conclusions in mouse developmental pathology experiments are drawn from fairly small data sets, acquired from a few animals or litters in a single experiment or from only a few replicate studies, and then applied to the entire population of similar mice. Interpretations made from such small data sets may be affected substantially if the investigator cannot limit the extent of experimental variation to the factor of interest (e.g., genotype or drug dose) by vigorously controlling other potential differences (developmental age at the time of exposure, environmental conditions, genetic background, maternal health, etc.). This section will review basic principles needed for appropriate design and statistical analysis of mouse developmental pathology studies in order to avoid unintended bias and minimize uncontrollable sources of error.

Variables

Experiments are conducted to obtain data in the form of qualitative observations or semiquantitative or quantitative measurements. Typically, one comparison between divergent conditions will be the focus of the study, although comparisons of two or more conditions simultaneously may be required in some cases to attain the sturdiest conclusion. Different types of data have distinct implications with respect to their underlying assumptions and the nature of the statistical tests that may be used to analyze them.

Biological data typically represent a series of effects associated with the presence of a specific manipulation (e.g., a genetic pattern or treatment). Such effects and manipulations are called *variables* because they can differ depending upon the design and conduct of a study. *Independent variables* (e.g., drug dose, route of exposure) are manipulations that are chosen in advance and can be controlled during the course of the study. *Dependent variables* (e.g., fetal weights, incidences of developmental malformations) are effects for which the quality (i.e., lesion severity) and/or quantity (lesion number) is dependent on the presence of a manipulation. For a given set of developmental pathology variables, the *population* of variables is the entire field of all potential measurements that might be observed in all possible individuals that ever have, currently have, or ever will share a similar set of demographic characteristics (e.g., male mouse embryos at gestational day [GD] 11). Because the entire population cannot be assessed, measurements are taken from a representative *sample* as an estimate of the effect that might be expected to occur within the population as a whole.

Data Types and Statistical Tests

The types of data encountered in mouse developmental pathology studies are of three basic kinds. The first is *nominal* (categorical) data, where different classes of observations are identified by specific names.

Each named category is mutually exclusive (i.e., only one name per category may be used for each event) and mutually exhaustive (i.e., one of the category names must be used for each event). The nature of nominal entries usually represents either a quantal state of being (i.e., *yes-or-no* propositions, like alive or dead, and normal or malformed) or a categorical diagnosis (i.e., multiple classes, such as exencephaly, encephalocele, and meningocele as categories of neural tube defects [NTD] affecting the dorsal skull). When segregating effects among the permissible categories, partial or *average* diagnoses cannot be made. Nominal data are tested using nonparametric statistical methods (i.e., statistical tests that do not require the data to be a measured parameter).

The second data type is *ordinal*, in which semiquantitative levels detail the nature of an ordered data set based on an arbitrary internal scale. A common example in the pathology setting is the subjective scoring scheme for lesion severity (e.g., 0 = within normal limits, 1+ = minimal, 2+ = mild, 3+ = moderate, and 4+ = marked abnormalities). Lesion scores assigned to the same specimen by different operators often will differ in their absolute scores since the two graders do not share a known, objective scale of measurement. Nonetheless, the trend of the scores provided by both individuals usually will resemble each other. Ordinal data also are analyzed using nonparametric statistical techniques.

The third kind of data is *interval continuous*, in which true quantitative measurements (termed parameters) are obtained objectively using a recognized external scale (e.g., mg, mm) to define the places of individual data points within the data set. Such measurements may occupy a truly continuous scale (e.g., crown-rump length, total body weight), where a data point may take on any possible value between two fixed points. Alternatively, measurements may constitute an interval scale (e.g., counts of proliferating or apoptotic cells within a given embryonic organ), where events may take on any whole number value but cannot assume any of the intervening noninteger values along the scale. Continuous data typically are evaluated by parametric statistical methods (i.e., statistical tests that require the data to be a quantitatively measured parameter). Interval data often approximate a continuous distribution when large numbers of events have been recorded (e.g., the numbers of cells from various leukocyte classes circulating in blood), and as such often may be assessed using parametric methods for analyzing continuous data.

Specific statistical procedures can only be used when certain underlying assumptions have been satisfied. Parametric statistical methods generally require more stringent assumptions to be met before they may be used, such as normal distribution of data points. Nonparametric tests typically are suitable whenever the assumptions needed to confirm that a data set is parametric cannot be verified or are unlikely to be true. In practice, however, the assumptions needed to employ nonparametric statistics often are not met as well, which serves to emphasize the need for biological researchers to develop a close collaborative relationship with a biostatistician.

Hypotheses

Conventional mouse developmental pathology studies are designed to differentiate between two competing hypotheses. The null hypothesis (designated H_0) represents a specific supposition about a population, which typically is framed in the form *The manipulation (i.e., a chemical treatment or genetic mutation) did not produce an effect*. The alternative hypothesis (designated H_1) generally is stated as *The manipulation produced an effect*. The purposes of statistical analyses are (1) to provide a means of deciding which of these two hypotheses is valid and (2) to be able to ascertain the degree of confidence one can have in reaching that decision.

Errors and Statistical Power

Four conclusions are possible when considering the potential outcomes of a hypothesis-based developmental pathology study (Table 18.1). The calculated statistical significance matches the anticipated biological relevance for case I (no statistical significance or biological response) and case IV (statistical significance accompanies a predicted biological response). However, for case II (*false-positive*) and case III (*false-negative*), the statistical calculation does not match the biological response. The existence

TABLE 18.1

Relationship of Biological Relevance to Statistical Significance

		Statistically Significant Difference? Phenotype Present? (Decision Regarding H_0)	
		No (Do Not Reject H_0)	**Yes** (Reject H_0)
Biological relevance? Test result abnormal? (veracity of H_0)	**No** (H₀ is true)	**Case I** *Correct decision* Confidence level $= 1 - \alpha$	**Case II** *Type I error* Significance level $= \alpha$
	Yes (H₀ is false)	**Case III** *Type II error* β	**Case IV** *Correct decision* Power $= 1 - \beta$

Notes: The null hypothesis H_0 represents a specific supposition about a population, which typically is framed in the form *The manipulation did not produce an effect.* The alternative hypothesis H_1 usually is stated as *The manipulation did produce an effect.*

of statistical significance in the absence of biological relevance (case II) is designated a *type I* error, while the lack of statistical significance in the presence of real biological significance (case III) is called a *type II* error. Erroneous false-positive and false-negative conclusions occur with some frequency during investigations of biological systems.

The probability of arriving at an erroneous conclusion may be controlled by an appropriate experimental design. The likelihood that a type I (false-positive) error will occur is termed the significance level or alpha (α). The confidence level is defined as $1 - \alpha$ (i.e., $100\% - \alpha\%$). The chance of making a type II (false-negative) error is designated beta (β). The power (also termed sensitivity) of a statistical test is calculated as $1 - \beta$. Therefore, power represents the probability that a test leads to rejection of the null hypothesis H_0 when the alternative hypothesis H_1 actually is correct. In other words, power is an estimation of the chance that a false-negative decision (type II error) can be avoided. The precise values for α and β are important in calculating the sample size for a given study, so both should be set in advance. The likelihood of committing type I and type II errors decreases as the size of the representative sample grows larger.

As with many biomedical experiments, mouse developmental pathology studies often set the significance level (α) at 0.05. This value means that the likelihood of an outcome occurring by chance in a group of manipulated animals is 5% or less. In our experience, investigators who perform mouse developmental pathology studies do not give sufficient attention to defining the power ($1 - \beta$) in advance, if they consider it at all. Instead, the usual practice is to set the group size arbitrarily using designs in published scientific reports. This unpremeditated approach may prevent the final data set from having sufficient power to make a conclusion with confidence. Accordingly, where possible mouse developmental pathology experiments should be planned so that the power will be at least 0.75, meaning that the probability of correctly deciding that a manipulation has had an effect is 75%.[6] In practice, the power for developmental pathology studies often is lower, sometimes by a substantial degree, especially when evaluating possible phenotypes in genetically engineered mice, due to the difficulty in acquiring large sample sizes of individuals with rare or lethal genotypes.

Functions of Statistical Analysis

Specific statistical tests each serve a distinct purpose. Descriptive statistics (i.e., exploratory or sensitivity analysis) provide an overview of what potential conclusions may be gleaned from a particular data set and may be utilized to generate new hypotheses and reveal unforeseen sources of bias. Hypothesis testing is employed to assess whether or not two or more groups of data diverge from each other by chance, and therefore is an effective means for minimizing false-positive (type I) errors. Model building is a means of predicting future outcomes using past data trends. Data simplification may be done to reduce the number of variables in a system while only minimally reducing the number of data. Mouse developmental pathology data may be used in principle in any of these fashions. However, in practice descriptive statistics and hypothesis testing typically are used when dealing with developmental pathology findings, and so will be emphasized here.

Descriptive Statistics

Numbers calculated to summarize the central tendency (i.e., an estimation of the most common response) and variability of the data are called descriptive statistics. These values are used to state both the central position of the data and assess the distribution of the data points about this central tendency. The most common numbers describing the central location of mouse developmental pathology data are the *mean* (i.e., arithmetic average); the *median* (i.e., the middle number, located at the 50th percentile); and the *mode* (i.e., the most common number). The usual means for defining data dispersion about the central location are the *standard deviation* (SD), *standard error of the mean* (SEM), and *coefficient of variation* (CV). The *range* falls between the lowest and highest data points, and thus defines the degree of data dispersion rather than describing the central trend.

The data distribution defines the appropriate choice of descriptive statistics. For instance, use of the mean for location with either the SD or SEM for dispersion implies that the sample data being summarized are from a population that at least approximates a *normal* (or Gaussian) distribution (i.e., a *bell-shaped curve*). Examples of mouse developmental pathology data exhibiting a normal distribution include body lengths and birth weights. However, it should be noted that the use of SD and SEM is not interchangeable in the strictest sense. In fact, the SD estimates the variability involved in acquiring the actual measurements from which a mean may be calculated (i.e., how the sample mean varies from the true mean of the population), while the SEM estimates the variability encountered when determining means for many possible samples. In the typical prospective studies performed in developing mice, the appropriate descriptive statistics for normally distributed pathology data will be the mean and SD.

The CV is a calculated ratio of the SD to the mean that defines the relative variability of a data set. A computed CV of 0.25 or 25% means that the SD represents 25% of the mean. The inherent variability in biological data often yields CV values between 25% and 50%, and values well over 100% are not uncommon.

The distribution of some pathology data sets is skewed (i.e., does not assume a normal bell-shaped distribution). An example of mouse developmental pathology data with a nonnormal distribution is the incidence of malformations induced by a particular toxic agent. The median is employed quite commonly for modestly shifted curves, while the mode and range are utilized less frequently for particularly skewed distributions.

Descriptive statistics are particularly effective when coupled with simple visual depictions of a data set. Graphs and data plots are a rapid, robust, and effective means of rendering complex data easier to understand. Mouse developmental pathology data sets usually are *univariate* in nature (i.e., affected by a single variable) and thus are best visualized using a 2D, rectangular graph. In general, nominal and ordinal data are depicted as bars, while interval continuous data are shown as uninterrupted lines. The horizontal (X) axis shows an independent variable (e.g., dose), while the vertical (Y) axis is assigned to a dependent variable (e.g., body weight or lesion score). The range of values for each axis scale is selected to best showcase the data trends.

If the data points represent a summary statistic like mean or median, an appropriate estimate of dispersion (e.g., one- or two-tailed error bars for the SD) should be included with each point in the graph to display the degree of variability. It is important to recall that error and variability actually represent distinct concepts: the degree of variability cannot be reduced due to the inherent variability of the experimental model and conditions, while the extent of error can be minimized using effective design and analytical procedures.

Hypothesis Testing

Hypothesis testing distinguishes between the competing possibilities that an observed effect was the result of an experimental manipulation or the consequence of random chance. The usual means of stating the conclusion at the end of a study is to indicate that a treatment-related effect was statistically significant, $p \leq 0.05$. This phrasing acknowledges the usual practice of setting α (the likelihood of a false-positive outcome) at 0.05. A p-value of 0.05 does not confirm that the effect resulted from the manipulation, but simply communicates that the likelihood the effect occurred by chance is 5% (1 case in 20) or less.

The choice of p-value will affect the choice of statistical procedure. In typical developmental pathology experiments with genetically engineered mice, p-values are two-tailed as they estimate the probability of observing a treatment-induced effect (an increase OR a decrease) by chance alone. This choice is dictated because the nature of the phenotype often is not known, so no *a priori* predictions are possible regarding the direction of the response relative to the normal state. Common examples of parameters requiring two-tailed tests are length and weight measurements. However, in developmental toxicity studies, many pathology endpoints may be evaluated using a one-tailed (one-sided) p-value as an estimate of the probability that chance will elicit a treatment effect in a specific direction (e.g., decreased weight, increased numbers of malformations). By convention, p-values are always of the two-tailed variety unless a one-tailed version is specified.

Hypothesis testing gives no direct indication with respect to how large the effect might be. Large and important effects may be unrecognized if the study is small, measurements are not precise, or the inter-individual response is large (type II error—inadequate power). On the other hand, in large studies small and unimportant biological effects may exhibit statistical significance (type I error—excessive power). Thus, the decision regarding whether or not a statistically significant result can be considered to support a hypothesis requires judgment.

Statistical Principles for Experimental Design

Mouse developmental pathology studies are typically designed to answer two questions. The first is whether or not a given manipulation (e.g., genotype or treatment) can be associated with a particular biological effect. The second question is to define the extent of the effect. Good developmental pathology experiments are devised to supply data that are suitable for statistical analysis. In return, statistical considerations are essential parameters to be considered when designing good experiments. Several facets must be envisaged when designing developmental pathology studies in mice, all of which are obvious to experienced researchers.

Experimental units in mouse developmental pathology studies typically are either individual animals or a litter of animals. The experimental unit is the entity that receives the manipulation and yields a response that can be observed and measured to yield a datum. For genetically engineered mouse lines, the individual usually represents the experimental unit as all animals of a given genotype are essentially equivalent subjects. In contrast, in the developmental toxicology setting, the litter is the better experimental unit as the pregnant dam receives the treatment and all conceptuses that she carries are assumed to have an equal opportunity for exposure. The entire collection of experimental units for a given group is the sample (N). The value of N defines the types of statistical tests that can be used.

Factors are sources of systematic difference other than the manipulation (i.e., mutation or treatment) that might influence an individual's response. Examples of common factors include differences in age,[2,10,13,16] sex,[13,14,18] and sometimes intrauterine position.[1,8,12,19,21] In some cases, factors need to be taken into account when designing studies so that a more powerful statistical test can be chosen. Interactions between the manipulation and each factor need to be assessed for each level of the predetermined factor (e.g., factor = sex; levels = male and female). If the levels of the factor do not impact the outcome, the data can be pooled to raise the sample size (N), thereby strengthening the power of the statistical analysis.

Randomization is a common practice to ensure that each manipulation (e.g., dose level for a test article) has an equal chance to produce its effects relative to the chances afforded to other manipulations in the study. For example, in developmental toxicity studies, a common practice is to segregate mice among treatment groups immediately after conception, or at the same stage of pregnancy, by using random number tables. Randomization reduces the likelihood of introducing unexpected bias into the study design, such as populating the vehicle control group entirely from one shipment of mice and the high-dose treatment group from a separate shipment. For a typical phenogenomic study, randomization is carried out on individuals of a given genotype taken from multiple litters representing at least two different lines that express the same genetic modification. Randomization should not be limited to the in-life experimental phase but also should be applied to *postmortem* evaluation (see Chapter 7) and processing steps for fluid (Chapter 8) and tissue (Chapter 9) samples.

Controls are essential to distinguishing between a manipulation and chance as the likely source for an observed effect. Inclusion of appropriate experimental controls is done in an attempt to remove systematic (i.e., known) factors that might influence the response of the experimental system so that the sole known source of variation among groups is the independent variable (i.e., the genotype or treatment). In some cases, it is desirable to incorporate not only negative control cohorts (e.g., +/+ genotype, vehicle treatments) but also positive controls (e.g., a group manipulated to exhibit a positive outcome, to confirm that subtle effects in the test groups can actually be detected). In general, control data for mouse developmental pathology studies should be derived from concurrent control groups sharing the same demographic characteristics (especially the genetic background and husbandry conditions) rather than from repositories of historical data.

Replication is required to adjust for the extensive degree of interindividual variability in biological responses. The dispersion of data resulting from experimental errors and biological variability tends to be minimized as the number of replicates is increased. Replicates require that a particular manipulation be applied to multiple experimental units (e.g., mouse litters, fetuses of a given genotype) for any given study, and often that two or more studies of the same design be performed serially or in parallel to gain confidence that the study results reflect the genuine outcome to be expected when a manipulation is applied to the entire population. In practical terms, replication in mouse developmental pathology research typically is handled by ensuring that the group size (N) is large enough to provide for a reasonably sensitive statistical analysis without having to incur the expense of repeating the whole study.

Balance in the group sizes typically is desirable to strengthen the statistical analysis, as mathematical comparisons work best when groups are of nearly the same size. Therefore, in mouse developmental pathology studies it generally is inappropriate to expand the size of the treatment groups but not the control cohorts. Balanced designs also are important if multiple factors are to be assessed in the course of a single study (e.g., the individual and combined impact of two mutant genes in the same animal). That said, conventional statistical methods can accommodate unequal group sizes, within limits. The most common reason for an unbalanced study design in mouse developmental pathology studies is to include more mutant or xenobiotic-exposed individuals in an attempt to heighten the sensitivity for detecting subtle phenotypes, or to enhance the confidence that an apparent absence of a phenotype is a genuine outcome of the manipulation.

Sampling is a crucial stage of designing any experiment. Most statistical tests assume that each member of the sample is collected independently and without bias so that any member of the population has an equal chance of being selected. A number of sampling methods may be utilized for mouse developmental pathology studies. Where numbers permit, the most common strategy is random sampling to avoid all known bias. This approach is suitable for genetically engineered mouse lines where individuals of a given genotype represent the experimental unit. However, in the developmental toxicology setting, the litter typically serves as the experimental unit as the pregnant dam receives the treatment and all conceptuses are assumed to have an

TABLE 18.2

Appropriate Sample (Group) Sizes for Standard Mouse Developmental Pathology Studies

Background Incidence of Developmental Defects	Statistical Power $(1-\beta)$	Confidence Level $(1-\alpha)$									
		0.95 (95%)	0.90 (90%)	0.80 (80%)	0.70 (70%)	0.60 (60%)	0.50 (50%)	0.40 (40%)	0.30 (30%)	0.20 (20%)	0.10 (10%)
0.30	0.90 (90%)	10	12	18	31	46	102	389	—	—	—
	0.50 (50%)	6	6	9	12	22	32	123	—	—	—
0.20	0.90 (90%)	8	10	12	18	30	42	88	320	—	—
	0.50 (50%)	5	5	6	9	12	19	28	101	—	—
0.10	0.90 (90%)	6	8	10	12	17	25	33	65	214	—
	0.50 (50%)	3	3	5	6	9	11	17	31	68	—
0.05	0.90 (90%)	5	6	8	10	13	18	25	35	75	464
	0.50 (50%)	3	3	5	6	7	9	12	19	24	147
0.01	0.90 (90%)	5	5	7	8	10	13	19	27	46	114
	0.50 (50%)	3	3	5	5	6	8	10	13	25	56

Notes: The significance level (α) represents the probability that a false-positive result (type I) error will occur, and $1 - \alpha$ defines the confidence level. The probability of obtaining a false-negative outcome (type II error) is designated beta (β), while $1 - \beta$ defines the statistical power. The mathematical meaning of these values with respect to the percentage likelihood of their happening is given in parentheses.

equal degree of exposure. Therefore, for developmental toxicity studies, the usual approach is cluster sampling, where the subject pool is divided into numerous separate groups (i.e., litters) from which a few individuals (fetuses) are selected at random for evaluation. The cluster paradigm is often supplemented by a stratified approach, where the pool is divided into subsets (i.e., strata) that share a similar trait—often treatment with a specific dose level of a chemical.

Group size (i.e., sample size, or *N*) is a major determinant of the precision and power of the study. In general, a large number of animals are required per group to detect a small effect, while obvious changes can be identified with relatively few animals. Similarly, the lower the desired significance level (α) or the higher the desired power ($1 - \beta$), the greater will be the necessary sample size. The group size in developmental pathology studies should not be chosen at random, but instead should be calculated purposefully so that the outcome of the experiment is amenable to statistical analysis. Appropriate sample size calculations for mouse developmental pathology studies are shown in Table 18.2.

Selecting Statistical Tests for Mouse Developmental Pathology Studies

Test selection generally should occur in advance of the study (*a priori*) rather than be defined after the fact (*post hoc*). The rationale is that the experimental design defines the nature of the analysis, so the appropriate statistical procedure for the predicted data set needs to be established up front to insure that the study is organized correctly.

Numerous statistical tests may be utilized for developmental pathology analysis. A detailed consideration of the various options and their assumptions is outside the scope of this chapter, but such information is readily available in many print and web-based resources. Thus, the remainder of this section will briefly consider common statistical methods used for mouse developmental pathology studies (Figure 18.1 and Table 18.3) so that researchers will have a solid foundation for selecting appropriate methods from among the many available possibilities.

In general, pathology analysis of developing mice should strive to provide quantitative data for analysis. The typical means for accomplishing this task is to record the presence or absence of a particular lesion, preferably with some semiquantitative grade or quantitative measurement regarding its severity and/or size. Consistent terminology (see Chapter 3) and scoring criteria need to be utilized for all

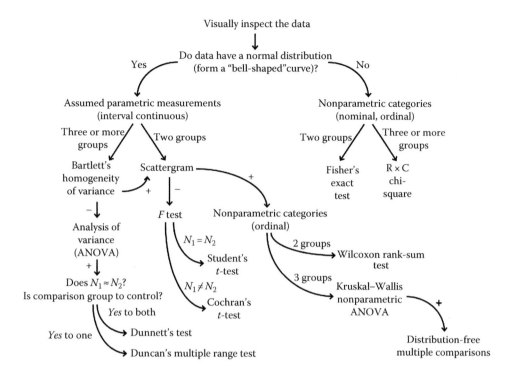

FIGURE 18.1 Flowchart for selecting among common statistical tests for analyzing mouse developmental pathology data. The diagram shows a decision tree by which various types of data may be used to determine the nature of the most appropriate statistical method. Symbols: + indicates that the outcome of the statistical test demonstrates a significant difference between the groups being compared, while − indicates that no significant difference was identified. N_1 and N_2 denote the sizes of two study groups. ≈ indicates *approximately equal*.

animals if a statistical analysis based on grade is to have meaning. Slightly different terms applied to the same lesion (*diagnostic drift*) should be reviewed and, where possible, merged prior to performing statistical calculations.

Where possible, the authors prefer to employ simple nonparametric methods, needing few assumptions, as they are robust and widely appropriate. Such broad techniques are particularly relevant when engaged in hypothesis testing, where many developmental pathology experiments have a single independent variable by design. Nonparametric statistical analysis of ranked data (e.g., lesion scores) exactly parallels parametric methods used for continuous data (e.g., body length and weight measurements). Furthermore, nonparametric procedures are as or more powerful than parametric techniques if underlying assumptions (e.g., homogeneous variability) cannot be confirmed. In all cases, the desired conclusion must have no part in the selection of the statistical test to be used for analysis of a given data set. Instead, the choice of statistical test should be dictated only by the nature of the data (Figure 18.1). The remainder of this section will define preferred statistical procedures for various common types of mouse developmental pathology data.

Lengths and Weights

Objective measurements of various lengths (e.g., crown-rump length) and weights (e.g., brain weight, total body weight) are often an early and sensitive indication that a manipulation has had an effect. These continuous data usually are assessed using an analysis of variance (ANOVA) followed, if necessary by a *post hoc* procedure such as Duncan's multiple range test. The nonparametric Kruskal–Wallis ANOVA is appropriate for many mouse developmental pathology studies due to its ability to detect effects in small sample sizes (<5/group) or if there is any uncertainty about whether or not the data follow a normal distribution.

TABLE 18.3

Summary of Assumptions and Limitations for Some Common Statistical Tests

Test	Assumptions	Limitations and Comments
Assessing the distribution of the data		
Bartlett's test for homogeneity of variance	Homogeneity of variance (also termed equality of variances or homoscedasticity) is an important assumption for Student's *t*-test, analysis of variance (ANOVA), and analysis of covariance. Bartlett's test is designed for three or more groups. Data must be normally distributed (i.e., the test verifies whether or not a data set follows a normal distribution).	Bartlett's test does not test for normality, but rather homogeneity of variance. Bartlett's test is very sensitive to departures from normality. As a result, a significant value in Bartlett's test may indicate nonnormality rather than heteroscedasticity (i.e., nonequal variances). Outliers can bring about such a finding. The sensitivity to such erroneous findings is extreme with small sample sizes ($N \leq 17$/group).
Parametric statistical tests		
Analysis of variance (ANOVA)	This test is parametric (i.e., a test assuming that data assume a particular distribution). A multiple-comparison *post hoc* method (e.g., Duncan's multiple range test, Scheffe's multiple comparison test) must be used to define group effects indicated by a statistically significant result from an ANOVA test. Samples are normally distributed, and exhibit homogeneity of variances. If the sample sizes are approximately equal (i.e., $N_1 \approx N_2$), ANOVA is robust for moderate inequality of variances (as determined by Bartlett's test). Data are independent.	The test is robust for moderate departures from normality if the sample sizes are large ($N \geq 30$/group). It is not appropriate to use a *t*-test to evaluate multiple groups in a *2-groups-at-a-time* comparison (instead of the appropriate method, which is ANOVA followed by a *post hoc* test) to identify where significant differences exist among three or more groups.
Student's *t*-test (rarely used due to limitations in experimental design)	The test assumes that the data are univariate (i.e., able to differ in only one variable), continuous, and normally distributed. Data are collected via random sampling. The test should be used when the aforementioned assumptions are met and only two groups are to be compared. The variances of the populations to be compared are equal. Where the sample size (N) exceeds 30, a *Z*-test is used rather than a *t*-test. The *Z*-test shares the same attributes as the *t*-test: data should be distributed normally (although this is less important for large N), and the variances should be equal.	Do not use when the data are ranked, are not approximately normally distributed, or when more than two groups are to be compared. Do not use for paired observations. This test is the single most commonly misused statistical method, except in those few cases where one truly is comparing only two groups of data and the group sizes are roughly equivalent. This test is not valid for multiple comparisons (because of resulting additive errors) or where group sizes are very unequal. The test is robust for moderate departures from normality and, when the sample sizes are approximately equal (i.e., $N_1 \approx N_2$), is robust for moderate departures from homogeneity of variances. The main difference between the *Z*-test and the *t*-test is that the *Z*-statistic is based on a known standard deviation, σ, while the *t*-statistic uses the sample standard deviation, s, as an estimate of σ. If data are distributed normally, the variance σ^2 is more closely estimated by the sample variance s^2 as N gets large.

Trend tests	Trend tests seek to evaluate whether or not a monotonic tendency exists in response to a change in treatment (i.e., that the direction of the dose response is absolute [e.g., as dose goes up, the incidence of congenital defects increases]).	The test loses power rapidly in response to the occurrences of reversals (e.g., a low dose group with a decreased incidence of birth defects). Trend tests will have higher power than the chi-square test when the suspected trend is correct, but the ability to detect unsuspected trends is sacrificed when a trend test is used. Trend tests exploit the suspected direction of an effect to increase power, but this does not affect the sampling distribution of the test statistic under the null hypothesis.

Post hoc analysis

Duncan's multiple range test	For dependent variables, the data are a random sample from a multivariate (i.e., able to differ in several variables), normally distributed population. In the population, the variance–covariance matrices for all cells are the same. Assumptions are checked with homogeneity of variances tests (e.g., Bartlett's).	Duncan's test assumes a set alpha (α) level (the false-positive or type I error rate) for all tests when means are separated by approximately regular intervals. Preserving this α level means that the test is less sensitive than other similar procedures, such as the Student–Newman–Keuls test. The test is inherently conservative and not resistant or robust.
Scheffe's multiple comparisons	This method tests all linear contrasts among the population means. It is not formulated on the basis of groups with equal numbers (as one of Duncan's procedures is), and if $N_1 \neq N_2$, there is no separate weighing procedure. In the population, the variance–covariance matrices for all cells are the same. Assumptions are checked with homogeneity of variances tests (e.g., Bartlett's).	The Scheffe's procedure is robust to moderate violations of the normality and homogeneity of variance assumptions. The Scheffe's procedure is powerful because of its robustness, yet it is very conservative. Type I error (the false-positive rate) is held constant at the selected probability for each comparison.
Dunnett's *t*-test	Treated group sizes must be approximately equal (i.e., $N_1 \approx N_2$). In the population, the variance–covariance matrices for all cells are the same. Assumptions are checked with homogeneity of variances tests (e.g., Bartlett's).	Dunnett's test seeks to ensure that the false-positive (type 1 error) rate will be fixed at the desired level by incorporating correction factors into the design of the test value table.

Nonparametric statistical tests

Chi-square	All expected frequencies should be 10 or greater. Data are univariate and categorical. Data are from a multinomial population (i.e., where the organism exhibits one response out of several possibilities). Data are compiled by random and independent sampling. Groups are of approximately the same size (i.e., $N_1 \approx N_2$), especially for small group sizes. If any expected frequencies are <10, but ≥5, Yates' Correction for continuity can be used. This is done by subtracting 0.5 from the absolute value of observed − expected values ($O - E$) before squaring.	*Use when:* The data are of a categorical (or frequency) nature. The data fit the aforementioned assumptions. The goodness-to-fit to a known form of distribution is to be tested. Cell sizes are large. *Do not use when:* The data are continuous rather than categorical. Sample sizes are small and very unequal. Sample sizes are too small (e.g., when total N [the sum of all animals in all groups] is <50 or if any expected group size is <5). For any 2×2 comparison, use Fisher's exact test instead.

(Continued)

TABLE 18.3 (CONTINUED)

Summary of Assumptions and Limitations for Some Common Statistical Tests

Test	Assumptions	Limitations and Comments
Contingency table	Fisher's exact test must be used in preference to the chi-square test when there are small cell numbers (i.e., <6). There are no interactions between row and column classifications. The test assumes no outliers.	Fisher's probabilities are not necessarily symmetric. Tables are available that provide individual exact probabilities for contingency tables made to test studies with small sample size. The probability resulting from a two-tailed chi-square test is exactly double that of a one-tailed probability from the same data.
Fisher's exact test	Fisher's exact test must be used in preference to the chi-square test when there are small cell numbers (i.e., <6). Fisher's probabilities are not necessarily symmetric. Although some analysts will double the one-tailed p-value to obtain the two-tailed result, this method is usually overly conservative. Samples are random and independent. The response is dichotomous (i.e., exhibiting a quantal response [such as *yes* vs. *no* for the presence of malformation]).	Tables are available that provide individual exact probabilities for contingency tables made to test studies with small sample size. Ghent has developed and proposed a good method extending the calculation of exact probabilities to 2×3, 3×3, and $R \times C$ (row-by-column) contingency tables. The probability resulting from a two-tailed chi-square test is exactly double that of a one-tailed probability from the same data.
Kruskal–Wallis nonparametric analysis of variance (ANOVA)	The test statistic H is used for both small and large samples. Data must be independent for the test to be valid. The effect of adjusting for tied ranks is to slightly increase the value of H. Omission of this adjustment results in a more conservative test. Within each sample, the observations are independent and identically distributed. The samples are independent of each other. The test makes no assumption about the distribution of the data.	When we find a significant difference, we do not know which groups are different. It is incorrect to then perform a Mann–Whitney U-test (see following text) on all possible combinations. Rather, a multiple comparison method must be used, such as the distribution-free multiple comparisons. Too many tied ranks will decrease the power of this test and also lead to false-positive results. When $k = 2$, the Kruskal–Wallis chi-square value has one degree of freedom (df). This test is identical to the normal approximation used for the Wilcoxon rank-sum test. A chi-square with one df can be represented by the square of a standardized normal random variable. In the case of $k = 2$, the H-statistic is the square of the Wilcoxon rank-sum (without the continuity correction).
Log-rank test	The endpoint of concern is in fact or is defined so that it is right censored (i.e., once it happens, it does not reoccur). Examples include death, a defect, and a minimum or maximum value of an enzyme activity or physiologic function (such as respiration rate). The method requires no assumption regarding distribution of event times. A continuity correction can also be used, in which the numerators are reduced by 1/2 before squaring. Use of such a correction leads to further conservatism and may be omitted when sample sizes are moderate or large (i.e., ≥15 per cell).	The Wilcoxon rank-sum test (see following text) could be used to analyze the event times in the absence of censoring. A *generalized Wilcoxon* test, sometimes called the Gehan test, based on an approximate chi-square distribution has been developed for use in the presence of censored observations. Both the log-rank and the Wilcoxon rank-sum tests are nonparametric tests, and require no assumptions regarding the distribution of event times. When the event rate is greater early in the trial than toward the end, the generalized Wilcoxon test is the more appropriate test since it gives greater weight to the earlier differences.

Test	Assumptions	Comments
	The subjects are randomly sampled from, or at least representative of, larger populations. The subjects were chosen independently. Consistent criteria are used. Baseline survival rate is not changing over time. The survival of the censored subjects would be the same, on average, as the survival of the remaining subjects.	Life tables can be constructed to provide estimates of the event time distributions. Estimates commonly used are known as the survival rate. Survival and failure times often follow the exponential distribution. If such a model can be assumed, a more powerful alternative to the log-rank test is the likelihood ratio test. This nonparametric test assumes that event probabilities are constant over time. In other words, the chance that a patient becomes event positive at time t given that he is event negative up to time t does not depend on t.
Mann–Whitney U-test	The test statistics are linearly related to those of the Wilcoxon rank-sum test (see later text). The two tests will always yield the same result. The Mann–Whitney U-test has been much favored in developmental and reproductive toxicology (DART) studies. It does not matter whether the observations are ranked from smallest to largest or vice versa. The samples are randomly taken from a population. The samples are independent within and mutually independent between samples. The measurement scale is at least ordinal.	This test should not be used for paired observations.
Row × column (R × C) chi-square	None of the expected frequency values should be <5. The chi-square test is always one tailed. The results from each additional column (group) are approximately additive. Due to this characteristic, chi-square can be readily used for evaluating any R × C combination. The results of the chi-square calculation must be a positive number, which is an inevitable outcome given the other conditions. Data must be independent.	Without the use of some form of correction, the test becomes less accurate as the differences between group sizes increase. The test is weak with either small sample sizes ($N \leq 5$/group) or when the expected frequency in any cell is <5. Pooling—combining cells can overcome these limitations. The test can be used to test the probability of validity of any distribution. Test results are independent of the ordering of the cells.
Wilcoxon rank-sum test	Differences are assumed to be independent. Data are ordinal in nature (e.g., within normal limits, or minimal, mild, moderate, or marked lesions) The test assumes that the distribution of signs is symmetric.	Too many tied ranks increase the likelihood of a false-positive result (i.e., type I error), leading to rejection of the null hypothesis at <5%, even though the α level is set at 0.05. The Wilcoxon rank-sum test can be highly biased in the presence of censored data.

Clinical Chemistry

These interval continuous data are best analyzed using rank-based nonparametric methods like the Wilcoxon rank-sum test and Kruskal–Wallis ANOVA rather than the more traditional parametric approaches like the *t*-test and ANOVA. The nonparametric strategy is preferred for clinical chemistry data for two reasons. First, alterations often occur together rather than in isolation, indicating that they are not truly independent of one another. Examples include the reciprocal deflections of electrolytes and the simultaneous increase in various enzyme activities following certain types of tissue injury. Second, the distribution of clinical chemistry data commonly is skewed rather normal, and thus requires the use of nonparametric procedures.

Hematology

The appropriate choice of statistical test depends on the analyte being assessed. Directly measured values such as the numbers of red blood cells (RBC), white blood cells (WBC), and platelets as well as the mean corpuscular volume (MCV) of RBCs represent continuous data that may be analyzed using traditional parametric methods: the *t*-test and ANOVA. The hematocrit (HCT) also may be evaluated using parametric procedures if it has been measured directly, but calculated HCT values (derived from RBC and MCV) should be examined using nonparametric approaches like the Wilcoxon rank-sum test or Kruskal–Wallis ANOVA. Hemoglobin content (HGB) is directly measured and has a continuous nature but typically exhibits a multimodal distribution due to the simultaneous presence of many forms (e.g., fetal and adult) and conformations (oxyhemoglobin, deoxyhemoglobin, etc.); the resulting nonnormal distribution warrants a nonparametric statistical technique. Counts of various leukocyte lineages are handled differently depending on the manner in which they were enumerated. Direct differential counts using automated cell counters provide continuous data that may be analyzed using parametric tests. In contrast, differentials calculated by manually tallying 100 or a few hundred cells and then multiplying the WBC value by the percentage for each cell type do not fulfill the assumption for a normal distribution (especially for rare cells like eosinophils and basophils) and thus should be examined by nonparametric methods.

The interdependence of hematological parameters (e.g., RBC, MCV, and HCT) means that genuine hematological effects should affect multiple values at once. A statistically significant alteration in any one parameter is unlikely to have a biologically useful meaning. Furthermore, hematologic values vary among different mouse strains, and also tend to drift over time within a single strain. Accordingly, control values for hematologic data should come from concurrent control groups representing animals of the same strain acquired from the same vendor to ensure the most robust performance by the chosen statistical test.

Histopathological Findings

Statistical analysis is essential to define whether or not the anatomic effects observed in manipulated mice really do differ in number and/or severity from the appearance of similar structures seen in control animals. Comparisons of lesion incidence between control and treatment groups in developmental pathology studies with balanced designs usually are made using parametric methods like Fisher's exact test (where $N \leq 6$/group) or a chi-square (χ^2) test (where $N \geq 6$/group). Unbalanced designs resulting from premature death or other effects caused by the manipulation necessitate that data be analyzed using a nonparametric procedure like the Wilcoxon rank-sum test. Ordinal data for lesion severity scores generally are examined by a nonparametric test as well.

Summary and Conclusions

Statistical analysis is a critical component of arriving at conclusions for data derived from mouse developmental pathology studies. That said, formal statistical analysis is a useful tool to permit comprehension of a data set rather than an absolute requirement for evaluation and interpretation.

Investigators should be familiar with fundamental concepts and assumptions of statistics so that they can develop study designs appropriate to answering particular questions. However, researchers should consult regularly with biostatisticians when designing experiments as expert counsel will conserve effort, money, time, and unique animal resources while ensuring that the final data set actually can address the stated objective(s) of the study. If planned correctly, the wonders of modern software make the actual statistical calculations a relatively innocuous task.

REFERENCES

1. Barr Jr M, Brent RL (1970). The relation of the uterine vasculature to fetal growth and the intrauterine position effect in rats. *Teratology* **3** (3): 251–260.
2. Chester N, Kuo F, Kozak C, O'Hara CD, Leder P (1998). Stage-specific apoptosis, developmental delay, and embryonic lethality in mice homozygous for a targeted disruption in the murine Bloom's syndrome gene. *Genes Dev* **12** (21): 3382–3393.
3. Cochran WG, Cox GM (1992). *Experimental Designs*, 2nd edn. New York: John Wiley & Sons.
4. Daniel WW (2009). *Biostatistics: A Foundation for Analysis in the Health Sciences*, 9th edn. Hoboken, NJ: John Wiley & Sons.
5. Diamond WJ (2001). *Practical Experimental Designs for Engineers and Scientists*. New York: John Wiley & Sons.
6. Dingus CA, Teuschler LK, Rice GE, Simmons JE, Narotsky MG (2011). Prospective power calculations for the Four Lab study of a multigenerational reproductive/developmental toxicity rodent bioassay using a complex mixture of disinfection by-products in the low-response region. *Int J Environ Res Public Health* **8** (10): 4082–4101.
7. Gad SC (2001). Statistics for toxicologists. In: *Principles and Methods of Toxicology*, 4th edn., Hayes AW (ed.). Philadelphia, PA: Taylor & Francis; pp. 285–364.
8. Hernandez-Tristan R, Leret ML, Almeida D (2006). Effect of intrauterine position on sex differences in the gabaergic system and behavior of rats. *Physiol Behav* **87** (3): 625–633.
9. Hicks CR, Turner KV (1999). *Fundamental Concepts in the Design of Experiments*, 5th edn. Oxford, U.K.: Oxford University Press.
10. Higashibata Y, Sakuma T, Kawahata H, Fujihara S, Moriyama K, Okada A, Yasui T, Kohri K, Kitamura Y, Nomura S (2004). Identification of promoter regions involved in cell- and developmental stage-specific osteopontin expression in bone, kidney, placenta, and mammary gland: An analysis of transgenic mice. *J Bone Miner Res* **19** (1): 78–88.
11. Hollander M, Wolfe DA (1999). *Nonparametric Statistical Methods*, 2nd edn. New York: John Wiley & Sons.
12. Hurd PL, Bailey AA, Gongal PA, Yan RH, Greer JJ, Pagliardini S (2008). Intrauterine position effects on anogenital distance and digit ratio in male and female mice. *Arch Sex Behav* **37** (1): 9–18.
13. Janmohamed A, Hernandez D, Phillips IR, Shephard EA (2004). Cell-, tissue-, sex- and developmental stage-specific expression of mouse flavin-containing monooxygenases (*Fmos*). *Biochem Pharmacol* **68** (1): 73–83.
14. Lu F, Bytautiene E, Tamayo E, Gamble P, Anderson GD, Hankins GD, Longo M, Saade GR (2007). Gender-specific effect of overexpression of sFlt-1 in pregnant mice on fetal programming of blood pressure in the offspring later in life. *Am J Obstet Gynecol* **197** (4): 418.e1–418.e5.
15. Motulsky H (2010). *Intuitive Biostatistics: A Nonmathematical Guide to Statistical Thinking*, 2nd edn. Oxford, U.K.: Oxford University Press.
16. Nakamura K, Sugawara Y, Sawabe K, Ohashi A, Tsurui H, Xiu Y, Ohtsuji M, Lin QS, Nishimura H, Hasegawa H, Hirose S (2006). Late developmental stage-specific role of tryptophan hydroxylase 1 in brain serotonin levels. *J Neurosci* **26** (2): 530–534.
17. Rousseaux CG, Gad SC (2013). Statistical assessment of toxicologic pathology studies. In: *Haschek and Rousseaux's Handbook of Toxicologic Pathology*, 3rd edn., Vol. 2, Haschek WM, Rousseaux CG, Wallig MA, Bolon B, Ochoa R, Mahler BW (eds.). San Diego, CA: Academic Press (Elsevier); pp. 894–988.
18. Silva IA, El Nabawi M, Hoover D, Silbergeld EK (2005). Prenatal $HgCl_2$ exposure in BALB/c mice: Gender-specific effects on the ontogeny of the immune system. *Dev Comp Immunol* **29** (2): 171–183.

19. Ward WF, Aceto Jr H, Karp CH (1977). The effect of intrauterine position on the radiosensitivity of rat embryos. *Teratology* **16** (2): 181–186.
20. Zar JH (2010). *Biostatistical Analysis*. Englewood Cliffs, NJ: Prentice-Hall.
21. Zielinski WJ, Vandenbergh JG (1991). Effect of intrauterine position and social density on age of first reproduction in wild-type female house mice (*Mus musculus*). *J Comp Psychol* **105** (2): 134–139.

Appendix: Common Abbreviations in Mouse Developmental Pathology Studies

2D	Two-dimensional
3D	Three-dimensional
ABC	Avidin-biotin complex
AEC	3-Amino-9-ethylcarbazole
AGD	Anogenital distance
ALP	Alkaline phophatase
ALT	Alanine aminotransferase
ANOVA	Analysis of variance
AS	Antisense
AST	Aspartate aminotransferase
BrdU	5-Bromo-2′-deoxyuridine
BSA	Bovine serum albumin
BUN	Blood urea nitrogen
BW	Body weight
CBC	Complete blood count
CL	Corpus luteum (plural: corpora lutea)
CLSM	Confocal laser scanning microscopy
CSF	Cerebrospinal fluid
CV	Coefficient of variation
DAB	3,3′-Diaminobenzidine
DIG	Digoxigenin
dpc	Days *post coitum*
E	Embryonic day
EDTA	Ethylenediaminetetraacetic acid.
EHC	Enzyme histochemistry
EM	Electron microscopy
ENU	*N*-Ethyl-*N*-nitrosourea
EPC	Ectoplacental cone
ES	Embryonic stem (cells)
F_1 (or F1)	First-generation progeny of two parental strains
FCS (or FBS)	Fetal calf serum (fetal bovine serum)
GD	Gestational day
GEM	Genetically engineered mouse/mice
GFP	Green fluorescent protein
H&E	Hematoxylin and eosin
HCT	Hematocrit
Heterozygous	(+/−)
HGB	Hemoglobin concentration
HRP	Horseradish peroxidase
ICM	Inner cell mass
IHC	Immunohistochemistry
IKMC	International Knockout Mouse Consortium
IMPC	International Mouse Phenotyping Consortium (https://www.mousephenotype.org/)
IP	Intraperitoneal (route of administration)
ISH	*In situ* hybridization

Knockout	(–/–)
lacZ	β-Galactosidase (bacterial origin)
LCM	Laser capture microdissection
LHC	Lectin histochemistry
μCT	Microscopic computer-assisted tomography
MCV	Mean corpuscular volume (of red blood cells)
MGI	Mouse Genome Informatics (http://www.informatics.jax.org/)
MPM	Multiphoton microscopy
MRM	Magnetic resonance microscopy
N	Group/sample size
NBF	Neutral buffered 10% formalin
NOAEL	No observed adverse effect level
NTD	Neural tube defect
OCT	Optical cutting temperature compound
OPT	Optical projection tomography
PBS	Phosphate-buffered saline
PCD	Programmed cell death
PCNA	proliferating cell nuclear antigen
PCR	Polymerase chain reaction
PEN	Polyethylene naphthalate
PET	Positron emission tomography
PFA (or PF)	Paraformaldehyde
PIPES	Piperazine-*N*,*N*'-*bis*(2-ethanesulfonic acid)
PND	Postnatal day
RBC	Red blood cells
RT	Room temperature
SD	Standard deviation
SEM	Scanning electron microscopy; also stands for Standard error of the mean
siRNA	Small interfering RNAs
SOP	Standard operating procedure (or protocol)
SPECT	Single photon emission computed tomography
TEM	Transmission electron microscopy
TGC	Trophoblast giant cell
tg	Transgenic
tm	Targeted mutation
TS	Theiler stage
TUNEL	Terminal deoxynucleotidyl transferase (TdT)-mediated 2'-deoxyuridine-5'-triphosphate (dUTP) nick-end labeling
UBM	Ultrasound biomicroscopy
US	Ultrasound (conventional)
WBC	White blood cells
WT	Wild-type (+/+)
X-gal	5-Bromo-4-chloro-3-indolyl-β-D-galactoside
YS	Yolk sac

Index

Printed and bound by CPI Group (UK) Ltd, Croydon, CR0 4YY

24/10/2024

01778285-0005